Lecture Notes in Computer Science   14433

Founding Editors

Gerhard Goos
Juris Hartmanis

Editorial Board Members

Elisa Bertino, *Purdue University, West Lafayette, IN, USA*
Wen Gao, *Peking University, Beijing, China*
Bernhard Steffen⬛, *TU Dortmund University, Dortmund, Germany*
Moti Yung⬛, *Columbia University, New York, NY, USA*

The series Lecture Notes in Computer Science (LNCS), including its subseries Lecture Notes in Artificial Intelligence (LNAI) and Lecture Notes in Bioinformatics (LNBI), has established itself as a medium for the publication of new developments in computer science and information technology research, teaching, and education.

LNCS enjoys close cooperation with the computer science R & D community, the series counts many renowned academics among its volume editors and paper authors, and collaborates with prestigious societies. Its mission is to serve this international community by providing an invaluable service, mainly focused on the publication of conference and workshop proceedings and postproceedings. LNCS commenced publication in 1973.

Qingshan Liu · Hanzi Wang · Zhanyu Ma ·
Weishi Zheng · Hongbin Zha · Xilin Chen ·
Liang Wang · Rongrong Ji

Editors

Pattern Recognition and Computer Vision

6th Chinese Conference, PRCV 2023
Xiamen, China, October 13–15, 2023
Proceedings, Part IX

Springer

Editors
Qingshan Liu (ID)
Nanjing University of Information Science
and Technology
Nanjing, China

Zhanyu Ma (ID)
Beijing University of Posts
and Telecommunications
Beijing, China

Hongbin Zha (ID)
Peking University
Beijing, China

Liang Wang
Chinese Academy of Sciences
Beijing, China

Hanzi Wang (ID)
Xiamen University
Xiamen, China

Weishi Zheng (ID)
Sun Yat-sen University
Guangzhou, China

Xilin Chen (ID)
Chinese Academy of Sciences
Beijing, China

Rongrong Ji (ID)
Xiamen University
Xiamen, China

ISSN 0302-9743   ISSN 1611-3349 (electronic)
Lecture Notes in Computer Science
ISBN 978-981-99-8545-6   ISBN 978-981-99-8546-3 (eBook)
https://doi.org/10.1007/978-981-99-8546-3

Preface

Welcome to the proceedings of the Sixth Chinese Conference on Pattern Recognition and Computer Vision (PRCV 2023), held in Xiamen, China.

PRCV is formed from the combination of two distinguished conferences: CCPR (Chinese Conference on Pattern Recognition) and CCCV (Chinese Conference on Computer Vision). Both have consistently been the top-tier conference in the fields of pattern recognition and computer vision within China's academic field. Recognizing the intertwined nature of these disciplines and their overlapping communities, the union into PRCV aims to reinforce the prominence of the Chinese academic sector in these foundational areas of artificial intelligence and enhance academic exchanges. Accordingly, PRCV is jointly sponsored by China's leading academic institutions: the Chinese Association for Artificial Intelligence (CAAI), the China Computer Federation (CCF), the Chinese Association of Automation (CAA), and the China Society of Image and Graphics (CSIG).

PRCV's mission is to serve as a comprehensive platform for dialogues among researchers from both academia and industry. While its primary focus is to encourage academic exchange, it also places emphasis on fostering ties between academia and industry. With the objective of keeping abreast of leading academic innovations and showcasing the most recent research breakthroughs, pioneering thoughts, and advanced techniques in pattern recognition and computer vision, esteemed international and domestic experts have been invited to present keynote speeches, introducing the most recent developments in these fields.

PRCV 2023 was hosted by Xiamen University. From our call for papers, we received 1420 full submissions. Each paper underwent rigorous reviews by at least three experts, either from our dedicated Program Committee or from other qualified researchers in the field. After thorough evaluations, 522 papers were selected for the conference, comprising 32 oral presentations and 490 posters, giving an acceptance rate of 37.46%. The proceedings of PRCV 2023 are proudly published by Springer.

Our heartfelt gratitude goes out to our keynote speakers: Zongben Xu from Xi'an Jiaotong University, Yanning Zhang of Northwestern Polytechnical University, Shutao Li of Hunan University, Shi-Min Hu of Tsinghua University, and Tiejun Huang from Peking University.

We give sincere appreciation to all the authors of submitted papers, the members of the Program Committee, the reviewers, and the Organizing Committee. Their combined efforts have been instrumental in the success of this conference. A special acknowledgment goes to our sponsors and the organizers of various special forums; their support made the conference a success. We also express our thanks to Springer for taking on the publication and to the staff of Springer Asia for their meticulous coordination efforts.

We hope these proceedings will be both enlightening and enjoyable for all readers.

October 2023

Qingshan Liu
Hanzi Wang
Zhanyu Ma
Weishi Zheng
Hongbin Zha
Xilin Chen
Liang Wang
Rongrong Ji

Organization

General Chairs

Hongbin Zha   Peking University, China
Xilin Chen   Institute of Computing Technology, Chinese
Academy of Sciences, China
Liang Wang   Institute of Automation, Chinese Academy of
Sciences, China
Rongrong Ji   Xiamen University, China

Program Chairs

Qingshan Liu   Nanjing University of Information Science and
Technology, China
Hanzi Wang   Xiamen University, China
Zhanyu Ma   Beijing University of Posts and
Telecommunications, China
Weishi Zheng   Sun Yat-sen University, China

Organizing Committee Chairs

Mingming Cheng   Nankai University, China
Cheng Wang   Xiamen University, China
Yue Gao   Tsinghua University, China
Mingliang Xu   Zhengzhou University, China
Liujuan Cao   Xiamen University, China

Publicity Chairs

Yanyun Qu   Xiamen University, China
Wei Jia   Hefei University of Technology, China

Local Arrangement Chairs

Xiaoshuai Sun	Xiamen University, China
Yan Yan	Xiamen University, China
Longbiao Chen	Xiamen University, China

International Liaison Chairs

Jingyi Yu	ShanghaiTech University, China
Jiwen Lu	Tsinghua University, China

Tutorial Chairs

Xi Li	Zhejiang University, China
Wangmeng Zuo	Harbin Institute of Technology, China
Jie Chen	Peking University, China

Thematic Forum Chairs

Xiaopeng Hong	Harbin Institute of Technology, China
Zhaoxiang Zhang	Institute of Automation, Chinese Academy of Sciences, China
Xinghao Ding	Xiamen University, China

Doctoral Forum Chairs

Shengping Zhang	Harbin Institute of Technology, China
Zhou Zhao	Zhejiang University, China

Publication Chair

Chenglu Wen	Xiamen University, China

Sponsorship Chair

Yiyi Zhou	Xiamen University, China

Exhibition Chairs

Bineng Zhong	Guangxi Normal University, China
Rushi Lan	Guilin University of Electronic Technology, China
Zhiming Luo	Xiamen University, China

Program Committee

Baiying Lei	Shenzhen University, China
Changxin Gao	Huazhong University of Science and Technology, China
Chen Gong	Nanjing University of Science and Technology, China
Chuanxian Ren	Sun Yat-Sen University, China
Dong Liu	University of Science and Technology of China, China
Dong Wang	Dalian University of Technology, China
Haimiao Hu	Beihang University, China
Hang Su	Tsinghua University, China
Hui Yuan	School of Control Science and Engineering, Shandong University, China
Jie Qin	Nanjing University of Aeronautics and Astronautics, China
Jufeng Yang	Nankai University, China
Lifang Wu	Beijing University of Technology, China
Linlin Shen	Shenzhen University, China
Nannan Wang	Xidian University, China
Qianqian Xu	Key Laboratory of Intelligent Information Processing, Institute of Computing Technology, Chinese Academy of Sciences, China
Quan Zhou	Nanjing University of Posts and Telecommunications, China
Si Liu	Beihang University, China
Xi Li	Zhejiang University, China
Xiaojun Wu	Jiangnan University, China
Zhenyu He	Harbin Institute of Technology (Shenzhen), China
Zhonghong Ou	Beijing University of Posts and Telecommunications, China

Contents – Part IX

Neural Network and Deep Learning II

Decoupled Contrastive Learning for Long-Tailed Distribution

Xiaohua Chen[1,2], Yucan Zhou[1(✉)], Lin Wang[1,2], Dayan Wu[1],
Wanqian Zhang[1], Bo Li[1,2], and Weiping Wang[1,2]

[1] Institute of Information Engineering, Chinese Academy of Sciences, Beijing, China
{chenxiaohua,zhouyucan,wanglin5812,wudayan,
zhangwanqian,libo,wangweiping}@iie.ac.cn
[2] School of Cyber Security, University of Chinese Academy of Sciences,
Beijing, China

Abstract. Self-supervised contrastive learning is popularly used to obtain powerful representation models. However, unlabeled data in the real world naturally exhibits a long-tailed distribution, making the traditional instance-wise contrastive learning unfair to tail samples. Recently, some improvements have been made from the perspective of model, loss, and data to make tail samples highly evaluated during training, but most of them explicitly or implicitly assume that the sample with a large loss is the tail. We argue that due to the lack of hard negatives, tail samples usually occupy a small loss at the initial stage of training, which will make them eliminated at the beginning of training. To address this issue, we propose a simple but effective two-stage learning scheme that decouples traditional contrastive learning to discover and enhance tail samples. Specifically, we identify the sample with a small loss in Stage I while a large loss in Stage II as the tail. With the discovered tail samples, we generate hard negatives for them based on their neighbors, which will balance the distribution of the hard negatives in training and help learn better representation. Additionally, we design the weight inversely proportional or proportional to the loss in each stage to achieve fairer training by reweighting. Extensive experiments on multiple unlabeled long-tailed datasets demonstrate the superiority of our DCL compared with the state-of-the-art methods. The code will be released soon.

Keywords: Long-tailed distribution · Unsupervised contrastive learning · Hard negative samples generation

1 Introduction

Deep learning have achieved great improvements on many tasks [10,26,28,33]. Among them, contrastive learning has been popularly explored for learning powerful data representations without requiring manually annotated datasets and achieved significant success [2,11,29]. Despite the great success, existing methods always significantly benefit from the well-curated balanced training data

Q. Liu et al. (Eds.): PRCV 2023, LNCS 14433, pp. 3–15, 2024.
https://doi.org/10.1007/978-981-99-8546-3_1

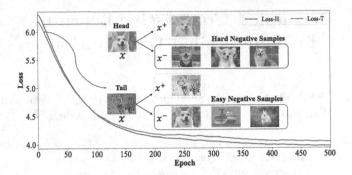

Fig. 1. The average loss of the head and tail in each epoch on CIFAR-100-LT.

[8,20,31,32]. In contrast, real-world datasets naturally exhibit a long-tailed [3,16] distribution, i.e., many tail classes contain extremely fewer training samples than a few head classes. Recent research [13] has shown that state-of-the-art contrastive learning is certainly vulnerable to long-tailed data. However, only less attention has been paid to this [13,21]. Therefore, how to achieve satisfying performance for long-tailed self-supervised learning needs more exploration.

Existing work on long-tailed self-supervised representation learning can be roughly divided into *loss adjustment*, *model improvement* and *data augmentation*. *Loss adjustment* mainly allocates larger weights to tail samples, e.g., focal loss [19] and SAM [21]. *Model improvement* [13,24] relies on a specific model to obtain hard positive or increase the proportion of hard negative samples for tail samples to attract more attention of the model, such as SDCLR via pruning and DnC ensemble. For *data augmentation*, based on RandAugment [5], BCL [34] proposes to assign heavier augmentation strength for tail samples. Although some progress has been achieved, these methods explicitly or implicitly suggest recognizing tail samples with large losses. However, this is not always the case.

For balanced datasets, each sample has the same number of hard negative samples, which are from the same class of the anchor. Whereas, the distribution of hard negatives is heavily imbalanced for long-tailed datasets. Figure 1 shows the average loss of the head and tail class in different epochs. Since there are sufficient samples in the class "Dog", for any dog sample, other dogs are its hard negatives in training. While, for a sample from the tail class "Tiger", there are only few tigers in its negatives. Therefore, the loss of a tail sample is inevitably much smaller than that of a head sample. Consequently, tail samples are out of favor at the beginning of the training, resulting in under-represented tail samples.

In this paper, we propose an effective **D**ecoupled **C**ontrastive **L**earning (**DCL**) method for long-tailed distribution, which is a two-stage scheme. Specifically, in Stage I, we argue that tail samples usually own small losses and can be identified with the top-K smallest contrastive losses. After mining the tail samples, we generate their hard negative samples by mixing their neighbors based on cosine similarity. Then, we use the original samples and generated hard negatives together for contrastive learning, which not only enlarges the loss for the tail but also enriches the diversity of tail classes. With the new losses, the weights are simply designed inversely proportional to the losses to achieve a more balanced training by reweighting. Note that, the generated samples also play as anchors for contrastive learning, which will increase the proportion of the tail in total loss. In Stage II, when the model fits most head samples, tail samples will own large losses, since head samples are usually memorized prior. Therefore, we select the samples with the top-K largest losses as the tail and generate hard negatives for them. This mainly provides more and higher quality hard negatives for helping the feature learning of the tail. Then, the weights are designed proportionally to the losses to further balance the training by enhancing the tail.

Contributions. (1) To our best knowledge, we are the first to analyse the different behaviors of tail samples in different learning stages, and propose to discover them in a two-stage way with the sorted contrastive losses. (2) We argue that tail samples are under-represented because of the lack of hard negative samples, and propose to generate hard negative samples for each tail sample by mixing with its neighbors. (3) The proposed DCL outperforms current state-of-the-art methods on three benchmark unlabeled long-tailed datasets.

2   Related Work

Supervised Long-Tailed Contrastive Learning. Long-tailed distribution is a long-term research problem in machine learning [4,23]. Recently, Supervised contrastive learning (SCL) [17] uses label information to extend self-supervised to supervised contrastive learning, which surpasses the performance of traditional cross-entropy loss on image recognition. Inspired by this, many methods based on SCL have emerged to assist supervised long-tailed learning. Hybrid-SC [25] designs a two-branch network that includes an SCL-based feature learning branch and a classifier branch based on the cross-entropy loss. However, SCL still suffers greatly from the bias of the head [6,15]. Consequently, PaCo [6] introduces a set of parametric class-wise learnable centers to rebalance the contrastive losses for each class. KCL [15] exploits the balanced number of positive pairs in each batch for all classes to generate a more balanced feature space. Recently, TSC [18] proposes to achieve fair training by forcing the features of different classes to converge to a set of corresponding uniformly distributed targets on the hypersphere based on SCL. However, the targets of TSC are learned without class semantics. Therefore, balanced-CL [35] proposes to balance the gradient contribution of negative classes and allows all classes to appear as prototypes in each batch. Unfortunately, the above methods rely on expensive label information and are not directly applicable to unsupervised representation learning.

Self-supervised Long-Tailed Contrastive Learning. Self-supervised contrastive learning has gained more attention because annotation is time-consuming and labor-expensive. However, it meets significant challenges when applying the practical data [9] that follows a long-tailed distribution. Since samples from the head class will dominate the training process, resulting in under-represented samples for the tail. So, it is intractable to learn a good representation with long-tailed data. Recent methods to improve long-tailed contrastive learning can be mainly divided into *loss adjustment, model improvement* and *data augmentation.*

Loss adjustment is the most commonly used strategy to achieve balanced training via reweighting the sample-wise loss according to their learning difficulty. For instance, focal loss [19] explicitly regards the hard samples as the tail and allocates large weights for them to achieve balanced training. *Model improvement* usually designs a specific model structure to obtain hard positive or increase the proportion of hard negative samples for the tail to make the tail attract more attention. SDCLR [13] explicitly assumes that the difficult-to-memorize samples with large losses are the tail. By constructing a larger prediction difference between the pruned and non-pruned branches, it can implicitly give tail samples more attention with the enlarged loss. DnC [24] designs a multi-expert framework to implicitly increase the proportion of hard negatives for the tail by contrasting locally between semantically-similar samples in the divided clusters. With the expanded loss, it captures more discriminative representation for tail samples. Recently, from the *data augmentation* perspective, BCL [34] proposes to explicitly regard the samples with large losses in historical memory as the tail and assign higher intensities of augmentation to them based on the RandAugment to improve the representation learning of tail samples.

Although previous studies have shown some improvement, they explicitly or implicitly assume that samples with large losses are the tail. In contrast to them, we argue that tail samples tend to occupy small losses due to the imbalanced distribution of hard negative samples, leading to their elimination in the initial stage of training. Consequently, we propose a two-stage learning scheme to mine tail samples online and enhance them.

3   Methods

In this section, we present DCL for unsupervised long-tailed learning, which contains three key components: (1) online tail samples discovery, (2) hard negatives generation, and (3) contrastive loss reweighting.

3.1   Online Tail Samples Discovery

Given a set of N unlabeled samples, we follow the popular self-supervised contrastive learning framework SimCLR [2] to learn visual representation by strengthening the similarity of positive sample pairs (x_i, x_i^+) and expanding the distance of negative sample pairs (x_i, x_i^-). Formally, the total contrastive loss is:

$$L_{CL} = \frac{1}{N} \sum_{i=1}^{N} -\frac{\exp(\frac{f(x_i)^T f(x_i^+)}{\tau})}{\sum_{x_i' \in X^- \cup \{x_i^+\}} \exp(\frac{f(x_i)^T f(x_i')}{\tau})}, \tag{1}$$

where (x_i, x_i^+) are the two augmentations of the same sample i while all other augmented samples are treated as negative samples and form the negative sample set X^-. τ is the temperature hyper-parameter. f denotes the feature extractor. The goal of contrastive learning is to learn a "good" representation, which can be easily used as a feature extractor for downstream lightweight tasks.

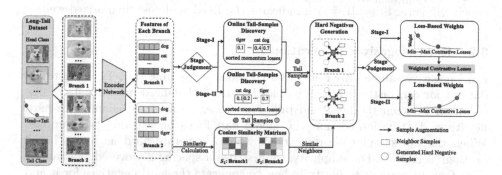

Fig. 2. Framework of DCL.

In our work, given an unlabeled long-tailed dataset, we use Eq. (1) to learn. As shown in Fig. 1, since the tail sample is short of hard negatives, its contrastive loss is relatively smaller than the head during the early training stage. In the later stage of training, the model has memorized most of the head samples while the tail samples are not well memorized, so the tail will produce a larger loss than the head. Therefore, our online tail-samples discovery contains two stages.

Online Tail Samples Discovery. For a specific positive pair (x_i, x_i^+), we record its contrastive loss in epoch e as $L_{i,e}$ and trace a sequence of the loss values in each epoch as $\{L_{i,0}, ..., L_{i,E}\}$. To alleviate the impact of the randomness of augmentation on the selection of tail samples, similar to BCL [34], we use the moving-average momentum loss to trace the historical contrastive loss as follows:

$$L_{i,0}^m = L_{i,0},$$
$$L_{i,e}^m = \beta L_{i,e-1}^m + (1 - \beta)L_{i,e}, \tag{2}$$

where β is the hyperparameter to control the degree smoothed by the historical losses. Then, we use $L_e^m = \{L_{1,e}^m, ..., L_{i,e}^m, ..., L_{N,e}^m\}$ to define the momentum losses of all N samples in epoch e, which represents the cumulative historical loss of N samples. In Stage I, we choose the samples corresponding to the top-K smallest momentum losses as the tail while in Stage II, we choose the top-K largest losses. Since the contrastive loss is calculated by a pair of positive samples which is the two views of the same sample, when we select top-K losses, it is equivalent to selecting K tail sample pairs. Therefore, we use $T = \{x_i^t, x_i^{t^+}\}_{i=1}^K$ to denote the tail set. For simplicity, we define $(x_i^t, x_i^{t^+}) \in T$ as a tail sample pair that belongs to Branch 1 and 2 of SimCLR in Fig. 2, respectively.

Stage Judgement. We use different strategies to discover tail samples online in different stages, so we need to determine which stage the current epoch belongs to. We use the difference value of the overall momentum loss at a certain epoch interval as the basis for judging the transition from Stage I to II. The overall average momentum loss at the epoch e can be obtained by $L_{all,e}^m = \frac{1}{N} \sum_{i=1}^{N} L_{i,e}^m$. When the $(L_{all,e-interval}^m - L_{all,e}^m)$ is less than a certain threshold t, the $(e+1)$-th epoch will enter Stage II, which indicates the head is almost memorized-well.

3.2 Hard Negatives Generation

Hard negative samples are critical for contrastive learning [14]. The long-tailed dataset for contrastive learning will lead to the imbalanced hard negatives distribution between the head and tail. Therefore, we need to balance the hard negatives distribution so that the model can focus on learning tail samples. When the tail samples are discovered, we can generate hard negatives based on their neighbors. For simplicity, we first use a specific discovered tail sample $x_i^t \in T$ as an example to illustrate how to generate the hard negatives for it and then extend this pipeline to all discovered tail samples.

Hard Negatives Generation for One Tail Sample. For a tail sample x_i^t, we first calculate the cosine similarity between x_i^t and other samples in the same branch. The cosine similarity is typically defined as:

$$s(x_i^t, x_j) = \frac{f(x_i^t)^{\mathrm{T}} \cdot f(x_j)}{|f(x_i^t)||f(x_j)|}, \tag{3}$$

where $j \in [1, N]$, and $x_i^t \neq x_j$. The x_i^t and x_j belong to the same branch. With Eq. (3), we can get its top-M neighbors' feature $\{n_1, n_2, ..., n_M\}$.

Previous work [14,30] shows that the data mixing technology operated at the pixel or feature level can help the model learn more robust features. Inspired by this, we also choose the data mixing to synthesize hard negatives at the feature level because of its simplicity and effectiveness. Specifically, we mix the feature of tail sample x_i^t with each negative neighbor in $\{n_1, n_2, ..., n_M\}$ to synthesize M new features. In formal, we define $\{h_1, h_2, ..., h_M\}$ as the set of synthetic hard negative features generated for x_i^t. By creating a convex linear combination of each neighbor with the feature of $f(x_i^t)$, the j-th synthetic hard negatives feature h_j for the tail sample x_i^t can be given by:

$$h_j = \frac{\tilde{h}_j}{|\tilde{h}_j|}, \text{ where } \tilde{h}_j = \alpha_j f(x_i^t) + (1 - \alpha_j) n_j, \tag{4}$$

where $j \in [1, M]$, n_j is the j-th neighbor of x_i^t. $\alpha_j \in (0, 1)$ is a randomly selected mixing coefficient.

Hard Negatives Generation for All Tail Samples. In the above, we mainly introduce the generation of negatives for a specific tail sample. So, for each tail in T, we can generate hard negatives according to the above pipeline.

Specifically, with Eq. (3), we can calculate the similarity matrix of all tail samples in Branch 1 and 2 to form $S_1 \in R^{K \times N}$ and $S_2 \in R^{K \times N}$, respectively. Then, each tail sample can generate hard negatives based on its top-M neighbors in its branch (S_1 or S_2) with Eq. (4). Therefore, we can generate hard negatives for all tail samples and we define them $H = \{h_i, h_i^+\}_{i=1}^{KM}$ with the size $2KM$ since we generate M hard negatives for each of all $2K$ tail samples respectively. When the hard negatives for the tail are generated, we recalculate the total contrastive loss with the expanded training feature set $F = \{f(x_1), f(x_1^+), ..., f(x_N), f(x_N^+), h_1, h_1^+, ..., h_{KM}, h_{KM}^+\}$. For simply, we redefine the expanded set $F = \{f(x_i), f(x_i^+)\}_{i=1}^{N+KM}$, where $\{f(x_i), f(x_i^+)\}_{i=N+1}^{N+KM}$ corresponds to $\{h_i, h_i^+\}_{i=1}^{KM}$. Then Eq. (1) can be modified as:

$$\mathcal{L}_{CL}' = -\frac{1}{N+KM}\left(\sum_{i=1}^{N+KM} \frac{\exp(\frac{f(x_i)^T f(x_i^+)}{\tau})}{\sum_{x_i' \in \tilde{X}^- \cup \{x_i^+\}} \exp(\frac{f(x_i)^T f(x_i')}{\tau})}\right)$$
$$= -\frac{1}{N+KM} \sum_{i=1}^{N+KM} L_i', \tag{5}$$

where $\tilde{X}^-$ with the size of $(2N + 2KM - 2)$ denotes the new negative samples set to instance i and L_i' is the i-th contrastive loss.

3.3 Contrastive Loss Reweighting

Although we take advantage of generated hard negative samples for the tail to achieve comparable instance-level losses with the head, the head samples still dominate the training process. Because the number of selected neighbors increases, the similarity between negatives and the tail samples will decrease, which will lead to easy negative samples.

To achieve more balanced training, we explicitly enhance the learning of the tail samples by strengthening the importance of their losses like the supervised reweighting strategy [1,7]. Detailedly, in Stage I, we assign larger weights to the small losses while in Stage II we assign larger weights to the large ones. We simply design the weight to each sample according to the recalculated contrastive losses of all samples $\{L_i'\}_{i=1}^{N+KM}$ in Eq. (5). The weight of sample i can be designed by:

$$w_i = \begin{cases} \frac{1/L_i'}{\sum_{j=1}^{N+KM} 1/L_j'} \times (N+KM) & \text{Stage I,} \\ \frac{L_i'}{\sum_{j=1}^{N+KM} L_j'} \times (N+KM) & \text{Stage II.} \end{cases} \tag{6}$$

With the simple weights in Eq. (6), we can achieve more balanced training for long-tailed samples with the reweighted loss based on Eq. (5) as follows:

$$\mathcal{L}_{WCL} = -\frac{1}{N+KM} \sum_{i=1}^{N+KM} w_i \times L_i', \tag{7}$$

4   Experiments

Datasets and Baseline. We conduct our experiments on three popular long-tailed datasets: CIFAR-100-LT [1], ImageNet-LT [22], and Places-LT [22]. We use λ as the imbalance ratio of the sample sizes between the most and least frequent classes (i.e., $\lambda = N_{max}/N_{min}$). The test sets are all balanced and the details of CIFAR-100-LT, ImageNet-LT and Places-LT are shown in Table 1.

Compared Methods. To verify the effectiveness of the proposed DCL, we compare it with several state-of-the-art (SOTA) methods under the unsupervised long-tailed representation learning, including (1) **Baseline**: contrastive learning based on SimCLR [2]; (2) **Loss adjustment**: Focal Loss [19]; (3) **Model improvement**: DnC [24] and SDCLR [13]; (4) **Data augmentation**: BCL [34].

Table 1. The details of CIFAR-100-LT, ImageNet-LT and Places-LT.

Datasets	Class Number	λ	Train Set	Test Set
CIFAR-100-LT [1]	100	100	9,754	10,000
ImageNet-LT [22]	1,000	256	115,846	100,000
Places-LT [22]	365	996	62,500	36,500

Implementation Details. Following BCL [34], we use the SGD with a momentum 0.9 and the cosine annealing schedule for all experiments. Our model is based on the SimCLR [2] and the augmentation follows the RandAugment [5]. Besides, the initial learning rate is 0.5 and the temperature is 0.2 for each dataset. For CIFAR-100-LT, we use ResNet-18 [12] as the backbone and train 2000 epochs with batch size 512. We set $K = 15$ and $M = 2$ to generate hard negatives for the tail. We calculate the difference between the average losses over a 100 epochs interval. When the difference value is less than 0.06, our DCL transition from Stage I to II. For ImageNet-LT and Places-LT, we use ResNet-50 as the backbone and train 500 epochs with batch size 256. For the linear probing evaluation on CIFAR-100-LT, we follow BCL [34] to train 500 epochs and use the learning rate decaying from 10^{-2} to 10^{-6}. Besides, we use the Adam optimizer with the weight decay 5×10^{-6}. For ImageNet-LT and Places-LT, we train 200 epochs.

Evaluation Protocol. We follow [15] to conduct the linear probing evaluation, which contains three steps: (1) *Representation learning*, pre-train the f on unlabeled long-tailed data with above L_{WCL} in Eq. (7); (2) *Classifier learning*, in order to eliminate the effect of the long-tailed distribution, we use the balanced dataset to train a linear classifier with cross-entropy [27] loss based on the feature extracted from the fixed f for classification. (3) *Representation evaluation*, evaluating the learned classifier and the representation from f on the balanced test dataset. Concretely, we mainly perform the *few-shot* accuracy. Same as BCL, we conduct 100-shot evaluations on CIFAT-100-LT, ImageNet-LT, and Places-LT.

We calculate the average accuracy to measure the representation. To visualize the fine-grained performance under the long-tailed setting, we split the test dataset into three groups based on the number of training samples in each class. Following [13], for CIFAR-100-LT, we split three groups as Many (more than 106 samples), Medium (105~20 samples), and Few (less than 20 samples). For the ImageNet-LT and Places-LT, we follow [13, 22] to split the groups as Many (more than 100 samples), Medium (100~20 samples) and Few (less than 20 samples).

4.1 Linear Probing Evaluations

Overall Performance. Table 2 summarizes results of the recent SOTA methods pre-trained on CIFAR-100-LT, ImageNet-LT, and Places-LT with linear probing evaluation, i.e. features are frozen and a linear classifier is trained. Our DCL can achieve outstanding performance compared with all SOTA on three different datasets. Compared with SDCLR and BCL, our DCL achieves a significant improvement of 2.11% and 2.66% on CIFAR-100-LT, respectively. For large-scale datasets ImageNet-LT and Places-LT, our DCL outperforms 1.12% and 1.37% compared with BCL. These results demonstrate the superiority of our method.

Table 2. Top-1 accuracy on CIFAR-100-LT, ImageNet-LT, and Places-LT.

Methods	CIFAR-100-LT				ImageNet-LT				Places-LT			
	Many	Medium	Few	All	Many	Medium	Few	All	Many	Medium	Few	All
SimCLR [2]	48.70	46.81	44.02	46.53	41.16	32.91	31.76	35.93	31.12	33.85	35.62	33.22
Focal [19]	48.46	46.73	44.12	46.46	40.55	32.91	31.29	35.63	30.18	31.56	33.32	31.41
DnC [24]	54.00	46.68	45.65	48.53	29.54	19.62	18.38	23.27	28.20	28.07	28.46	28.19
SDCLR [13]	51.22	49.22	45.85	48.79	41.24	33.62	32.15	36.35	32.08	35.08	35.94	34.17
BCL [34]	50.45	48.23	45.97	48.24	42.53	35.66	33.93	38.07	32.27	34.96	38.03	34.59
DCL	**54.20**	**50.97**	**47.62**	**50.90**	**42.99**	**37.16**	**35.62**	**39.19**	**33.38**	**36.56**	**39.35**	**35.96**

Fine-Grained Analysis. For a comprehensive evaluation, we further report the fine-grained accuracy on three groups Many, Medium, and Few of each dataset. Table 2 summarizes the results and we can conclude that:

(1) Our DCL is superior to the existing SOTA methods in all groups of each dataset. It outperforms BCL and SDCLR on Many, Medium, and Few by a large margin at {3.75%, 2.74%, 1.65%} and {2.98%, 1.75%, 1.77%} on CIFAR-100-LT, respectively. For ImagNet-LT and Places-LT, DCL outperforms the second-best method by {0.46%, 1.50%, 1.69%} and {1.11%, 1.48%, 1.32%} on Many, Medium and Few, respectively. (2) Note that our method can achieve a significant consistent improvement not only on the Few but also on Many and Medium groups. For instance, compared with BCL on CIFAR-100-LT, ImageNet-LT and Places-LT, DCL obtains {1.65%, 1.69%, 1.32%} and {2.74%, 1.50%, 1.6%} improvements on Few and Medium on corresponding three datasets, respectively. Furthermore, it gains {3.75%, 0.46%, 1.11%} on Many of these three datasets. A possible reason is that when generating hard negatives for the tail, most of their neighbors are head samples, which also indirectly enhances the learning for the head.

4.2 Analysis

To study the effectiveness of each component of the proposed DCL, we conduct a detailed ablation study and discuss the parameter sensitivity on CIFAR-100-LT.

Influence of Each Component. To verify the effectiveness of each component in our DCL, we conduct experiments including (1) using randomly selected samples (+random aug) VS online tail samples discovery (+tail-discovery aug) for hard negative samples generation; (2) only reweighting the contrastive loss (+reweighting); (3) our DCL. As shown in Table 3, it demonstrates that:

(1) Tail-discovery aug is significant to DCL. Compared with SimCLR, +tail-discovery aug can boost the performance by {0.88%,1.32%,1.30%} on 50/100/ Full-shot. However, when using +random aug, the results will decrease by {0.61%, 0.57%,0.47% }, which shows that randomly selecting samples cannot ensure the tail be chosen. Therefore, using the discovered tail samples and generating hard negatives for them can alleviate the imbalanced hard negatives distribution.
(2) Reweighting is also important for our DCL. Simply combining reweighting with SimCLR can obtain {0.96%, 1.2%, 0.86%} improvements on 50/100/Full-shot, respectively. It shows that a simple two-stage weighting can explicitly strengthen the learning of the tail by enlarging the importance of their losses.

Table 3. Ablation study on CIFAR-100-LT.

Methods	50-shot	100-shot	Full-shot
SimCLR	43.42	47.26	53.14
+random aug	43.69	48.01	53.97
+tail-discovery aug	44.30	48.58	54.44
+reweighting	44.38	48.46	54.00
DCL	**46.98**	**50.78**	**56.32**

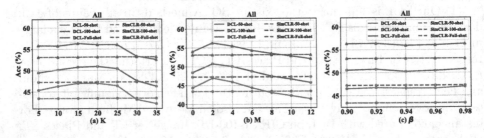

Fig. 3. (a) Effect of tail sample numbers K when neighbor $M = 2$. (b) Effect of neighbor numbers M when tail number $K = 15$. (c) Effect of $\beta \in \{0.90, 0.92, 0.94, 0.96, 0.98\}$.

(3) Cooperating reweighting with tail-discovery aug is our DCL, which can consistently have a significant improvement in all cases compared with baseline SimCLR. For instance, DCL improves a large margin by $\{3.56\%, 3.52\%, 3.18\%\}$ in terms of 50/100/Full-shot, respectively. Therefore, our DCL is not only suitable for many-shot but also for few-shot downstream tasks, which strongly confirms the effectiveness of our DCL on long-tailed representation learning.

Effect of K Tail Samples. An important step of DCL is to discover the K tail samples. Selecting tail samples as much as possible can better balance the training. Therefore, we explore the effect of tail samples by varying the $K \in \{5, 10, 15, 20, 25, 30, 35\}$. In Fig. 3(a), we can see that $K = 15$ can achieve the best with neighbors $M = 2$. When K is too large, the proportion of selected tail may be reduced, resulting in generating hard negatives for the head, which further aggravates the imbalance distribution of hard negatives. For instance, when $K = 30$, the results on 50/100/Full-shot will be worse than the SimCLR. Therefore, it is unreasonable to blindly select too many samples for enhancement. On the contrary, if K is too small, some tail samples will be missed.

Effect of M Tail Neighbors. After obtaining K-discovered tail samples, the heavily imbalanced distribution of hard negatives in long-tailed datasets can be balanced by generating hard negatives for the tail based on their neighbors. We analyze the influence of various numbers of neighbors $M \in \{0, 2, 4, 6, 8, 10, 12\}$. From Fig. 3(b), we can see that $M = 2$ consistently has achieved the best performance with $K = 15$. When M increases, the similarity of the neighbors will decrease, resulting in easy negatives. For instance, when $M = 8$, the accuracy on 50/100/Full-shot will be worse than baseline SimCLR.

Effect of β. We can see that β is robust to DCL in Fig. 3(c). The results on 50/100-shot achieves the best when $\beta = 0.98$ while on Full-shot $\beta = 0.92$. Besides, the biggest difference between the best and worst results when the β takes different values is about 0.5%. Therefore, in general, we set $\beta = 0.98$.

5   Conclusion

In this paper, we have proposed an online decoupled loss-driven contrastive learning method for long-tailed distribution. By selecting tail samples with a small or large loss in Stage I and Stage II respectively, we can generate hard negative samples for the tail samples based on their neighbors. Thus, the problem of imbalanced hard negative samples caused by the long-tailed distribution is alleviated. Besides, we further design simple weights to strengthen the importance of the tail according to the contrastive losses to achieve fairer training. Through extensive experiments, we have demonstrated the effectiveness of our DCL in different settings. We believe that our two-stage learning strategy can offer inspiration for future work on self-supervised contrastive learning.

Acknowledgements. This work was supported by the National Natural Science Foundation of China under Grants 62006221, 62106258, 62006242, 62202459 and 61925602, the National Key R&D Program of China under Grant 2022YFB31 03500, the Grant No. XDC02050200, the China Postdoctoral Science Foundation under Grant 2022M713348 and 2022TQ0363, and Young Elite Scientists Sponsorship Program by BAST (No. BYESS2023304).

References

1. Cao, K., Wei, C., Gaidon, A., Arechiga, N., Ma, T.: Learning imbalanced datasets with label-distribution-aware margin loss. In: NeurIPS (2019)
2. Chen, T., Kornblith, S., Norouzi, M., Hinton, G.: A simple framework for contrastive learning of visual representations. In: ICML, pp. 1597–1607 (2020)
3. Chen, X., et al.: Area: adaptive reweighting via effective area for long-tailed classification. In: ICCV (2023)
4. Chen, X., et al.: Imagine by reasoning: a reasoning-based implicit semantic data augmentation for long-tailed classification. In: AAAI, pp. 356–364 (2022)
5. Cubuk, E.D., Zoph, B., Shlens, J., Le, Q.V.: Randaugment: practical automated data augmentation with a reduced search space. In: CVPR, pp. 702–703 (2020)
6. Cui, J., Zhong, Z., Liu, S., Yu, B., Jia, J.: Parametric contrastive learning. In: ICCV, pp. 715–724 (2021)
7. Cui, Y., Jia, M., Lin, T.Y., Song, Y., Belongie, S.: Class-balanced loss based on effective number of samples. In: CVPR, pp. 9268–9277 (2019)
8. Deng, J., Dong, W., Socher, R., Li, L.J., Li, K., Fei-Fei, L.: Imagenet: a large-scale hierarchical image database. In: CVPR, pp. 248–255 (2009)
9. Doersch, C., Gupta, A., Efros, A.A.: Unsupervised visual representation learning by context prediction. In: ICCV, pp. 1422–1430 (2015)
10. Hao, X., Zhang, W., Wu, D., Zhu, F., Li, B.: Dual alignment unsupervised domain adaptation for video-text retrieval. In: CVPR, pp. 18962–18972 (2023)
11. He, K., Fan, H., Wu, Y., Xie, S., Girshick, R.: Momentum contrast for unsupervised visual representation learning. In: CVPR, pp. 9729–9738 (2020)
12. He, K., Zhang, X., Ren, S., Sun, J.: Deep residual learning for image recognition. In: CVPR, pp. 770–778 (2016)
13. Jiang, Z., Chen, T., Mortazavi, B.J., Wang, Z.: Self-damaging contrastive learning. In: ICML, pp. 4927–4939 (2021)
14. Kalantidis, Y., Sariyildiz, M.B., Pion, N., Weinzaepfel, P., Larlus, D.: Hard negative mixing for contrastive learning. In: NeurIPS, pp. 21798–21809 (2020)
15. Kang, B., Li, Y., Xie, S., Yuan, Z., Feng, J.: Exploring balanced feature spaces for representation learning. In: ICLR (2020)
16. Kang, B., et al.: Decoupling representation and classifier for long-tailed recognition. In: ICLR (2020)
17. Khosla, P., et al.: Supervised contrastive learning. In: NeurIPS (2020)
18. Li, T., et al.: Targeted supervised contrastive learning for long-tailed recognition. In: CVPR, pp. 6918–6928 (2022)
19. Lin, T.Y., Goyal, P., Girshick, R., He, K., Dollár, P.: Focal loss for dense object detection. In: Proceedings of the IEEEICCV, pp. 2980–2988 (2017)
20. Lin, T.-Y., et al.: Microsoft COCO: common objects in context. In: Fleet, D., Pajdla, T., Schiele, B., Tuytelaars, T. (eds.) ECCV 2014. LNCS, vol. 8693, pp. 740–755. Springer, Cham (2014). https://doi.org/10.1007/978-3-319-10602-1_48

21. Liu, H., HaoChen, J.Z., Gaidon, A., Ma, T.: Self-supervised learning is more robust to dataset imbalance. In: ICLR (2022)
22. Liu, Z., Miao, Z., Zhan, X., Wang, J., Gong, B., Yu, S.X.: Large-scale long-tailed recognition in an open world. In: CVPR, pp. 2537–2546 (2019)
23. Sinha, S., Ohashi, H.: Difficulty-net: learning to predict difficulty for long-tailed recognition. In: WACV, pp. 6433–6442 (2023)
24. Tian, Y., Henaff, O.J., van den Oord, A.: Divide and contrast: self-supervised learning from uncurated data. In: ICCV, pp. 10063–10074 (2021)
25. Wang, P., Han, K., Wei, X., Zhang, L., Wang, L.: Contrastive learning based hybrid networks for long-tailed image classification. In: CVPR, pp. 943–952 (2021)
26. Wu, D., Dai, Q., Liu, J., Li, B., Wang, W.: Deep incremental hashing network for efficient image retrieval. In: CVPR, pp. 9069–9077 (2019)
27. Xie, S., Girshick, R.B., Dollár, P., Tu, Z., He, K.: Aggregated residual transformations for deep neural networks. In: CVPR, pp. 5987–5995 (2017)
28. Yang, Z., Wu, D., Zhang, W., Li, B., Wang, W.: Handling label uncertainty for camera incremental person re-identification. In: MM (2023)
29. You, Y., Chen, T., Sui, Y., Chen, T., Wang, Z., Shen, Y.: Graph contrastive learning with augmentations. In: NeurIPS, pp. 5812–5823 (2020)
30. Zhang, H., Cisse, M., Dauphin, Y.N., Lopez-Paz, D.: mixup: beyond empirical risk minimization. In: ICLR (2018)
31. Zhang, W., Wu, D., Zhou, Y., Li, B., Wang, W., Meng, D.: Binary neural network hashing for image retrieval. In: SIGIR, pp. 317–326 (2021)
32. Zhao, S., Wu, D., Zhang, W., Zhou, Y., Li, B., Wang, W.: Asymmetric deep hashing for efficient hash code compression. In: MM, pp. 763–771 (2020)
33. Zhao, S., Wu, D., Zhou, Y., Li, B., Wang, W.: Rescuing deep hashing from dead bits problem. In: IJCAI, pp. 1338–1344 (2021)
34. Zhou, Z., Yao, J., Wang, Y.F., Han, B., Zhang, Y.: Contrastive learning with boosted memorization. In: ICML, pp. 27367–27377 (2022)
35. Zhu, J., Wang, Z., Chen, J., Chen, Y.P., Jiang, Y.: Balanced contrastive learning for long-tailed visual recognition. In: CVPR, pp. 6898–6907 (2022)

MFNet: A Channel Segmentation-Based Hierarchical Network for Multi-food Recognition

Kelei Jin⬮, Jing Chen(✉)⬮, and Tingting Song⬮

Jiangnan University, Wuxi 214122, China
{6213113072,6213113018}@stu.jiangnan.edu.cn, chenjing@jiangnan.edu.cn

Abstract. Due to significant differences between foods from different regions, there is a relative lack of publicly available multi-food image datasets, which is particularly evident in the field of Chinese cuisine. To alleviate this issue, we create a large-scale Chinese food image dataset, JNU FoodNet, which contains 17,128 original images. Although a considerable amount of prior research has focused on single food image recognition, such methods are not suitable for recognizing multiple food items in one image. Moreover, in a food image, there are usually multiple food regions, and the key features of each region tend to gradually disperse from the center to the edges, leading to cumulative errors. To overcome this difficulty, we design a selective discriminative feature constrained module, SGC, which restricts model attention to regions from a global information perspective. Furthermore, we propose a progressive hierarchical network, MFNet, based on channel segmentation from both the whole image and local region perspectives, combined with the SGC branch. Experimental results show that MFNet achieves state-of-the-art mAP values on JNU FoodNet, UEC Food-100, and UEC Food-256.

Keywords: Food computing · Hierarchical structure · Multi-food recognition

1 Introduction

The main task of food image recognition is to automatically identify the category, quantity, and other information of food by processing and analyzing food images using computer vision technology [21]. The accuracy of food image recognition results will have a decisive impact on downstream tasks related to it, such as calorie and energy estimation [12]. Single food recognition, as shown in Fig. 1(a), refers to the identification of a single food item in a single image [23]. Multi-food recognition, as shown in Figs. 1(b), (c), and (d), refers to the identification of multiple food items in a single image.

Most research has focused on the recognition of single food images based on global features [18,23]. However, multi-food images are also common in real-life dining scenarios, and existing research has mostly focused on specific regional

Q. Liu et al. (Eds.): PRCV 2023, LNCS 14433, pp. 16–28, 2024.
https://doi.org/10.1007/978-981-99-8546-3_2

Fig. 1. Examples of food images. (a) is a single food image. (b) represents a multi-food image in which each type of food is placed in a different dish. (c) is also a multi-food image in which different types of food are placed in the same dish. (d) is another type of multi-food image in which the boundaries between different foods are less clear.

foods, either by first localizing the regions of each food and then classifying the food category within the region, or by using co-occurrence matrices of dataset labels to identify the food in images [2,17,19]. Nevertheless, recognizing multiple foods has long been a challenge due to the significant differences between foods from different regions, the inter-class similarities and intra-class differences of food categories [20], and the overlapping boundaries of food regions in images. These problems are particularly prominent in the field of Chinese cuisine, and there is a lack of publicly available targeted datasets. Therefore, we create a new and challenging Chinese multi-food image dataset, JNU FoodNet, which includes 17,128 original images of the most popular food categories.

In Figs. 1(c) and (d), the key discriminative features of each food region in the image gradually become more dispersed as they move from the center to the boundary, which poses a challenge for annotation and model development [23]. To address this issue, we design a selective discriminative feature constraint module, SGC, based on global information to limit the accumulation of errors. Considering the limitations of existing food image recognition methods due to dataset constraints (different cultural regions or food image categories), they lack a comprehensive consideration of food images in various dining scenes [2,5, 17,19]. Therefore, based on JNU FoodNet, we propose a comprehensive multi-food image recognition network, MFNet, that considers the importance of global features of a single food image, regional feature information of different food areas in multiple food images, and combines them with the SGC module to limit the accumulation of errors. In summary, our contributions are as follows:

1. We create a new food image dataset named JNU FoodNet that includes bounding boxes and category annotations. This dataset aims to address the lack of available datasets in this field and provide a platform for relevant experiments to be conducted.
2. To constrain the model's focus to food regions, we design a selective discriminative feature constrained module, SGC, which limits the selective discriminative features based on global information to reduce model errors.
3. We propose a progressive hierarchical structure network for channel segmentation called MFNet, which leverages both channel and spatial-level features to understand food images. Extensive experiments on multiple datasets show that MFNet achieves state-of-the-art results.

2   Related Work

Currently, the construction of single-food datasets has relatively matured, including medium to large-scale datasets such as Food-101, VIREO food 172, ChinaMartFood-109, and Food2K. However, there is a lack of publicly available multi-food image datasets, with only a few existing datasets such as UEC Food-100 [19], UEC Food-256 [11], Mixed-dish dataset [2], and VIPER-FoodNet [17]. The number of datasets for multi-food recognition is still inadequate, especially for publicly available multi-food datasets that focus on Chinese cuisine, which are almost nonexistent.

The early research employed traditional methods, namely fusing four region recognizers to output the candidate regions for recognition, including using the Deformable Part Model method (DPM), circular recognizer, and JSEG region segmentation method [18,19] to identify the food regions in the entire image. This method considered the different characteristics of different foods and modeled them from multiple perspectives, but lacked subsequent research.

Recent studies have used CNN-based deep learning networks to model different food image datasets [8,24,30]. These methods have achieved encouraging results on specific food image datasets, such as using Affinity Propagation (AP) clustering and label-level co-occurrence matrices with Faster R-CNN for food image recognition [2,17]. Other works have achieved good experimental results by adding a CA attention mechanism on top of YOLOv5, which has better performance [5]. However, most of these methods focus on specific scenario datasets, such as Western and mixed food datasets, and their applicable scenarios are relatively narrow and do not fully represent real dining scenarios. In contrast, our model is capable of processing food images in Chinese dining scenarios that are closer to reality.

3   Food Image Datasets Construction

3.1   Food Images Collection

The sources of food image collection include various occasions such as school canteens, restaurants, and home dining, as well as some public food datasets and online resources. Through the collection, screening, and classification by 96 volunteers, the most popular 92 categories are selected, totaling 17,128 images. To restore the real scenes as much as possible, data augmentation techniques such as horizontal and vertical flipping are used to expand the dataset, resulting in a total of 22,801 images.

3.2   Annotation and Statistics

Multi-food images often present challenges in terms of boundary blur, occlusion, and variations in shape, which make annotation difficult. To address these issues, we develop the following annotation rules: 1) marking the upper-left and lower-right coordinates of the bounding box that encloses the smallest food area; 2)

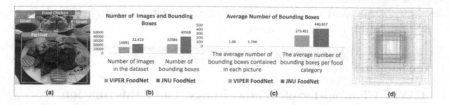

Fig. 2. The annotation details of JNU FoodNet and the comparison of its data with VIPER-FoodNet.

selecting the food category that is most similar to the enclosed area, as shown in Fig. 2(a); 3) conducting cross-validation by multiple annotators. In total, we annotate 40,568 bounding boxes, with an average of 1.796 bounding boxes per image and 440.957 bounding boxes per food category. We compare the differences in quantity between VIPER-FoodNet [17] and JNU FoodNet, as shown in Fig. 2(b–c). As Fig. 2(d) demonstrates, the bounding boxes in JNU FoodNet are concentrated around the center of the image, and most images contain multiple foods.

4   Method

Inspired by the design idea of ShuffleNet V2 [16], the overall network structure of MFNet is shown in Fig. 3, which comprises three branches: CWF, SGC, and SPF. These branches model the input feature maps at different visual levels and learn information at different scales [28]. Specifically, CWF captures the global information of the food image at the channel level, while SGC and SPF obtain the food image information from the global and local levels, respectively. First, we aggregate the outputs Y_{SGC} and Y_{SPF} of SGC and SPF, and then concatenate them with the output Y_{CWF} of CWF to obtain the final output result, as shown in Eq. 1. The model can better attend to both the "where" and

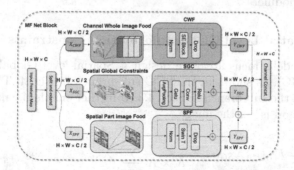

Fig. 3. The overall overview of MFNet proposed for multi-food image recognition is presented herein.

the "what" of the food [31]. The implementation details of each branch will be described in detail in the following sections.

$$Y = Concat(Y_{CWF}, Y_{SGC} + Y_{SPF}) \tag{1}$$

4.1   CWF: Channel-Level Whole Image Food Information Acquisition

Early studies focus on designing classifiers to identify food information in the entire image region [18]. Inspired by these, we focus on the global information at the channel level, and explicitly modeled the food information correlation between channels. The main part of the Channel-level Whole image Food (CWF), which acquires the global food information at the channel level, is implemented using a Squeeze-and-Excitation (SE) block network structure [7], and the specific implementation is shown in the CWF sub-branch in Fig. 4.

$$Z_{X_{CWF}} = F_{sq}(X_{CWF}) = \frac{1}{H \times W} \sum_{i=1}^{H} \sum_{j=1}^{W} X_{CWF}(i,j) \tag{2}$$

$Z \in R^C$, F_{sq} denotes the squeeze transformation, while $X_{CWF}(i,j)$ represents the feature value of X_{CWF} at the (i,j) position. $H \times W$ denotes the spatial dimension of the feature map, C representing the channel dimension of the feature map.

$$S_{X_{CWF}} = F_{ex}(Z_{CWF}, W) = \sigma(g(Z_{X_{CWF}}, W)) = \sigma(W_2 \delta(W_1 Z_{CWF})) \tag{3}$$

$S \in R^C$, F_{ex} represents the expand transformation, with $W_1 \in R^{\frac{C}{r} \times C}$ and $W_2 \in R^{C \times \frac{C}{r}}$, where r is the reduction factor. δ represents the ReLu activation function, and a sigmoid gate mechanism is used at the end.

$$\tilde{X}_{CWF} = F_{scale}(X_{CWF}, S_{X_{CWF}}) = X_{CWF} S_{X_{CWF}} \tag{4}$$

F_{scale} represents the matrix multiplication between the input X_{CWF} and $S_{X_{CWF}}$ in the original module.

4.2   SGC: Spatial-Level Global Information Constraints

To limit the model's focus area, we design a Spatial-level Global Information Constraints (SGC), as shown in the SGC sub-branch in Fig. 4. The purpose is to combine with the sub-branch in Sect. 4.3 to further enable the model to focus on "where the food is" in the image.

$$\tilde{X}_{SGC} = F_{GAP}(X_{SGC}) = \frac{1}{K \times K} \sum_{p,q=1}^{K} X_{SGC:,i+p-\frac{K+1}{2},i+q-\frac{K+1}{2}} \tag{5}$$

For a given input X_{SGC} of the SGC branch, as described in Formula 5, we first perform an average pooling on X_{SGC} aimed at capturing the global contextual

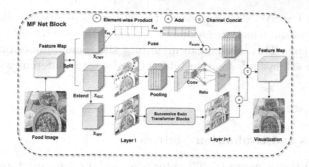

Fig. 4. The MFNet block comprises three sub-branch modules, namely CWF, SGC, and SPF, which perform specific processing steps on the input features.

information of the input features. K represents the size of the pooling kernel, and this operation partly compensates for the deficiency of the lack of inter-window contextual information interaction in the W-MSA operation of the first continuous sub-block of the Swin-Transformer block in the next branch module.

$$Z_{\tilde{X}_{SGC}} = \text{MLP}(\xi(\text{Norm}(\tilde{X}_{SGC}))) \tag{6}$$

Norm represents the normalization operation, where $\tilde{X}_{SGC}$ represents the output of the previous step's F_{GAP} operation, ξ denotes the GELU activation function. MLP represents Multi-Layer Perceptron, which is used here for extracting discriminative optimal features, subsequently limiting the accumulation of errors, and improving the discriminative power of the branching model [1].

4.3   SPF: Spatial-Level Part Image Food Information Acquisition

The main body of the Spatial-level Part image Food information acquisition (SPF) branch is composed of a continuous Swin-Transformer block [15], aiming to capture the features of multiple food regions distributed in the image. The implementation details are illustrated in the SPF sub-branch shown in Fig. 4.

$$\hat{Z}^l = \text{W} - \text{MSA}\left(\text{LN}\left(\text{X}_{\text{SPF}}\right)\right) + \text{X}_{\text{SPF}} \tag{7}$$

$$Z^l = \text{MLP}\left(\text{LN}\left(\hat{Z}^l\right)\right) + \hat{Z}^l \tag{8}$$

$$\hat{Z}^{l+1} = \text{SW} - \text{MSA}\left(\text{LN}\left(Z^l\right)\right) + Z^l \tag{9}$$

$$Z^{l+1} = \text{MLP}\left(\text{LN}\left(\hat{Z}^{l+1}\right)\right) + \hat{Z}^{l+1} \tag{10}$$

Formulas 7 and 8, along with Formulas 9 and 10, respectively constitute the first and second sub-blocks' self-attention mechanisms used in the Swin-Transformer block [15]. The W-MSA denotes the window-based multi-head self-attention mechanism, which performs local self-attention in each non-overlapping image window. The SW-MSA represents the shift-window-based multi-head self-attention mechanism, which merges non-adjacent sub-windows obtained through shifting operations. This mechanism maintains the same number of windows as the W-MSA to achieve information interaction between different windows.

5   Experiments

5.1   Datasets and Evaluation Metrics

UEC Food-100 [19] consists of 100 categories and 12,740 annotated images of Japanese food, where each food item is annotated with a corresponding bounding box. UEC Food-256 [11] is an extension of the UEC Food-100 dataset, containing 256 categories and 25,088 images of Japanese food. We use mAP [4] as a comprehensive evaluation metric in the experiments.

5.2   Implementation Details

We pretrain the model on a large-scale image dataset MS COCO. The input image size is set at 640×640 pixels, and the model is trained for 200 epochs using a batch size of 80 on an NVIDIA GeForce RTX GPU. The initial learning rate is set as 0.01. The loss function used is a combination of BCE loss and CIOU loss [32]. Similar to the approach used in [19], we use a training-to-testing ratio of 1:9 for the experiments.

5.3   Performance Comparison with Other Method

Table 1. The performance comparison of MFNet and other methods on publicly available datasets UEC Food-100 and UEC Food-256 is presented.

Method	mAP_0.5/ UEC Food-100	mAP_0.5/UEC Food-256
Faster R-CNN	42.00%	-
MobileNet [6]	68.25%	-
TLA-MobileNet [25]	76.37%	-
MobileNetV2 [22]	59.51%	-
TLA-MobileNetV2 [25]	78.29%	-
ResNet18 [25]	53.19%	-
TLA-ResNet18 [25]	61.66%	-
BP-based model [24]	49.90%	-
DeepFood [8]	17.50%	10.50%
Visual Aware Hierarchy [17]	60.63%	56.73%
Baseline [9]	86.24%	66.99%
MFNet	89.21%	72.48%

Comparison on UEC Food-100 and UEC Food-256. We compare the mAP of MFNet and other methods on the UEC Food-100 and UEC Food-256 datasets, as shown in Table 1. The experimental results demonstrate that MFNet achieves the highest average precision compared with lightweight models such as MobileNet and MobileNetV2, as well as with more complex models such as Faster R-CNN and Visual Aware Hierarchy [17].

Comparison on JNU FoodNet. We compare MFNet with existing methods for food recognition and object detection [2,5,10,13,26,27] on JNU Food-Net, a food image dataset. In addition, some attention models [3,14,29,31] with state-of-the-art performance are also included in the comparison experiment. As shown in Table 2, overall experimental results indicate that MFNet achieves state-of-the-art performance in food image recognition compared to other methods. Specifically, compared with the DISH-YOLOv5 [5] model which uses the same baseline algorithm [9], MFNet achieves a 15.48% increase in mAP value.

Table 2. The performance comparison of MFNet and other methods on the new food image dataset, JNU FoodNet, is presented.

Method	precision	recall	mAP_0.5
Faster R-CNN(ResNet-50) [2]	-	-	44.56%
Faster R-CNN(VGG) [2]	-	-	54.37%
EffcientDet [26]	-	-	63.60%
Baseline [9]	70.06%	67.20%	69.70%
Baseline +ViT [3]	68.43%	66.78%	69.44%
Baseline +S^{2}MLPV2 [29]	73.29%	67.47%	71.34%
Baseline+GAM [14]	71.59%	67.20%	69.61%
Baseline+ShuffleAttention [31]	70.00%	67.33%	69.24%
Dish-Yolov5 [5]	73.02%	56.77%	60.49%
Dish-Yolov5+CBAM [5]	55.07%	50.46%	53.36%
Dish-Yolov5+SE [5]	65.58%	60.13%	62.43%
Yolov6 [13]	59.10%	62.00%	62.20%
Yolov7 [27]	71.48%	72.53%	74.37%
Yolov8 [10]	71.85%	67.99%	72.10%
MFNet	75.73%	71.88%	75.97%

5.4 Ablation Study

Table 3 lists the contributions of the three branches, CWF, SGC, and SPF, in MFNet. With the addition of each branch, the mAP value increases, indicating that the three branches are complementary and indispensable to each other.

Table 3. Ablation study of sub-branch structures.

	CWF	SGC	SPF	precision	recall	mAP_0.5
MFNet				70.06%	67.20%	69.70%
			√	71.45%	66.31%	71.12%
		√	√	72.10%	67.90%	71.39%
	√	√	√	74.81%	72.56%	75.94%

Figure 5 further demonstrates the improvement of 20 food categories after the incorporation of three sub-branches, with a significant increase in their mAP values. For example, "Shredded Pork Noodles With Mushrooms" increased from 14.20% to 49.70% in terms of recognition results. "Watermelon" is often presented in image as either cut into pieces or as a whole spherical shape. Despite this, its recognition results still significantly improve upon the inclusion of our designed modules, with an increase in mAP from 67.40% to 79.00%.

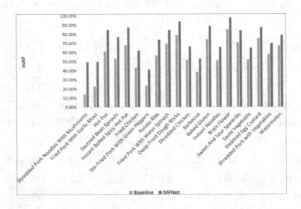

Fig. 5. The mAP values for 20 food categories on MFNet and Baseline models.

5.5   Visualization Result

As shown in (b), (e), and (i) in Fig. 6, without using our designed module, the regions in the feature maps of various food categories do not show clear distinctions, and are relatively dark. However, after being processed by our designed module, as shown in (c), (f), and (j) in Fig. 6, the key features of regions in the feature maps of various food categories become more distinguishable, as evidenced by the clear boundaries between light and dark areas in the figure.

Figure 7 further illustrates a visual comparison of recognition results between MFNet and Baseline model. As the figure shows, our model usually achieves higher confidence and more comprehensive recognition results. Especially for the recognition of mixed foods, the recognition results of MFNet are significantly

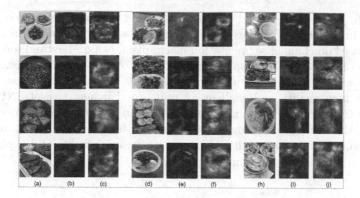

Fig. 6. (a), (d), and (h) are examples of original input food images. (b), (e), and (i) show the distribution of attention in Baseline. (c), (f), and (j) depict the distribution of attention in MFNet.

improved compared to Baseline. For example, as shown in (c) of Fig. 7, the mixed food combination "Sausage" and "Rice" at the bottom right of the image is recognized by our model with a confidence score of 0.72, but Baseline model only recognizes "Sausage".

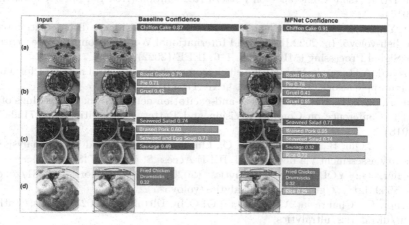

Fig. 7. Partial food image recognition results on the MFNet method are shown as examples.

6   Conclusion

In this paper, we first design a selective discriminative feature restriction module (SGC) based on the characteristics of feature distribution of each food area in multi-food images to restrict the model error accumulation. Then, we propose a new multi-food recognition method MFNet by combining the importance of global features of a single food image (CWF) and feature information of each food area in multi-food images (SPF), together with the SGC module. This method, based on channel segmentation mechanism, can understand food image information from multiple perspectives, including overall and local image information as well as error accumulation restriction. Moreover, we create a new large-scale, high-quality and challenging multi-food image dataset JNU Food-Net in the Chinese food domain, which aims to address the lack of food image datasets in this domain and provide a platform for future related work. In future work, we will continue to enrich and improve existing datasets and explore new perspectives for learning food image information.

References

1. Alaeddine, H., Jihene, M.: Wide deep residual networks in networks. Multimedia Tools Appl. **82**(5), 7889–7899 (2023)
2. Deng, L., et al.: Mixed-dish recognition with contextual relation networks. In: Proceedings of the 27th ACM International Conference on Multimedia, pp. 112–120 (2019)
3. Dosovitskiy, A., et al.: An image is worth 16x16 words: transformers for image recognition at scale. arXiv preprint arXiv:2010.11929 (2020)
4. Ege, T., Yanai, K.: Estimating food calories for multiple-dish food photos. In: 2017 4th IAPR Asian Conference on Pattern Recognition (ACPR), pp. 646–651. IEEE (2017)
5. Gu, Y., Cai, L., Wang, J., Chen, Y., Zhu, P., Gao, M.: Chinese dish detection based on dish-yolov5. In: 2022 IEEE 32nd International Workshop on Machine Learning for Signal Processing (MLSP), pp. 1–6. IEEE (2022)
6. Howard, A.G., et al.: Mobilenets: efficient convolutional neural networks for mobile vision applications. arXiv preprint arXiv:1704.04861 (2017)
7. Hu, J., Shen, L., Sun, G.: Squeeze-and-excitation networks. In: Proceedings of the IEEE Conference on Computer Vision and Pattern Recognition, pp. 7132–7141 (2018)
8. Jiang, L., Qiu, B., Liu, X., Huang, C., Lin, K.: Deepfood: food image analysis and dietary assessment via deep model. IEEE Access **8**, 47477–47489 (2020)
9. Jocher, G.: YOLOv5 by Ultralytics (2020). https://doi.org/10.5281/zenodo.3908559. https://github.com/ultralytics/yolov5
10. Jocher, G., Chaurasia, A., Qiu, J.: YOLO by Ultralytics (2023). https://github.com/ultralytics/ultralytics
11. Kawano, Y., Yanai, K.: FoodCam-256: a large-scale real-time mobile food recognition system employing high-dimensional features and compression of classifier weights. In: Proceedings of the 22nd ACM International Conference on Multimedia, pp. 761–762 (2014)
12. Kim, J., Lee, Y.K., Herr, P.M.: The impact of menu size on calorie estimation. Int. J. Hosp. Manag. **100**, 103083 (2022)
13. Li, C., et al.: Yolov6: a single-stage object detection framework for industrial applications. arXiv preprint arXiv:2209.02976 (2022)
14. Liu, Y., Shao, Z., Hoffmann, N.: Global attention mechanism: Retain information to enhance channel-spatial interactions. arXiv preprint arXiv:2112.05561 (2021)

15. Liu, Z., et al.: Swin transformer: hierarchical vision transformer using shifted windows. In: Proceedings of the IEEE/CVF International Conference on Computer Vision, pp. 10012–10022 (2021)
16. Ma, N., Zhang, X., Zheng, H.T., Sun, J.: ShuffleNet V2: practical guidelines for efficient CNN architecture design. In: Proceedings of the European Conference on Computer Vision (ECCV), pp. 116–131 (2018)
17. Mao, R., He, J., Shao, Z., Yarlagadda, S.K., Zhu, F.: Visual aware hierarchy based food recognition. In: Del Bimbo, A., et al. (eds.) ICPR 2021. LNCS, vol. 12665, pp. 571–598. Springer, Cham (2021). https://doi.org/10.1007/978-3-030-68821-9_47
18. Matsuda, Y., Hoashi, H., Yanai, K.: Recognition of multiple-food images by detecting candidate regions. In: 2012 IEEE International Conference on Multimedia and Expo, pp. 25–30. IEEE (2012)
19. Matsuda, Y., Yanai, K.: Multiple-food recognition considering co-occurrence employing manifold ranking. In: Proceedings of the 21st International Conference on Pattern Recognition (ICPR 2012), pp. 2017–2020. IEEE (2012)
20. Metwalli, A.S., Shen, W., Wu, C.Q.: Food image recognition based on densely connected convolutional neural networks. In: 2020 International Conference on Artificial Intelligence in Information and Communication (ICAIIC), pp. 027–032. IEEE (2020)
21. Min, W.Q., Liu, L.H., Liu, Y.X., Luo, M.J., Jiang, S.Q.: A survey on food image recognition. Chin. J. Comput. 45(3) (2022)
22. Sandler, M., Howard, A., Zhu, M., Zhmoginov, A., Chen, L.C.: Mobilenetv 2: inverted residuals and linear bottlenecks. In: Proceedings of the IEEE Conference on Computer Vision and Pattern Recognition, pp. 4510–4520 (2018)
23. Sheng, G., Sun, S., Liu, C., Yang, Y.: Food recognition via an efficient neural network with transformer grouping. Int. J. Intell. Syst. 37(12), 11465–11481 (2022)
24. Shimoda, W., Yanai, K.: CNN-based food image segmentation without pixel-wise annotation. In: Murino, V., Puppo, E., Sona, D., Cristani, M., Sansone, C. (eds.) ICIAP 2015. LNCS, vol. 9281, pp. 449–457. Springer, Cham (2015). https://doi.org/10.1007/978-3-319-23222-5_55
25. Sun, J., Radecka, K., Zilic, Z.: Exploring better food detection via transfer learning. In: 2019 16th International Conference on Machine Vision Applications (MVA), pp. 1–6. IEEE (2019)
26. Tan, M., Pang, R., Le, Q.V.: Efficientdet: scalable and efficient object detection. In: Proceedings of the IEEE/CVF Conference on Computer Vision and Pattern Recognition, pp. 10781–10790 (2020)
27. Wang, C.Y., Bochkovskiy, A., Liao, H.Y.M.: Yolov7: trainable bag-of-freebies sets new state-of-the-art for real-time object detectors. In: Proceedings of the IEEE/CVF Conference on Computer Vision and Pattern Recognition, pp. 7464–7475 (2023)
28. Wang, J., Lv, P., Wang, H., Shi, C.: SAR-U-Net: squeeze-and-excitation block and atrous spatial pyramid pooling based residual U-Net for automatic liver segmentation in computed tomography. Comput. Methods Programs Biomed. 208, 106268 (2021)
29. Yu, T., Li, X., Cai, Y., Sun, M., Li, P.: S^2-MLPV2: improved spatial-shift MLP architecture for vision. arXiv preprint arXiv:2108.01072 (2021)
30. Zhang, H., et al.: Resnest: split-attention networks. In: Proceedings of the IEEE/CVF Conference on Computer Vision and Pattern Recognition, pp. 2736–2746 (2022)

31. Zhang, Q.L., Yang, Y.B.: SA-Net: shuffle attention for deep convolutional neural networks. In: ICASSP 2021-2021 IEEE International Conference on Acoustics, Speech and Signal Processing (ICASSP), pp. 2235–2239. IEEE (2021)
32. Zheng, Z., Wang, P., Liu, W., Li, J., Ye, R., Ren, D.: Distance-IoU loss: faster and better learning for bounding box regression. In: Proceedings of the AAAI Conference on Artificial Intelligence, vol. 34, pp. 12993–13000 (2020)

Improving the Adversarial Robustness of Object Detection with Contrastive Learning

Weiwei Zeng[1], Song Gao[1(✉)], Wei Zhou[1], Yunyun Dong[2], and Ruxin Wang[1]

[1] National Pilot School of Software, Engineering Research Center of Cyberspace, Yunnan University, Kunming, China
zww@mail.ynu.edu.cn, {gaos,zwei}@ynu.edu.cn
[2] National Pilot School of Software, School of Information Science and Engineering, Yunnan University, Kunming, China
dongyy929@ynu.edu.cn

Abstract. Object detection plays a crucial role and has wide-ranging applications in computer vision. Nevertheless, object detectors are susceptible to adversarial examples. Some works have been presented to improve the adversarial robustness of object detectors, which, however, often come at the loss of some prediction accuracy. In this paper, we propose a novel adversarial training method that integrates the contrastive learning into the training process to reduce the loss of accuracy. Specifically, we add a contrastive learning module to the primary feature extraction backbone of the target object detector to extract contrastive features. During the training process, the contrastive loss and detection loss are used together to guide the training of detectors. Contrastive learning ensures that clean and adversarial examples are more clustered and are further away from decision boundaries in the high-level feature space, thus increasing the cost of adversarial examples crossing decision boundaries. Numerous experiments on PASCAL-VOC and MS-COCO have shown that our proposed method achieves significantly superior defense performance.

Keywords: Object detection · adversarial training · contrastive learning · adversarial robustness

1 Introduction

Deep learning has achieved good results in many computer vision tasks, such as image classification [25], semantic segmentation [15], and object detection [22]. However, deep learning models are susceptible to adversarial examples which are generated by adding imperceptible perturbations to clean examples [1,9,17,18, 27]. Adversarial examples bring higher requirements to the applications of deep learning in security-related fields. In this paper, we focus on object detection and explore ways to enhance the robustness of detectors against adversarial examples.

Q. Liu et al. (Eds.): PRCV 2023, LNCS 14433, pp. 29–40, 2024.
https://doi.org/10.1007/978-981-99-8546-3_3

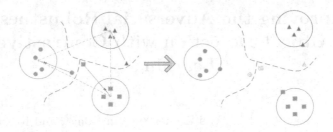

Fig. 1. Contrastive learning can shorten the distance between similar images and expand the distance between different types of images in the feature space. This increases the distance from the example to the decision boundary and increases the difficulty of attacks.

Lu et al. [16] first introduced that adding specific noise to traffic signs could fool an advanced detector (Faster RCNN [24] or YOLO9000 [23]). Since then, many adversarial attacks against object detection tasks have been proposed, such as DAG [30], RAP [13], TOG [4] and UEA [28], which seriously threaten the security of object detectors.

Some researches have been presented to defend against adversarial examples like input reconstruction [7,26], adversarial training [2,9,17,32] and detection only [6,8]. Among these defenses, adversarial training has been proven to be a highly effective way to promote the adversarial robustness of deep learning models. By incorporating adversarial examples into the training data, deep models become more resilient to attacks and perform well in real-world scenarios. Zhang et al. [32] first introduced adversarial training to enhance the adversarial robustness of object detectors. Although their method improves the robustness of object detectors against adversarial samples, it loses too much accuracy on clean examples. Chen et al. [2] proposed CWAT that improves the generation method of adversarial examples but still does not solve Zhang's problem. Xie et al. [30] leveraged knowledge distillation and feature alignment to reduce the loss on clean samples, but the effectiveness is limited.

In this paper, we propose a novel method that incorporates contrastive learning into the adversarial training process. As shown in Fig. 1, the motivation behind our method is to increase the similarity of high-level features between clean examples and their corresponding adversarial examples, while reducing the similarity of examples from different categories. In detail, we add a contrastive learning module following the primary feature extraction backbone to extract contrastive features of inputs. During training, we first generate the adversarial examples of input images with white-box settings. Then clean images and their corresponding adversarial examples are passed into the modified detector to get contrastive features and detection results. Finally, by optimizing the proposed contrastive adversarial loss to obtain the adversarial robust detector. Our proposed method clusters clean and adversarial examples of the same class together, and keeps the clustering centers of different categories away from each other, which increases the cost of adversarial attacks.

We summarize the main contributions of our work as follows:

- We propose a novel defense method that integrates contrastive learning into the adversarial training of object detectors. Our method can improve the similarity of the high-level representations of clean examples and adversarial examples, diffuse cluster centres of different classes and reduce the loss of accuracy of clean examples.
- Our approach requires only simple modifications to the architecture of the target network, and the modifications exist only during the training process, without consuming additional computing resources in the testing phase.
- We test our proposed method on PASCAL-VOC and MS-COCO, and the experimental results indicate that our approach has outstanding performance on both adversarial examples and clean examples.

2   Related Work

2.1   Adversarial Attacks

Szegedy et al. [27] first proposed to maximize the prediction error of the target network to generate adversarial examples, denoted as:

$$x_{adv} = argmax(l(f(x, \theta), y)), \tag{1}$$

where l is the target function, f is a deep model with parameters θ, x is an input, and y is the true label of x. x_{adv} is the generated adversarial example.

Fast Gradient Sign Method (FGSM) [9] is an earlier method to produce adversarial examples, it involves a single-step gradient update along the direction of the gradient sign at each pixel. Moosavi-Dezfooli et al. [18] introduced a technique known as DeepFool to determine the closest distance from the original input to the decision boundary. C&W [1] generates adversarial examples by optimizing low-perturbation adversarial examples through iteration. Madry et al. [17] suggested a highly effective method termed PGD for first-order attacks, which iteratively adds perturbations to the input image along the direction of the gradient. Lu et al. [16] first demonstrated that adversarial examples also exist in object detection. They proposed DFool to attack Faster-RCNN and showed that the generated adversarial samples can be transferred to YOLO9000 [23]. Xie et al. [30] proposed DAG that attacks the classification loss of the two-stage object detector. Li et al. [13] proposed RAP by destroying the unique RPN structure in the two-stage detector. Chow et al. [4] proposed TOG to attack both two-stage and one-stage detectors.

2.2   Adversarial Defenses

To improve the adversarial robustness of deep learning models, researchers have proposed many defense approaches, such as defensive distillation [20], adversarial training [2,9,32], input reconstruction [7], and detection only [6,8]. Among these strategies, adversarial training is considered one of the most effective methods. Adversarial training involves adding images deliberately altered to the training data to improve the deep model's ability to handle adversarial examples. For example, Goodfellow et al. [9] utilized FGSM to generate adversarial examples and added them into the training data to train deep models. Madry et al. [17] adopted PGD to generate adversarial examples for adversarial training. In object detection, Zhang and Wang [32] divided the object detection task into class and location and generated different adversarial examples for different task domains. Chen et al. [2] proposed CWAT that balances the relationship between adversarial examples and clean examples through a class-aware method for class-aware adversarial training. Xu et al. [31] proposed FA that uses the feature alignment to align clean samples and adversarial samples in the feature space. Although these methods can improve the robustness of object detectors against adversarial examples, they may lead to a decrease in the accuracy of detectors on clean examples.

2.3   Contrastive Learning

The central idea behind contrastive learning is to shorten the distance between similar images and expand the distance between different types of images in the feature space to obtain a better feature extraction model. Hadsell et al. [10] used two images for contrastive learning. Wu et al. [29] proposed InstDisc that encodes the low-dimensional features of input images and distinguishes them in the feature space as much as possible. He et al. [11] proposed MoCo that encodes images into query vectors and key vectors. Chen et al. [3] proposed SimCLR to generate different data perspectives through various data enhancement method, and trained an encoder to ensure that the different perspectives of an image are similar and the different images are dissimilar. Zhang et al. [33] proposed a novel spectral feature enhancement method for contrastive learning. Pan et al. [19] proposed a novel contrastive combination clustering method for adjusting combination features by dynamically establishing similar and dissimilar pairs. In this study, we incorporate contrastive learning into adversarial training and adopt adversarial examples as positive examples. In this way, the target detector can learn the potential connection between clean examples and adversarial examples, and increase the distance between different examples in the high-level feature space (Fig. 2).

3   Proposed Method

In this section, we first introduce the contrastive learning module used in our work, and then describe the details of different object detectors with the contrastive learning module.

3.1   Contrastive Learning Module

The contrastive learning module is mainly composed of a global mean pooling (GAP) and two fully connected layers, in which ReLU is used as the activation

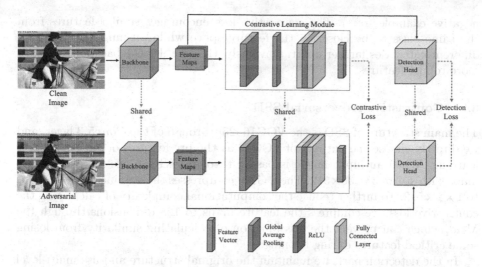

Fig. 2. Overall framework of our proposed Contrastive Adversarial Training for one-stage object detector. A clean example and its corresponding adversarial example are passed through the backbone to get feature maps, and then the feature maps are input into the contrastive learning module and the detection head for the contrastive loss and the detection loss, respectively.

function. The primary function of the contrastive learning module is to reduce the high-dimensional features of the backbone and obtain the low-dimensional contrastive features for comparison similarity calculation.

In the training phase, given an input image x, we need to obtain another perspective of the image. To achieve this, we utilize an adversarial attack (PGD in this work) to generate the corresponding adversarial example of x, and consider the adversarial example as an image from a different viewpoint of the original image. Subsequently, the original image and its corresponding adversarial example are passed through the backbone of a detector to obtain feature representations. We then employ the GAP operation to handle the complex-dimensional features. GAP compresses the data dimension while preserving background information, which is crucial in detection tasks. Next, we utilize a two-layer MLP to obtain contrastive vectors, and the objective of contrastive learning is

$$\mathcal{L}_{cl} = \frac{1}{2N} \sum_{k=1}^{N} \left[\ell\left(x_{2k}, x_{2k-1}\right) + \ell\left(x_{2k-1}, x_{2k}\right) \right], \tag{2}$$

$$\ell_{i,j} = -\log \frac{exp\left(sim\left(z_i, z_j\right)\right)/\tau}{\sum_{k=1}^{2N} \mathbb{1}_{[k \neq i \ and \ j]} exp\left(sim\left(z_i, z_j\right)\right)/\tau}, \tag{3}$$

where N is the batch size. For N clean examples randomly selected, we generate N adversarial examples by an attack, and calculate the cosine similarity between each image in $2N$ relative to its own positive and negative examples. The positive sample of z_i is its adversarial version z_j, and the other $2N - 2$ images are all negative examples of z_i. The contrastive loss encourages similar features from the same class to be closer in the feature space while pushing features from different categories farther apart. Obviously, this enables the backbone to learn more robust features.

3.2 Contrastive Adversarial SSD

The main structure of SSD is the VGG16 [25] through of the $Conv5\_3$ layer, and we directly choose the outputs of VGG16 as the pre-features of our contrastive learning. When an input image is passed through VGG16, we will get feature maps with size of $38 \times 38 \times 512$. Then, GAP compresses the obtained feature maps to $1 \times 1 \times 512$. To further reduce the computational complexity of calculating the contrastive loss, we compress the feature maps to 128 dimensions through the MLP, which can prevent the loss function of calculating similarity from losing some critical features during training.

In the detection part, we maintain the original structure and use multi-level feature maps to ensure the detection effect of objects with different scales. Finally, the objects of interest and their location information are obtained through the non-maximum suppression algorithm. Overall, the objective for our method with SSD is

$$\mathcal{L}_{ssd} = \mathcal{L}_{cl} + \mathcal{L}_{det}, \tag{4}$$

$$\mathcal{L}_{det} = \begin{cases} \dfrac{1}{M}(\mathcal{L}_{conf} + \alpha \mathcal{L}_{loc}) & M > 0 \\ 0 & M = 0, \end{cases} \tag{5}$$

$$\mathcal{L}_{conf} = -\sum_{i \in Pos}^{M} m_{ij}^p log(\hat{c}_i^p) - \sum_{i \in Neg} log(\hat{c}_i^b), \tag{6}$$

$$\mathcal{L}_{loc} = \sum_{i \in Pos}^{M} \sum m_{ij}^k smooth_{L_1}(P - T), \tag{7}$$

where M is the number of matched default boxes. P represents the predicted results, $P = \{x', y', w', h'\}$. x' and y' are the predicted position of the centre point of the default box, w' is the width of the default box, and h' is the height of the default box. T represents the ground truth of the input image, $T = \{x, y, w, h\}$, $m_{ij}^c = \{0, 1\}$ indicates that matching the i-th default box to the j-th ground truth box of category c. α is a hyper-parameter.

3.3 Contrastive Adversarial YOLO

YOLO consists of three main components: backbone, neck, and head. The backbone usually uses ResNet [12] or DarkNet [21] that have been pre-trained on large-scale datasets like ImageNet or COCO, and three information scales, 76×76, 38×38, and 19×19, are mainly used to extract the feature information of different scale objects in the detection. The neck is primarily accountable for fusing information from multiple scales. The head is responsible for extracting classification and localization information for outputs. Finally, the detection results are obtained through non-maximum suppression (NMS).

Since the backbone of YOLO produces three different sized feature maps to capture objects of different sizes in the image, we should consider the information from these three scales simultaneously. Firstly, we concatenate the high-dimensional features from the three scales and then utilize the contrastive learning module to extract contrastive features. We define the features obtained by the YOLO backbone as h_0, h_1, h_2, and the contrastive feature of an input is $f_{cl} = M(cat(G(h_0), G(h_1), G(h_2)))$, where cat is the concatenate operation, G denotes global average pooling, and M is the MLP. The total loss is

$$\mathcal{L}_{yolo} = \mathcal{L}_{loc} + \mathcal{L}_{conf} + \mathcal{L}_{cls} + \mathcal{L}_{cl}, \tag{8}$$

$$\mathcal{L}_{loc} = if^{obj}(2 - w_i \times h_i)(1 - CIOU), \tag{9}$$

$$\mathcal{L}_{conf} = if^{obj}BCE(\hat{C}_i, C_i) + if^{noobj}BCE(\hat{C}_i, C_i), \tag{10}$$

$$\mathcal{L}_{cls} = if^{obj} \sum_{c \in classes} BCE(\hat{c}_i, c_i), \tag{11}$$

where K is the number of predict boxes. $if^{obj} = \sum_{i=0}^{K \times K} \sum_{j=0}^{M} \mathbb{1}_{ij}^{obj}$. $\hat{C}^i$ represents the result of the model and C^i representing the ground truth.

3.4   Adversarial Training with Contrastive Learning

For an object detector, we add a contrastive learning module to it and use an attack to generate adversarial examples in the training phase. The total objective function is Eq. (12), and the specific training process of our method is summarized in Algorithm 1.

$$\mathcal{L}_{total} = \mathcal{L}_{cl} + \beta \mathcal{L}_{det}^{clean} + \gamma \mathcal{L}_{det}^{adv}. \tag{12}$$

After training, the contrastive learning module is useless. This means that our method does not incur any additional computational overhead during the testing phase, which is consistent with the traditional adversarial training.

4   Experiments

In this section, we first present the settings used in our experiments, and then conduct robustness tests on PASCAL-VOC and MS-COCO to evaluate the performance of our proposed method.

Algorithm 1: Contrastive Adversarial Training

Require: dataset $\mathcal{D}$, train epochs $\mathcal{T}$
1: Initialize θ
2: **for** epoch $t = 1$ to $\mathcal{T}$ **do**
3:      sample a batch images X from $\mathcal{D}$
4:      generate adversarial examples X_{adv}
5:      forward X and X_{adv}
6:      compute loss $\mathcal{L}_{total} = \mathcal{L}_{cl} + \beta \mathcal{L}_{det}^{clean} + \gamma \mathcal{L}_{det}^{adv}$
7:      update network parameter θ with stochastic gradient descent
8: **end for**

4.1 Experimental Settings

We test our method on PASCAL-VOC [5] and MS-COCO [14]. The attacks, including PGD [17], DAG [30], CWA [2], TOD [32], are set as default. And the metric for different methods is mAP@0.5. For SSD, we resize input images to 300×300, and set the initial learning rate to 0.002, the momentum to 0.9, and the weight decay to 0.0005. The learning rate drop method is cosine. The training batch size is 32. For YOLO, we resize input images to 416×416, and set the initial learning rate to 0.01, the momentum to 0.9, and the weight decay to 0.0005. The learning rate drop method is cosine. The training batch size is 8.

Table 1. The mAP(%) of different defenses with SSD on PASCAL-VOC.

	CLEAN	A_{cls}	A_{loc}	DAG	CWA	AVG
AT-CLS	46.7	21.8	32.2	-	-	27
AT-LOC	51.9	23.7	26.5	-	-	25.1
AT-CON	38.7	18.3	27.2	26.4	-	23.97
MTD	48	29.1	31.9	28.5	-	29.83
MTD-fast	46.6	31.1	41.8	48.6	18.2	34.93
TOAT-6	43	30	21.8	46.6	19.7	29.53
OWAT	51.8	32.7	43.4	50.4	20.3	36.7
CWAT	51.3	32.5	43.3	50.3	19.9	36.5
Ours	**74.7**	**47.4**	**53.2**	**69.47**	**45.99**	**54.01**

4.2 Defense Capability Evaluation

Verification on SSD. Table 1 shows the defense performances of different defense methods with SSD on PASCAL-VOC. We can see that our method achieves the highest scores on both clean examples and adversarial examples. Especially, our method can get 74.7% mAP on clean examples, which outperforms other defense methods by a big margin. Figure 3 provides an intuitive

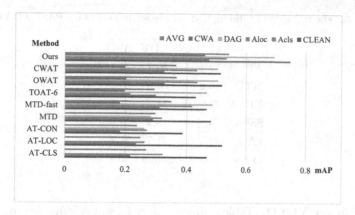

Fig. 3. The intuitive display of mAP scores of different defense methods on PASCAL-VOC when protecting SSD.

display, it can be seen that our method is superior to other baselines in all cases. Table 2 displays the defense performances of different methods with SSD on MS-COCO. We can see that our method also achieves the highest mAP scores on both adversarial examples and clean examples.

AT-CLS, AT-LOC and AT-CON only conduct adversarial training for specific object detection domains. Although these single-domain adversarial training methods can improve the adversarial robustness of the object detector, their forced learning of adversarial characteristics leads to drastic changes in decision boundaries, resulting in unsatisfactory performances on clean examples. MTD uses examples in the composite domain to train the object detector to learn the features of the training, making the object detector learn the features of pieces to improve the robustness of model. We can see that MTD has a good mAP on clean examples. CWAT uses the class-wise method to balance the unbalanced effects among various classes. It can effectively and uniformly improve the robustness of the target model against all target classes. Our approach takes advantage of contrastive learning to increase the robustness of the target detector by narrowing the distance between similar examples while rejecting heterogeneous sample centres. In this way, the cost of clean examples becoming adversarial will increase. Therefore, our method can achieve good defense performances on both adversarial examples and clean examples.

Table 2. The mAP(%) of different defenses with SSD on MS-COCO.

	CLEAN	A_{cls}	A_{loc}	CWA	AVG
MTD	19	12.7	14.6	8.2	11.83
MTD-FAST	24.2	16.7	18.2	7.7	14.2
TOAT-6	18.2	12	14.8	7.4	11.4
OWAT	21.1	12.9	16.9	7.4	12.4
CWAT	23.7	16.8	18.9	9.2	14.97
Ours	**30.14**	**20.29**	**19.84**	**18.71**	**19.61**

Table 3. The mAP(%) of different defenses with YOLO on PASCAL-VOC.

	CLEAN	ADV	AVG
AT	72.4	20.1	20.1
MTD	70.2	20.8	20.8
KDFA	73.9	21.7	21.7
SSFA	73	18.9	18.9
FA	74.5	21.7	21.7
Ours	**85.67**	**35.8**	**35.8**

Verification on YOLO. Table 3 shows the defense performances of defenses with YOLO on PASCAL-VOC. We can see that our method achieves the highest mAP scores in all cases. Specially, our method reaches an mAP of 85.67% on clean examples. Table 4 displays the defense performances of defenses with YOLO on MS-COCO. It can be seen that our method also achieves the highest mAP scores on both clean examples and mean adversarial robustness.

KDFA and SSFA [31] force the alignment in the feature space, thus significantly improving the robustness of the target detector on adversarial examples and clean examples. In contrast, our method still exhibits significant advantages, indicating that contrastive learning can effectively alleviate the problems existing in vanilla adversarial training and further enhance the adversarial robustness of the target detector.

Table 4. The mAP(%) of different defenses with YOLO on MS-COCO.

	CLEAN	ADV	AVG
AT	49	8.9	8.9
MTD	48.8	8.8	8.8
KDFA	50.6	11	11
SSFA	49.9	10.3	10.3
FA	51	12	12
Ours	**55.2**	**22.19**	**22.19**

5 Conclusion

We propose a method of adversarial training coupled with contrastive learning. Through simple and few modifications to the target detector, our proposed adversarial training method can enable the model to learn robust feature representations against adversarial examples and increase the cost of adversarial attack. We verify our approach on PASCAL-VOC and MS-COCO. The experimental results show that our method has a significant improvement compared to previous works, and is not as effective on YOLO as on SSD. YOLO's backbone is more complex than that of SSD, in order to obtain more robust feature space on YOLO, we adopt the multi-scale contrastive feature fusion. However, we only use simple concatenate operation to fuse this multi-scale information. In our future work, we will explore more effective way to fuse multi-scale information and further enhance the defense ability of our method on complex detectors.

Acknowledgement. This work is supported in part by the National Natural Science Foundation of China under Grant No. 62101480, the Yunnan Foundational Research Project under Grant No. 202201AT070173 and No. 202201AU070034, Yunnan Province Education Department Foundation under Grant No.2022j0008, in part by the National Natural Science Foundation of China under Grant 62162067, Research and Application of Object detection based on Artificial Intelligence, in part by the Yunnan Province expert workstations under Grant 202205AF150145.

References

1. Carlini, N., Wagner, D.A.: Towards evaluating the robustness of neural networks. In: S&P, pp. 39–57 (2017)
2. Chen, P., Kung, B., Chen, J.: Class-aware robust adversarial training for object detection. In: CVPR, pp. 10420–10429 (2021)
3. Chen, T., Kornblith, S., Norouzi, M., Hinton, G.E.: A simple framework for contrastive learning of visual representations. In: ICML, pp. 1597–1607 (2020)
4. Chow, K.H., et al.: Adversarial objectness gradient attacks in real-time object detection systems. In: TPS-ISA, pp. 263–272 (2020)
5. Everingham, M., Gool, L.V., Williams, C.K.I., Winn, J.M., Zisserman, A.: The pascal visual object classes (VOC) challenge. Int. J. Comput. Vis. **88**(2), 303–338 (2010)
6. Gao, S., et al.: Detecting adversarial examples on deep neural networks with mutual information neural estimation. IEEE Trans. Depend. Secure Comput. (2023). https://doi.org/10.1109/TDSC.2023.3241428
7. Gao, S., Yao, S., Li, R.: Transferable adversarial defense by fusing reconstruction learning and denoising learning. In: INFOCOMW, pp. 1–6 (2021)
8. Gao, S., Yu, S., Wu, L., Yao, S., Zhou, X.: Detecting adversarial examples by additional evidence from noise domain. IET Image Process. **16**(2), 378–392 (2022)
9. Goodfellow, I.J., Shlens, J., Szegedy, C.: Explaining and harnessing adversarial examples. In: ICLR (2015)
10. Hadsell, R., Chopra, S., LeCun, Y.: Dimensionality reduction by learning an invariant mapping. In: CVPR, pp. 1735–1742 (2006)
11. He, K., Fan, H., Wu, Y., Xie, S., Girshick, R.B.: Momentum contrast for unsupervised visual representation learning. In: CVPR, pp. 9726–9735 (2020)
12. He, K., Zhang, X., Ren, S., Sun, J.: Deep residual learning for image recognition. In: CVPR, pp. 770–778 (2016)
13. Li, Y., Bian, X., Lyu, S.: Attacking object detectors via imperceptible patches on background. arXiv preprint arXiv:1809.05966 (2018)
14. Lin, T.-Y., et al.: Microsoft COCO: common objects in context. In: Fleet, D., Pajdla, T., Schiele, B., Tuytelaars, T. (eds.) ECCV 2014. LNCS, vol. 8693, pp. 740–755. Springer, Cham (2014). https://doi.org/10.1007/978-3-319-10602-1_48

15. Liu, Q., et al.: Learning part segmentation through unsupervised domain adaptation from synthetic vehicles. In: CVPR, pp. 19118–19129 (2022)
16. Lu, J., Sibai, H., Fabry, E.: Adversarial examples that fool detectors. arXiv preprint arXiv:1712.02494 (2017)
17. Madry, A., Makelov, A., Schmidt, L., Tsipras, D., Vladu, A.: Towards deep learning models resistant to adversarial attacks. In: ICLR (2018)
18. Moosavi-Dezfooli, S., Fawzi, A., Frossard, P.: Deepfool: a simple and accurate method to fool deep neural networks. In: CVPR, pp. 2574–2582 (2016)
19. Pan, Z., Chen, Y., Zhang, J., Lu, H., Cao, Z., Zhong, W.: Find beauty in the rare: contrastive composition feature clustering for nontrivial cropping box regression. In: AAAI, pp. 2011–2019 (2023)
20. Papernot, N., McDaniel, P.D., Wu, X., Jha, S., Swami, A.: Distillation as a defense to adversarial perturbations against deep neural networks. In: S&P, pp. 582–597 (2016)
21. Redmon, J.: Darknet: open source neural networks in C (2013-2016). http://pjreddie.com/darknet/
22. Redmon, J., Divvala, S.K., Girshick, R.B., Farhadi, A.: You only look once: unified, real-time object detection. In: CVPR, pp. 779–788 (2016)
23. Redmon, J., Farhadi, A.: Yolo9000: better, faster, stronger. In: CVPR, pp. 6517–6525 (2017)
24. Ren, S., He, K., Girshick, R.B., Sun, J.: Faster R-CNN: towards real-time object detection with region proposal networks. In: NeurIPS, pp. 91–99 (2015)
25. Simonyan, K., Zisserman, A.: Very deep convolutional networks for large-scale image recognition. In: ICLR (2015)
26. Sun, M., et al.: Can shape structure features improve model robustness under diverse adversarial settings? In: ICCV, pp. 7506–7515 (2021)
27. Szegedy, C., et al.: Intriguing properties of neural networks. In: ICLR (2014)
28. Wei, X., Liang, S., Chen, N., Cao, X.: Transferable adversarial attacks for image and video object detection. In: IJCAI, pp. 954–960 (2019)
29. Wu, Z., Xiong, Y., Yu, S.X., Lin, D.: Unsupervised feature learning via non-parametric instance discrimination. In: CVPR, pp. 3733–3742 (2018)
30. Xie, C., Wang, J., Zhang, Z., Zhou, Y., Xie, L., Yuille, A.L.: Adversarial examples for semantic segmentation and object detection. In: ICCV, pp. 1378–1387 (2017)
31. Xu, W., Huang, H., Pan, S.: Using feature alignment can improve clean average precision and adversarial robustness in object detection. In: ICIP, pp. 2184–2188 (2021)
32. Zhang, H., Wang, J.: Towards adversarially robust object detection. In: ICCV, pp. 421–430 (2019)
33. Zhang, Y., Zhu, H., Song, Z., Koniusz, P., King, I.: Spectral feature augmentation for graph contrastive learning and beyond. In: AAAI, pp. 11289–11297 (2023)

CAWNet: A Channel Attention Watermarking Attack Network Based on CWABlock

Chunpeng Wang[1], Pengfei Tian[1], Ziqi Wei[2], Qi Li[1], Zhiqiu Xia[1], and Bin Ma[1($\boxtimes$)]

[1] Qilu University of Technology, Jinan 250353, China
sddxmb@126.com
[2] Institute of Automation Chinese Academy of Sciences, Beijing 100190, China

Abstract. In recent years, watermarking technology has been widely used as a common information hiding technique in the fields of copyright protection, authentication, and data privacy protection in digital media. However, the development of watermark attack techniques has lagged behind. Improving the efficiency of watermark attack techniques and effectively attacking watermarks has become an urgent problem to be solved. Therefore, this paper proposes a watermark attack network called CAWNet. Firstly, this paper designs a convolution-based watermark attack module (CWABlock), which introduces channel attention mechanism. By replacing fully connected layers with global average pooling layers, the parameter quantity of the network is reduced and the computational efficiency is improved, enabling effective attacks on watermark information. Secondly, in the training phase, we utilize a large-scale real-world image dataset for training and employ data augmentation strategies to enhance the robustness of the network. Finally, we conduct ablation experiments on CWABlock, attention mechanism, and other modules, as well as comparative experiments on different watermark attack methods. The experimental results demonstrate significant improvements in the effectiveness of the proposed watermark attack approach.

Keywords: watermarking attack · deep learning · CWABlock · attention mechanism · Imperceptible · Robustness

1 Introduction

Watermark attack [1] refers to the manipulation, removal, or destruction of watermark information in digital images or other media through various means, in order to bypass watermark protection or steal original data. With the widespread dissemination and application of digital media, watermark technology [2,3] has been widely used as a means of digital copyright protection and authentication, and its development has become relatively mature. However, the

Q. Liu et al. (Eds.): PRCV 2023, LNCS 14433, pp. 41–52, 2024.
https://doi.org/10.1007/978-981-99-8546-3_4

development of watermark attack techniques has struggled to keep pace with the progress of digital watermarking technology.

The background and significance of studying watermark attacks lie in the fact that watermark attacks, as a means of digital media tampering, can potentially render digital image watermarking techniques ineffective, thereby threatening the goals of copyright protection, authentication, and data privacy protection in digital media. Therefore, in-depth research on the types, methods, and counter-measures of watermark attacks is of great importance. By understanding watermark attack techniques, it becomes possible to enhance the anti-attack capabilities of digital image watermarking techniques. This is crucial for ensuring the secure dissemination and protection of digital media [4].

Currently, traditional watermark attacks mainly include several common types, such as image processing attacks [5,6], geometric attacks [7], signal processing attacks [8], etc. These attack methods often involve adding noise, cropping images, or applying geometric transformations to digital images in order to disrupt or remove watermark information. However, traditional watermark attack methods often have some disadvantages, such as low success rates, unstable attack effects, limitations in the face of resistance measures, and significant impact on the visual aesthetics and practicality of image carriers.

In response to the disadvantages of traditional watermark attacks, we propose a novel watermark attack approach that utilizes advanced deep learning and image processing techniques. Through analysis and research on a large amount of experimental data and real attack scenarios, we design and implement an efficient, accurate, and imperceptible [9] watermark attack method. This method allows for efficient and covert tampering of watermark information in digital images without compromising the original image quality. It counters traditional watermark detection and removal methods and improves the success rate and stability of watermark attacks.

The contributions of our research are mainly in the following aspects:

- A novel watermark attack scheme is proposed, which is characterized by high efficiency, accuracy, and strong imperceptibility. It can effectively attack and test watermark algorithms, thereby promoting the development of watermark algorithms in a reverse manner.
- Extensive experiments and validation in practical attack scenarios have been conducted, demonstrating the superior attack effectiveness and stability of the proposed watermark attack scheme in various contexts.
- The CWABlock module is introduced, combined with channel attention mechanism [10,11], to optimize and adjust the attack algorithm, thereby further improving the attack effectiveness and imperceptibility of watermark attacks.
- In-depth analysis and research are conducted on the disadvantages of traditional watermark attack methods, providing new ideas and solutions for the security and reliability of digital image watermarking techniques.

In conclusion, our research holds significant research significance and practical application value in the field of watermark attacks, providing new technical support and solutions for copyright protection, authentication, and data privacy protection in digital media.

2 Related Work

2.1 Attacked Watermarking Algorithm

The watermark algorithm based on Quaternionic Polar Harmonic Fourier Matrices (QPHFMs) [12] is a geometrically invariant watermarking algorithm widely used in pattern recognition, watermarking, and image reconstruction. The QPHFMs watermark algorithm embeds the watermark into the image invariants (QPHFM) by calculating the invariants of the carrier image, thereby ensuring geometric invariance. Compared to other watermark algorithms, QPHFMs exhibit stronger robustness [13] against image processing attacks and geometric attacks. Therefore, this study will conduct experiments using the QPHFMs watermark algorithm.

2.2 Watermarking Attack Techniques

Conventional image processing attacks and geometric attacks are relatively mature at present, but these conventional attacks inevitably result in a loss of visual quality in the original carrier image. In recent years, due to the widespread application of deep learning in image processing, computer vision, and other fields, researchers have begun to explore the use of deep learning in the development of watermark attack algorithms [14]. Geng et al. [15] proposed a watermark attack algorithm using convolutional neural networks. In 2020, Hatoum et al. [16] introduced the concept of denoising attacks and used fully convolutional neural networks (FCNN) for watermark attacks. In 2023, Wang et al. [17] proposed a covert watermark attack network for removing low-frequency watermark information from watermarked images.

3 The Proposed Method

A watermark attack network refers to a network structure or algorithm designed to counter digital image watermarking techniques. It primarily aims to reduce the visibility, integrity, or extractability of watermarks by altering, tampering with, or attacking digital images, thereby posing a threat to the effectiveness and security of watermarks.

The definition of a watermark attack network encompasses two aspects. Firstly, it serves as a means or algorithm to undermine the visibility, integrity, or extractability of digital image watermarks. Secondly, it can be an independent network structure that learns and optimizes to analyze digital watermarks, thus serving as an evaluation tool for digital watermarking techniques and promoting their development [18].

3.1 CAWNet

The algorithm combines the advantages of Convolutional Neural Networks (CNN) feature extraction and attention mechanisms to create the CWABlock module for constructing the CAWNet watermark attack network. The network architecture consists of convolutional layers, max pooling layers, the CWABlock module, and attention mechanisms. Please refer to Fig. 1 for a visual representation.

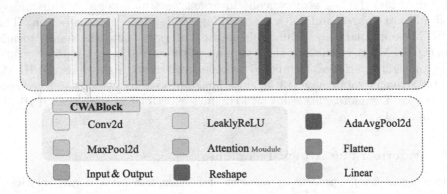

Fig. 1. CAWNet network structure diagram

At this stage, we have defined a neural network model called CAWNet for image de-watermarking tasks in watermark attacks. The network consists of a series of CWABlock modules that extract features from the watermark information and transform them using global average pooling layers and attention mechanisms.

The main structure of the network includes three CWABlock modules, each containing convolutional layers, LeakyReLU activation functions, max pooling layers, and attention modules. These modules aid in extracting higher-level features from the input image and enhance the importance of the focused regions through attention mechanisms. Finally, a convolutional layer is used to convert the feature maps into the final output.

In watermark attacks, the role and effectiveness of this network mainly lie in the feature extraction and attention enhancement through the CWABlock modules. This allows for better identification and removal of watermarks from images, thereby improving the effectiveness of de-watermarking. Convolutions aid the network in recognizing different image details. The channel attention mechanism, on the other hand, is used to compete for the representational capacity of features. It guides the model to focus on the more important parts of the image while reducing the negative impact of interfering factors such as redundant components. As a result, the network's detection and tracking capabilities for embedded watermarks are improved.

3.2   Attention Mechanism

The attention mechanism is a mechanism used to enhance the focus of neural networks on different parts of the input. In deep learning, the attention mechanism allows the network to automatically weight important features during processing, thereby improving the performance and generalization ability of the network.

As shown in Fig. 1 and Fig. 2. In our model, attention mechanisms are added to four convolutional layers of the CWABlock model and incorporated into the analysis after feature extraction. Specifically, each attention module receives the output of a convolutional layer as input. After a convolution operation and activation with the sigmoid function, an attention weight is obtained, indicating the importance of different positions in the input. This weight is multiplied with the output of the convolutional layer, weighting the features at different positions and obtaining a weighted feature representation. The output feature maps preserve the original feature information while enhancing the representation capability of local important features, improving the performance of downstream tasks.

Through the attention mechanism, the network can automatically emphasize important features in the input image, thereby enhancing the network's ability to recognize and remove watermarks. The attention weights can dynamically adjust based on the content and contextual information of the input image, allowing the network to accurately focus on the location and features of the watermark under different input conditions. The attention mechanism also helps the network suppress distracting features in the input image, improving the accuracy of watermark recognition and removal, as well as enhancing the network's generalization ability. By adjusting the attention weights according to different characteristics of the input image, the network can effectively focus on the location and features of the watermark in various scenarios.

In summary, the attention mechanism helps the network to better focus on the location and features of watermarks in a watermark attack network. This improves the network's ability to recognize and remove watermarks, enhancing its generalization capability to effectively handle watermarks in different scenarios.

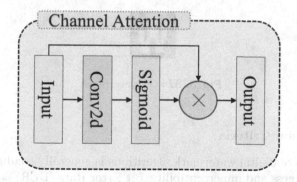

Fig. 2. Structure of attention mechanism network

3.3   CWABlock

In our network, as shown in Fig. 1, we have constructed the CWABlock module, which is an enhanced convolutional block used for image processing. It consists of a convolutional layer, LeakyReLU activation function, max pooling layer, and an attention module. The feature extraction is performed by the convolutional layer, followed by non-linear activation through the activation function, and spatial downsampling via the max pooling layer. Then, the attention module enhances the feature maps to boost the importance of the focused regions.

In watermark attacks, the CWABlock can be used for image de-watermarking tasks. The convolutional layer and pooling layer are utilized for feature extraction and downsampling of the image, aiming to capture higher-level feature representations. The LeakyReLU activation function introduces non-linearity and enhances the expressive power of the network. Combined with the self-attention mechanism, it further improves the feature extraction capability of the watermark attack network.

4   Experiments and Results Analysis

The experiment was conducted in the environment of PyTorch 1.1 and Python 3.6. The experiments were conducted using an NVIDIA Tesla V100 GPU for training. For convenience, the VOC2012 [19] dataset was resized to a unified size of 256 × 256, and a total of 16,700 images with embedded watermarks were selected as the training set. In the experiment, we compared and analyzed different network models and different attack methods. The QPHFMs watermark algorithm was used in this experiment. The 32 × 32 watermark information shown in Fig. 3 was embedded in the images with an embedding strength of 0.5 to obtain watermarked images with high imperceptibility and robustness.

Fig. 3. 32 × 32 watermark

4.1   Evaluation Criteria

The evaluation of digital watermark algorithms is generally conducted from two aspects: robustness and imperceptibility. Bit Error Rate (BER) is used to assess the robustness of digital watermark algorithms, while Peak Signal-to-Noise Ratio (PSNR) is used for comparing imperceptibility. Watermark attack techniques are a research branch of digital watermark technology. Starting from digital watermark algorithms, we utilize PSNR and BER as evaluation criteria for watermark attacks.

Peak Signal-to-Noise Ratio (PSNR). [20] is a metric used to evaluate the quality of images or videos. It is typically calculated by comparing the mean squared error between the original image and the compressed image. PSNR is used to represent the extent of information loss that occurs during compression and transmission of an image. A higher value indicates better image quality and less loss. The formula is as follows:

$$PSNR = 10 \cdot \log_{10} \left(\frac{(2^n - 1)^2}{MSE} \right)$$
(1)

where n denotes the bit depth of each pixel of the original image, which is generally 8 bits (i.e., $n = 8$); MSE denotes the mean square error between the original image and the compressed image, defined as

$$MSE = \frac{1}{m \cdot n} \sum_{i=0}^{m-1} \sum_{j=0}^{n-1} (I(i,j) - K(i,j))^2$$
(2)

where $I(i,j)$ and $K(i,j)$ denote the grayscale values of the corresponding pixels in row i and column j of the original image and the compressed image, respectively, and m and n denote the number of rows and columns of pixels in the original image and the compressed image, respectively.

Bit Error Rate (BER). [21] is a metric used to measure the error rate of data transmission in a digital communication system. It is typically calculated by comparing the number of different bits between the bit stream transmitted by the sender and the bit stream received by the receiver. BER is used to represent the level of difference between the information sent by the sender and the information received by the receiver in digital communication. A lower value indicates higher reliability of data transmission. The formula is as follows:

$$BER = \frac{N_{error}}{N_{total}}$$
(3)

Where N_{error} represents the number of erroneous bits received by the receiving end, and N_{tatal} represents the total number of bits received by the receiving end.

4.2   Ablation Experiment

In this section of the experiment, we conducted an ablation study on the number of layers in the CWABlock module and the attention mechanism. We designated the network without the four-layer dense CWABlock module as model1, the network without the attention mechanism as model2, and the normally trained network as CAWNet. We also performed attack effect experiments as shown in Table 1. In the ablation study, by comparing the PSNR and BER metrics of Model1, Model2, and CAWNet, it can be observed that CAWNet demonstrates more robust performance in image de-watermarking under complex scenes and achieves higher BER values compared to the other two models. Although CAWNet's PSNR value is slightly lower than that of Model1, the addition of CWABlock and attention mechanism brings several advantages: Improved feature extraction capability: CWABlock enhances the network's ability to perceive features at different scales, thereby improving the efficiency and accuracy of feature extraction. Additionally, the attention mechanism allows the network to focus on important information, further enhancing the precision and robustness of feature extraction. Better de-watermarking effects: With the addition of CWABlock and attention mechanism, CAWNet exhibits superior de-watermarking results, effectively removing watermarks while preserving the visual quality of the image. In contrast, models without these two modules (such as Model1 and Model2) have limited de-watermarking effects.

In conclusion, CAWNet, with the addition of CWABlock and attention mechanism, exhibits superior watermark attack performance, imperceptibility, robustness, stability, and promising application prospects.

4.3   Effects of Traditional Attack and Different Deep Learning Attack Methods

In this section, we compare CAWNet with some traditional image processing attacks, geometric attacks, and several deep learning watermark attack methods.

Detailed attack effect data is shown in Table 2 and Table 3. In Table 2, we can observe that the watermark attack method based on the CAWNet network outperforms other attack methods in various types of attacks. CAWNet achieves high PSNR values compared to most traditional statistical methods such as Gaussian noise, multiplicative noise, and salt-and-pepper noise. Especially compared to compression attack types like JPEG, CAWNet exhibits significant advantages in terms of PSNR. This indicates that the method can effectively protect the image quality while removing watermarks, demonstrating high imperceptibility. CAWNet also achieves relatively high BER scores under most attack types. Although JPEG20 has a slightly higher BER score, its PSNR value is relatively lower. Overall, CAWNet still demonstrates good performance.

Table 1. Effect diagram of CWABlock and attention mechanism ablation experiment

	Model1	Model2	CAWNet
Original image			
Attacked image			
PSNR	34.0212	33.7076	33.8686
Watermark Information			
BER	0.2031	0.2021	0.2158

Therefore, in traditional attack types, CAWNet shows high watermark attack effectiveness and imperceptibility, making it a highly effective de-watermarking solution with potential for practical applications. In Table 3, we compare the proposed CAWNet network with FCNNDA, GENG et al., and CWAN. As some algorithms can only process grayscale images, we performed color channel decomposition to make comparisons. Based on experimental data, we have a more significant advantage in terms of both PSNR and BER, indicating that our proposed watermark attack method possesses better imperceptibility and attack capability.

Table 2. Effect diagram of CWABlock and attention mechanism ablation experiment

Evaluation Criteria	No Attack	Gaussian Noise	JPEG 20	Speckle Noise	Salt&pepper	Rotate 45°	CAWNet
Original image							
Attacked image							
PSNR		32.9939	27.1949	25.7618	25.3063	10.5966	33.8686
Watermark Information							
BER	0	0.0127	0.2891	0.1357	0.1123	0.0918	0. 2158

Table 3. Effect diagram of CWABlock and attention mechanism ablation experiment

Evaluation Criteria	No attack	FCNNDA [16]	Geng et al. [15]	CWAN [17]	CAWNet
PSNR		25.30	26.74	30.40	33.87
BER	0	0.2120	0.1563	0.1797	0. 2158

4.4 Stability and Suitability Testing

In this experiment, we used the QPHFMs watermark algorithm to generate a test set of 1000 color images with a size of $256 \times 256 \times 3$. The purpose was to analyze the experimental data structure of the watermark attack method based on CAWNet after attacking a large number of watermarked images. The experiment aimed to evaluate the performance of the method when attacking different watermarked images and compare the differences in the original carrier images to demonstrate the stability and universality of the method. We subjected the 1000 watermarked images to different watermark attacks and summarized and analyzed the resulting PSNR and BER. The results are presented in line charts shown in Fig. 4 and Fig. 5.

In common attack types such as multiplicative noise, salt-and-pepper noise, rotation by 45°, the CAWNet watermark attack method achieves very high PSNR and BER scores. Particularly, compared to attack types like JPEG20,

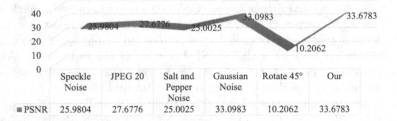

	Speckle Noise	JPEG 20	Salt and Pepper Noise	Gaussian Noise	Rotate 45°	Our
■ PSNR	25.9804	27.6776	25.0025	33.0983	10.2062	33.6783

Fig. 4. 1000 line chart Diagram of PSNR Value after Watermark Image Attack

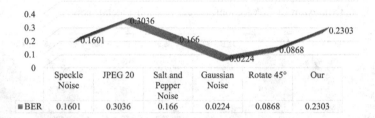

	Speckle Noise	JPEG 20	Salt and Pepper Noise	Gaussian Noise	Rotate 45°	Our
■ BER	0.1601	0.3036	0.166	0.0224	0.0868	0.2303

Fig. 5. BER value line chart after 1000 watermarked images are attacked

there is minimal decrease in CAWNet's PSNR score. This indicates that the CAWNet watermark attack method can preserve the quality of the original image while effectively attacking the watermark information.

The superiority of this method is demonstrated by its high PSNR, BER, strong robustness, and high imperceptibility. These characteristics make CAWNet a highly effective watermark attack solution with potential for practical applications.

5   Conclusion

In order to address the limitations of current watermark attack techniques, in this study, we developed a channel-wise attention-based watermark attack network called CAWNet, based on the CWABlock. We designed the CWABlock module to enhance the feature representation capability in watermark attack tasks, thereby improving the overall performance of watermark attacks. The attention mechanism was incorporated to strengthen the neural network's focus on different parts of the input. Experimental results demonstrate that the inclusion of the CWABlock module and attention mechanism effectively enhances the sensitivity of the neural network watermark attack method towards watermark information. This method achieves better watermark attack performance on various carrier images, validating the stability and applicability of the watermark attack algorithm. Additionally, in comparative experiments with other watermark attack algorithms, we found that this method exhibits superior performance in terms of imperceptibility and watermark attack effectiveness.

For future research prospects, we suggest exploring the following aspects: improving and optimizing existing watermark attack methods by integrating fields such as image restoration and image processing to further enhance the accuracy and imperceptibility of watermark attacks; exploring more effective approaches for dataset construction, model training, and model evaluation to enhance the credibility and practicality of experiments and extend the application of watermark attack methods to wider domains.

Acknowledgements. This work was funded by Taishan Scholar Program of Shandong (tsqn202306251); Youth Innovation Team of Colleges and Universities in Shandong Province (2022KJ124); National Natural Science Foundation of China (62302249, 62272255, 62302248); The "Chunhui Plan" Cooperative Scientific Research Project of Ministry of Education (HZKY20220482); National Key Research and Development Program of China (2021YFC3340602); Shandong Provincial Natural Science Foundation (ZR2023QF032, ZR2022LZH011, ZR2023QF018, ZR2020MF054); Ability Improvement Project of Science and technology SMES in Shandong Province (2023TSGC0217, 2022TSGC2485); Project of Jinan Research Leader Studio (2020GXRC056); Project of Jinan Introduction of Innovation Team (202228016); Science, Education and Industry Integration Project (2023PY060, 2023PX071, 2023PX006).

References

1. Tanha, M., Torshizi, S.D.S., Abdullah, M.T., Hashim, F.: An overview of attacks against digital watermarking and their respective countermeasures. In: Proceedings Title: 2012 International Conference on Cyber Security, Cyber Warfare and Digital Forensic (CyberSec), pp. 265–270. IEEE (2012)
2. Kadian, P., Arora, S.M., Arora, N.: Robust digital watermarking techniques for copyright protection of digital data: a survey. Wireless Pers. Commun. **118**, 3225–3249 (2021)
3. Qian, Z., Zhang, X.: Reversible data hiding in encrypted images with distributed source encoding. IEEE Trans. Circuits Syst. Video Technol. **26**(4), 636–646 (2015)
4. Li, Q., et al.: Concealed attack for robust watermarking based on generative model and perceptual loss. IEEE Trans. Circuits Syst. Video Technol. **32**(8), 5695–5706 (2021)
5. Xiao, B., Luo, J., Bi, X., Li, W., Chen, B.: Fractional discrete tchebyshev moments and their applications in image encryption and watermarking. Inf. Sci. **516**, 545–559 (2020)
6. Rohilla, T., Kumar, M., Kumar, R.: Robust digital image watermarking in YCbCr color space using hybrid method. Inf. Technol. Ind. **9**(1), 1200–1204 (2021)
7. Wang, H., et al.: Detecting aligned double jpeg compressed color image with same quantization matrix based on the stability of image. IEEE Trans. Circuits Syst. Video Technol. **32**(6), 4065–4080 (2021)
8. Cheddad, A., Condell, J., Curran, K., Mc Kevitt, P.: Digital image steganography: survey and analysis of current methods. Signal Process. **90**(3), 727–752 (2010)
9. Agarwal, N., Singh, A.K., Singh, P.K.: Survey of robust and imperceptible watermarking. Multimedia Tools Appl. **78**, 8603–8633 (2019)
10. Guo, M.-H., et al.: Attention mechanisms in computer vision: a survey. Comput. Vis. Media **8**(3), 331–368 (2022)

11. Hu, J., Shen, L., Sun, G.: Squeeze-and-excitation networks. In: Proceedings of the IEEE Conference on Computer Vision and Pattern Recognition, pp. 7132–7141 (2018)

12. Wang, C., Wang, X., Li, Y., Xia, Z., Zhang, C.: Quaternion polar harmonic fourier moments for color images. Inf. Sci. **450**, 141–156 (2018)

13. Wang, C., Wang, X., Zhang, C., Xia, Z.: Geometric correction based color image watermarking using fuzzy least squares support vector machine and Bessel K form distribution. Signal Process. **134**, 197–208 (2017)

14. Wang, C., et al.: RD-IWAN: residual dense based imperceptible watermark attack network. IEEE Trans. Circuits Syst. Video Technol. **32**(11), 7460–7472 (2022)

15. Geng, L., Zhang, W., Chen, H., Fang, H., Yu, N.: Real-time attacks on robust watermarking tools in the wild by CNN. J. Real-Time Image Proc. **17**, 631–641 (2020)

16. Hatoum, M.W., Couchot, J.-F., Couturier, R., Darazi, R.: Using deep learning for image watermarking attack. Signal Process. Image Commun. **90**, 116019 (2021)

17. Wang, C., et al.: CWAN: covert watermarking attack network. Electronics **12**(2), 303 (2023)

18. Voloshynovskiy, S., Pereira, S., Iquise, V., Pun, T.: Attack modelling: towards a second generation watermarking benchmark. Signal Process. **81**(6), 1177–1214 (2001)

19. Everingham, M., Eslami, S.A., Van Gool, L., Williams, C.K., Winn, J., Zisserman, A.: The pascal visual object classes challenge: a retrospective. Int. J. Comput. Vision **111**, 98–136 (2015)

20. Korhonen, J., You, J.: Peak signal-to-noise ratio revisited: Is simple beautiful? In: 2012 Fourth International Workshop on Quality of Multimedia Experience, pp. 37–38. IEEE (2012)

21. Jeruchim, M.: Techniques for estimating the bit error rate in the simulation of digital communication systems. IEEE J. Sel. Areas Commun. **2**(1), 153–170 (1984)

Global Consistency Enhancement Network for Weakly-Supervised Semantic Segmentation

Le Jiang[1], Xinhao Yang[1], Liyan Ma[1,2]($\boxtimes$), and Zhenglin Li[2]

[1] School of Computer Engineering and Science, Shanghai University, Shanghai, China
liyanma@shu.edu.cn
[2] School of Artificial Intelligence, Shanghai University, Shanghai, China

Abstract. Generation methods for reliable class activation maps (CAMs) are essential for weakly-supervised semantic segmentation. These methods usually face the challenge of incomplete and inaccurate CAMs due to intra-class inconsistency of final features and inappropriate use of deep-level ones. To alleviate these issues, we propose the Global Consistency Enhancement Network (GCENet) that consists of Middle-level feature Auxiliary Module (MAM), Intra-class Consistency Enhancement Module (ICEM), and Critical Region Suppression Module (CRSM). Specifically, MAM uses middle-level features which carry clearer edges information and details to enhance output features. Then, for the problem of incomplete class activation maps caused by the high variance of local context of the image, ICEM is proposed to enhance the representation of features. It takes into account the intra-class global consistency and the local particularity. Furthermore, CRSM is proposed to solve the problem of excessive CAMs caused by the over-activation of features. It activates the low-discriminative regions appropriately, thus improving the quality of class activation maps. Through our comprehensive experiments, our method outperforms all other competitors and well demonstrates its effectiveness on the PASCAL VOC2012 dataset.

Keywords: Weakly-supervised semantic segmentation · Semantic segmentation · Intra-class consistency

1 Introduction

Semantic segmentation is a fundamental task in the region of computer vision and plays an essential role in applications, such as autonomous driving and medical diagnosis. In recent years, significant success has been achieved in the research field of fully supervised semantic segmentation [1–3, 21] due to the large-scale datasets with dense pixel-level annotations. However, preparing such datasets

---

L. Jiang and X. Yang—These authors contributed equally to this work and share first authorship.

Q. Liu et al. (Eds.): PRCV 2023, LNCS 14433, pp. 53–65, 2024.
https://doi.org/10.1007/978-981-99-8546-3_5

is laborious. Instead, weakly-supervised semantic segmentation (WSSS) is proposed to alleviate this issue by employing weak supervisions, such as image-level tags, bounding-box annotations, scribbles or points. In this paper, we concentrate on WSSS under image-level class labels because it is of the lowest labeling demand and is less likely to lead to incorrect labels among these supervision options.

Most existing WSSS methods [4,5,19,24] follow a two-stage learning process. First, a classification network is trained to generate class activation maps (CAMs) under the image-level supervision and then generates pseudo masks through refined CAMs. Finally, these pseudo ground-truth masks are utilized to train a segmentation network. CAMs are used to present activation degree of objects in images. However, most existing methods for CAM generation [4,5,19,20] only utilize deep-level features of the backbone. Thus, inaccurate CAMs and inconsistent final features are obtained.

As a momentous research field, many large language models (LLMs) utilize the Chain-of-Thought (CoT) technique [6] that simulates the way our brain works. In short, CoT splits a inference problem into multiple intermediate steps, allocates more consideration for them and integrates all results for the final answer. CoT successfully sparks emergent capacity of LLMs since it fully exploits middle-level features to equip models with stronger inferential ability. Besides, vision transformers enlarge the receptive field of CNNs to a global extent through fully connections, but suffer from high costs both in time and space.

Inspired by these facts, we propose a novel WSSS method named Global Consistency Enhancement Network (GCENet) based on the Cross Language Image Matching framework (CLIMS) [22]. To tackle with the problem of inaccurate CAMs caused by misuse of deep-level features, we propose Middle-level feature Auxiliary Module (MAM) that is different from most existing WSSS methods. It exploits middle-level features which contain clearer edges and details to supplement deep-level features. Since local context is highly unstable from a global perspective, focusing on restricted regions may omit parts of the entire object. To alleviate the issue of inconsistent final features due to the intra-class inconsistency caused by the high variance of local context of the image, we propose Intra-class Consistency Enhancement Module (ICEM) that utilizes local and global feature centers to improve the global consistency with low costs while considering the local particularity. Furthermore, we propose Critical Region Suppression Module (CRSM) to suppress the excess activation regions of features used to generate CAMs so that highly activated regions in CAMs can be preciser.

The main contributions of this paper can be summarized as follows:

- We propose a framework of first enlarging, then suppressing, which is called Global Consistency Enhancement Network (GCENet) for WSSS.
- For problems of ambiguous details, intra-class inconsistency, and excessive regions of features, Middle-level feature Auxiliary Module (MAM), Intra-class Consistency Enhancement Module (ICEM), Critical Region Suppression Module (CRSM) are designed as corresponding solutions.
- We validate the effectiveness of the network through extensive experiments on PASCAL VOC2012.

2   Related Work

In previous works, the pipeline of weakly supervised semantic segmentation mainly consists of two steps: pseudo mask generation with class activation maps and semantic segmentation model training. Most existing studies merely concentrate on using deep-level features to generate CAMs. PSA [4] proposes an affinity network to refine CAMs by looking for foreground regions with high confidence and background ones. Wang et al. [20] designed a self-supervised equivariant attention mechanism to narrow the gap between full and weak supervisions. Sun et al. [19] focused on semantic relations between images through two co-attentions which leverage context from images to boost the localization of common objects from two images. Chang et al. [5] introduced a self-supervised task to enforce the model to focus on more parts of each object through sub-category exploration. Yao et al. [24] proposed a non-salient region object mining method which utilizes a graph-based unit to boost the representation of global relations so that features of non-salient objects can be activated well. Xu et al. [23] proposed a multi-task framework called AuxSegNet. It utilizes a multi-label image classification task with saliency detection to assist the semantic segmentation task. Xie et al. [22] proposed a CLIP-based [18] cross-language image matching method which introduces natural language supervisions to activate more object regions and suppress background regions. Du et al. [11] proposed the pixel-to-prototype contrast that imposes the cross-view feature semantic consistency regularization to narrow the gap between classification and segmentation tasks. Ru et al. [9] first proposed an end-to-end WSSS framework based on a transformer which leverages learned semantic affinity to propagate pseudo masks. Chen et al. [7] designed a semantic-guided attention mechanism to complete the segmentation task. However, above methods can not capture effective features, since they only use deep-level features and the high variance of local context of images.

3   Method

The overview network architecture is shown in Fig. 1 that is built on the baseline CLIMS [22]. CLIMS introduces language supervisions to activate complete object regions and suppress background regions. In order to obtain more representative features, we first utilize Middle-level feature Auxiliary Module to employ the middle-level features. After that, Intra-class Consistency Enhancement Module is utilized to aggregate the local and global feature centers appropriately in order to improve the representation of features by taking into account the intra-class global consistency while considering the local particularity. Critical Region Suppression Module suppressing the excessive regions of features is used to generate more accurate CAMs. At last, the text-driven evaluator from the baseline is attached to activate more foreground regions and restrain background ones.

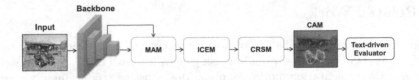

Fig. 1. An overview of proposed GCENet.

3.1 Middle-Level Feature Auxiliary

Previous weakly-supervised semantic segmentation methods only take partial advantages of middle-level features. Therefore, low-quality features further harm the quality of generated CAMs. Inspired by CoT that sparks emergent capacity of LLMs through sufficient exploitation of middle-level features, we propose Middle-level feature Auxiliary Module to refine deep-level features with middle-level ones which contain more clearer edges and details. The architecture of this module is displayed in Fig. 2. In this module, we utilize the third block of ResNet-50 [12] to refine the fourth block.

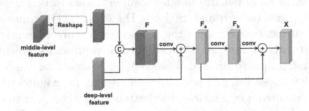

Fig. 2. Process of Middle-level feature Auxiliary Module.

First, we reshape the spatial size of the middle-level feature to the same one as the deep-level feature. Then we concatenate them together as follows:

$$F = Concat(F_{middle}, F_{deep}), \tag{1}$$

where F_{middle} and F_{deep} describe outputs of the third block and the fourth one of ResNet-50, respectively. After we apply a 1×1 convolution layer and ReLU activation function on the output feature for complete knowledge, a residual connection is used to reserve information from the deep-level feature as follows:

$$F_a = ReLU(Conv_{1 \times 1}(F)) + F_{deep}. \tag{2}$$

Two 3×3 convolution layers and a ReLU activation function are added to extract effective features further as follows:

$$F_b = ReLU(Conv_{3 \times 3}(F_a)), \tag{3}$$

$$X = ReLU(Conv_{3\times3}(F_b)) + F_a, \tag{4}$$

where X denotes the output of the module. Eventually we obtain the refined features boosted by the middle-level feature.

3.2   Intra-class Consistency Enhancement

Intra-class Consistency Enhancement Module improves global consistency of final features while considering the local context specificity. For the purpose of generating more precise CAMs, ICEM introduces spatial information. As shown in Fig. 3, we split input features into several non-overlapping patches. Then we generate initial CAMs and probability score maps for these patches as follows:

$$A_c = Conv_{1\times1}(X), \tag{5}$$

$$S_c = Sigmoid(AvgPool^{S\times S}(A_c)), \tag{6}$$

where S_c and A_c describe the probability score map and initial CAM, respectively. $AvgPool$ means the adaptive average pooling. $S \times S$ means the number of split patches, and S is empirically chosen as 2.

First we generate a local feature center X^i for each initial CAM of different patches. Then we calculate the similarity map between each pixel and local feature center. After that, we aggregate all local feature centers into a global feature center to enhance global consistency. Given an image of size $H \times W$, we split it into $N_p \times h \times w$ by adaptive pooling, where $h = H/S$, $w = W/S$ and $N_p = S \times S$ denotes the number of split patches. For each initial CAM generated by corresponding patches, we calculate the local feature center through the patch feature. The probability map is calculated in the following:

$$F_l^i = S_c^i \cdot [\sigma_s(A_c^i)^T \times X^i], \tag{7}$$

where ".", "$\times$" stands for matrix dot product and matrix multiplication, respectively. $X^i \in \mathbb{R}^{N\times C}$, $N = h \times w$, $i \in \{0, ..., N_{p-1}\}$ denotes the index of each patch after the input is split into N_p patches. σ_s denotes the Softmax operation along the spatial dimension. Moreover, ICEM employs the probability score map to filter out part of the local feature center that belongs to nonexistent classes so that the feature of relevant classes can be clustered appropriately.

To enforce local feature centers to interact with each other, this module utilizes a local interaction operation to aggregate separate local feature centers into a global feature center. Thus, the semantic correlation of local feature centers can be boosted and key regions in the global feature center are activated to a larger extent. We treat every local feature center as a node in a graph and then build the structure of different nodes. The process of this operation is as follows:

$$F_l^{i'} = Linear(Conv_{1\times1}(F_l^i)). \tag{8}$$

Since local feature centers are calculated in different regions, their feature representations for each class may vary to a large extent. Thus, we leverage them to generate the global feature center so that the intra-class consistency can be enhanced while considering the local specificity. During the generation process of the global feature center, ICEM applies a weighted-sum operation on local feature centers to capture the global feature center F_g:

$$F_g = \sum_i \alpha_i F_l^{i'}, \tag{9}$$

where α_i denotes the learnable weight of the i-th local feature center.

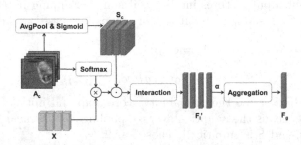

Fig. 3. Illustration of generating local feature centers and global feature centers.

After local feature centers and the global feature center are obtained, ICEM calculates attention maps by applying them on the original feature for enhancement, which is shown in Fig. 4. First, ICEM leverages local feature centers to calculate the inter-pixel similarity map of each split as follows:

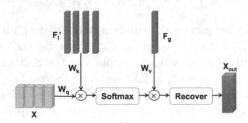

Fig. 4. Illustration of feature enhancement through the attention mechanism.

$$P^i = \sigma_c(W_q(X^i) \times W_k(F_l^{i'})^T), \tag{10}$$

where P denotes the relation between each pixel and the local feature center, σ_c denotes the Softmax operation along class dimension. The boosted feature is as follows:

$$X_{out}^i = P^i \times W_v(F_g), \tag{11}$$

where X_{out} denotes the final feature that is boosted by local feature centers and the global feature center. The size of the final feature is recovered to $H \times W \times C$, finally. ICEM leverages different local feature centers and the global feature center to enhance the original feature so that the intra-class inconsistency of the local context can be alleviated.

3.3 Critical Region Suppression

The quality of generated CAMs plays an important role in weakly-supervised semantic segmentation. However, numerous CAMs in previous works just focus on the critical region of the image while ignoring the activation of less critical regions. Thus, incomplete CAMs may be generated. To alleviate the problem, we propose Critical Region Suppression Module that alleviates the problem of excess activation caused by Intra-class Consistency Enhancement Module. Besides, it also diffuses these highly-activated regions to neighbouring weakly-activated ones since target regions are not merely located at regions with high activation degrees. The specific process of CRSM is shown in Fig. 5.

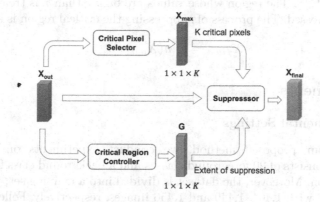

Fig. 5. The process of Critical Region Suppression Module.

CRSM consists of a critical pixel selector, critical region controller, and critical region suppressor. We define the input feature of this module as $X_{out} \in \mathbb{R}^{H \times W \times K}$, where H, W and K denote the height, width and number of channels of the input, respectively. The critical pixel selector is designed to extract K most critical pixels as follows:

$$X_{max} = GMP(ReLU(X_{out})). \tag{12}$$

Specifically, we first apply the ReLU function on the input, then the global max pooling GMP is applied. These K elements are defined as the critical region and suppressed in the next step so that more critical regions which contain the target object can be activated.

The following critical region controller G decides the extent of the suppression to the critical region. The k-th value of the controller determines the extent of the suppression to the responding critical value. The controller is as follows:

$$G = f(GAP(X_{out}); \theta), \tag{13}$$

where f denotes a fully-connected layer, GAP denotes global average pooling and θ denotes the learnable parameter of the critical region controller. While the critical pixel selector aims to find out most critical spatial information from a pixel view, the critical region controller focuses on the overall activation degree of a region of pixels.

After generating the critical region controller, CRSM finally suppresses the feature by the critical region suppressor that demands previous K selected critical pixels and a critical region controller as inputs. Firstly, we conduct element-wise multiplication between K selected critical pixels and the critical region controller. Next, the result is treat as the upper-bound of critical values that is as follows:

$$\tau = X_{max} \cdot G, \tag{14}$$

where $\tau \in \mathbb{R}^{1 \times 1 \times K}$. The region whose values are bigger than τ is treated as the one to be suppressed. The process of suppressing the critical region is as follows:

$$X_{final} = min(X_{out}, \tau). \tag{15}$$

4   Experiments

4.1   Experimental Settings

To evaluate our proposed method, we conduct experiments on PASCAL VOC2012. It consists of 20 foreground classes and a background class for semantic segmentation. Moreover, the dataset is divided into a training set, validation set and test set with 1,464, 1,449 and 1,456 images, respectively. Following previous works, we utilize the augmented dataset for training that contains 10582 images. We use mean Intersection-over-Union (mIoU) for evaluation.

4.2   Implementation Details

All experiments in this paper are conducted in the PyTorch framework [17] and trained on a NVIDIA RTX 3090 GPU. ResNet-50 is adopted as the backbone. Input images are randomly rescaled and then randomly cropped to 512×512. Moreover, horizontal flipping is leveraged to augment training data. We choose 16 as the batch size for training and utilize the text label descriptions to fine-tune the CLIP model including the image encoder and the text encoder for 20 epochs, with the initial learning rate of 0.00005 and a weight decay of 0.003. For the whole GCENet, we train it for 10 epochs with an initial learning rate of 0.00025 and a weight decay of 0.0001.

4.3   CAM Performance

We compare our approach with other SOTA weakly-supervised semantic segmentation methods. We show the comparison for the quality of CAMs that
are generated by different methods. As shown in Table 1, our proposed method
outperforms all other methods. In the case of using ResNet-50 as backbone,
our method achieves 58.8% mIoU. The result is 0.7% higher than our baseline
CLIMS. We can conclude that the Global Consistency Enhancement Network
that is proposed by us is effective to improve the quality of generated CAMs.
Moreover, MAM, ICEM and CRSM can generate more accurate CAMs. We also
show some qualitative results of initial CAMs generated by the baseline and our
proposed method in Fig. 6.

Table 1. Comparison for the quality of CAMs on PASCAL VOC2012. The result of
CLIMS is reproduced by us.

Method	Backbone	mIoU
L2G [13]	ResNet-38	56.8
RIB [14]	ResNet-50	56.5
ESOL [16]	ResNet-50	53.6
AdvCAM [15]	ResNet-50	55.6
CLIMS [22]	ResNet-50	58.1
GCENet (ours)	ResNet-50	58.8

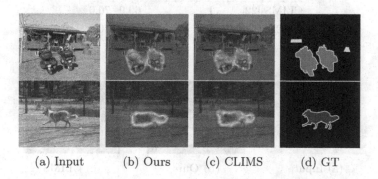

(a) Input   (b) Ours   (c) CLIMS   (d) GT

Fig. 6. Qualitative comparison of CAMs on PASCAL VOC2012. (a): the input image.
(b): the results of our method. (c): the results of CLIMS. (d): the ground-truth label.

We can observe that our method can activate more complete target regions
than the baseline, which benefits from the improvement of global consistency and
expansion of less critical regions. In the second column, our CAMs have more
accurate edges and details because of the utilization of middle-level features.

4.4   Segmentation Performance

Typically, pseudo masks generated by CAMs are treated as ground-truth labels to train a fully-supervised semantic segmentation network. In order to verify the effectiveness of our method on semantic segmentation, we conduct segmentation experiments and fairly compare it with other existing mainstream methods on the validation set and test set of PASCAL VOC2012 in Table 2. We train a DeepLabV2 network with ResNet-101 as the backbone. The supervision information that we use is just pseudo masks generated by CAMs, not any auxiliary saliency maps. As shown in Table 2, our method achieves 70.5% and 70.2% mIoU on the validation set and test set of PASCAL VOC2012 with the same setting as the baseline, 0.6% and 0.2% higher than CLIMS. Moreover, our method outperforms all other methods on PASCAL VOC2012 without using auxiliary saliency maps. Some qualitative results of segmentation are presented in Fig. 7.

Table 2. Segmentation performance on PASCAL VOC2012 validation and test sets. I denotes image-level supervision. S denotes saliency map supervision.

Method	Supervision	Val	Test
RIB [14]	I	68.3	68.6
AdvCAM [15]	I	68.1	68.0
ReCAM [10]	I	68.5	68.4
L2G [13]	I+S	68.5	68.9
SIPE [8]	I	68.8	69.7
ESOL [16]	I	69.9	69.3
CLIMS [22]	I	69.9	70.0
GCENet (ours)	I	70.5	70.2

(a) Input   (b) Ours   (c) CLIMS

Fig. 7. Qualitative comparison of segmentation on PASCAL VOC2012. (a): the input image. (b): the results of our methods. (c): the results of CLIMS.

4.5   Ablation Study

In this section, we present extensive ablation studies on PASCAL VOC2012 to verify the effectiveness of each component.

Table 3. Ablation studies of our proposed method.

MAM	ICEM	CRSM	mIoU
			58.1
✓			58.3
✓	✓		58.6
✓	✓	✓	58.8

As shown in Table 3, only with MAM, the mIoU of CAMs is 0.2% higher than the baseline. It can be seen that the utilization of middle-level features improves the representation of deep-level features. Furthermore, after applying ICEM, we observe that the mIoU of CAMs increases to 58.6%, which is 0.5% higher than the baseline. We can conclude that ICEM improves the global consistency while considering the local particularity. Moreover, after applying CRSM, the model finally achieves 58.8% mIoU. With these three modules, our model increases the mIoU of baseline by 0.7%, showing the effectiveness of our method.

5   Conclusion

In this paper, we discuss a common problem in weakly-supervised semantic segmentation that most methods still face the challenge of generating incomplete and inaccurate CAMs. To solve this problem, we propose a novel framework called Global Consistency Enhancement Network which mainly consists of Middle-level feature Auxiliary Module, Intra-class Consistency Enhancement Module and Critical Region Suppression Module. Moreover, extensive experiments conducted on PASCAL VOC2012 dataset illustrate the effectiveness of GCENet. Compared with other SOTA methods, the initial CAMs generated by our method and the final segmentation results are more complete and accurate.

Acknowledgments. This work was supported in part by the National Key R&D Program of China (No. 2021YFA1003004), in part by the Shanghai Municipal Natural Science Foundation (No. 21ZR1423300), in part by National Natural Science Foundation of China (No. 62203289).

References

1. Long, J., Shelhamer, E., Darrell, T.: Fully convolutional networks for semantic segmentation. In: Proceedings of the IEEE Conference on Computer Vision and Pattern Recognition, pp. 3431–3440 (2015)
2. Chen, L.C., Papandreou, G., Kokkinos, I., Murphy, K., Yuille, A.L.: DeepLab: semantic image segmentation with deep convolutional nets, atrous convolution, and fully connected CRFs. IEEE Trans. Pattern Anal. Mach. Intell. **40**(4), 834–848 (2017)

3. Zhao, H., Shi, J., Qi, X., Wang, X., Jia, J.: Pyramid scene parsing network. In: Proceedings of the IEEE Conference on Computer Vision and Pattern Recognition, pp. 2881–2890 (2017)
4. Ahn, J., Kwak, S.: Learning pixel-level semantic affinity with image-level supervision for weakly supervised semantic segmentation. In: Proceedings of the IEEE Conference on Computer Vision and Pattern Recognition, pp. 4981–4990 (2018)
5. Chang, Y.T., Wang, Q., Hung, W.C., Piramuthu, R., Tsai, Y.H., Yang, M.H.: Weakly-supervised semantic segmentation via sub-category exploration. In: Proceedings of the IEEE/CVF Conference on Computer Vision and Pattern Recognition, pp. 8991–9000 (2020)
6. Wei, J., et al.: Chain of thought prompting elicits reasoning in large language models. arXiv Preprint arXiv:2201.11903 (2022)
7. Chen, J., Zhao, X., Luo, C., Shen, L.: Semformer: semantic guided activation transformer for weakly supervised semantic segmentation. arXiv preprint arXiv:2210.14618 (2022)
8. Chen, Q., Yang, L., Lai, J.H., Xie, X.: Self-supervised image-specific prototype exploration for weakly supervised semantic segmentation. In: Proceedings of the IEEE/CVF Conference on Computer Vision and Pattern Recognition, pp. 4288–4298 (2022)
9. Ru, L., Zhan, Y., Yu, B., Du, B.: Learning affinity from attention: end-to-end weakly-supervised semantic segmentation with transformers. In: Proceedings of the IEEE/CVF Conference on Computer Vision and Pattern Recognition, pp. 16846–16855 (2022)
10. Chen, Z., Wang, T., Wu, X., Hua, X.S., Zhang, H., Sun, Q.: Class re-activation maps for weakly-supervised semantic segmentation. In: Proceedings of the IEEE/CVF Conference on Computer Vision and Pattern Recognition, pp. 969–978 (2022)
11. Du, Y., Fu, Z., Liu, Q., Wang, Y.: Weakly supervised semantic segmentation by pixel-to-prototype contrast. In: Proceedings of the IEEE/CVF Conference on Computer Vision and Pattern Recognition, pp. 4320–4329 (2022)
12. He, K., Zhang, X., Ren, S., Sun, J.: Deep residual learning for image recognition. In: Proceedings of the IEEE Conference on Computer Vision and Pattern Recognition, pp. 770–778 (2016)
13. Jiang, P.T., Yang, Y., Hou, Q., Wei, Y.: L2G: a simple local-to-global knowledge transfer framework for weakly supervised semantic segmentation. In: Proceedings of the IEEE/CVF Conference on Computer Vision and Pattern Recognition, pp. 16886–16896 (2022)
14. Lee, J., Choi, J., Mok, J., Yoon, S.: Reducing information bottleneck for weakly supervised semantic segmentation. Adv. Neural. Inf. Process. Syst. **34**, 27408–27421 (2021)
15. Lee, J., Kim, E., Yoon, S.: Anti-adversarially manipulated attributions for weakly and semi-supervised semantic segmentation. In: Proceedings of the IEEE/CVF Conference on Computer Vision and Pattern Recognition, pp. 4071–4080 (2021)
16. Li, J., Jie, Z., Wang, X., Wei, X., Ma, L.: Expansion and shrinkage of localization for weakly-supervised semantic segmentation. arXiv preprint arXiv:2209.07761 (2022)
17. Paszke, A., et al.: Pytorch: an imperative style, high-performance deep learning library. In: Advances in Neural Information Processing Systems, vol. 32 (2019)
18. Radford, A., et al.: Learning transferable visual models from natural language supervision. In: International Conference on Machine Learning, pp. 8748–8763. PMLR (2021)

19. Sun, G., Wang, W., Dai, J., Van Gool, L.: Mining cross-image semantics for weakly supervised semantic segmentation. In: Vedaldi, A., Bischof, H., Brox, T., Frahm, J.-M. (eds.) ECCV 2020. LNCS, vol. 12347, pp. 347–365. Springer, Cham (2020). https://doi.org/10.1007/978-3-030-58536-5_21

20. Wang, Y., Zhang, J., Kan, M., Shan, S., Chen, X.: Self-supervised equivariant attention mechanism for weakly supervised semantic segmentation. In: Proceedings of the IEEE/CVF Conference on Computer Vision and Pattern Recognition, pp. 12275–12284 (2020)

21. Xie, E., Wang, W., Yu, Z., Anandkumar, A., Alvarez, J.M., Luo, P.: Segformer: simple and efficient design for semantic segmentation with transformers. Adv. Neural. Inf. Process. Syst. **34**, 12077–12090 (2021)

22. Xie, J., Hou, X., Ye, K., Shen, L.: Cross language image matching for weakly supervised semantic segmentation. arXiv preprint arXiv:2203.02668 (2022)

23. Xu, L., Ouyang, W., Bennamoun, M., Boussaid, F., Sohel, F., Xu, D.: Leveraging auxiliary tasks with affinity learning for weakly supervised semantic segmentation. In: Proceedings of the IEEE/CVF International Conference on Computer Vision, pp. 6984–6993 (2021)

24. Yao, Y., et al.: Non-salient region object mining for weakly supervised semantic segmentation. In: Proceedings of the IEEE/CVF Conference on Computer Vision and Pattern Recognition, pp. 2623–2632 (2021)

Enhancing Model Robustness Against Adversarial Attacks with an Anti-adversarial Module

Zhiquan Qin[1], Guoxing Liu[2], and Xianming Lin[3(✉)]

[1] School of Electronic Science and Engineering, Xiamen University,
Xiamen 361005, People's Republic of China
[2] Xiamen University, Xiamen 361005, People's Republic of China
[3] Key Laboratory of Multimedia Trusted Perception and Efficient Computing,
Ministry of Education of China, Xiamen University,
Xiamen 361005, People's Republic of China
linxm@xmu.edu.cn

Abstract. Due to the rapid development of artificial intelligence technologies, such as deep neural networks in recent years, the subsequent emergence of adversarial samples poses a great threat to the security of deep neural network models. In order to defend threats brought by adversarial attacks, the recent mainstream method is to use adversarial training methods to add adversarial samples into model's training process. Although such a type of method can defend against adversarial attacks, it requires increased computing resources and time, and reduces the accuracy of the original samples. Adversarial defense method that conduct anti-adversity in the inference stage. Inspired by adversarial example generation methods, we propose a defense against adversarial in the inference phase of the model. The adversarial sample generation is to add disturbances in the direction of maximizing the loss function after obtaining the sample gradient. Therefore, we add a perturbation in the opposite direction of the adversarial example generated before the sample is fed into the network. The main advantages of our method are that the method requires less computing resources and time; and our method can effectively improve the robust accuracy of the model against adversarial attacks. As a summary, the research content in this paper can stabilize adversarial training, alleviate the high resource consumption of adversarial training, and improve the overall robust performance of the model, which is of great significance to adversarial defense.

Keywords: Deep neural network · Adversarial training · Adversarial attack · Adversarial defense

1 Introduction

Owing to the versatility and significance of deep neural networks, their robustness is increasingly valued. The prevailing research mainly involves augmenting adversarial samples during training to improve the robustness of deep neural networks themselves. Furthermore, most of the prior research employed adversarial training approaches, such as AT [20], TRADES [35], MART [31] and other methods, which have some efficacy in withstanding adversarial attacks, but these methods also have some shortcomings, namely, these methods are relatively resource-intensive and time-consuming, expensive, and degrade the recognition accuracy of the original clean samples. Besides, these methods aim to fend off most white-box attacks, but are helpless against black-box attacks, which may entail serious losses in the real world. There are also some research methods that are oriented towards input adversarial samples, which differ from adversarial sample detection methods. They manipulate the input samples to filter out perturbations or disrupt gradients. For instance, the paper [17] suggests that adding Gaussian noise to the activation mapping during training and testing can successfully fend off various black-box attacks, and the effect is satisfactory; moreover, Small Input Noise (SND) method [4] is to superimpose a layer of random perturbation on the input samples, augmenting the robustness against black-box attacks, but this stochastic method has conspicuous drawbacks, they can be bypassed by expectation over transformation (EOT) [2]technique.

Hence, it is especially vital to seek a more economical and more intuitive and efficient way to attain the research of enhancing robustness. Adversarial samples are the primary defense targets of defense methods. We need to reevaluate their generation and mechanism of action and examine them inversely. Adversarial attacks are directed at a well-trained deep neural network model, producing an adversarial sample along the adversarial direction (the direction of maximizing the loss function), shifts the adversarial sample towards the decision boundary as much as possible, and then feeds it into the network, lowering the confidence of the original correct sample label, in order to accomplish the goal of attack.

As a defender, the prior methods were to augment the robustness of the model itself, so as to attain the goal of defending against attacks, and motivated by the manner of generating adversarial samples, we conceived of applying a specific treatment to the input adversarial samples, for the adversarial samples produced in the direction of adversarial, before they feed into the deep neural network, we can update them in the reverse direction of adversarial, so that they can eliminate the perturbation carried by the adversarial samples themselves, thereby boosting the model recognition accuracy and enhancing robustness.

Therefore, we proposes a viable solution, that is, to introduce an anti-adversarial module prior to the deep neural network model. The role of this module is to remove the perturbation information carried by the adversarial samples by generating the inverse gradient of the adversarial samples, thereby diminishing the influence of the adversarial samples and improving the accuracy of the ultimate outcome of the model. Simultaneously, in order to withstand both black-box and white-box attacks, and amplify the robustness of the model as much as possible, this paper also incorporates SND method [82], which combines them effectively to optimize the model performance and achieve the best result. The main contributions of this paper are as follows:

- The whole deep neural network processing does not require much computing resources, and the time impact is also small.
- There is almost no impact on the classification accuracy of the original clean samples.
- This method has very simple operability, and can be easily combined with other adversarial training methods at this stage, with flexibility.
- This method can defend against black-box and white-box attacks, and combine it with SND method, which has a very significant effect on black-box attacks.

2   Related Works

Since there are attacks, there must be defenses. They are the opposing sides of a phenomenon, and it is a process of mutual contest, mutual advancement and mutual enhancement. In order to investigate the harm of adversarial attacks and reduce their harm as much as possible, relevant researchers have also proposed many efficacious methods. These methods can be roughly categorized into three types: defense based on gradient masking, defense based on adversarial sample detection and defense based on optimization.

2.1   Gradient Masking

The gradient masking defense mechanism aims to render gradients ineffective for the targeted model. By ensuring smoothness in the loss function around the input clean samples, it becomes challenging for attackers to find meaningful directions in the training data and generate adversarial samples.

Papernot et al. [24] introduced the defense distillation method to smooth the gradients of sample data, reducing sensitivity to perturbations carried by adversarial samples. However, Ross et al. [27] proposed input gradient regularization as a more effective approach for optimizing deep neural network models by smoothing gradient information through input and output regularization, despite its high training complexity. Lee et al. [16] This work presents new theoretical robustness bounds based on local gradient information and implements a scalable version of input gradient regularization, showing its competitiveness with adversarial training and its ability to avoid gradient obfuscation or masking.

Despite the impact of gradient masking-based defense methods, they have limitations. They are only effective against attack methods that rely on gradient information and offer no protection against other attack methods.

2.2   Adversarial Examples Detection

An alternative approach to defending against adversarial samples is through sample detection. Metzen et al. [22] trained a sub-network to detect adversarial samples based on intermediate layer features, but it is not effective against unknown attack methods. SafetyNet [18] exploits differences in the output distribution of ReLU activation functions to detect adversarial samples. Similarly, the I-defender method [36] uses the output distribution of hidden layers as a criterion for detection. Ma et al. [19] leverage the invariance of deep neural network features to enhance detection. Meng et al. [21] propose MagNet, a defense mechanism against adversarial samples that combines a detector and a reconstructor. Mao et al. [5] employ control theory to design robust defense strategies by treating the neural network as a closed-loop control system.

2.3 Robust Optimization

Adversarial Training. Adversarial training has emerged as a prominent defense mechanism against adversarial examples [29]. Methods such as Fast Gradient Sign Method (FGSM) [9] and Lagrange relaxation [28] were introduced to enhance model robustness. GANs were also employed to generate perturbations [3,32]. Kannan et al. proposed Adversarial Logit Pairing using M-PGD [12,13]. Wang et al. [30] proposed a dynamic training strategy that gradually improves the convergence quality of adversarial examples, based on the first-order stationary condition of constrained optimization (FOSC), resulting in significant enhancement of robustness. Input transformation methods, including bit-depth reduction, JPEG compression, and image stitching, have been explored. Guo et al. [11] utilized JPEG compression to mitigate pixel displacement. Raff et al. [25] proposed an ensemble of weak transformations for improved robustness, encompassing color precision reduction, JPEG noise, contrast group, grayscale group, denoising, and more.

Model Regularization. Regularization methods reduce the impact of small input perturbations on decisions. Gu et al. [10] introduced gradient regularization using a hierarchical shrinkage penalty. Katz et al. [14] analyzed generalization error and derived bounds based on the Jacobian Matrix. Cisse et al. proposed Parseval Networks for robustness against input perturbations [6]. Zhang et al. [35] addressed the trade-off between robustness and accuracy with the TRADES defense mechanism. Yang et al. [34] presents a novel adversarial training algorithm that combines data-adaptive regularization to enhance both the generalization and robustness of deep neural networks. These methods balance accuracy and robustness through regularization techniques.

Certified Defense. Certified defenders aim to prove the robustness of deep CNN models against adversaries. Katz et al. [14] used SMT solvers for ReLU activation functions to verify local adversarial robustness. Gehr et al. [8] introduced logical formulas to approximate models and provided a comprehensive framework for reasoning on deep CNNs. Singh et al. [28] extended the approach to consider multiple activation functions and implemented parallelized layer transformations for faster verification.

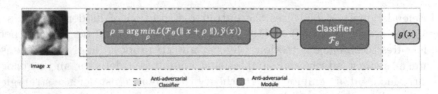

Fig. 1. Adversarial module. The adversarial module introduces perturbations ρ to the input samples of the classifier by minimizing the loss function $\mathcal{L}$.

3   Methods

The anti-adversarial method proposed in this article defends against adversarial attacks by applying a counter-adversarial treatment to input samples. It maximizes the network's resistance to such attacks by taking an opposite direction to the adversarial perturbation.

3.1   Counter-Adversarial Module

The general approach for generating adversarial examples involves computing the sample gradient and perturbing it to maximize the loss function. This moves the sample away from the classification boundary, causing misclassification by the network.The expression for this process is shown as follows:

$$\max_{\delta} \mathcal{L}(\mathcal{F}_\theta(x + \delta), y),$$
$$\text{s.t.} \left\|\delta\right\|_p \le \epsilon. \tag{1}$$

Our method draws inspiration from the generation process of adversarial examples and adopts a similar approach. We incorporate a preprocessing layer before inputting samples into the deep neural network model. This layer aims to reduce the influence of adversarial examples, thereby maximizing the classification accuracy of the model. The formulation of this layer is as follows:

$$g(x) = \mathcal{F}_\theta(\left\|x + \rho\right\|)$$
$$\text{s.t.}\quad \rho = arg \min_{\rho} \mathcal{L}\big(\mathcal{F}_\theta(x + \rho), \check{y}(x)\big), \tag{2}$$
$$\text{s.t.} \left\|\rho\right\|_p \le \epsilon.$$

Our anti-adversarial approach involves a classifier $g(x)$ with a preceding counter-adversarial module. We apply a counter-adversarial perturbation ρ to input samples and consider the prediction result $\check{y}(x)$ instead of the true label. This enhances classification accuracy and defends against adversarial attacks. Unlike other methods, our approach does not require retraining of the original classifier $\mathcal{F}_\theta$. It preserves the accuracy of clean samples and improves the performance of $\mathcal{F}_\theta$. See Fig. 1 for an illustration.

In our method, we utilize a single-step PGD operation as the anti-adversarial module. Unlike the adversarial sample generation process, we subtract the perturbation from the input sample before feeding it into the network. This perturbation is distinct from the ρ perturbation used during the attack. To minimize the impact of adversarial samples, we increase the step size and perturbation constraint. The corresponding equation is shown as Eq. (3).

$$\rho = \epsilon \cdot sign\left(\nabla_x \mathcal{L}\big(\mathcal{F}_\theta(x), \check{y}(x)\big)\right),^2$$
$$\text{s.t. } \|\rho\|_p \leq \epsilon.$$
$$(3)$$

Algorithm 1 presents a pseudocode for our method. Since our method primarily utilizes the single-step PGD method, the parameter K in the algorithm is set to 1.

Algorithm 1. The counter-adversarial classifier $g(x)$

Input: Data samples x, model $\mathcal{F}_\theta$, parameter α, stide τ
Output: The classification result of the anti-adversarial classifier $g(x)$
1: $\check{y}(x) = arg\max_i \mathcal{F}_\theta^i(x)$;
2: $x^0 = x$;
3: **for** $k = 0,\dots,K-1$ **do**
4:    $x^{k+1} = \|x^k - \alpha \cdot sign\left(\nabla_{x^k} \mathcal{L}\big(\mathcal{F}_\theta(x^k), \check{y}(x)\big)\right)\|$
5:    $x^{k+1} = clip(x^{k+1})$
6: **end for**
7: **return** $\mathcal{F}_\theta(x^{k+1})$

3.2 Enhancing Defense Against Black-Box Attacks

To enhance defense against black-box attacks, we introduce the Small Input Noise (SND) [4]. Black-box attacks pose practical challenges due to data diversity, limiting complete knowledge about them. In [4], a solution is proposed: random input noise is added to disrupt query-based attacks, hindering gradient estimation and local search. This approach effectively strengthens defense against black-box attacks, while maintaining minimal impact on recognition accuracy for clean samples, distinguishing it from methods designed for white-box attacks.

$$\mathcal{F}_\eta = \mathcal{F}(x + \eta), \text{where} \quad \eta \sim \mathcal{N}(0, \sigma^2 I) \quad \text{and} \quad \sigma \ll 1. \qquad (4)$$

The Small Input Noise (SND) method involves manipulating the samples that are about to be fed into the deep neural network. It simply adds a randomly obtained value to the input sample, and the magnitude of this value is controlled by a hyperparameter η, as shown in Eq. (4). As depicted in Fig. 2(a), the small noise introduced by the SND method has minimal impact on the predictions for clean samples when compared to the effects of larger noise.

The introduction of additional noise disrupts the effectiveness of local search attacks, which aim to minimize the adversarial target loss through iterative

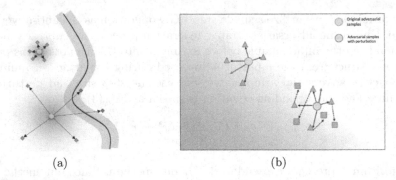

(a)   (b)

Fig. 2. (a) Impact of small perturbations vs. adversarial perturbations [4]. Yellow circles represents original samples, green squares are same-class samples, red squares are adversarial samples and solid line is decision boundary. (b) Blue arrow is gradient estimation for original adversarial sample. After adding a small perturbation, the gradient estimation (red arrow) changes direction to opposite. (Color figure online)

image updates. With the presence of a small perturbation η, the input adversarial sample x_n^{adv} is modified as $x_n^{adv} = x_{(n-1)}^{adv} + \rho + \eta$, where ρ represents the perturbation added at each update and $\eta \sim N(0, \sigma^2 I)$. This perturbation alters the direction of the attack, causing the loss of the attack objective to increase instead. As a result, the attack algorithm enters an ineffective iterative process.

Figure 2(b) provides a visual representation of this mechanism. The central circle represents the adversarial sample. In query-based attacks, updates are made to reduce the target loss (blue arrows). However, with the introduction of Small Input Noise (SND), the estimated gradient information is altered. The gradient estimate shifts from the original green triangle to the orange square, changing the direction of updates. Instead of the original blue arrows, the direction changes to red arrows, opposite to the target loss reduction. This alteration effectively hinders adversarial sample generation, defending against attacks.

4   Experiments

4.1   Experiment Settings

Datasets. CIFAR-10 [15], CIFAR-100 [15], and SVHN [23] are popular computer vision datasets. CIFAR-10 consists of 60,000 32×32 color images divided into ten classes, while CIFAR-100 has 100 classes grouped into 20 superclasses. SVHN focuses on street view house numbers and contains a large number of training and testing images. These datasets serve as valuable resources for computer vision research and algorithm development.

Compared Methods. To validate the effectiveness of our proposed method, we conducted comprehensive comparisons with popular defense methods, including standard adversarial training [20], TRADES [35], and MART [31]. For evaluating defense performance, we employed white-box attacks: PGD-20 and PGD-100 [20], as well as APGD [7], which is an advanced gradient-based app-roach called Auto-PGD. Additionally, we considered black-box attacks using the Square Attack [1]. For deep neural networks, we adopted ResNet18 as it enables us to conduct fast and accurate comparative experiments.

Implementation Details. In our anti-adversarial module, we have two main parameters: α and the step size in the anti-adversarial module. We set α to 0.1 and the step size τ to 0.5. For the parameters in SND, we followed the original paper's settings and set σ to 0.01. To mitigate overfitting [26] in 200-epoch training, we employed the Adversarial Weight Perturbation (AWP) [33]. AWP not only addresses overfitting but also enhances model robustness, providing a dual benefit. It seamlessly integrates with existing adversarial training methods. We set the AWP hyperparameter γ to 0.005 as recommended in the original paper.

4.2 Main Results

White-Box Attacks. To validate the practical effectiveness of our method, we conducted white-box attack evaluations on three datasets (Table 2 and 3).

Table 1. Comparing the white-box adversarial robustness accuracy (%) of different defense methods on CIFAR-10 (%)

Method		Original	PGD attack		
			PGD-20	PGD-100	APGD
AT+AWP		85.26	56.13	53.85	53.16
	+ anti-adversarial	85.2	**85.1**	**85**	**84.76**
TRADES+AWP		82.01	56.5	55.52	55.31
	+ anti-adversarial	81.96	**81.87**	**81.88**	**81.54**
MART+AWP		81.41	57.9	56.43	55.81
	+ anti-adversarial	81.38	**81.12**	**81.14**	**80.89**

Table 2. Comparing the white-box adversarial robustness accuracy (%) of different defense methods on CIFAR-100.

Method		Original	PGD attack		
			PGD-20	PGD-100	APGD
AT+AWP		61.58	29.9	28.55	28
	+ anti-adversarial	61.41	**61.47**	**61.35**	**60.79**
TRADES+AWP		59.4	32.04	31.44	31.15
	+ anti-adversarial	59.37	**59.4**	**59.32**	**58.8**
MART+AWP		57.3	33.61	32.95	32.59
	+ anti-adversarial	57.09	**57.12**	**57.18**	**56.48**

Table 1 displays the performance of our method on CIFAR-10. By incorporating the adversarial module, our model achieves significantly improved robustness against both PGD and APGD attacks. For example, with the AT+AWP method, the robust accuracy increases from 53.85% to 85% against PGD-100 attacks, with only a 0.2% difference compared to clean samples. The adversarial module has minimal impact on the accuracy of clean samples, as indicated in Table 1. Similar trends are observed on CIFAR-100 and SVHN, where the inclusion of the adversarial module leads to substantial improvements in robust accuracy, particularly against PGD-100 attacks.

Table 3. Comparing the white-box adversarial robustness accuracy (%) of different defense methods on SVHN (%)

Method		Original	PGD attack		
			PGD-20	PGD-100	APGD
AT+AWP		86.72	32.76	32.31	19.66
	+ anti-adversarial	60.48	**48.27**	**48.36**	**46.68**
TRADES+AWP		85.37	51.4	50.64	50.07
	+ anti-adversarial	74.32	**65.67**	**65.68**	**65.35**
MART+AWP		81.69	52.76	51.38	50.24
	+ anti-adversarial	76.3	**69.25**	**69.25**	**68.87**

Based on the analysis above, it is evident that our adversarial module can be easily applied to different datasets and various adversarial training methods. It significantly improves the robustness against white-box attacks without increasing additional computational resources. The performance improvement achieved is highly significant.

Black-Box Attacks. We conducted experiments using the Square Attack method on CIFAR-10, as shown in Table 4. Our anti-adversarial module was effective against black-box attacks, especially when combined with SND. With our module, accuracy of AT+AWP increased by 14.31% to reach 72.89%. Combining it with SND resulted in 78.88% accuracy, a substantial 20.03% improvement in robust accuracy. Moreover, our anti-adversarial module combined with SND, with only a 7% difference from clean sample accuracy against Square attacks.

Table 4. Comparing Square black-box attacks robustness accuracy (%) of different defense methods on CIFAR-10

Method		Original	Robust
AT+AWP		85.26	58.58
	+ anti-adversarial	85.2	72.89
	+anti-adversarial + SND	85.14	**78.88**
TRADES+AWP		82.01	57.53
	+ anti-adversarial	82.01	72.13
	+anti-adversarial + SND	82.01	**75.44**
MART+AWP		81.41	57.11
	+ anti-adversarial	81.38	70.17
	+anti-adversarial + SND	81.19	**74.81**

4.3 Ablation Study

We examine the effects of different parameters on the results. Specifically, we conducted experiments on CIFAR-10 using the TRADES+AWP model with the addition of the anti-adversarial module.

Table 5. (a) Ablation study on the step size parameter in the counter-adversarial module. (b) The impact of iteration count in the anti-adversarial module on the results.

(a)

Step	Original	APGD	Square
0.05	82.01	81.57	69.19
0.1	82.01	**81.58**	**72.13**
0.15	81.96	81.42	70.2
0.2	81.76	81.01	62.82
0.25	81.05	80.14	53.17
0.3	79.53	78.16	43.48

(b)

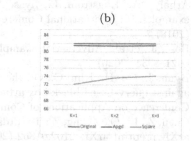

Table 5(a) presents the results obtained by fixing the iteration count to 1 and varying the step size from 0.05 to 0.3. Notably, the optimal performance for both white-box (APGD) and black-box (Square) attacks is achieved when the step size is set to 0.1. Hence, we selected this value for our experiments.

In Table 5(b), the impact of the iteration count on the results is depicted. Notably, increasing the iteration count has minimal effect on defending white-box attacks, as even with $K = 1$, high accuracy is maintained. However, for black-box attacks, the iteration count significantly influences the results and necessitates additional computational resources. To enhance defense against black-box attacks, we incorporated the SND method, which yielded a substantial improvement. Consequently, we set the parameter K to 1 in our experiments.

5   Conclusion

We commence by providing an introductory overview of the proposed method, followed by a comprehensive comparison from various experimental perspectives. The experimental results consistently demonstrate the remarkable effectiveness of our approach. Additionally, we conduct ablation experiments to examine the impact of different parameters within the anti-adversarial module, aiming to identify the optimal configuration. In conclusion, the anti-adversarial module can be seamlessly integrated into the model's inference stage, effectively enhancing the model's defense against both black-box and white-box attacks.

Acknowledgement. This work was supported by National Key R&D Program of China (No. 2022ZD0118202), the National Science Fund for Distinguished Young Scholars (No. 62025603), the National Natural Science Foundation of China (No. U21B2037, No. U22B2051, No. 62176222, No. 62176223, No. 62176226, No. 62072386, No. 62072387, No. 62072389, No. 62002305 and No. 62272401), and the Natural Science Foundation of Fujian Province of China (No. 2021J01002, No. 2022J06001).

References

1. Andriushchenko, M., Croce, F., Flammarion, N., Hein, M.: Square attack: a query-efficient black-box adversarial attack via random search. In: Vedaldi, A., Bischof, H., Brox, T., Frahm, J.-M. (eds.) ECCV 2020. LNCS, vol. 12368, pp. 484–501. Springer, Cham (2020). https://doi.org/10.1007/978-3-030-58592-1_29
2. Athalye, A., Engstrom, L., Ilyas, A., Kwok, K.: Synthesizing robust adversarial examples. In: International Conference on Machine Learning, pp. 284–293. PMLR (2018)
3. Bose, J., et al.: Adversarial example games. Adv. Neural. Inf. Process. Syst. **33**, 8921–8934 (2020)
4. Byun, J., Go, H., Kim, C.: On the effectiveness of small input noise for defending against query-based black-box attacks. In: Proceedings of the IEEE/CVF Winter Conference on Applications of Computer Vision, pp. 3051–3060 (2022)
5. Chen, Z., Li, Q., Zhang, Z.: Towards robust neural networks via close-loop control. arXiv preprint arXiv:2102.01862 (2021)
6. Cisse, M., Bojanowski, P., Grave, E., Dauphin, Y., Usunier, N.: Parseval networks: improving robustness to adversarial examples. In: International Conference on Machine Learning, pp. 854–863. PMLR (2017)
7. Croce, F., Hein, M.: Reliable evaluation of adversarial robustness with an ensemble of diverse parameter-free attacks. In: International Conference on Machine Learning, pp. 2206–2216. PMLR (2020)
8. Gehr, T., Mirman, M., Drachsler-Cohen, D., Tsankov, P., Chaudhuri, S., Vechev, M.: AI2: safety and robustness certification of neural networks with abstract interpretation. In: 2018 IEEE Symposium on Security and Privacy (SP), pp. 3–18. IEEE (2018)
9. Goodfellow, I.J., Shlens, J., Szegedy, C.: Explaining and harnessing adversarial examples. arXiv preprint arXiv:1412.6572 (2014)
10. Gu, S., Rigazio, L.: Towards deep neural network architectures robust to adversarial examples. arXiv preprint arXiv:1412.5068 (2014)

11. Guo, C., Rana, M., Cisse, M., Van Der Maaten, L.: Countering adversarial images using input transformations. arXiv preprint arXiv:1711.00117 (2017)
12. Huang, Q., Katsman, I., He, H., Gu, Z., Belongie, S., Lim, S.N.: Enhancing adversarial example transferability with an intermediate level attack. In: Proceedings of the IEEE/CVF International Conference on Computer Vision, pp. 4733–4742 (2019)
13. Kannan, H., Kurakin, A., Goodfellow, I.: Adversarial logit pairing. arXiv preprint arXiv:1803.06373 (2018)
14. Katz, G., Barrett, C., Dill, D.L., Julian, K., Kochenderfer, M.J.: Reluplex: an efficient SMT solver for verifying deep neural networks. In: Majumdar, R., Kunčak, V. (eds.) CAV 2017. LNCS, vol. 10426, pp. 97–117. Springer, Cham (2017). https://doi.org/10.1007/978-3-319-63387-9_5
15. Krizhevsky, A., Hinton, G., et al.: Learning multiple layers of features from tiny images (2009)
16. Lee, H., Bae, H., Yoon, S.: Gradient masking of label smoothing in adversarial robustness. IEEE Access **9**, 6453–6464 (2020)
17. Liu, X., Cheng, M., Zhang, H., Hsieh, C.J.: Towards robust neural networks via random self-ensemble. In: Proceedings of the European Conference on Computer Vision (ECCV), pp. 369–385 (2018)
18. Lu, J., Issaranon, T., Forsyth, D.: Safetynet: detecting and rejecting adversarial examples robustly. In: Proceedings of the IEEE International Conference on Computer Vision, pp. 446–454 (2017)
19. Ma, S., Liu, Y., Tao, G., Lee, W.C., Zhang, X.: NIC: detecting adversarial samples with neural network invariant checking. In: 26th Annual Network And Distributed System Security Symposium (NDSS 2019). Internet Soc (2019)
20. Madry, A., Makelov, A., Schmidt, L., Tsipras, D., Vladu, A.: Towards deep learning models resistant to adversarial attacks (2017)
21. Meng, D., Chen, H.: Magnet: a two-pronged defense against adversarial examples. In: Proceedings of the 2017 ACM SIGSAC Conference on Computer and Communications Security, pp. 135–147 (2017)
22. Metzen, J.H., Genewein, T., Fischer, V., Bischoff, B.: On detecting adversarial perturbations. arXiv preprint arXiv:1702.04267 (2017)
23. Netzer, Y., Wang, T., Coates, A., Bissacco, A., Wu, B., Ng, A.Y.: Reading digits in natural images with unsupervised feature learning (2011)
24. Papernot, N., McDaniel, P., Wu, X., Jha, S., Swami, A.: Distillation as a defense to adversarial perturbations against deep neural networks. In: 2016 IEEE Symposium on Security and Privacy (SP), pp. 582–597. IEEE (2016)
25. Raff, E., Sylvester, J., Forsyth, S., McLean, M.: Barrage of random transforms for adversarially robust defense. In: Proceedings of the IEEE/CVF Conference on Computer Vision and Pattern Recognition, pp. 6528–6537 (2019)
26. Rice, L., Wong, E., Kolter, Z.: Overfitting in adversarially robust deep learning. In: International Conference on Machine Learning, pp. 8093–8104. PMLR (2020)
27. Ross, A., Doshi-Velez, F.: Improving the adversarial robustness and interpretability of deep neural networks by regularizing their input gradients. In: Proceedings of the AAAI Conference on Artificial Intelligence, vol. 32 (2018)
28. Sinha, A., Namkoong, H., Volpi, R., Duchi, J.: Certifying some distributional robustness with principled adversarial training. arXiv preprint arXiv:1710.10571 (2017)
29. Szegedy, C., et al.: Intriguing properties of neural networks. arXiv preprint arXiv:1312.6199 (2013)

30. Wang, Y., Ma, X., Bailey, J., Yi, J., Zhou, B., Gu, Q.: On the convergence and robustness of adversarial training. arXiv preprint arXiv:2112.08304 (2021)
31. Wang, Y., Zou, D., Yi, J., Bailey, J., Ma, X., Gu, Q.: Improving adversarial robustness requires revisiting misclassified examples. In: International Conference on Learning Representations (2020)
32. Weng, T.W., et al.: Evaluating the robustness of neural networks: an extreme value theory approach. arXiv preprint arXiv:1801.10578 (2018)
33. Wu, D., Xia, S.T., Wang, Y.: Adversarial weight perturbation helps robust generalization. Adv. Neural. Inf. Process. Syst. **33**, 2958–2969 (2020)
34. Yang, D., Kong, I., Kim, Y.: Adaptive regularization for adversarial training. arXiv preprint arXiv:2206.03353 (2022)
35. Zhang, H., Yu, Y., Jiao, J., Xing, E., El Ghaoui, L., Jordan, M.: Theoretically principled trade-off between robustness and accuracy. In: International Conference on Machine Learning, pp. 7472–7482. PMLR (2019)
36. Zheng, Z., Hong, P.: Robust detection of adversarial attacks by modeling the intrinsic properties of deep neural networks. In: Advances in Neural Information Processing Systems, vol. 31 (2018)

FGPTQ-ViT: Fine-Grained Post-training Quantization for Vision Transformers

Caihua Liu[1,2](✉), Hongyang Shi[1,2], and Xinyu He[3]

[1] The College of Computer Science and Technology,
Civil Aviation University of China, Tianjin 300300, China
chliu@cauc.edu.cn
[2] Key Laboratory of Intelligent Airport Theory and System, CAAC,
Tianjin 300300, China
[3] College of Computer Science, Nankai University, Tianjin 300350, China

Abstract. The complex architecture and high training cost of Vision Transformers (ViTs) have prompted the exploration of post-training quantization (PTQ). However, introducing previous PTQ into ViTs performed worse because the activation values after processing by the softmax and GELU functions were extremely unbalanced distributed and not the common Gaussian distribution. To solve this problem, we propose a fine-grained ViT quantization method to fit this special distribution and reduce the quantization error of the activation values. We also design an adaptive piecewise point search algorithm that can automatically find the optimal piecewise point. Both the piecewise point and its search process are in the form of a power of two, making it possible to be implemented on general-purpose hardware with a simple shift operation. Experiments show that the quantization algorithm requires only 32 calibration images and achieves nearly lossless prediction accuracy in the classification task of ImageNet dataset. (The accuracy degradation for 8-bit quantization does not exceed 0.45%, and the average degradation is 0.17%).

Keywords: Vision transformer quantization · Piecewise quantization · Post-training quantization · Adaptive search algorithm

1 Introduction

In recent years, transformer-based vision models have achieved remarkable performance in various computer vision tasks, including image classification [21], object detection [4], semantic segmentation [5], and video recognition [6]. However, compared with traditional Convolutional Neural Networks (CNNs), ViTs suffer from issues such as large storage requirements, high computational complexity, and significant power consumption. These limitations hinder their deployment and real-time inference on resource-constrained edge devices. Therefore, compression methods for ViTs have attracted significant attention.

This work was supported by the Scientific Research Project of Tianjin Educational Committee under Grant 2021KJ037 and the Fundamental Research Funds for the Central Universities of Civil Aviation University of China under Grant 3122021063.

Q. Liu et al. (Eds.): PRCV 2023, LNCS 14433, pp. 79–90, 2024.
https://doi.org/10.1007/978-981-99-8546-3_7

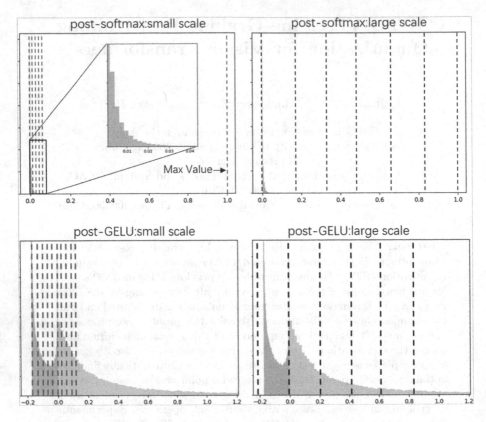

Fig. 1. Demonstration of different scaling factors to quantize the post-softmax and post-GELU activation values. The dotted lines represent the quantization points. Each value is quantized to the nearest quantization point.

Model quantization is one of the most widely used and effective compression methods. Although PTQ has achieved tremendous success in CNNs, directly applying it to ViTs results in a significant decrease in accuracy. Because the activation values in ViT show a special distribution after processing by softmax and GELU functions, as shown in Fig. 1. Most of the values are close to zero, which does not follow the common Gaussian distribution. This poses the following three problems, (1) Directly applying uniform quantization with a small scaling factor to a given range can result in important values being lost, as larger values are mapped to the same value as smaller values. Conversely, using a large scaling factor would result in small values all being quantized to zero, so direct use of uniform quantization is not suited to this particular distribution. (2) Most hardware devices cannot handle non-uniform quantized values efficiently, so non-uniform quantization is also not suitable. (3) Existing methods use using binary segmentation to fit this special distribution, but this coarse interval division lead to a fit that differs significantly from the original distribution.

To resolve the above problems, we propose an efficient quantization framework called FGPTQ-ViT, which can complete the entire quantization process within minutes. Furthermore, the piecewise point and its search process are both based on powers of two, enabling them to be implemented through simple shift operations on general-purpose hardware. Experiments show that our method requires only 32 calibration images and achieves near lossless prediction accuracy in the classification task of ImageNet-1k. (Accuracy degradation is no more than 0.45% for 8-bit quantization, and 0.17% on average)

The main contributions of our work can be summarized as follows:

- We propose a fine-grained ViT quantization method that fits this special distribution through multiple piecewise to achieve improved quantization performance.
- We design an adaptive piecewise point search algorithm that automatically identifies optimal piecewise points, improving the accuracy and efficiency of quantization.
- Extensive experimental validation has been conducted on various Transformer-based architectures, and the results demonstrate that our method achieves accuracy comparable to that of full-precision models.

2   Related Work

2.1   CNN Quantization

PWLQ [13] proposed piecewise linear quantization to uniformly quantize within different quantization regions, targeting bell-shaped distributions. Pot [14] used powers-of-two as quantization factors to achieve hardware-friendly non-uniform quantization. Apot [15] represents the quantization factor as the addition of multiple PoTs, solving the problem that increasing the bit width in a PoT only increases the accuracy of the representation near the zero value, further improving accuracy. EasyQuant [17] used cosine distance to improve the quantization performance of CNNs. AdaRound [2] proposed a novel rounding mechanism to adapt to data and task requirements. AdaQuant [16] further improved the performance of quantized models by incorporating hierarchical verification and integer encoding based on AdaRound. BRECQ [24] leveraged Gaussian-Newton matrix analysis of second-order errors and introduced block reconstruction to balance inter-layer dependencies and generalization errors. Additionally, some notable works adopted more advanced quantization strategies, such as non-uniform quantization [15], channel quantization [11], and mixed-precision quantization [9] to further improve quantization accuracy. Due to the special architecture and numerical distribution of ViTs, it is not feasible to apply these methods directly to ViTs.

2.2   Vision Transformer Quantization

Recently, several quantization methods have been proposed specifically for ViT structures. Liu et al. [12] first introduced a PTQ method for quantizing ViTs,

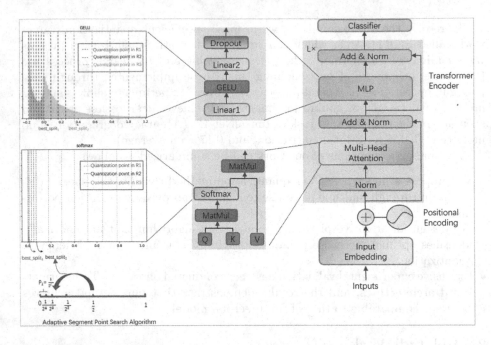

Fig. 2. Overview of the proposed methods.

using pearson correlation coefficient and ranking loss as metrics to determine the scaling factors. PSAQViT [18,19] presented a quantization framework based on patch similarity, suitable for data-free scenarios. PTQ4ViT [10] proposes twin uniform quantization to reduce the quantization error on the activation values of Softmax and GELU, and uses a Hessian guided metric to evaluate different scaling factors. I-ViT [20] employed only integer operations and shift operations during inference. BiViT [22] introduced Softmax-aware binarization and dynamic adaptation to data distribution to reduce binarization-induced errors. NoisyQuant [23] introduced quantization noise before quantization to make the activation values relatively uniform, reducing quantization loss. These methods have achieved certain advancements in quantizing ViTs. However, we observe that current methods still have limitations and challenges, such as insufficient handling of special distributions and significant accuracy loss. Therefore, we propose a new quantization method to address these issues and further enhance the quantization effectiveness of ViTs (Fig. 2).

3   Method

3.1   FGPTQ-ViT Framework

To achieve fast quantization and deployment, we have developed an efficient post-training quantization framework called FGPTQ-ViT for Vision Transformers, as shown in Algorithm 1. This framework consists of two quantization stages:

1) The first stage is the preprocessing stage. Before quantization, we collect the outputs O^l and output gradients $\frac{\partial L}{\partial O^l}$ for each layer. By performing forward propagation on a calibration dataset, we obtain the lth layer outputs O^l and compute the output gradients $\frac{\partial L}{\partial O^l}$ using backward propagation. To reduce GPU memory consumption, we store these outputs and gradients in the main memory.

2) The second stage is the quantization stage. In the lth layer, we transfer the obtained O^l and $\frac{\partial L}{\partial O^l}$ from the first stage to the GPU. Utilizing Hessian-guided metrics, we calculate the optimal scaling factors to quantize the activation values and weight values separately. We then compute the quantized output $\hat{O}^l$ for that layer and discard O^l and $\frac{\partial L}{\partial O^l}$.

For the specific distribution observed after applying softmax and GELU functions, we propose a fine-grained ViT quantization method to fit this distribution, reducing quantization errors in activation values. Additionally, we design an adaptive piecewise point search algorithm that automatically find the optimal piecewise points. The remaining parts of the ViT are uniformly quantized, except for the activation values with the specific distribution mentioned above.

3.2   Fine-Grained ViT Quantization

The distribution of activation values in ViTs after processing by softmax and GELU functions is extremely unbalanced, and as analysed in Sect. 1, all existing methods can cause serious loss of accuracy, so we propose a fine-grained ViT quantization method to solve this problem.

For the softmax function, when quantizing the model to k bits, we first find a point p within the $[0,1]$ range and determine two piecewise points p and $2^{-k}p$. This divides the range into three quantization regions (denoted as R_1, R_2, and R_3 from left to right). Let the scale factors for the three regions be $r_i (i = 1, 2, 3)$, with r_2 fixed as $\frac{A_{max}}{2^k}$, where A is the matrix to be quantized. Then, the quantization factor for the left region is $r_1 = r_2 p$. This allows for handling smaller values effectively within the range $R_1 = [0, 2^{-k}p]$. In the range $R_3 = (p, 1]$, larger values can be quantized using a larger scale factor $r_3 = r_2 p^{-1}$. Uniform quantization is performed separately within these three regions.

For the GELU function, we quantified it into three parts, R_1, R_2 and R_3. Let the scale factor of the three parts be set as $r_i (i = 1, 2, 3)$, r_2 is fixed as $\frac{A_{max}}{2^k}$, then $r_1 = 2^{-m} r_2$, $r_3 = 2^{-m} r_2$, where m is the parameter that we need to optimize, and here the Hessian guided metric is chosen to determine the value of m for each layer. Also the three quantization regions of the GELU function can be determined as $R_1 \in (\infty, 0]$, $R_2 \in [0, 2^k r_2)$, $R_3 \in (2^k r_2, \infty)$, respectively, to do uniform quantization in these three regions, and the quantization equation can be expressed as

$$Q(x, r_1, r_2, r_3, k) = \begin{cases} \text{uni } (x, r_1, k)\, x \in R_1 \\ \text{uni } (x, r_2, k)\, x \in R_2 \\ \text{uni } (x, r_3, k)\, x \in R_3. \end{cases} \tag{1}$$

Similarly we also divide the quantized region into four and five pieces and verify the effect in Sect. 4.3.

During the quantization process, the selection of parameters p and m for each layer is guided by the Hessian guided Metric. Our optimization objective is to minimize the difference between the feature values before and after quantization. For classification tasks, we define the task loss function as $L = \Omega(y, \hat{y})$, where y is the output of the full-precision model at the current layer/block, and $\hat{y}$ is the output of the quantized model. When considering weights as variables, the task loss function can be represented as $L(W)$, where the expected weight loss is $E[L(W)]$. During quantization, we introduce a small perturbation $\hat{w} = w + \epsilon$ on the weights and analyze its impact on the task loss using Taylor series expansion.

$$\mathbb{E}[L(\hat{W})] - \mathbb{E}[L(W)] \approx \epsilon^T \bar{g}^{(W)} + \frac{1}{2}\epsilon^T \bar{H}^{(W)}\epsilon, \tag{2}$$

where g(W) represents the gradient and H(W) represents the Hessian matrix. Since the pretrained model has already converged to a local optimum, the gradient g(W) is close to zero, allowing us to neglect the first-order term and focus only on the impact of the second-order term. Based on the analysis presented in [10], we only consider the elements on the diagonal, i.e., $\left(\frac{\partial L}{\partial O_1}\right)^2, \cdots, \left(\frac{\partial L}{\partial O_m}\right)^2$, which only requires the first-order gradients with respect to the output O. Therefore, our optimization formulation is:

$$\min_{\Delta w} \mathbb{E}\left[(\hat{O} - O)^T \operatorname{diag}\left(\left(\frac{\partial L}{\partial O^1}\right)^2, \cdots, \left(\frac{\partial L}{\partial O^l}\right)^2\right)(\hat{O} - O)\right], \tag{3}$$

where O^l and $\hat{O}^l$ are the outputs of the l-th layer before and after quantization.

3.3   Adaptive Piecewise Point Search Algorithm

To automatically search for the optimal piecewise points during the quantization process, we propose an adaptive piecewise point search algorithm, referred to as Algorithm 2.

In the first step, we set the candidates of p_1 to the power of two form, i.e., $p_1 \in \left\{\frac{1}{2}\cdot, \frac{1}{2^2}\cdot\frac{1}{2^3}, \cdots, \cdot\frac{1}{2^t}\cdot\right\} = G$, and find the optimal point p_1 by iteration.

In the second step, we set the left endpoint of the search range to low $= \frac{p_1}{2}$ and the right endpoint to hight $= p_1 \times 2$. It is worth noting that both low and hight are values in the set G and are to the left and right of p_1, respectively. We use the highest non-zero term of the low order (maximum number of digits) as the search step size. (e.g., if $p_1 = 0.00068$, then low $= 0.00034$ and step $= 0.0001$).

Finally, in the third step, we determine the piecewise point p by further iterations, using the same method as in the second step to search for the best point p. Where in each step, finding the optimal point p is specifically explained as follows: the point p has many candidate terms, and the point p_i is used as a piecewise point to obtain the current scaling factor as described in Sect. 3.2. The current layer is quantized and the output $\hat{O}$ is obtained, a loss value is obtained using Eq. 3 and stored. After traversing the point p once, take the smallest loss value corresponding to p i.e. the optimal point p.

With these three steps, we are able to automatically find the optimal piece-wise point p to optimize the accuracy loss during quantization.

Algorithm1: FGPTQ-ViT	**Algorithm2:** Adaptive piecewise
for k in 1 to L **do**	point search algorithm
Forward-propagation $O^k = A^k B^k$	$P \in \Omega = \left\{ \frac{1}{2}, \frac{1}{2^2} \frac{1}{2^3}, \cdots, \frac{1}{2^{15}} \right\}$
end	**for** i in 1 to 15 **do**
for k in L to 1 **do**	Search for P_{best} using Eq. 3
backward-propagation $grad = \frac{\partial L}{\partial O^k}$	**end**
end	**for** i in 1 to 3 **do**
for k in 1 to L **do**	low $= p/2$ hight $= p*2$
initialize serarch spaces of S_A and S_A	s $=$ maxium_number_of_digits(low)
for r in 1 to #round **do**	**for** j in low to hight using step $=$ s **do**
search for S_A using Eq. 3	Search for p_{best} using Eq. 3
search for S_B using Eq. 3	**end**
end	return p_{best}
end	**end**

4   Experiments

4.1   Implementation Details

Models and Dataset. We evaluate our proposed quantization methods on multiple transformer-based visual models, including ViT [1], Deit [3], and Swin [8]. Experiments on the publicly available ImageNet (ILSVRC-2012) [7], which consists of 1000 classes of images. For our experiments, we randomly select 32 images from the dataset as the calibration images. The pretrained models used in our experiments are obtained from the timm package.

Experimental Settings. We quantize all the weights and inputs, including those in the first and last layers, unless otherwise specified. Unless stated otherwise, our experimental results are based on dividing the quantization range into three pieces. We do not quantize the softmax and normalization layers, as the parameters in these layers can be disregarded, and quantizing them may lead to significant accuracy degradation. Note that the method proposed in this paper quantizes the softmax function in the mutil-head attention module and does not quantize the final softmax layer.

We used our own full-accuracy model accuracy results obtained in our experiments instead of publicly available accuracy values. This is because different hardware environments may lead to slight differences in accuracy when testing pre-trained models locally. In order to exclude the effect of hardware differences on the results, we chose results from our own runs to ensure fairness and credibility of the comparison.

Table 1. Results on ViT Backbones. W8A6 represents that the weights are quantized to 8 bits and the activation values are quantized to 6 bits. ViT-S/224 indicates that the input image resolution is 224×224 by default. The values in parentheses are the accuracy loss compared with the full precision model.

Model	FP32	FGPTQ-ViT(drop)	
		W8A8	W6A6
ViT-S/224	80.460	80.210(−0.250)	78.232(−2.280)
ViT-B/224	83.546	83.57(+0.024)	81.552(−1.994)
ViT-B/384	85.010	84.984(−0.026)	82.948(−2.062)
DeiT-S/224	79.308	78.990(−0.318)	76.798(−2.510)
DeiT-B/224	81.160	80.858(−0.302)	79.640(−1.520)
DeiT-B/384	82.458	82.226(−0.232)	81.218(−1.204)
Swin-T/224	79.952	79.910(−0.042)	79.410(−0.542)
Swin-S/224	81.756	81.726(−0.030)	80.958(−0.798)
Swin-B/224	82.042	84.042(+2.00)	83.092(+1.05)

4.2 Experimental Results

To verify the effectiveness of the proposed method, we evaluated various ViTs on the ImageNet, and the results are shown in Table 1. We can observe that the accuracy of FGPTQ-ViT degrades less than 0.45% under 8-bit quantization, achieving almost lossless post-training quantization. In the case of using 6-bit quantization, the average accuracy drop is 1.74%. Despite this decrease, the accuracy loss remains within an acceptable range. Notably, the prediction accuracy of ViT-B/224 is even 0.024% higher than that of the full-precision model under 8-bit quantization. And Swin-B/224 outperforms the full-accuracy model by 2% and 1.05% under 8-bit and 6-bit quantization, respectively, which fully demonstrates the effectiveness and robustness of our proposed method.

In Table 2, we compared our method with existing post-training quantization methods, and it can be observed that existing methods exhibit significant accuracy degradation on Transformer-based vision models. Taking Deit-S/224 as an example, Percentile shows a 5.328% Top-1 accuracy drop under 8-bit quantization, while EasyQuant, PTQ-Liu and Bit-Split all have accuracy drops exceeding 1%. Furthermore, all methods experience more pronounced accuracy degradation under 6-bit quantization, with PTQ4ViT being the state-of-the-art method, but still showing significant accuracy loss compared to the full-precision model. The PTQ4ViT article states that parallel quantisation gives better results than sequential quantisation, however my device doesn't support parallel quantisation, and to ensure a fair comparison, the results from a local run were chosen. In summary, we have achieved new state-of-the-art results in post-training quantization for ViTs.

In terms of dataset requirements, Percentile, Bit_Split, while EasyQuant and PTQ-Liu use 1024 images. In contrast, our method only requires 32 images,

Table 2. Comparison of Top-1 Accuracy for different post-training quantization methods. "*" denotes results obtained locally, while other results are reported in other literature. "#ims" represents the number of calibration images, and the accuracy results below the model names are for full-precision models.

Model	Method	#ims	Top-1 Accuracy	
			W8A8	W6A6
ViT-B/224 (83.546)	Percentile	all	74.100	71.580
	PTQ4ViT *	32	83.420	81.276
	our*	32	**83.570**	**81.552**
Deit-S/224 (79.308)	Percentile [11]	all	73.980	70.490
	EasyQuant [17]	1024	76.590	73.260
	PTQ-Liu [12]	1024	77.470	74.580
	Bit_Split [3]	all	77.060	74.040
	PTQ4ViT [10]*	32	78.844	76.658
	our*	32	**78.990**	**76.798**
Deit-B/224 (81.160)	Percentile	all	75.210	73.990
	EasyQuant	1024	79.360	75.860
	PTQ-Liu	1024	80.480	77.020
	Bit_Split	all	79.420	76.390
	PTQ4ViT*	32	80.816	**79.822**
	our*	32	**80.858**	79.640

significantly reducing the dataset requirements. This makes it suitable for scenarios with limited datasets or high data confidentiality, such as medical imaging and defense applications.

4.3 Ablation Study

To verify the effectiveness of our proposed method in fitting the distributions of softmax and GELU functions, we conducted ablation experiments and the results are presented in Table 3. From the results, we can observe that applying our method to these two specific distributions improves the model accuracy, indicating that better quantization results can be achieved by applying our method on specific distribution functions. It is worth noting that our method has a more significant effect under 6-bit quantization. For example, taking the ViT-B/224 as an example, when our method is applied to the softmax and GELU functions, 6-bit quantization improves the accuracy by 2.35 compared to not applying the method, while 8-bit quantization only improves the accuracy by 0.114.

The dataset size and quantization speed are also important considerations in our method. We compared the influence of 32, 64, and 128 calibration images on the accuracy and time. From Table 4, we can see that when the number of calibration images is 32, most ViT can complete the quantization process in a few minutes and achieve good accuracy results. However, when the number of calibration images increases to 64 or 128, the quantization time significantly increases,

Table 3. Ablation study of the effect of the proposed a Fine-grained ViT Quantization method and Adaptive piecewise point search algorithm. We mark a ✓ if the proposed method is used.

Model	SoftMax	GELU	Top-1 Accuracy	
			W8A8	W6A6
ViT-S/224 (80.460)			80.068	76.402
	✓		80.020	76.138
		✓	80.300	77.988
	✓	✓	80.210	78.232
ViT-B/224 (83.546)			83.456	79.202
	✓		83.522	80.054
		✓	83.490	80.676
	✓	✓	83.570	81.552
ViT-B/384 (85.010)			84.802	78.798
	✓		84.846	80.514
		✓	84.928	80.962
	✓	✓	84.984	82.948

Table 4. Comparisons of quantization time and accuracy for different numbers of calibration images. T represents the quantization time, measured in minutes.

Model	#ims=32			#ims=64			#ims=128		
	W8A8	W6A6	T	W8A8	W6A6	T	W8A8	W6A6	T
ViT-S/16/224	80.210	78.232	7	80.114	77.968	12	80.142	78.174	11
ViT-S/32/224	75.050	72.450	2	75.146	72.352	3	75.054	72.242	5
ViT-B/16/224	83.570	81.552	12	83.510	81.478	23	83.558	81.498	22
DeiT-T/16/224	70.872	69.354	4	70.898	69.232	5	70.898	69.414	7
DeiT-S/16/224	78.990	76.798	7	79.030	76.344	8	79.000	76.452	12
DeiT-B/16/224	80.858	79.640	12	80.962	79.410	15	80.928	79.406	23
Swin-T/224	79.910	79.410	8	79.908	79.352	11	79.920	79.406	16
Swin-S/224	81.726	80.958	10	81.798	81.076	15	81.794	81.160	19
Swin-B/224	84.042	83.092	18	84.098	83.030	24	84.034	82.792	35

and the Top-1 accuracy of some models slightly differs or even decreases. This indicates that FGPTQ-ViT is not sensitive to the number of calibration images.

Additionally, we explored the number of search rounds, as shown in Fig. 3 (left). We found that the accuracy has reached the optimum when round = 3. We also compared the accuracy results when quantization ranges were divided into three, four, or five pieces, as shown in Table 3 (right), having more quantization piece does not improve the model accuracy, but rather increases the computational overhead.

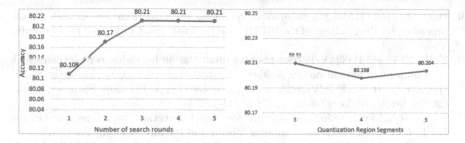

Fig. 3. The accuracy values of ViT-S/224 during W8A8 quantization are shown in the left image with different search rounds. The right image divides the quantization area into 3, 4, and 5 segments.

5   Conclusion

In this paper, we proposed a quantization framework called FGPTQ-ViT, which enables quantization of Vision Transformer in just a few minutes using only 32 calibration images. Specifically, We propose a fine-grained ViT quantization method to more accurately fit the special distribution of activation values processed by softmax and GELU functions. Additionally, we designed an adaptive piecewise point search algorithm that automatically find the optimal positions for piecewise points. The piecewise point and its search process both adopt powers of two, effectively reducing the hardware resource requirements. The experimental results demonstrate that FGPTQ-ViT achieves comparable accuracy to the full-precision baseline model.

References

1. Dosovitskiy, A., et al.: An image is worth 16x16 words: transformers for image recognition at scale. arXiv preprint arXiv:2010.11929 (2020)
2. Nagel, M., et al.: Up or down? Adaptive rounding for post-training quantization. In: International Conference on Machine Learning. PMLR (2020)
3. Wang, P., et al.: Towards accurate post-training network quantization via bit-split and stitching. In: International Conference on Machine Learning. PMLR (2020)
4. Carion, N., Massa, F., Synnaeve, G., Usunier, N., Kirillov, A., Zagoruyko, S.: End-to-end object detection with transformers. In: Vedaldi, A., Bischof, H., Brox, T., Frahm, J.-M. (eds.) ECCV 2020. LNCS, vol. 12346, pp. 213–229. Springer, Cham (2020). https://doi.org/10.1007/978-3-030-58452-8_13
5. Chen, L.Y., Yu, Q.T.U.N.: Transformers Make Strong Encoders for Medical Image Segmentation. arXiv (2021). arXiv preprint arXiv:2102.04306
6. Arnab, A., et al.: Vivit: a video vision transformer. In: Proceedings of the IEEE/CVF International Conference on Computer Vision (2021)
7. Deng, J., et al.: Imagenet: a large-scale hierarchical image database. In: 2009 IEEE Conference on Computer Vision and Pattern Recognition. IEEE (2009)
8. Liu, Z., et al.: Swin transformer: hierarchical vision transformer using shifted windows. In: Proceedings of the IEEE/CVF International Conference on Computer Vision (2021)

9. Dong, Z., et al.: HAWQ: hessian aware quantization of neural networks with mixed-precision. In: Proceedings of the IEEE/CVF International Conference on Computer Vision (2019)

10. Yuan, Z., et al.: PTQ4ViT: post-training quantization for vision transformers with twin uniform quantization. In: Avidan, S., Brostow, G., Cissé, M., Farinella, G.M., Hassner, T. (eds.) ECCV 2022. LNCS, vol. 13672, pp. 191–207. Springer, Cham (2022). https://doi.org/10.1007/978-3-031-19775-8_12

11. Li, R., et al.: Fully quantized network for object detection. In: Proceedings of the IEEE/CVF Conference on Computer Vision and Pattern Recognition (2019)

12. Liu, Z., et al.: Post-training quantization for vision transformer. In: Advances in Neural Information Processing Systems, vol. 34, pp. 28092–28103 (2021)

13. Fang, J., Shafiee, A., Abdel-Aziz, H., Thorsley, D., Georgiadis, G., Hassoun, J.H.: Post-training piecewise linear quantization for deep neural networks. In: Vedaldi, A., Bischof, H., Brox, T., Frahm, J.-M. (eds.) ECCV 2020. LNCS, vol. 12347, pp. 69–86. Springer, Cham (2020). https://doi.org/10.1007/978-3-030-58536-5_5

14. Miyashita, D., Lee, E.H., Murmann, B.: Convolutional neural networks using logarithmic data representation. arXiv preprint arXiv:1603.01025 (2016)

15. Li, Y., Dong, X., Wang, W.: Additive powers-of-two quantization: an efficient non-uniform discretization for neural networks. arXiv preprint arXiv:1909.13144 (2019)

16. Hubara, I., et al.: Improving post training neural quantization: layer-wise calibration and integer programming. arXiv preprint arXiv:2006.10518 (2020)

17. Wu, D., et al.: EasyQuant: post-training quantization via scale optimization. arXiv preprint arXiv:2006.16669 (2020)

18. Li, Z.: Patch similarity aware data-free quantization for vision transformers. In: Avidan, S., Brostow, G., Cissé, M., Farinella, G.M., Hassner, T. (eds.) ECCV 2022. LNCS, vol. 13671, pp. 154–170. Springer, Cham (2022). https://doi.org/10.1007/978-3-031-20083-0_10

19. Li, Z., et al.: PSAQ-ViT V2: Towards Accurate and General Data-Free Quantization for Vision Transformers. arXiv preprint arXiv:2209.05687 (2022)

20. Li, Z., Gu, Q.: I-ViT: integer-only quantization for efficient vision transformer inference. arXiv preprint arXiv:2207.01405 (2022)

21. Graham, B., et al.: LeViT: a vision transformer in convnet's clothing for faster inference. In: Proceedings of the IEEE/CVF International Conference on Computer Vision (2021)

22. He, Y., et al.: BiViT: Extremely Compressed Binary Vision Transformer. arXiv preprint arXiv:2211.07091 (2022)

23. Liu, Y., et al.: NoisyQuant: Noisy Bias-Enhanced Post-Training Activation Quantization for Vision Transformers. arXiv preprint arXiv:2211.16056 (2022)

24. Li, Y., et al.: BRECQ: pushing the limit of post-training quantization by block reconstruction. arXiv preprint arXiv:2102.05426 (2021)

Learning Hierarchical Representations in Temporal and Frequency Domains for Time Series Forecasting

Zhipeng Zhang[1], Yiqun Zhang[1(✉)], An Zeng[1], Dan Pan[2], and Xiaobo Zhang[1]

[1] Guangdong University of Technology, Guangzhou, China
2112105278@mail2.gdut.edu.cn, {yqzhang,zengan,zxb_leng}@gdut.edu.cn
[2] Guangdong Polytechnic Normal University, Guangzhou, China
pandan@gpnu.edu.cn

Abstract. Long-term time series forecasting is a critical task in many domains, including finance, healthcare, and weather forecasting. While Transformer-based models have made significant progress in time series forecasting, their high computational complexity often leads to compromises in model design, limiting the full utilization of temporal information. To address this issue, we propose a novel hierarchical decomposition framework that disentangles latent temporal variation patterns. Specifically, we decompose time series into trend and seasonal modes and further decompose seasonal temporal changes into coarse- and fine-grained states to capture different features of temporal sequences at different granularities. We use linear layers to embed local information for capturing fine-grained temporal changes and Fourier-domain attention to capture multi-periodic seasonal patterns to extract coarse-grained temporal dependency information. This forms a time series forecasting modeling from fine to coarse, and from local to global. Extensive experimental evaluation demonstrates that the proposed approach outperforms state-of-the-art methods on real-world benchmark datasets.

Keywords: Time series forecasting · hierarchical decomposition · fourier-domain attention · deep learning · supervised learning

1 Introduction

Supervised learning is one of the most popular tasks in pattern recognition, which includes the conventional classification [1], challenging concept-drift adaptation [2], and many cutting-edge applications [3]. Time series forecasting, also being an important pattern recognition task, has been widely applied to time series datasets in various domains, e.g., finance, healthcare, weather, etc. These applications rely on historical time series to make predictions for future time series, assisting decision-making and planning. As distribution of time series data is easily influenced by missing values [4], concept-drifts [5], and unforseen external

Q. Liu et al. (Eds.): PRCV 2023, LNCS 14433, pp. 91–103, 2024.
https://doi.org/10.1007/978-981-99-8546-3_8

factors, the patterns become complex, and are usually composed of seasonality and trend components. An intuitive example is the overall growth (trend) and 24-h changes (seasonal) in electricity consumption. Most existing methods directly consider the stack of these two patterns as a whole for time series prediction, which cannot capture the potential temporal changes, and thus results in unsatisfactory prediction results, especially for long-term forecasting.

Although deep learning methods, e.g., convolutional neural networks (CNN) [6], recurrent neural networks (RNN) [7], and temporal convolutional networks (TCN) [8], have achieved better results compared to traditional methods, they possess fixed receptive fields, constraining their ability to capture long-term patterns. Recently, transformer models [9] have achieved tremendous success, owing to their ability to capture long-range dependencies in sequence data. Therefore, an increasing number of transformer-based models have been proposed demonstrating remarkable performance in long-term time series forecasting. Since their high computational cost becomes bottlenecks, the current research efforts, e.g., Logtrans [10], Reformer [11], Triformer [12] and Informer [13], mainly focus on improving the efficiency. However, these linear complexity models may limit the performance as they do not fully exploit the information and lack built-in prior structures. Moreover, they are incapable of learning necessary periodic and seasonal dependencies in complex and diverse time patterns.

To tackle the issues mentioned above, we focus on analyzing and utilizing the inherent patterns of variation in time series data. Figure 1(a) shows that the original input sequence changes without specific regularity. If we learn representations directly from raw sequences, the learned representations may not generalize well. If the observed sequence is composed of the trend and seasonality modules shown in Fig. 1(b) and Fig. 1(c), and we know that the distribution of the seasonality module changes due to different periods leading to different fine-grained and coarse-grained time changes, we can still make reasonable predictions based on the invariant trend module. Therefore, it is promising to deduce a modular architecture for modeling complex temporal variations through a multi-stage decomposition approach.

In this paper, we aim to tackle the complexity and diversity of temporal patterns in multivariate long-term time series data by decomposing the important factors and variables in multiple stages. Accordingly, a novel LHRTF architecture has been designed to model the dependency relationships across the entire temporal range, from fine-grained to coarse-grained time scales. It is based on the multi-level decomposition framework, which combines the trend extraction ability of multi-layer perceptron and the seasonal information expression ability of frequency domain attention to form a model to predict time series with complex and diverse time patterns. Experimental results demonstrate that the proposed model can effectively capture the complex temporal patterns and underlying factors in multi-level time series data, and thus achieves more accurate long-term time series forecasting. The main contributions are summarized below:

- This work addresses the intricacy and variability of temporal patterns in multi-dimensional long-term time series data. By decomposing time series into

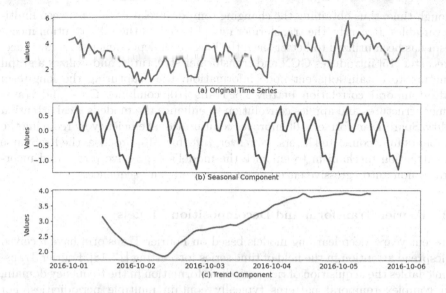

Fig. 1. Seasonal (b) and trend (c) components of a time series (a).

multiple stages, significant information can be comprehensively extracted. As a result, inherent multiple periodic and seasonal time patterns in the seasonal sequences can be captured.

- A new mechanism has been designed to extract and integrate seasonal temporal dependencies, which involves capturing fine-grained temporal dependencies in the time domain and coarse-grained temporal dependencies in the frequency domain, which forms an information extraction and integration process of seasonal temporal dependencies from local to global.
- Extensive experiments have been conducted on real-world multivariate time series datasets from various domains, including energy, economy, weather, and disease. The empirical results show that the prediction performance of the proposed model is very competitive, reducing the average MSE and MAE by 14.9% and 13.1%, respectively, compared to the state-of-the-art methods.

2 Related Work

2.1 Convolutional and Transformer Models

Deep learning has achieved remarkable success in the field of time series modeling, and various deep models have been proposed for time series modeling. Typical models include RNN and its variants [14] adopting RNN as a modeling framework and simulating the distribution of future sequences. LSTNet [7] further combines RNN and CNN to better capture the long- and short-term dependencies of time series. MTCN uses multiple convolutional layers to slide through time, thus obtaining the changing temporal relationship between multiple variables. Recently, the transformer model based on the self-attention mechanism has been utilized to better learn the long-term dependencies. For instance, Forecaster [15] introduces GCN and transformer structures, and utilizes a graph structure to obtain adjacent node information, thus capturing the long-term spatio-temporal correlation better. LogTrans [10] combines CNN and transformer structures and applies convolution to enhance the model's local attention ability. Spare Transformer [16] improves computational efficiency by reducing the number of fully connected layers. Moreover, Informer [13] proposes the ProSparse Self-Attention mechanism to enhance the model's expressive power and incorporates non-autoregressive encoding to generate future sequences.

2.2 Fourier Transform and Decomposition Models

In recent years, deep learning models based on Fourier Transform have received widespread attention in the field of time series forecasting [17,18]. Fourier Transform enables the acquisition of rich periodic information in the frequency domain, and complex temporal patterns typically contain multiple periodicities. For example, Autoformer [19] utilizes fast Fourier transform to design an autocorrelation mechanism that better aggregates information. FEDformer [20] introduces a frequency domain information enhancement mechanism that effectively captures periodic information in long-term time series, while ESTformer [21] combines exponential smoothing and transformer models to simultaneously process various time series components, including trends and seasonality. With the development of time series forecasting, there has been an increasing awareness of the need to focus on and utilize complex temporal patterns for modeling. DeepFS [22] decomposes the sequence and extracts the features of each component into an attention network for interaction and fusion, enabling the extraction of temporal features. Furthermore, MICN [23] is based on a decomposition model of convolutional neural networks and performs convolution operations in the time domain to extract multi-scale temporal dependencies of seasonal time series.

3 Proposed Approach

In this section, we first introduce our problem definition and then describe our LHRTF framework that can capture both seasonal and trend dependencies. The LHRTF overall framework is illustrated in Fig. 2, where we propose a modular architecture to capture the time dependencies of trends and seasonality separately. In Sect. 3.1, we present our multi-level decomposition strategy. In Sect. 3.2, we provide detailed information on the trend prediction module, and in Sect. 3.3, we specifically describe our approach to modeling seasonality. We begin by presenting some basic definitions to facilitate the research, discussion, and analysis in subsequent sections. The definition of the long-term sequence forecasting problem is as follows: Given a historical input sequence $X^{(h)} = [x_1, x_2, x_3, \ldots, x_n] \in \mathbb{R}^{n \times d}$ with a length of the historical sequence is n

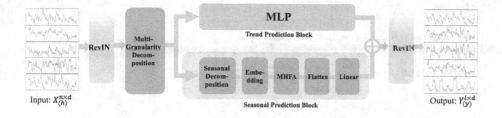

Fig. 2. Overall architecture of LHRTF.

and dimensionality of d, the forecasting task is to forecast the values for the next l steps: $X^{(y)} = [x_{n+1}, x_{n+2}, x_{n+3}, \ldots, x_{n+l}] \in \mathbb{R}^{l \times d}$ (i.e., learning the mapping function as $f : H^{n \times d} \mapsto Y^{l \times d}$).

Remark 1. To address the non-stationarity problem of time series data, recent studies [24] have proposed instance normalization methods to reduce the influence of distribution shift by normalizing time series instances using mean and standard deviation. Similarly, we normalize the input time series data by reversible instance normalization (RevIN) before forecasting and finally add back the mean and deviation to the predicted sequence.

3.1 Time Series Hierarchical Decomposition

We propose a multi-level decomposition framework to address the challenge of modeling the complex patterns of time series forecasting and to obtain more latent factors for improved accuracy. Firstly, we performed a first-level decomposition by utilizing multiple averaging filters of varying scales to separate the trend and seasonal information of different patterns. The trend component was obtained by averaging the resulting patterns, while the seasonal component was obtained by subtracting the trend from the original sequence. Separating the seasonal and trend components allows for improved transferability and generalization of non-stationary time series data, which is critical for effectively modeling non-stationary time series data, the process is:

$$X_{trend} = \frac{\sum_{i=1}^{n} f(x_i)}{n}, \quad X_{season} = X - X_{trend} \tag{1}$$

where $X_{trend}, X_{season} \in \mathbb{R}^{n \times d}$, and $f(x_i)$ is an average filter with a certain scale.

To facilitate better modeling and forecasting of the seasonal component, we performed a second decomposition on the seasonal time series obtained from the first-level decomposition, resulting in multiple sub-sequences. This approach enables the extraction of deeper levels of periodic and seasonal information, the process is:

$$\sum_{i=1}^{m} X_{s(i)} = X_s^{n \times d} \tag{2}$$

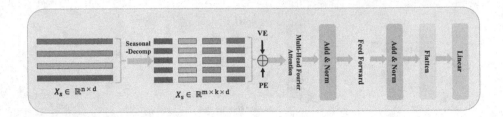

Fig. 3. Seasonal Forecasting Module.

where $X_s^{(i)} \in \mathbb{R}^{\frac{n}{m} \times d}$, $X_s^{(i)}$ represents the i-th subsequence, and m represents the number of subsequences.

After the second seasonal decomposition, we can better capture the different fluctuations or frequencies in seasonal data. For example, for seasonal data with one peak per month in a year, we can decompose each month into a subsequence to better understand the seasonal fluctuations of each month. During prediction, we can utilize the seasonal pattern of each month instead of simply using the seasonal pattern of the entire year. Moreover, decomposing the seasonal component into multiple identical subsequences can more precisely characterize the seasonality, which improves the accuracy of the model in predicting seasonal changes, which is crucial for the accurate modeling of complex time series patterns.

3.2 Trend Forecasting Module

We propose to use a multi-layer perceptron (MLP) for trend forecasting. By mapping the decomposed trend sequence through an MLP, we obtain the predicted trend sequence, which can be written as:

$$Z_{trend} = MLP(X_{trend}) \tag{3}$$

where $Z_{trend} \in \mathbb{R}^{l \times d}$ represents the trend part of the predicted sequence after passing through the multi-layer perceptron network.

3.3 Seasonal Forecasting Module

The seasonal forecasting module, as depicted in Fig. 3, is designed to predict complex and diverse seasonal time variations. We leverage the advantages of the Fourier transform in obtaining a spectrum of rich multi-periodicity information and apply Fourier attention to the seasonal module to capture periodic changes at different time scales, which can be expressed as:

$$X_s = X_s^{d \times m \times k}, \quad X_s^{embed} = Embedding(X_s^{d \times m \times k}), \quad \tilde{X}_s = MHFA(X^{embed})$$
$$Z_{season} = Linear(Flatten(\tilde{X}_s)), \quad Y_{pred} = Z_{season} + Z_{trend}$$

$$\tag{4}$$

where k is the subsequence length and $Y_{pred} \in \mathbb{R}^{l \times d}$ is the prediction. The Fourier transform converts time-domain signals to frequency-domain signals by decomposing them into a combination of sinusoidal and cosinusoidal waves with varying frequencies. As a result, it excels at capturing time series with periodic changes on different time scales. Moreover, frequency-domain attention is with a smaller scale of parameters, which ensures efficient training and testing.

Embedding. We project the length of each subsequence resulting from the second decomposition into a K-dimensional hidden space using a linear layer in the time domain. This allows us to capture fine-grained temporal dependencies within each subsequence while incorporating learnable position embedding to model the temporal positions of the time series data, the process is:

$$X_d^{(i)} = W_{linear} X_s^{(i)} + W_{pos}, \quad X_s^{embed} = W_{linear} (\sum_{i=1}^{m} X_s^{(i)}) + W_{pos} \quad (5)$$

where W_{pos} and W_{linear} represent encoding of position and value, respectively.

MHFA and Flatten. We first provide Remark 2 below.

Remark 2. The convolution theorem states that the Fourier transform of the cyclic convolution of two signals is equal to the point-wise multiplication of their Fourier transforms in the frequency domain. Given a signal $x[n]$ and a filter $h[n]$, the convolution theorem can be expressed as follows:

$$\mathfrak{F}[f_1(t) * f_2(t)] = \mathfrak{F}_1(w) \bullet \mathfrak{F}_2(w) \quad (6)$$

where $f_1(t) \leftrightarrow \mathfrak{F}_1(w)$, $\mathfrak{F}$ for Fourier transform, and $*$ for convolution operation.

According to Remark 2, we know that the pointwise product of the spectra of two sequences is equivalent to their circular convolution in the time domain. Multiplying sequences with larger receptive fields reflects more global features (such as periodicity) and requires less computational cost. Although transformer-based models typically use self-attention mechanisms to capture long-term dependencies in time series, they do not consider the local dependencies among subsequences within the sequence during modeling, which may limit the utilization of local structural information in the sequence.

In contrast, we input each embedded subsequence of every dimension into the multi-head Fourier attention mechanism in sequence, which calculates the Fourier attention mechanism in the frequency domain to capture the global seasonal sequence, forming a local-to-global information extraction and integration for temporal dependencies by:

$$\mathfrak{F}\left(Q_h^{(i)}, Q_K^{(i)}, Q_V^{(i)}\right) = \mathfrak{F}\left(\left(X_d^{(i)}\right)\left(W_h^Q, W_h^K, W_h^V\right)\right) \quad (7)$$

and

$$\left(O_h^{(i)}\right)^T = FA\left(Q_h^{(i)}, K_h^{(i)}, V_h^{(i)}\right) = \mathfrak{F}^{-1}\left(Softmax\left(\frac{\mathfrak{F}\left(Q_h^{(i)}\right)\mathfrak{F}\left(K_h^{(i)}\right)^T}{\sqrt{d_k}}\right)\right)\mathfrak{F}V_h^{(i)} \quad (8)$$

where $W_h^Q, W_h^K \in \mathbb{R}^{K \times d_k}$, $W_h^V \in \mathbb{R}^{K \times K}$, and FA indicates the Fourier Attention operation. Accordingly, attention output $O_h^{(i)} \in \mathbb{R}^{K \times d_k}$ can be obtained by reducing the number of input tokens from n to m, and a linear complexity can be achieved for the frequency-domain attention mechanism. $O_h^{(i)}$ is then processed by a norm layer and a FeedForward Network (FFN) with residual connections by:

$$\tilde{X}_s^{(i)} = FFN\left(\left(O_h^{(i)}\right)^T + X_d^{(i)}\right). \quad (9)$$

Finally, the output $\tilde{X}_s \in \mathbb{R}^{K \times m \times d}$ is flattened and then mapped through a linear layer to generate the final seasonal forecasting sequence $Z_{trend} \in \mathbb{R}^{l \times d}$.

4 Experiments

4.1 Dataset

We have extensively experimented with seven real-world publicly available benchmark datasets. The ETT [13] dataset includes four subsets, namely, two hourly datasets (ETTh1 and ETTh2) and two minutely datasets (ETTm1 and ETTm2), which record six power load features and the target variable "oil temperature" collected from power transformers. The Electricity dataset [25] records the hourly electricity consumption of 321 users from 2012 to 2014. The Weather dataset [26] contains 21 weather indicators, such as humidity, air pressure, and rainfall, recorded every 10 min from July 2020 to July 2021, from nearly 1600 locations in the United States. The ILI dataset [27] records the weekly patient data of influenza-like illness (ILI) from the Centers for Disease Control and Prevention in the United States from 2002 to 2021, describing the ratio of observed ILI patients to the total number of patients. Following the same standard protocol as before, we split all the forecasting datasets into training, validation, and testing sets with ratios of 6:2:2 for ETT datasets and 7:1:2 for other datasets.

4.2 Baselines and Setup

The proposed method is compared with seven baseline methods, including five transformer-based models: LogTrans [10], Pyraformer [28], Informer [13], Autoformer [19], and FEDformer [20], and two non-transformer models: SCINet [29] and MICN [23]. All baselines follow the same evaluation protocol to ensure a fair comparison. The forecasting horizon for the ILI dataset is set to $T \in \{24, 36, 48, 60\}$, and for other datasets, it is set to $T \in \{96, 192, 336, 720\}$, consistent with the settings in the original papers. We collect the results of these time series forecasting baselines from their respective papers. For the state-of-the-art non-transformer MICN and the transformer-based FEDformer, we compare with their improved versions, i.e., MICN-regre and FEDformer-f, respectively.

4.3   Implement Details and Evaluation Metrics

Our approach employs the Adam optimizer. Our method is trained with $L2$ loss. The batch size of the ETT dataset is set to 128, and the batch size of other datasets is set to 16 and 32 respectively. Our model contains 1 frequency-domain attention with the number of heads $H = 8$, and the latent space dimension $D = 512$. All experiments use dropout with a probability of 0.05. We adopt early stopping by terminating training if the MAE on the validation set does not decrease for three consecutive rounds. The training process is stopped prematurely within 30 epochs. MSE and MAE are used as evaluation metrics for all benchmarks. All models are implemented in PyTorch and trained and tested on a single Nvidia GeForce RTX 3090 GPU.

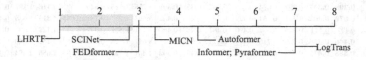

Fig. 4. Significance test using Bonferroni-Dunn test at confidence interval 95% (i.e. $\alpha = 0.05$). The light green region stands for the right side of the critical difference interval. The proposed method performs significantly better than the comparison methods that rank outside this interval. (Color figure online)

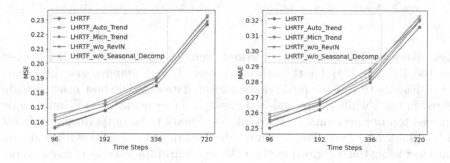

Fig. 5. Ablation results in terms of (a) MSE and (b) MAE on Electricity dataset.

4.4   Main Results

To ensure a fair comparison, we followed the same evaluation protocol where the historical range length was set to 36 for influenza-like ILI and 96 for

Table 1. Multivariate long-term series forecasting results.

Method		LHRTF		MICN		FEDformer		Autoformer		Informer		Pyraformer		LogTrans		SCINet	
Metric		MSE	MAE	MSE	MAE	MSE	MAE	MSE	MAE	MSE	MAE	MSE	MAE	MSE	MAE	MSE	MAE
Weather	96	**0.161**	**0.210**	**0.161**	0.229	0.217	0.296	0.266	0.336	0.300	0.384	0.622	0.556	0.458	0.490	0.239	0.271
	192	**0.209**	**0.252**	0.220	0.281	0.276	0.336	0.307	0.367	0.598	0.544	0.739	0.624	0.658	0.589	0.283	0.303
	336	**0.268**	**0.295**	0.278	0.331	0.339	0.380	0.359	0.395	0.578	0.523	1.004	0.753	0.797	0.652	0.330	0.335
	720	0.325	**0.344**	**0.311**	0.356	0.403	0.428	0.419	0.428	1.059	0.741	1.420	0.934	0.869	0.675	0.400	0.379
	Avg.	**0.241**	**0.275**	0.243	0.299	0.309	0.360	0.338	0.382	0.634	0.548	0.946	0.717	0.696	0.602	0.313	0.322
Electricity	96	**0.156**	**0.250**	0.164	0.269	0.193	0.308	0.201	0.317	0.274	0.368	0.386	0.449	0.258	0.357	0.205	0.312
	192	**0.168**	**0.262**	0.177	0.285	0.201	0.315	0.222	0.334	0.296	0.386	0.378	0.443	0.266	0.368	0.197	0.308
	336	**0.185**	**0.280**	0.193	0.304	0.214	0.329	0.231	0.338	0.300	0.394	0.376	0.443	0.280	0.380	0.202	0.312
	720	0.227	**0.316**	0.212	0.321	0.246	0.355	0.254	0.361	0.373	0.439	0.376	0.445	0.283	0.376	0.234	0.338
	Avg.	**0.184**	**0.277**	0.187	0.295	0.214	0.327	0.227	0.338	0.311	0.397	0.379	0.445	0.272	0.370	0.210	0.318
ILI	24	**1.848**	**0.847**	2.684	1.112	3.228	1.260	3.483	1.287	5.764	1.677	7.394	2.012	4.480	1.444	2.782	1.106
	36	**1.947**	**0.888**	2.667	1.068	2.679	1.080	3.103	1.148	4.755	1.467	7.551	2.031	4.799	1.467	2.689	1.064
	48	**1.558**	**0.799**	2.558	1.052	2.622	1.078	2.669	1.085	4.763	1.469	7.662	2.057	4.800	1.468	2.324	0.999
	60	**1.917**	**0.892**	2.664	1.086	2.857	1.157	2.770	1.125	5.264	1.564	7.931	2.100	5.278	1.560	2.802	1.112
	Avg.	**1.817**	**0.856**	2.664	1.086	2.847	1.144	3.006	1.161	5.137	1.544	7.635	2.050	4.839	1.485	2.649	1.070
ETTh1	96	0.388	0.412	0.421	0.431	**0.376**	0.419	0.449	0.459	0.865	0.713	0.664	0.612	0.878	0.740	0.404	0.415
	192	0.450	0.443	0.474	0.487	**0.420**	0.448	0.500	0.482	1.008	0.792	0.790	0.681	1.037	0.824	0.456	0.445
	336	0.490	0.465	0.569	0.551	**0.459**	**0.465**	0.521	0.496	1.107	0.809	0.891	0.738	1.238	0.932	0.519	0.481
	720	0.484	0.474	0.770	0.672	0.506	0.507	0.514	0.512	1.181	0.865	0.963	0.782	1.135	0.852	0.564	0.528
	Avg.	0.453	0.448	0.559	0.535	**0.440**	**0.460**	0.496	0.487	1.040	0.795	0.827	0.703	1.072	0.837	0.486	0.467
ETTh2	96	**0.298**	**0.347**	0.299	0.364	0.358	0.397	0.358	0.397	3.755	1.525	0.645	0.597	2.116	1.197	0.312	0.355
	192	**0.379**	**0.402**	0.441	0.454	0.429	0.439	0.456	0.452	5.602	1.931	0.788	0.683	4.315	1.635	0.401	0.412
	336	**0.429**	**0.444**	0.654	0.567	0.496	0.487	0.482	0.486	4.721	1.835	0.907	0.747	1.124	1.604	0.413	0.432
	720	**0.446**	**0.456**	0.956	0.716	0.463	0.474	0.515	0.511	3.647	1.625	0.963	0.783	3.188	1.540	0.490	0.483
	Avg.	**0.388**	**0.412**	0.588	0.525	0.437	0.449	0.453	0.462	4.431	1.729	0.826	0.703	2.686	1.494	0.404	0.421
ETTm1	96	0.322	**0.361**	**0.316**	0.362	0.379	0.419	0.505	0.475	0.672	0.571	0.543	0.510	0.600	0.546	0.350	0.385
	192	0.362	**0.382**	0.363	0.390	0.426	0.441	0.553	0.496	0.795	0.669	0.557	0.537	0.837	0.700	0.382	0.400
	336	**0.391**	**0.403**	0.408	0.426	0.445	0.459	0.621	0.537	1.212	0.871	0.754	0.655	1.124	0.832	0.419	0.425
	720	0.458	**0.442**	0.481	0.476	0.543	0.490	0.671	0.561	1.166	0.823	0.908	0.724	1.153	0.820	0.494	0.463
	Avg.	**0.383**	**0.397**	0.392	0.414	0.448	0.452	0.588	0.517	0.961	0.734	0.691	0.607	0.929	0.725	0.411	0.418
ETTm2	96	**0.179**	**0.265**	**0.179**	0.275	0.203	0.287	0.255	0.339	0.365	0.453	0.435	0.507	0.768	0.642	0.201	0.280
	192	**0.240**	**0.303**	0.307	0.376	0.269	0.328	0.281	0.340	0.533	0.563	0.730	0.673	0.989	0.757	0.283	0.331
	336	**0.294**	**0.339**	0.325	0.388	0.325	0.366	0.339	0.372	1.363	0.887	1.201	0.845	1.334	0.872	0.318	0.352
	720	**0.392**	**0.397**	0.502	0.490	0.421	0.415	0.422	0.419	3.379	1.338	3.625	1.451	3.048	1.328	0.439	0.423
	Avg.	**0.276**	**0.326**	0.328	0.382	0.305	0.349	0.324	0.368	1.410	0.810	1.498	0.869	1.535	0.900	0.310	0.347

other datasets. The forecasting horizons were $\{24, 36, 48, 60\}$ for ILI dataset and $\{96, 192, 336, 720\}$ for the other datasets. Table 1 summarizes the results of multivariate time series prediction for seven datasets. The best result is highlighted in bold, while the second-best result is underlined. LHRTF consistently achieved top-tier performance across all benchmark tests, outperforming almost all baselines. Moreover, We evaluated the experimental results with a significance test using the described method [30] to confirm the validity of our method, as shown in Fig. 4. Compared to the previously best-performing model MICN, LHRTF showed significant improvements on the ILI and ETTh2 datasets, with our results achieving a relative reduction in MSE and MAE averages of 31.7% and 21.2% for ILI, and 34% and 21.5% for ETTh2, respectively. In particular, compared to the previous state-of-the-art results, we achieved an overall reduction in MSE by 14.9% and a decrease in MAE by 13.1%.

4.5 Ablation Study

To demonstrate the necessity and effectiveness of trend, seasonal modular, and multilevel decomposition modeling, we conduct related ablation studies. LHRTF-Micn-Trend replaces our MLP with a single linear layer to predict the trend module, while LHRTF-Auto-Trend uses the mean to predict the trend module in Autoformer instead of our MLP. LHRTF w/o RevIN removes instance normalization, and LHRTF w/o Seasonal Decomp indicates that the second-stage seasonal decomposition module has been removed. We conduct experiments on one large electricity dataset, and the experimental results are shown in Fig. 5. It is evident that our multilayer perceptron network performs the best, further demonstrating its effectiveness in trend modeling. The removal of instance normalization shows some errors, but we still achieve the best results compared to MICN methods, indicating the importance of normalization for non-stationary time series data. When the two-stage decomposition module is removed, the performance of the model is poor, which fully demonstrates the effectiveness and rationality of the two-stage decomposition.

5 Conclusion

We propose a multi-hierarchical decomposition framework based on a frequency-domain attention mechanism, which leverages the powerful information extraction capability of deep learning to obtain information on potential variables for time series forecasting, and provides more accurate and reliable solutions for time series forecasting. We emphasize the importance of respecting and utilizing the complex patterns of time series and propose a multi-decomposition architecture for trend and season module forecasting. In the seasonal forecasting module, we model potential different patterns at different granularities, decompose the input seasonal sequence into multiple identical subsequences, capture fine-grained local time dependencies within each subsequence through high-dimensional embeddings, and then model global time dependencies at different scales using Fourier attention. Extensive experiments demonstrate the effectiveness of our model in long-term time series forecasting, and it achieves linear complexity.

Acknowledgements. This work was supported in part by the National Natural Science Foundation of China (NSFC) under grants 62102097, 61976058, and 92267107, the Science and Technology Planning Project of Guangdong Province under grants: 2023A1515012855 and 2022A1515011592, 2021B0101220006, 2019A050510041, and 2021A1515012300, the Key Science and Technology Planning Project of Yunnan Province under grant 202102AA100012, and the Science and Technology Program of Guangzhou under grants 202201010548 and 202103000034.

References

1. Zhang, Y., Cheung, Y.M.: Discretizing numerical attributes in decision tree for big data analysis. In: ICDMW, pp. 1150–1157. IEEE (2014)

2. Zhao, L., Zhang, Y., et al.: Heterogeneous drift learning: classification of mix-attribute data with concept drifts. In: DSAA, pp. 1–10. IEEE (2022)
3. Zeng, A., Rong, H., et al.: Discovery of genetic biomarkers for Alzheimers disease using adaptive convolutional neural networks ensemble and genome-wide association studies. Interdiscip. Sci. **13**(4), 787–800 (2021)
4. Zhang, Z., Zhang, Y., et al.: Time-series data imputation via realistic masking-guided tri-attention Bi-GRU. In: ECAI, pp. 1–9 (2023)
5. Zhao, M., Zhang, Y., et al.: Unsupervised concept drift detection via imbalanced cluster discriminator learning. In: PRCV, pp. 1–12 (2023)
6. Mittelman, R.: Time-series modeling with undecimated fully convolutional neural networks. arXiv preprint arXiv:1508.00317 (2015)
7. Lai, G., Chang, W.C., et al.: Modeling long-and short-term temporal patterns with deep neural networks. In: SIGIR, pp. 95–104 (2018)
8. He, Y., Zhao, J.: Temporal convolutional networks for anomaly detection in time series. In: Journal of Physics: Conference Series, p. 042050 (2019)
9. Vaswani, A., Shazeer, N., et al.: Attention is all you need. In: NeurIPS, pp. 5998–6008 (2017)
10. Li, S., Jin, X., et al.: Enhancing the locality and breaking the memory bottleneck of transformer on time series forecasting. In: NeurIPS, pp. 5244–5254 (2019)
11. Kitaev, N., et al.: Reformer: the efficient transformer. In: ICLR (2020)
12. Cirstea, R., Guo, C., et al.: Triformer: triangular, variable-specific attentions for long sequence multivariate time series forecasting. In: IJCAI, pp. 1994–2001 (2022)
13. Zhou, H., Zhang, S., et al.: Informer: beyond efficient transformer for long sequence time-series forecasting. In: AAAI, pp. 11106–11115 (2021)
14. Flunkert, V., Salinas, D., et al.: Deepar: probabilistic forecasting with autoregressive recurrent networks. arXiv preprint arXiv:1704.04110 (2017)
15. Li, Y., Moura, J.M.F.: Forecaster: a graph transformer for forecasting spatial and time-dependent data. In: ECAI, vol. 325, pp. 1293–1300 (2020)
16. Child, R., Gray, S., et al.: Generating long sequences with sparse transformers. arXiv preprint arXiv:1904.10509 (2019)
17. Xu, K., Qin, M., et al.: Learning in the frequency domain. In: CVPR, pp. 1740–1749 (2020)
18. Guibas, J., Mardani, M., et al.: Adaptive fourier neural operators: efficient token mixers for transformers. arXiv preprint arXiv:2111.13587 (2021)
19. Wu, H., Xu, J., et al.: Autoformer: decomposition transformers with auto-correlation for long-term series forecasting. In: NeurIPS, pp. 22419–22430 (2021)
20. Zhou, T., Ma, Z., et al.: Fedformer: frequency enhanced decomposed transformer for long-term series forecasting. In: ICML, pp. 27268–27286 (2022)
21. Woo, G., Liu, C., et al.: Etsformer: exponential smoothing transformers for time-series forecasting. arXiv preprint arXiv:2202.01381 (2022)
22. Jiang, S., Syed, T., et al.: Bridging self-attention and time series decomposition for periodic forecasting. In: CIKM, pp. 3202–3211 (2022)
23. Wang, H., Peng, J., et al: MICN: multi-scale local and global context modeling for long-term series forecasting. In: ICLR (2023)
24. Kim, T., Kim, J., et al.: Reversible instance normalization for accurate time-series forecasting against distribution shift. In: ICLR (2021)
25. UCI: Electricity. https://archive.ics.uci.edu/dataset/321/electricityloaddiagrams 20112014
26. Wetterstation: Weather. https://www.bgc-jena.mpg.de/wetter/
27. CDC: Illness. https://gis.cdc.gov/grasp/fluview/fluportaldashboard.html

28. Liu, S., Yu, H., et al.: Pyraformer: low-complexity pyramidal attention for long-range time series modeling and forecasting. In: ICLR (2021)
29. Liu, M., Zeng, A., et al.: Scinet: time series modeling and forecasting with sample convolution and interaction. In: NeurIPS, pp. 5010–5828 (2022)
30. Demsar, J.. Statistical comparisons of classifiers over multiple data sets. J. Mach. Learn. Res. 7, 1–30 (2006)

DeCAB: Debiased Semi-supervised Learning for Imbalanced Open-Set Data

Xiaolin Huang[1], Mengke Li[2], Yang Lu[1(✉)], and Hanzi Wang[1]

[1] Fujian Key Laboratory of Sensing and Computing for Smart City, School of
Informatics, Xiamen University, Xiamen, China
luyang@xmu.edu.cn

[2] Guangdong Laboratory of Artificial Intelligence and Digital Economy (SZ),
Shenzhen, China

Abstract. Semi-supervised learning (SSL) has received significant
attention due to its ability to use limited labeled data and various unlabeled data to train models with high generalization performance. However, the assumption of a balanced class distribution in traditional SSL
approaches limits a wide range of real applications, where the training
data exhibits long-tailed distributions. As a consequence, the model is
biased towards head classes and disregards tail classes, thereby leading
to severe class-aware bias. Additionally, since the unlabeled data may
contain out-of-distribution (OOD) samples without manual filtering, the
model will be inclined to assign OOD samples to non-tail classes with
high confidence, which further overwhelms the tail classes. To alleviate
this class-aware bias, we propose an end-to-end semi-supervised method
*De*bias *C*lass-*A*ware *B*ias (DeCAB). DeCAB introduces positive-pair
scores for contrastive learning instead of positive-negative pairs based on
unreliable pseudo-labels, avoiding false negative pairs negatively impacts
the feature space. At the same time, DeCAB utilizes class-aware thresholds to select more tail samples and selective sample reweighting for feature learning, preventing OOD samples from being misclassified as head
classes and accelerating the convergence speed of the model. Experimental results demonstrate that DeCAB is robust in various semi-supervised
benchmarks and achieves state-of-the-art performance. Our code is tem-
porarily available at https://github.com/xlhuang132/decab.

Keywords: Semi-supervised Learning · Imbalanced Learning ·
Contrastive Learning

This work was supported in part by the National Natural Science Foundation of
China under Grants 62002302, 62306181; in part by the FuXiaQuan National Independent Innovation Demonstration Zone Collaborative Innovation Platform under Grant
3502ZCQXT2022008; in part by the China Fundamental Research Funds for the Central Universities under Grants 20720230038.

Q. Liu et al. (Eds.): PRCV 2023, LNCS 14433, pp. 104–119, 2024.
https://doi.org/10.1007/978-981-99-8546-3_9

1 Introduction

Deep supervised learning models have attracted significant interest from both industrial and academia owing to their exceptional performance. However, this outstanding performance is mainly attributed to the abundance of human-annotated data, which can be relatively expensive to obtain [8,9,15]. As a solution, semi-supervised learning (SSL) has emerged as a viable method for leveraging limited labeled data and copious amounts of unlabeled data to achieve models with high generalization performance [1,12,17,19,28].

In the standard SSL paradigm, the assumption is made that the distribution of target data is balanced across all classes, with an equal number of labeled and unlabeled samples for each class. However, it is noteworthy that the class distribution of naturally collected data may potentially exhibit a long-tailed characteristic [24]. In situations where the labeled data is imbalanced, the performance of conventional SSL methods can be adversely affected as the unlabeled data is also likely to be imbalanced. The imbalanced nature of unlabeled data primarily enhances the performance of head classes, while simultaneously exacerbating the performance degradation of the tail classes. This phenomenon is referred to as class-aware bias, which is a significant challenge in semi-supervised learning. As a result, researchers have shown a growing interest in developing imbalance-robust SSL models [6,14,21] to mitigate the effect of this bias.

In addition to the challenge of imbalanced data in SSL, another problem is the existence of out-of-distribution (OOD) samples in the unlabeled data [7]. Without label information and manual filtering, the unlabeled data may have a high probability of containing OOD samples that do not belong to the target distribution. Common SSL methods will treat OOD samples in the same manner as in-distribution (ID) samples, resulting in introducing noisy samples during training. To address this problem, several open-set SSL methods have been proposed [2,7,23]. To mitigate the negative impact of OOD data in SSL, a common strategy is to identify and then filter out or reduce the weight of these samples [7,26]. However, these methods tend to classify unlabeled tail samples as OOD data [20] under long-tailed scenarios, further exacerbating the class-aware bias and diminishing performance. Thus, developing robust SSL methods that can handle both class imbalance and OOD data is crucial for achieving high generalization performance in practical scenarios.

This paper thereby investigates the problem of imbalanced SSL with OOD data, which is a more practical and general scenario in real-world applications. The long-tail distribution commonly leads to the erroneous classification of OOD samples as head samples, consequently intensifying the long-tail problem and perpetuating a detrimental loop. To alleviate this issue, we propose an end-to-end semi-supervised method named *De*bias *C*lass-*A*ware *B*ias (DeCAB). DeCAB leverages contrastive learning to mitigate the class-aware bias by introducing positive-pair scores to replace positive-negative pairs based on unreliable pseudo-labels, thereby avoiding the detrimental impact of false negative pairs on the model. In addition, DeCAB utilizes class-aware thresholds and selective sample reweighting to emphasize feature learning of unconfident unlabeled data.

Our main contributions can be summarized as follows:

- We study a novel problem of SSL on imbalanced open-set data and reveal the fact that the OOD samples exacerbate the long-tail problem in existing SSL methods.
- We propose a simple but effective semi-supervised method DeCAB to tackle more realistic scenarios, by evaluating the feature learning requirements of samples and sample pairs to facilitate feature training.
- DeCAB shows superior performance compared with various state-of-the-art SSL methods on more realistic scenarios, improving the performance of the tail classes while maintaining the performance of the head classes.

2   Related Work

2.1   General SSL Methods

General semi-supervised learning methods can be divided into three categories: pseudo-label-based methods [12,22], consistency regularization-based methods [10,18], and hybrid methods that combine the two [1,16,17]. Specifically, pseudo-label-based semi-supervised methods [12] use the self-training strategy to train the model with labeled data, then pseudo-labels unlabeled data for further training. Consistency regularization-based methods [10,18] encourage the model to output consistent results for the same input data with different forms of augmentation [3,4]. Hybrid methods [1,17] combine pseudo-label-based and consistency regularization-based methods to further improve performance. However, these methods both assume that the target data distribution is balanced and there are no OOD samples in unlabeled data. In more realistic scenarios, data often has two characteristics: 1) ID data follows a natural long-tailed distribution, and 2) unlabeled data may contain OOD data. These characteristics make existing methods vulnerable to distribution shifts and noisy data, thereby reducing performance. Therefore, further research and development of semi-supervised learning are needed to address the challenges in real-world scenarios.

2.2   Imbalanced SSL Methods

Significant progress has been made in developing imbalanced semi-supervised methods to address the first data characteristic. CReST [21] identifies more accurate tail samples and then performs probability sampling biased towards tail classes, leading to a significant improvement in tail class performance. DASO [14] combines similarity-based semantic pseudo-labels with linear classifier pseudo-labels in a self-adaptive manner, and utilizes semantic alignment loss to establish balanced feature representations, reducing biased predictions from the classifier. CoSSL [6] designs a tail class feature enhancement module to alleviate the unbalance problem.

2.3 Open-Set SSL Methods

To address the second data characteristic, open-set SSL has been proposed, which allows the model to identify and reject OOD data, thereby enhancing its robustness and performance on unseen data. OpenMatch [16] introduces the OVA classifier and proposes a soft consistency regularization loss to improve the smoothness of the ova classifier relative to the input transformation. There are currently many SSL methods resorting to contrastive learning to deal with open-set data. For example, CCSSL [23] uses class-aware contrastive learning to enhance the SSL model's ability to handle OOD samples, making the model more robust.

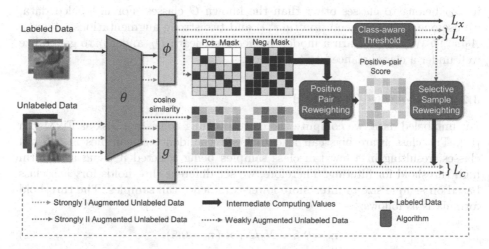

Fig. 1. Overview of the proposed DeCAB. Unlabeled data is filtered through class-aware thresholds to obtain pseudo-labels, and the selected samples undergo feature learning needs evaluation. Meanwhile, for each sample pair, positive-pair scores are computed to conduct weighted contrastive learning.

3 Proposed Method

To address the issue of class-aware bias, we propose Debias Class-Aware Bias (DeCAB), which is an end-to-end semi-supervised method that incorporates contrastive information. It consists of three core components, that is, class-aware threshold, selective sample reweighting, and positive-pair reweighting, with positive-pair reweighting being the core part. The overall framework of DeCAB is shown in Fig. 1.

3.1   Problem Setting and Notations

In a setting of SSL, we have labeled data $\mathcal{D}_L$ and unlabeled data $\mathcal{D}_U$. Each sample x_i^l in the labeled data $\mathcal{D}_L = \{(x_i^l, y_i^l)\}_{i=1}^N$ is associated with a label $y_i^l \in \{1, ..., C\}$. C is the total number of known classes. $N = \sum_{i=1}^C N_i$ is the total number of labeled data and N_i is the number of samples in class i. We assume the classes are imbalanced and sorted in non-ascending order, i.e., $N_i \geq N_j$ for $i < j$. The imbalance factor of a dataset is measured by the imbalance factor $IF = N_1/N_C$. Unlabeled data $\mathcal{D}_U = \{(x_i^u)\}_{i=1}^M$ contains M samples without annotation. We assume that unlabeled data consists of both ID data $\mathcal{D}_I$ and OOD data $\mathcal{D}_O$, i.e., $\mathcal{D}_U = \mathcal{D}_I \cup \mathcal{D}_O$. The samples in $\mathcal{D}_I$ follow the same class distribution as $\mathcal{D}_L$, and therefore IF is the same for both $\mathcal{D}_L$ and $\mathcal{D}_I$. Samples in $\mathcal{D}_O$ belong to classes other than the known C classes. For unlabeled data, we use one weak augmentation $(a(x^u))$ and two strong augmentations $(A_1(x^u), A_2(x^u))$. We aim to learn a model to effectively learn $\mathcal{D}_L$ and $\mathcal{D}_U$ to generalize well under a class-balanced test criterion.

3.2   Class-Aware Threshold

For unlabeled data, we compute the corresponding consistent loss as FixMatch [17]. The class-aware bias can lead to lower confidence of samples in the tail classes, resulting in a few tail class samples being selected if using a uniform high threshold for filtering. Therefore, we set different thresholds for each class based on sample size to filter more lower-confidence tail samples. The threshold formula is as follows:

$$\tau_c = 0.5 + N_c/N_1 \times (\tau_0 - 0.5), \tag{1}$$

where τ_0 denotes base threshold, and $c \in \{1, 2, 3, \dots, C\}$. For selected data, we compute the predicted class distribution of the weakly augmented version $a(x_i^u)$ of the unlabeled data point x_i^u: $q_i = f(a(x_i^u))$, and use $\hat{y}_i = \arg\max(q_i)$ as its pseudo-label. Then, we train the model to produce predicted class distribution $\tilde{q} = f(A_1(x_i^u))$ of its strongly augmented [4] version $A_1(x_i^u)$ and constrain it to be consistent with the pseudo-labels $\hat{y}$ by the following loss item:

$$L_u = \frac{1}{\mu B} \sum_{i=1}^{\mu B} \mathbb{1}(\max(q_i) > \tau_{\hat{y}_i}) H(\hat{y}_i, f(A_1(x_i^u))), \tag{2}$$

where $\tau_{\hat{y}_i}$ is the threshold related to class $\hat{y}_i$, μ is a hyper-parameter determining the relative ratio between the mini-batch size of labeled data, and that of unlabeled data, B is the batch size of mini-batch.

By using this mechanism, more tail-class samples are selected explicitly. Moreover, as is shown in Fig. 3, more OOD samples tended to be classified as non-head classes, implicitly mitigating the problem of exacerbated class perception bias due to OOD data. Therefore, the class-aware threshold can alleviate class-aware bias in both explicit and implicit ways.

3.3 Selective Sample Reweighting

For the samples that are filtered by the class-aware threshold, we think that the potential value of each sample in feature learning is not the same. In SSL methods, typically the same weight is assigned for feature learning on each sample. However, in situations with severe class-aware bias, the model may excessively prioritize learning confident head class samples and ignore unconfident tail class samples, resulting in low efficiency in feature learning. To overcome this problem, we utilize a sample reweighting mechanism that emphasizes the feature learning of unconfident samples. Specifically, the weight for feature contrastive learning of sample i is calculated as:

$$s_i = \mathbb{1}(\max(q_i) > \tau_{\hat{y}_i})(1 - \max(q_i)). \tag{3}$$

This mechanism, in combination with the class-aware threshold, can make the model focus more on the feature learning of less confident tail-class samples in the early stage of training, thus accelerating the convergence speed of the model.

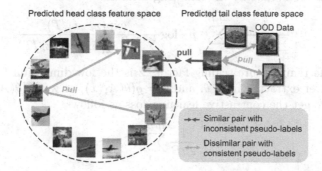

Fig. 2. A schematic shows how positive-pair scores work. The class-aware threshold tends to classify OOD samples as tail classes, and by using positive-pair scores, it pulls tail classes away from similar head classes while correcting misclassified tail samples.

3.4 Positive-Pair Reweighting

Class-aware contrastive learning approaches [13,27] have demonstrated remarkable performance in supervised learning tasks. However, in the context of imbalanced semi-supervised learning with OOD data, unreliable pseudo-labels can introduce false positive and false negative pairs for class-aware contrastive learning, potentially undermining the effectiveness of semantic feature learning. Therefore, we introduce positive-pair scores to address this issue, and Fig. 2 shows a schematic of how it works. We propose that the feature learning requirements of each sample pair may potentially differ, particularly in the semi-supervised scenario investigated in this paper. For sample pairs with consistent pseudo-labels but dissimilar features and sample pairs with inconsistent pseudo-labels but similar features, we argue they have more learning value, because the former larger feature differences contain more learnable information, and the latter may be potential positive sample pairs. Such a mechanism can avoid the negative impact of false negative pairs on the model.

Specifically, given a sample pair x_i^u and x_j^u, the positive-pair score is calculated as follows:

$$p_{ij} = \begin{cases} 1 - M_{ij}^p \cdot sim(\mathbf{v}_i, \mathbf{v}_j), & \text{if } \hat{y}_i = \hat{y}_j \\ 1 + M_{ij}^n \cdot sim(\mathbf{v}_i, \mathbf{v}_j), & \text{otherwise} \end{cases}, \tag{4}$$

where $\mathbf{v} = \theta(a(x))$ is the high-dimension feature of the weakly augmented version of x extracted by the feature extractor θ, $sim(\cdot, \cdot)$ is cosine similarity computation function, $\hat{y}_i$ and $\hat{y}_j$ denote the pseudo-labels of x_i^u and x_j^u respectively. M_{ij}^p and M_{ij}^n represent the positive-pair mask and negative-pair mask generated through pseudo-labeling (Pos. Mask and Neg. Mask in Fig. 2). If $\hat{y}_i = \hat{y}_j$, M_{ij}^p is set to 1, otherwise, 0, and M_{ij}^n follows the opposite logic.

Thus, for a given sample x_i^u, we calculate the contrastive learning loss of its strongly augmented views by the following formula:

$$L_{c_i} = \frac{1}{\sum_{j \neq i} p_{ij}} \sum_{j=1, j \neq i}^{2\mu B} p_{ij} \log \frac{\exp(\mathbf{z}_i \cdot \mathbf{z}_j / T)}{\sum_{k=1, k \neq i}^{2\mu B} \exp(\mathbf{z}_i \cdot \mathbf{z}_k / T)}, \tag{5}$$

where T is the temperature scaling factor, $\mathbf{z}$ is the low-dimension feature projected by g after extracted from θ, and $\mathbf{z} = g(\theta(A_1(x))$ or $\mathbf{z} = g(\theta(A_2(x))$. In a mini-batch, we get the contrastive learning loss as follows:

$$L_c = \frac{1}{2\mu B} \sum_{i=1}^{2\mu B} s_i L_{c_i}, \tag{6}$$

where s_i is the corresponding sample weight obtained from its weak augmentation by Eq. (3). Although this mechanism has no explicit negative pairs, it does not pose a significant risk of feature collapse. This is because it is combined with the class-aware threshold and sample reweighting mechanisms, where the feature learning intensity is reduced for highly confident samples. During the early stage of training, DeCAB prioritizes the learning of features from less confident tail samples while mitigating the impact of false negative pairs by employing positive-pair scores. In the later stages of training, the model assigns increasingly high confidence to samples, leading to a gradual weakening of the corresponding feature learning, reducing the risk of feature collapse.

3.5 Overall Training Objective

For the labeled data, we adopt the cross-entropy loss to utilize supervised information by the following:

$$L_x = \frac{1}{B} \sum_{i=1}^{B} l(y_i, f(x_i)), \tag{7}$$

where $l(\cdot, \cdot)$ is the common cross-entropy loss. The overall training objective function is as follows:

$$l_{total} = L_x + \lambda_u L_u + \lambda_c L_c, \tag{8}$$

where λ_u and λ_c are hyper-parameters, indicating the weight of each loss item.

4 Experimental Results

4.1 Experimental Settings

We compare DeCAB with several related SSL methods including general SSL methods FixMatch [17], MixMatch [1], OpenMatch [16], and imbalanced SSL method CReST [21] and DASO [14].

Imbalanced Dataset with OOD Data. The data used in the experiments consists of three subsets for the purpose of training, validation, and testing. **CIFAR-10/100** [11] are used as ID datasets, which are commonly adopted in the SSL literature [17]. We denote N_1 and M_1 as the number of head class samples in labeled data and unlabeled ID data, $N_i = (IF)^{-\frac{i-1}{C-1}} \cdot N_1$ and the same for M_i, while $N_1 = 1,500$, $M_1 = 3,000$ for CIFAR-10, $N_1 = 150$, $M_1 = 300$ for CIFAR-100. The testing sets of **Tiny ImageNet** (TIN) [5] and **LSUN** [25] are used as the OOD dataset. We mix these two OOD datasets into the unlabeled ID data and train the model on the mixed dataset.

Implementation Details. We employ Wide ResNet-28-2 for CIFAR-10 and Wide ResNet-28-8 for CIFAR-100 as backbone architecture, respectively. The standard training is performed for a total of 250,000 iterations, and validation is conducted every 500 iterations. The labeled data batch size is set to 64 for CIFAR-10, and 16 for CIFAR-100, while the batch size for unlabeled data is twice that, with μ set to 2. τ_0 in Eq. (1) is set to 0.95 for all experiments. The temperature scaling factor T in L_c is set to 0.007. λ_u and λ_c in Eq. (8) are set to 1.0 and 0.2 respectively for all experiments. We utilize the SGD optimizer with a basic learning rate of 0.03, momentum of 0.9, and weight decay of 1e-4. For experimental reproducibility, all experiments fix the random seed to 7.

Evaluation Criteria. In all experiments, the average top-1 accuracy (%) of each class is used for performance evaluation [17]. Additionally, we split the categories of the ID dataset into three groups (Head, Medium, and Tail) according to the class size, with the number of categories per group {3, 3, 4}, and {30, 35, 35} for CIFAR-10 and CIFAR-100, respectively.

4.2 Numerical Comparison

Experiments on the CIFAR-10 Dataset. We conduct a comparative analysis of model performance across different settings of CIFAR-10 with two OOD datasets (TIN and LSUN). As presented in Table 1, DeCAB exhibits remarkable

Table 1. Comparisons of group average accuracy with SSL methods on CIFAR-10 with two different OOD datasets (TIN and LSUN). The best results are shown in bold and the second-best ones are underlined.

Method	CIFAR-10-LT ($IF = 100$, TIN)				CIFAR-10-LT ($IF = 100$, LSUN)			
	Head	Medium	Tail	Avg. Acc	Head	Medium	Tail	Avg. Acc
General SSL Methods								
FixMatch	91.70	**77.10**	51.70	71.32	**94.20**	<u>73.17</u>	52.33	71.14
MixMatch	92.63	66.50	43.72	65.23	91.40	66.13	39.95	63.24
OpenMatch	<u>92.80</u>	70.53	40.42	65.17	91.77	64.37	41.87	63.59
Imbalanced SSL Methods								
CReST	92.13	72.97	51.35	70.07	91.57	70.07	49.98	68.48
DASO	91.67	<u>76.47</u>	<u>57.67</u>	<u>73.51</u>	<u>92.83</u>	72.17	<u>56.10</u>	<u>71.94</u>
DeCAB (Ours)	**94.10**	74.03	**62.42**	**75.41**	88.87	**74.30**	**60.60**	**73.19**

Table 2. Comparisons with SSL methods on CIFAR-100-LT under two different settings with OOD datasets (TIN and LSUN). The best results are shown in bold and the second-best ones are underlined.

Method	CIFAR-100-LT ($IF = 100$, TIN)				CIFAR-100-LT ($IF = 100$, LSUN)			
	Head	Medium	Tail	Avg. Acc.	Head	Medium	Tail	Avg. Acc
General SSL Methods								
FixMatch	<u>67.13</u>	36.69	6.37	35.21	66.97	**39.03**	5.89	35.81
MixMatch	61.20	32.14	5.77	31.63	60.20	29.94	6.63	30.86
OpenMatch	65.53	31.03	6.00	32.62	65.77	31.60	5.03	32.55
Imbalanced SSL Methods								
CReST	59.60	30.17	5.97	30.53	61.57	32.40	6.09	31.94
DASO	**71.70**	<u>36.83</u>	<u>7.06</u>	**36.87**	**71.13**	36.94	<u>7.29</u>	<u>36.82</u>
DeCAB (Ours)	66.33	**40.71**	**7.17**	<u>36.66</u>	<u>68.23</u>	<u>38.46</u>	**9.40**	**37.22**

performance and surpasses other methods in overall performance. Specifically, DeCAB outperforms MixMatch and OpenMatch by nearly 10% in overall accuracy when subjected to TIN or LSUN as OOD data. In the case of CIFAR-10-LT (IF=100, TIN), it surpasses FixMatch by almost 5%, CReST by 4%, and DASO by 2% in overall accuracy. In the group average accuracy, DeCAB outperforms MixMatch and OpenMatch by 20%, FixMatch and CReST by 10%, and DASO by approximately 4% in tail classes. From the results presented in the table, we can see that DeCAB achieves a more robust performance than other methods and is more friendly to non-head classes, particularly the tail classes.

Experiments on the CIFAR-100 Dataset. The experimental results on CIFAR-100 are represented on Table 2. It can be seen that DeCAB shows better performance on medium and tail classes compared to other methods, while there is a slight decrease in performance on head classes compared with FixMatch, which we think is acceptable. In the case of CIFAR-100-LT (IF=100, LSUN), DeCAB exhibits superior performance over FixMatch by 3.51%, MixMatch by 2.77%, OpenMatch by 4.37%, CReST by 3.31%, and DASO by 2.11% on the tail class. In the case of CIFAR-100-LT (IF=100, TIN), DeCAB is still more friendly to non-head classes, exhibiting a competitive advantage in terms of overall performance.

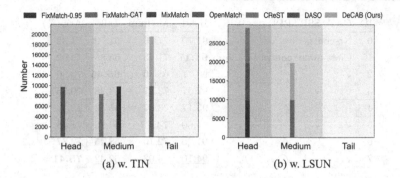

(a) w. TIN   (b) w. LSUN

Fig. 3. Comparison of different methods on the CIFAR-100-LT(IF=100, TIN) dataset in terms of the number of OOD misclassification with high confidence (0.95). FixMatch-CAT is a variation of the FixMatch algorithm that utilizes class-aware thresholds.

4.3 Analysis on Impact of OOD Data

In order to understand the role of OOD in the process of training, Fig. 3 depicts a comparison of various methods on the CIFAR-100-LT (IF=100, TIN) dataset in the number of OOD misclassification with high confidence (above 0.95). FixMatch-CAT is a variation of the FixMatch algorithm that utilizes class-aware thresholds. As shown in the figure, OpenMatch, CReST, and FixMatch tend to misclassify OOD samples as non-tail classes, and MixMatch and DASO demonstrate robustness to OOD data and avoid high-confidence misclassification, while FixMatch-CAT and DeCAB prefer to misclassify OOD data as tail classes. It is obvious that the utilization of class-aware thresholds tends to misclassify OOD data into non-head classes with high confidence, which implicitly mitigates the exacerbation of the class-aware bias caused by OOD data. In addition, although MixMatch and DASO did not misclassify OOD samples into ID classes with high confidence, their performance on tail classes is still inferior to that of DeCAB. This observation highlights that the OOD data in DeCAB serves as a beneficial source of information.

Table 3. Results of ablation experiments on CIFAR-10-LT under the setting of $IF = 100$ with Tiny ImageNet as OOD data. CAT, SSR, and PPR denote class-aware threshold, selective sample reweighting, and positive-pair reweighting, respectively. Bold values are the best and underlined values come next.

ID	CAT	SSR	PPR	Head	Medium	Tail	Acc.(%)
0	No contrastive loss			91.70	77.10	51.70	71.32
1	Class-aware contrastive loss			**94.17**	71.43	55.85	72.02
2	✓	–	–	90.77	68.30	**65.45**	<u>73.90</u>
3	–	✓	–	92.37	72.77	53.28	70.85
4	–	–	✓	93.17	73.37	55.53	72.17
5	✓	✓	–	91.60	**74.77**	53.92	71.48
6	✓	–	✓	91.20	72.10	58.45	72.37
7	✓	✓	✓	<u>94.10</u>	<u>74.03</u>	<u>62.42</u>	**75.41**

4.4 Ablation Experiments

We present ablation experiments on three major components of DeCAB, namely class-aware threshold (CAT), selective sample reweighting (SSR), and positive-pair reweighting (PPR). The results are presented in Table 3. **Experiment ID-0** refers to the base model FixMatch and **Experiment ID-1** utilizes class-aware contrastive loss with pseudo-labels, where the model performs best on the head group but poor on the tail group. The results of **Experiment ID-2**, **Experiment ID-3** and **Experiment ID-4** demonstrate that individually employing each module fails to yield significant performance improvements to the model. Comparing the results of **Experiments ID-5**, **ID-6**, and **ID-7**, it is apparent that PPR contributes the most to the model's performance improvement in DeCAB. However, solely relying on PPR may not be sufficient as the model's performance is still suboptimal. The overall results of ablation experiments reveal that the individual components of DeCAB do not independently result in a notable improvement in model performance. However, when integrated into a cohesive framework, the three modules effectively tackle the difficulties posed by imbalanced semi-supervised learning with OOD data.

5   Conclusion

In this paper, we proposed an end-to-end method DeCAB to alleviate serious class-aware bias under an imbalanced open-set scenario. DeCAB introduces positive-pair scores instead of positive-negative pairs in contrastive learning to avoid the detrimental effect of unreliable pseudo-labels. Moreover, by integrating class-aware thresholds, selective sample reweighting, and positive-pair scores, the model can focus on learning features of less confident tail class samples in the early stage and gradually reduce sample feature learning in later stages to avoid feature collapse, thereby improving the performance of tail classes and enhancing generalization performance. Overall, our method provides a simple but effective approach to address the challenge of feature learning in a semi-supervised environment and has the potential to advance state-of-the-art techniques in various machine learning tasks.

Appendix

A   Analysis of the Effect of OOD Data to SSL Methods

To reveal the fact that the OOD samples exacerbate the long-tail problem in existing SSL methods, we conduct a quick experiment. In detail, we consider a scenario in which the labeled data is characterized by a long-tailed distribution, with a small number of classes containing a disproportionate number of samples, while the vast majority of classes have a limited number of samples. In addition, the unlabeled data comprises both ID and OOD samples. The unlabeled ID samples follow the same class distribution and a similar long-tailed distribution to the labeled data, while the OOD samples do not belong to any of the ID classes. We conduct a quick experiment to demonstrate that the performance of existing SSL methods deteriorates when confronted with OOD data. To evaluate the model performance under various scenarios, we manipulate the imbalanced factor as well as the inclusion of OOD samples to simulate different settings. Figure 4 compares the confusion matrices of SSL methods on imbalanced training data with and without OOD data. The ID data uses the training set of CIFAR-100 with an imbalance factor of 100, while OOD data uses the testing set of Tiny ImageNet (TIN). It can be seen from the figure that the long-tailed problem leads to performance degradation on the tail class because many tail class samples are misclassified as a head class. The presence of OOD samples exacerbates the long-tailes problem for existing SSL methods.

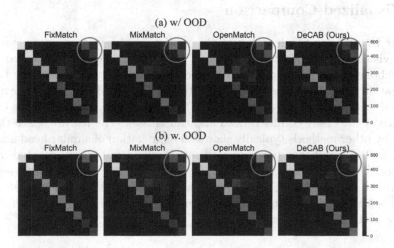

Fig. 4. Confusion matrices of SSL methods on the testing set of CIFAR-10 under two scenarios. The training set of CIFAR-10 is utilized as the labeled and unlabeled ID data with an imbalance factor of 100, and the testing set of Tiny ImageNet (TIN) is used as OOD data. (a) shows the case without OOD data, while (b) shows the case with OOD data.

B   Algorithm Flowchart

The algorithm of DeCAB is shown in Algorithm 1.

Algorithm 1. The Proposed DeCAB Algorithm

Input: Labeled data $\mathcal{D}_L$, unlabeled data $\mathcal{D}_U$, feature extractor θ, classifier ϕ, projector g, number of epochs E, number of iterations per epoch I and learning rate η.

Output: Feature extractor θ, classifier ϕ;

1: Initialize θ, ϕ and projector g;
2: **for** each epoch $t = 1, \ldots, E$ **do**
3:     **for** each iteration $i = 1, 2, \ldots, I$ **do**
4:         $S \leftarrow$ SAMPLEREWEIGHTING(X, θ, ϕ); // Obtain the selective sample weight by Eq. (3);
5:         $W \leftarrow$ PAIRREWEIGHTING(X, θ); // Obtain the posi- tive-pair score of sample pairs by Eq. (4);
6:         Compute L_{total} by Eq. (8) with S and W;
7:         Update θ, ϕ, g by L_{total} and η;
8:     **end for**
9: **end for**
10: **return** θ, ϕ.

C   Visualized Comparison

In order to evaluate the learning of the model on the feature space, we perform a visualized comparative analysis of the test set features extracted from the backbone of the model obtained by each method. Figure 5 shows the t-SNE visualization of feature space about the testing set of CIFAR-10, where the model is trained on CIFAR-10-LT (IF=100, TIN). In the figure, the black circle circles the space of the easily confused head and tail classes. The feature space that is learned by other methods typically shows an aggregation of similar head and tail classes, with a significant proportion of misclassified tail classes located in the middle of the black circles. In contrast, DeCAB exhibits a much clearer separation between head and tail classes in the feature space, resulting in fewer samples being misclassified in the middle. These results illustrate that our method can obtain a better feature space.

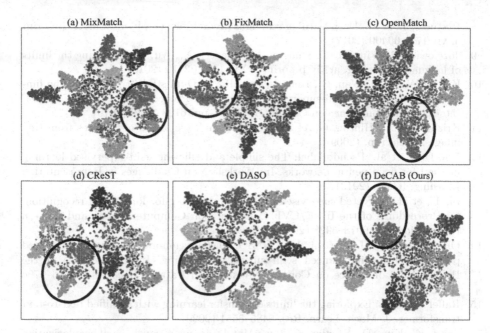

Fig. 5. The t-SNE visualization of feature space of CIFAR-10-LT test set, trained on CIFAR-10-LT with IF=100 and Tiny ImageNet as OOD data. The black circle circles the easily confused head and tail samples.

References

1. Berthelot, D., Carlini, N., Goodfellow, I., Papernot, N., Oliver, A., Raffel, C.A.: MixMatch: a holistic approach to semi-supervised learning. In: Advances in Neural Information Processing Systems, pp. 5050–5060 (2019)
2. Chen, Y., Zhu, X., Li, W., Gong, S.: Semi-supervised learning under class distribution mismatch. In: Proceedings of the AAAI Conference on Artificial Intelligence, pp. 3569–3576 (2020)
3. Cubuk, E.D., Zoph, B., Mane, D., Vasudevan, V., Le, Q.V.: AutoAugment: learning augmentation policies from data. In: Proceedings of the IEEE/CVF Conference on Computer Vision and Pattern Recognition, pp. 113–123 (2019)
4. Cubuk, E.D., Zoph, B., Shlens, J., Le, Q.: RandAugment: practical automated data augmentation with a reduced search space. In: Advances in Neural Information Processing Systems, pp. 18613–18624 (2020)
5. Deng, J., Dong, W., Socher, R., Li, L.J., Li, K., Fei-Fei, L.: ImageNet: a large-scale hierarchical image database. In: Proceedings of the IEEE/CVF Conference on Computer Vision and Pattern Recognition, pp. 248–255 (2009)
6. Fan, Y., Dai, D., Kukleva, A., Schiele, B.: CoSSL: co-learning of representation and classifier for imbalanced semi-supervised learning. In: Proceedings of the IEEE/CVF Conference on Computer Vision and Pattern Recognition, pp. 14574–14584 (2022)
7. Guo, L.Z., Zhang, Z.Y., Jiang, Y., Li, Y.F., Zhou, Z.H.: Safe deep semi-supervised learning for unseen-class unlabeled data. In: International Conference on Machine Learning, pp. 3897–3906 (2020)

8. Hestness, J., et al.: Deep learning scaling is predictable, empirically. arXiv preprint arXiv:1712.00409 (2017)

9. Jozefowicz, R., Vinyals, O., Schuster, M., Shazeer, N., Wu, Y.: Exploring the limits of language modeling. arXiv preprint arXiv:1602.02410 (2016)

10. Ke, Z., Wang, D., Yan, Q., Ren, J., Lau, R.W.: Dual student: Breaking the limits of the teacher in semi-supervised learning. In: Proceedings of the IEEE/CVF International Conference on Computer Vision, pp. 6728–6736 (2019)

11. Krizhevsky, A., Hinton, G., et al.: Learning multiple layers of features from tiny images. Tech. rep. (2009)

12. Lee, D.H., et al.: Pseudo-label: The simple and efficient semi-supervised learning method for deep neural networks. In: Workshop on Challenges in Representation Learning, ICML (2013)

13. Li, T., et al.: Targeted supervised contrastive learning for long-tailed recognition. In: Proceedings of the IEEE/CVF Conference on Computer Vision and Pattern Recognition, pp. 6918–6928 (2022)

14. Oh, Y., Kim, D.J., Kweon, I.S.: DASO: Distribution-aware semantics-oriented pseudo-label for imbalanced semi-supervised learning. In: Proceedings of the IEEE/CVF Conference on Computer Vision and Pattern Recognition,786–9796 (2022)

15. Raffel, C., et al.: Exploring the limits of transfer learning with a unified text-to-text transformer. J. Mach. Learn. Res. 5485–5551 (2020)

16. Saito, K., Kim, D., Saenko, K.: OpenMatch: open-set consistency regularization for semi-supervised learning with outliers. arXiv preprint arXiv:2105.14148 (2021)

17. Sohn, K., et al.: FixMatch: simplifying semi-supervised learning with consistency and confidence. In: Advances in Neural Information Processing Systems, pp. 596–608 (2020)

18. Tarvainen, A., Valpola, H.: Mean teachers are better role models: weight-averaged consistency targets improve semi-supervised deep learning results. In: Advances in Neural Information Processing Systems, pp. 1195–1204 (2017)

19. Van Engelen, J.E., Hoos, H.H.: A survey on semi-supervised learning. Mach. Learn. 373–440 (2020)

20. Wang, H., et al.: Partial and asymmetric contrastive learning for out-of-distribution detection in long-tailed recognition. In: International Conference on Machine Learning, pp. 23446–23458 (2022)

21. Wei, C., Sohn, K., Mellina, C., Yuille, A., Yang, F.: CReST: a class-rebalancing self-training framework for imbalanced semi-supervised learning. In: Proceedings of the IEEE/CVF Conference on Computer Vision and Pattern Recognition, pp. 10857–10866 (2021)

22. Xu, Y., et al.: Dash: semi-supervised learning with dynamic thresholding. In: International Conference on Machine Learning, pp. 11525–11536 (2021)

23. Yang, F., et al.: Class-aware contrastive semi-supervised learning. In: Proceedings of the IEEE/CVF Conference on Computer Vision and Pattern Recognition, pp. 14421–14430 (2022)

24. Yang, L., Jiang, H., Song, Q., Guo, J.: A survey on long-tailed visual recognition. Int. J. Comput. Vis. 1837–1872 (2022)

25. Yu, F., Seff, A., Zhang, Y., Song, S., Funkhouser, T., Xiao, J.: LSUN: construction of a large-scale image dataset using deep learning with humans in the loop. arXiv preprint arXiv:1506.03365 (2015)

26. Yu, Q., Ikami, D., Irie, G., Aizawa, K.: Multi-task curriculum framework for open-set semi-supervised learning. In: European Conference on Computer Vision, pp. 438–454 (2020)
27. Zhu, J., Wang, Z., Chen, J., Chen, Y.P.P., Jiang, Y.G.: Balanced contrastive learning for long-tailed visual recognition. In: Proceedings of the IEEE/CVF Conference on Computer Vision and Pattern Recognition, pp. 6908–6917 (2022)
28. Zhu, X.J.: Semi-supervised learning literature survey (2005)

An Effective Visible-Infrared Person Re-identification Network Based on Second-Order Attention and Mixed Intermediate Modality

Haiyun Tao, Yukang Zhang, Yang Lu[(⊠)], and Hanzi Wang

Fujian Key Laboratory of Sensing and Computing for Smart City, School of
Informatics, Xiamen University, Xiamen, China
{23020211153967,zhangyk}@stu.xmu.edu.cn,
{hanzi.wang,luyang}@xmu.edu.cn

Abstract. Visible-infrared person re-identification (VI-ReID) is a challenging cross-modality pedestrian retrieval problem. Due to the significant cross-modality discrepancy, it is difficult to learn discriminative features. Attention-based methods have been widely utilized to extract discriminative features for VI-ReID. However, the existing methods are confined by first-order structures that just exploit simple and coarse information. The existing approach lacks the sufficient capability to learn both modality-irrelevant and modality-relevant features. In this paper, we extract the second-order information from mid-level features to complement the first-order cues. Specifically, we design a flexible second-order module, which considers the correlations between the common features and learns refined feature representations for pedestrian images. Additionally, the visible and infrared modality has a significant gap. Therefore, we propose a plug-and-play mixed intermediate modality module to generate intermediate modality representations to reduce the modality discrepancy between the visible and infrared features. Extensive experimental results on two challenging datasets SYSU-MM01 and RegDB demonstrate that our method considerably achieves competitive performance compared to the state-of-the-art methods.

Keywords: Person re-identification · VI-ReID · Attention Mechanism · Intermediate Modality · Second-Order Information

1 Introduction

Person re-identification (Re-ID) aims to match the same person across disjoint cameras. Most of the existing methods consider images of people collected by visible cameras in the daytime. The infrared camera can capture valid appearance information during nighttime or in low-light conditions. Therefore, the applications are limited if only using visible cameras [8,18]. Cross-modality visible-

© The Author(s), under exclusive license to Springer Nature Singapore Pte Ltd. 2024
Q. Liu et al. (Eds.): PRCV 2023, LNCS 14433, pp. 120–132, 2024.
https://doi.org/10.1007/978-981-99-8546-3_10

Fig. 1. Five sample image pairs of the SYSU-MM01 dataset, where the images in each pair share the same identity. The visible and infrared images have great modality discrepancies.

infrared person re-identification (VI-ReID) task [40,41] aims to match the visible (infrared) images of the corresponding person given target infrared (visible) images.

The wavelengths between the two modalities are different. The visible images are composed of three channels, but the infrared modality images consist of a single channel containing invisible electromagnetic radiation. The two modalities are inherently different as shown in Fig. 1. VI-ReID task requires bridging the modality gap between the visible and infrared pedestrian images. Therefore, some existing VI-ReID methods [32,37] mainly use spatial attention, channel attention, or both of them [23,25,29] to obtain discriminative feature representations. But these commonly attention-based methods are either coarse or first-order. Some works [26,27] also use GAN to generate images to reduce the modality discrepancy. However, the image generation process usually brings additional computational costs and unavoidable noise. The above challenges result in less discriminative cross-modality features and unstable training.

To address the above limitations, inspired by MHN [2], we propose a second-order attention and mixed intermediate modality learning (SAIM) method, which consists of two main modules: second-order attention module (SAM) and mixed intermediate modality module (MIM). We consider that the second-order relationships between different modalities could better represent the person's features. Going further along with this method, the proposed SAM extracts second-order information from the deep features provided by the visible, intermediate, and infrared modalities. Many works [32] use spatial or channel-based attention networks which just exploit simple and coarse information and cannot be well applied to the challenging task of VI-ReID. Considering the complex real-world scenarios, the first-order features are insufficient to capture the interactions of visual parts. The second-order attention module identifies and captures the correlations among various body parts, resulting in enhanced and more detailed feature representations. It enhances the quality of features by incorporating second-order information into the representations of input images.

Furthermore, we generate intermediate modality representations to effectively capture complex interactions between different modalities. Unlike reducing domain adaptation differences [6], the mixed intermediate modality module aims to explicitly model appropriate intermediate modality to bridge the modal-

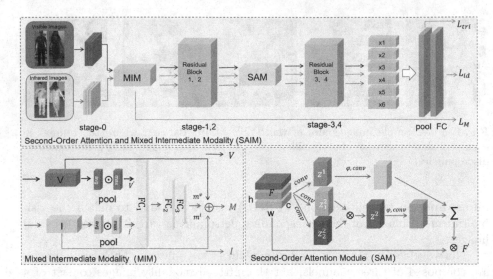

Fig. 2. The framework of our proposed approach. It involves two components: second-order attention module (SAM) and mixed intermediate modality (MIM).

ity discrepancy between the visible and infrared images. Specifically, it mixes the hidden stage's representations which can represent the characteristics of intermediate modality without complex generative and discriminative networks. To balance the visible and infrared modalities and avoid overfitting to either of them, we propose a balance loss, which constrains the standard deviation of two mixed weights. Therefore, it models the characteristics of appropriate intermediate modality to improve the model's discriminability.

Our contributions are summarized as follows:

- We propose a novel second-order attention module to capture and use second-order attention distributions among the visible, intermediate, and infrared modalities to facilitate feature learning for VI-ReID.
- We design a mixed intermediate modality module to reduce the visible and infrared modalities gap and transform cross-modality learning into three-modalities learning. Besides, we introduce loss function constraints to generate appropriate intermediate modality and improve the model's discriminability.
- Extensive experiments show that the proposed method achieves competitive performance compared with the state-of-the-art methods under various settings on both the SYSU-MM01 and RegDB datasets.

2   Related Work

VI-ReID aims to associate the same pedestrian images across the visible and infrared modalities. The image-level methods [26,28] address the modality discrepancy by efficiently exploring transformation methods. Specifically,

JSIA [27] performs set-level and instance-level alignment and generates cross-modality paired images. X-modality [16] utilizes an auxiliary modality to facilitate the cross-modality search and MMN [33] generates non-linear middle-modality images to reduce the modality discrepancy. The feature-level methods aim to exploit the modality-invariant features. MCLNet [11] introduces learning the modality-irrelevant features by confusing two modalities. MPANet [31] considers the cross-modality nuances. However, most of the mentioned methods primarily concentrate on extracting global or local features by only using first-order information and overlook the significance of second-order information. **Attention Mechanism** enables the network to concentrate on essential important information while suppressing less relevant information. Current convolution operations extract informative features by utilizing cross-channel or spatial information. Based on second-order information [2,4], some works [1,14,17] about fine-grained categorization have demonstrated improved performance by utilizing high-order information rather than only relying on first-order cues. We adopt our module to exploit meaningful features through second-order information. Using an attention mechanism enhances the model's discriminability and makes the network focus on important features and suppress irrelevant ones.

3 Proposed Method

The proposed method (SAIM) overview is illustrated in Fig. 2. SAIM contains a Second-Order Attention Module (SAM) in Sect. 3.1, and a Mixed Intermediate Modality (MIM) module in Sect. 3.2. Finally, we show the design of the loss function for the overall framework in Sect. 3.3.

3.1 Second-Order Attention Module

The attention mechanism helps the network focus on the important partial information. Firstly, we extract a deep feature map $F \in \mathbb{R}^{C \times H \times W}$ using a CNN where H and W are the height and width of the feature map and C is the channel features dimension. The overall attention process can be summarized as:

$$F' = S(F) \otimes F, \tag{1}$$

where $\otimes$ denotes element-wise multiplication, $S(\cdot)$ denotes a attention module and F' denotes the results of the proposed SAM. The spatial position in convolutional activations denotes $x \in \mathbb{R}^C$. Different from extracting local features, we use a polynomial predictor $f(\cdot)$ to model the second-order information and obtain a fine-grained discriminative representation. The polynomial predictor denotes as follows:

$$f(x) = \langle w, \phi(x) \rangle, \tag{2}$$

where w is a parameter to be learned and ϕ is a suitable local mapping function. The non-linear map is expressed as $\phi \to \otimes_r x$ and the r-order outer-product of x

is denoted by $\otimes_r$. The polynomial predictor can model all the degree r-th order information and can be written as:

$$f(x) = \sum_{k_1,\dots,k_r} W^r_{k_1,\dots,k_r}(\prod_{s=1}^{r} x_{k_s}),\qquad(3)$$

where $W^r_{k_1,\dots,k_r}$ denotes the weights of degree r-th order variable combinations. Specifically, when $r = 2$, $\sum_{i,j} W_{i,j}x_i x_j$ extracts all second-order interactions between variables. The polynomial predictor can be further reformulated as a tensor form:

$$f(x) = \langle w, x \rangle + \sum_{r=2}^{R} \langle w^r, \otimes_r x \rangle,\qquad(4)$$

where w^r is a r-th order tensor to be learned. The inner product of two tensors is denoted by $\langle \cdot, \cdot \rangle$. When $r > 1$, we use Tensor Tucker Decomposition [15] to break the independence of interaction parameters. Through this method, we save computing resources and reduce the number of trainable parameters.

$$f(x) = \langle w, x \rangle + \sum_{r=2}^{R} \sum_{d=1}^{D^r} \langle \sum \gamma^{r,d} u^{r,d}_1 \otimes \cdots \otimes u^{r,d}_r, \otimes_r x \rangle$$

$$= \langle w, x \rangle + \sum_{r=2}^{R} \sum_{d=1}^{D^r} \gamma^{r,d} \prod_{s=1}^{r} \langle u^{r,d}_s, x \rangle\qquad(5)$$

$$= \langle w, x \rangle + \sum_{r=2}^{R} \langle \gamma^r, z^r \rangle,$$

where w is approximated by D^r tensors, $u^{r,d}_r \in \mathbb{R}^C$ are vectors, $\gamma^{r,d}$ is the weight for d-th rank-1 tensor, and the d-th element of the vector $z^r \in \mathbb{R}^{D^r}$ is $\prod_{s=1}^{r} < u^{r,d}_s, x >$, $\gamma^r = [\gamma^{r,1},\dots,\gamma^{r,D^r}]^T$ is the weight vector of all D^r tensors. The computation of high-order statistics is very complex due to the high dimension, so we set the value of r as 2. Next, we use the sigmoid function $\varphi(z)$ to obtain the second-order attention map.

$$S(F) = sigmoid(f(F)).\qquad(6)$$

At last, the output is fed to the next convolutional layer. In contrast, the existing attention methods are not effective in capturing the subtle differences due to the significant cross-modality discrepancy. However, the proposed SAM uses second-order information to obtain discriminative and detailed attention maps.

3.2   Mixed Intermediate Modality Module

The structure of the proposed MIM is shown in Fig. 2. Inspired by [6], we obtain two intermediate feature maps $V \in \mathbb{R}^{C \times H \times W}, I \in \mathbb{R}^{C \times H \times W}$. Then, we utilize both max-pooling ($MaxPool$) outputs and average-pooling ($AvgPool$) outputs. According to the CBAM [29], we confirmed that exploiting both features improves the representation power of networks.

$$V' = AvgPool(V) \odot MaxPool(V), \qquad (7)$$

$$I' = AvgPool(I) \odot MaxPool(I), \qquad (8)$$

where $\odot$ means those are concatenated. Then the results V' and I' convolved by the FC_1 layers. Next, we merge the output feature vectors using element-wise summation for two modalities branches and then feed them into the FC_2 and FC_3 layers, which is computed in the following form:

$$m = \sigma(FC_3(FC_2(FC_1(V') + FC_1(I')))), \qquad (9)$$

$$M = m^v \cdot V + m^i \cdot I, \qquad (10)$$

which $\sigma(\cdot)$ is a softmax function. Finally, we achieve two mixed weights ($m = [m^v, m^i]$) and the mixed intermediate modality denoted by M. Then, we use the L_M to enforce that the intermediate modality generates better and trains stable.

$$L_M = -[\omega(\{m_i^v\}_{i=1}^N) + \omega(\{m_i^i\}_{i=1}^N)], \qquad (11)$$

where N is the batch size and the standard deviation is calculated by $\omega(\cdot)$.

3.3 Optimization

As shown in Fig. 2, we use a combination of identity loss L_{id} and triplet loss L_{tri} [38] to jointly optimize the proposed method. The identity loss L_{id} learns identity-invariant features. The triplet loss L_{tri} optimizes the relationships among triplet person images. The overall loss function can be defined as follows:

$$L = L_{id} + L_{tri} + \lambda L_M, \qquad (12)$$

where λ is the hyper-parameter that balances the contributions of the loss terms.

4   Experiments

4.1   Datasets

We use two publicly available VI-ReID datasets (SYSU-MM01 [30] and RegDB [21]) for the experiments. Moreover, we use mean average precision (mAP) and the mean Inverse Negative Penalty (mINP [38]) as our evaluation metrics.

The SYSU-MM01 dataset has 491 valid identities, out of which 296 identities are for training, 99 identities are for verification, and 96 identities are for testing. This dataset contains two different testing settings, which are all-search and indoor-search mode. The query set contains 96 identities with 3,803 infrared images. The gallery set contains all the images captured by four visible cameras in the all-search mode and only the images captured by two indoor visible cameras in the indoor-search mode. A single identity image under each visible camera to

Table 1. Comparison with the state-of-the-arts on the SYSU-MM01 dataset. The R1, R10, R20 denote Rank-1, 10 and 20 accuracies(%), respectively. The mean Average Precision (mAP) is also used as the evaluation metric. The best, second and third best results are indicated by red, blue, and green fonts.

Settings		All Search					Indoor Search				
Method	Venue	R1	R10	R20	mAP	mINP	R1	R10	R20	mAP	mINP
AGW [38]	TPAMI 21	47.50	84.39	92.14	47.65	35.30	54.17	91.14	95.98	62.97	59.23
X-Modal [16]	AAAI 20	49.92	89.79	95.96	50.73	–	–	–	–	–	–
DDAG [37]	ECCV 20	54.75	90.39	95.81	53.02	39.62	61.02	94.06	98.41	67.98	62.61
HAT [39]	TIFS 20	55.29	92.14	97.36	53.89	–	62.10	95.75	99.20	69.37	–
LBA [22]	ICCV 21	55.41	91.12	–	54.14	–	58.46	94.13	–	66.33	–
NFS [5]	CVPR 21	56.91	91.34	96.52	55.45	–	62.79	96.53	99.07	69.79	–
CICL [42]	AAAI 21	57.20	94.30	98.40	59.30	–	66.60	98.80	99.70	74.70	–
MSO [10]	ACMMM 21	58.70	92.06	97.20	56.42	42.04	63.09	96.61	–	70.31	–
CM-NAS [9]	ICCV 21	60.83	92.14	96.79	58.92	–	67.99	94.76	97.90	52.37	–
MID [13]	AAAI 22	60.27	92.90	–	59.40	–	64.86	96.12	–	70.12	–
SPOT [3]	TIP 22	65.34	92.73	97.04	62.25	48.86	69.42	96.22	99.12	74.63	70.48
MCLNet [11]	ICCV 21	65.40	93.33	97.14	61.98	47.39	72.56	96.98	99.20	76.58	72.10
SMCL [28]	ICCV 21	67.39	92.87	96.76	61.78	–	68.84	96.55	98.77	75.56	–
PMT [19]	AAAI 23	67.53	95.36	98.64	64.98	51.86	71.66	96.73	99.25	76.52	72.74
DART [34]	CVPR 22	68.72	96.39	98.96	66.29	53.26	72.52	97.84	99.46	78.17	74.94
CAJ [36]	ICCV 21	69.88	95.71	98.46	66.89	53.61	76.26	97.88	99.49	80.37	76.79
MPANet [31]	CVPR 21	70.58	96.21	98.80	68.24	–	76.64	98.21	99.57	80.95	–
DCLNet [24]	ACMMM 22	70.79	–	–	65.18	–	73.15	–	–	76.80	–
SAIM	Ours	72.44	97.05	99.15	70.35	58.21	80.23	99.20	99.90	83.87	80.49

form a gallery set is called a single shot. For the query set, all images will be used. We adopted a single-shot all-search mode evaluation protocol because it is the most challenging method.

The RegBD dataset contains 412 person identities, each with 10 visible and 10 infrared images. It includes 254 females and 158 males. Among the 412 individuals, 156 were photographed from the front and 256 from the back. Following the common settings [35], we randomly select all images of 206 identities for training and use the remaining 206 identities for testing. There are also two evaluation modes, which are visible-to-infrared and infrared-to-visible query modes. The performance is based on the average of ten trials on random training or testing splits.

4.2   Implementation Details

We implement our proposed method with the PyTorch framework. Following the AGW [38], we adopt a two-stream network with ResNet-50 [12] as our backbone. The network parameters are initialized using the pre-trained weights from ImageNet [7]. For the data augmentation, we apply the random horizontal flip and the random erasing [43] during the training process. The initial learning rate is set at 1×10^{-2} and then increases to 1×10^{-1} in 10 epochs. Then, we decay the learning rate to 1×10^{-2} at 20 epoch, and then we further decay it to 1×10^{-3}

Table 2. Comparison with the state-of-the-arts methods on the RegDB dataset. The best, second and third best results are indicated by red, blue, and green fonts.

Settings		Visible to Infrared					Infrared to Visible				
Method	Venue	R1	R10	R20	mAP	mINP	R1	R10	R20	mAP	mINP
AGW [38]	TPAMI 21	70.05	86.21	91.55	66.37	50.19	70.49	87.12	94.59	69.49	51.24
X-Modal [16]	AAAI 20	62.21	83.13	91.72	60.18	–	–	–	–	–	–
DDAG [37]	ECCV 20	69.34	86.19	91.49	63.46	49.24	68.06	85.15	90.31	61.80	48.62
HAT [39]	TIFS 20	71.83	–	–	67.56	–	70.02	–	–	66.30	–
LBA [22]	ICCV 21	74.17	–	–	67.64	–	72.43	–	–	65.46	–
NFS [5]	CVPR 21	80.54	91.96	95.07	72.10	–	77.95	90.45	93.62	69.79	–
CICL [42]	AAAI 21	78.80	–	–	69.40	–	77.90	–	–	69.40	–
MSO [10]	ACMMM 21	73.60	88.60	–	66.90	–	74.60	88.70	–	67.50	–
CM-NAS [9]	ICCV 21	82.79	95.06	97.74	79.25	–	81.68	94.06	96.91	77.58	–
MID [13]	AAAI 22	87.45	95.73	–	84.85	–	84.29	93.44	–	81.41	–
SPOT [3]	TIP 22	80.35	93.48	96.44	72.46	56.19	79.37	92.79	96.01	72.26	56.06
MCLNet [11]	ICCV 21	80.31	92.70	96.03	73.07	57.39	75.93	90.93	94.59	69.49	52.63
SMCL [28]	ICCV 21	83.93	–	–	79.83	–	83.05	–	–	78.57	–
PMT [19]	AAAI 23	84.83	–	–	76.55	–	84.16	–	–	75.13	–
DART [34]	CVPR 22	83.60	–	–	75.67	60.60	81.97	–	–	73.78	56.70
CAJ [36]	ICCV 21	85.03	95.49	97.52	79.14	65.33	84.75	95.33	97.51	77.82	61.56
MPANet [31]	CVPR 21	82.80	–	–	80.70	–	83.70	–	–	80.90	–
DCLNet [24]	ACMMM 22	81.20	–	–	74.30	–	78.00	–	–	70.60	–
SAIM	Ours	93.01	98.64	99.03	84.96	71.09	91.08	97.49	98.62	83.12	68.71

Table 3. Evaluation of each augmentation component on the SYSU-MM01 and RegDB datasets.

Method	Settings			SYSU-MM01					RegDB				
	M	S	L_M	R1	R10	R20	mAP	mINP	R1	R10	R20	mAP	mINP
B				66.63	94.59	98.03	63.41	49.16	90.43	97.21	98.52	82.78	68.06
B+M	✓			69.68	96.17	98.83	67.94	55.69	91.00	97.45	98.63	83.32	68.67
B+S		✓		70.54	96.47	99.07	68.64	56.51	91.26	97.53	98.66	83.87	69.65
B+M+S	✓	✓		70.93	96.55	99.05	68.39	56.60	91.43	97.54	98.71	84.07	70.40
Ours	✓	✓	✓	**72.44**	**97.05**	**99.15**	**70.35**	**58.21**	**93.01**	**98.64**	**99.03**	**84.96**	**71.09**

at epoch 80. Finally, we decay it to 1×10^{-4} at epoch 120. We also adopt a warm-up strategy [20]. The training process consists of a total of 150 epochs. At each training step, we randomly sample 6 identities, of which 4 visible and 4 infrared images in a mini-batch.

4.3   Comparison with State-of-the-Art Methods

In this section, we compare the proposed SAIM with state-of-the-art approaches published in the last two years to demonstrate the effectiveness of our method. We conduct the quantitative comparison on the SYSU-MM01 [30] and RegDB [21] datasets. The results can be seen in Table 1 and Table 2, our proposed method achieves competitive performance compared with the other existing methods on both two challenging datasets.

Table 4. Effectiveness of the SAM on the SYSU-MM01 and RegDB datasets.

1	SYSU-MM01					RegDB				
	R1	R10	R20	mAP	mINP	R1	R10	R20	mAP	mINP
B	66.63	94.59	98.03	63.41	49.16	90.43	97.21	98.52	82.78	68.06
1	67.14	95.78	98.42	64.46	52.95	90.91	97.42	98.29	83.51	69.20
2	**70.54**	**96.47**	**99.07**	**68.64**	**56.51**	**91.26**	**97.53**	**98.66**	**83.87**	**69.65**

The experiments on the SYSU-MM01 [30] dataset (Table 1) show that our method significantly improves the existing methods in both all-search and indoor search scenarios. We achieve 72.44% rank-1,70.35% mAP accuracy for the all-search settings. The experiments on the RegDB [21] dataset (Table 2) indicate that our proposed method exhibits robustness when confronted with various query settings. As shown in Table 2, our method can significantly outperform the existing methods by 5.56% and 6.33% in terms of R-1 scores with visible-to-infrared and infrared-to-visible query modes, respectively.

4.4   Ablation Study

In this section, we conduct some ablation experiments to show the influence of each component in our proposed method on both the SYSU-MM01 [30] and RegDB [21] datasets. As shown in Table 3, "B" indicates the baseline without the three components. "S" represents the second-order attention module. "M" denotes the mixed intermediate modality module. L_M can make the network learn a more appropriate intermediate modality. Especially, when introducing the second-order attention module, the performance is improved by 3.91% R-1 accuracy and 7.35% mINP on the SYSU-MM01 dataset. Table 3 shows the contribution of each component. When comparing our proposed method with the baseline, the integration of all three components results in the highest performance. This demonstrates the superior performance of our method.

Analysis on Different Order. As shown in Table 4, the experimental results show that second-order information benefits the ability to recognize the person identities on both the SYSU-MM01 and RegDB datasets. It significantly improves the baseline in all the metrics.

Different Stage Plugin. We insert our module after different stages of ResNet-50 to study how different stages will affect performance. The ResNet-50 backbone contains five stages, where stage-0 consists of two-stream structures and stage-1/2/3/4 corresponds to the other four convolutional blocks. Based on Table 5, we plug the SAM after stage-2 and plug the MIM module after stage-0 of the backbone in all our experiments.

Analysis on the Balance Parameter λ. We evaluate the effect of hyperparameter λ on both the SYSU-MM01 and RegDB datasets, as shown in Fig. 3

Table 5. Effectiveness on which stage to plug the SAM and MIM modules on the SYSU-MM01 dataset.

stage	SAM					MIM				
	R1	R10	R20	mAP	mINP	R1	R10	R20	mAP	mINP
0	67.05	95.35	98.67	65.42	52.93	**69.68**	**96.17**	**98.83**	**67.94**	**55.69**
1	70.40	95.83	98.30	67.79	55.42	68.65	95.75	98.85	65.34	51.25
2	**70.54**	**96.47**	**99.07**	**68.64**	**56.51**	69.02	95.83	98.81	66.95	55.37
3	70.03	95.69	98.02	67.57	55.94	67.96	95.19	98.42	63.67	51.54
4	66.39	94.78	98.46	63.13	49.19	68.51	95.84	98.16	66.82	54.72

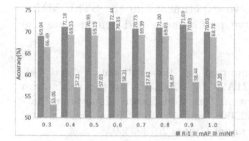

Fig. 3. Effect of hyperparameter λ in all-search mode on the SYSU-MM01 dataset.

Fig. 4. Effect of hyperparameter λ on the RegDB dataset.

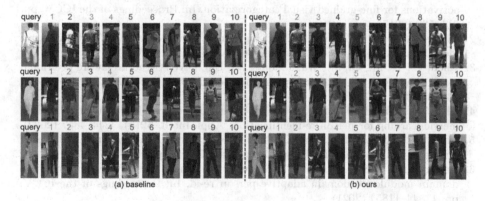

(a) baseline

(b) ours

Fig. 5. Visualization of the rank-10 retrieval results on the SYSU-MM01 dataset.

and Fig. 4. Experiments show that the best Rank-1 accuracy is achieved when λ is set to 0.6 on the SYSU-MM01 dataset and 0.5 on the RegDB dataset.

Visualization Analysis. Example results of three query images on the SYSU-MM01 dataset. Each row shows the rank-10 retrieved results. The same person is marked with green numbers and the different person is marked with red numbers.

Figure 5 demonstrates the effectiveness of the proposed method for identifying the same pedestrians and distinguishing different pedestrians compared with the baseline method.

5 Conclusion

In this paper, we propose an effective visible-infrared person re-identification network, which is composed of two modules. The second-order attention module for exploring the second-order information in VI-ReID. Our motivation here is to ensure the network can efficiently guarantee to capture a discriminative representation. We also present a mixed intermediate modality to bridge the gap between the two modalities. Extensive experiments on the SYSU-MM01 and RegDB datasets manifest competitive performance and demonstrate the effectiveness of our proposed method.

Acknowledgements. This work was supported by the National Natural Science Foundation of China under Grant U21A20514, 62002302, by the FuXiaQuan National Independent Innovation Demonstration Zone Collaborative Innovation Platform Project under Grant 3502ZCQXT2022008, and by the China Fundamental Research Funds for the Central Universities under Grants 20720230038.

References

1. Cai, S., Zuo, W., Zhang, L.: Higher-order integration of hierarchical convolutional activations for fine-grained visual categorization. In: Proceedings of the ICCV, pp. 511–520 (2017)
2. Chen, B., Deng, W., Hu, J.: Mixed high-order attention network for person re-identification. In: Proceedings of the ICCV, pp. 371–381 (2019)
3. Chen, C., Ye, M., Qi, M., Wu, J., Jiang, J., Lin, C.W.: Structure-aware positional transformer for visible-infrared person re-identification. In: IEEE TIP, pp. 2352–2364 (2022)
4. Chen, D., Wu, P., Jia, T., Xu, F.: Hob-net: high-order block network via deep metric learning for person re-identification. Appl. Intell. **52**(5), 4844–4857 (2022)
5. Chen, Y., Wan, L., Li, Z., Jing, Q., Sun, Z.: Neural feature search for RGB-infrared person re-identification. In: Proceedings of the CVPR, pp. 587–597 (2021)
6. Dai, Y., Liu, J., Sun, Y., Tong, Z., Zhang, C., Duan, L.Y.: IDM: an intermediate domain module for domain adaptive person re-id. In: Proceedings of the ICCV, pp. 11844–11854 (2021)
7. Deng, J., Dong, W., Socher, R., Li, L.J., Li, K., Fei-Fei, L.: Imagenet: a large-scale hierarchical image database. In: Proceedings of the CVPR, pp. 248–255 (2009)
8. Fan, X., Zhang, Y., Lu, Y., Wang, H.: Parformer: transformer-based multi-task network for pedestrian attribute recognition. In: IEEE TCSVT, p. 1 (2023)
9. Fu, C., Hu, Y., Wu, X., Shi, H., Mei, T., He, R.: CM-NAS: cross-modality neural architecture search for visible-infrared person re-identification. In: Proceedings of the ICCV, pp. 11803–11812 (2021)
10. Gao, Y., et al.: MSO: multi-feature space joint optimization network for RGB-infrared person re-identification. In: Proceedings of the 29th ACM MM, pp. 5257–5265 (2021)

11. Hao, X., Zhao, S., Ye, M., Shen, J.: Cross-modality person re-identification via modality confusion and center aggregation. In: Proceedings of the ICCV, pp. 16383–16392 (2021)
12. He, K., Zhang, X., Ren, S., Sun, J.: Deep residual learning for image recognition. In: Proceedings of the CVPR, pp. 770–778 (2016)
13. Huang, Z., Liu, J., Li, L., Zheng, K., Zha, Z.J.: Modality-adaptive mixup and invariant decomposition for RGB-infrared person re-identification. In: Proceedings of the AAAI, pp. 1034–1042 (2022)
14. Jacob, P., Picard, D., Histace, A., Klein, E.: Metric learning with HORDE: high-order regularizer for deep embeddings. In: Proceedings of the ICCVw, pp. 6539–6548 (2019)
15. Kolda, T.G., Bader, B.W.: Tensor decompositions and applications. SIAM Rev. **51**(3), 455–500 (2009)
16. Li, D., Wei, X., Hong, X., Gong, Y.: Infrared-visible cross-modal person re-identification with an x modality. In: Proceedings of the AAAI, pp. 4610–4617 (2020)
17. Li, P., Xie, J., Wang, Q., Zuo, W.: Is second-order information helpful for large-scale visual recognition? In: Proceedings of the ICCV, pp. 2089–2097 (2017)
18. Liu, L., Zhang, Y., Chen, J., Gao, C.: Fusing global and semantic-part features with multiple granularities for person re-identification. In: 2019 IEEE ISPA/BDCloud/SocialCom/SustainCom, pp. 1436–1440 (2019)
19. Lu, H., Zou, X., Zhang, P.: Learning progressive modality-shared transformers for effective visible-infrared person re-identification. In: Proceedings of the AAAI, pp. 1835–1843 (2022)
20. Luo, H., et al.: A strong baseline and batch normalization neck for deep person re-identification. IEEE Trans. Multim. **22**(10), 2597–2609 (2020)
21. Nguyen, D., Hong, H., Kim, K., Park, K.: Person recognition system based on a combination of body images from visible light and thermal cameras. Sensors **17**(3), 605 (2017)
22. Park, H., Lee, S., Lee, J., Ham, B.: Learning by aligning: Visible-infrared person re-identification using cross-modal correspondences. In: Proceedings of the ICCV, pp. 12026–12035 (2021)
23. Shao, R., Lan, X., Li, J., Yuen, P.C.: Multi-adversarial discriminative deep domain generalization for face presentation attack detection. In: Proceedings of the CVPR, pp. 10015–10023 (2019)
24. Sun, H., et al.: Not all pixels are matched: dense contrastive learning for cross-modality person re-identification. In: Proceedings of the ACM MM, pp. 5333–5341 (2022)
25. Tay, C.P., Roy, S., Yap, K.H.: Aanet: attribute attention network for person re-identifications. In: Proceedings of the CVPR, pp. 7127–7136 (2019)
26. Wang, G., Zhang, T., Cheng, J., Liu, S., Yang, Y., Hou, Z.: RGB-infrared cross-modality person re-identification via joint pixel and feature alignment. In: Proceedings of the ICCV, pp. 3622–3631 (2019)
27. Wang, G.-A., et al.: Cross-modality paired-images generation for RGB-infrared person re-identification. Proc. AAAI Conf. Artif. Intell. **34**(07), 12144–12151 (2020)
28. Wei, Z., Yang, X., Wang, N., Gao, X.: Syncretic modality collaborative learning for visible infrared person re-identification. In: Proceedings of the ICCV, pp. 225–234 (2021)
29. Woo, S., Park, J., Lee, J., Kweon, I.S.: CBAM: convolutional block attention module. In: Proceedings of the ECCV, pp. 3–19 (2018)

30. Wu, A., Zheng, W.S., Yu, H.X., Gong, S., Lai, J.: RGB-infrared cross-modality person re-identification. In: Proceedings of the ICCV, pp. 5390–5399 (2017)
31. Wu, Q., et al.: Discover cross-modality nuances for visible-infrared person re-identification. In: Proceedings of the CVPR, pp. 4328–4337 (2021)
32. Xu, J., Zhao, R., Zhu, F., Wang, H., Ouyang, W.: Attention-aware compositional network for person re-identification. In: Proceedings of the CVPR, pp. 2119–2128 (2018)
33. Yan, Y., Lu, Y., Wang, H.: Towards a unified middle modality learning for visible-infrared person re-identification. In: Proceedings of the ACM MM, pp. 788–796 (2021)
34. Yang, M., Huang, Z., Hu, P., Li, T., Lv, J., Peng, X.: Learning with twin noisy labels for visible-infrared person re-identification. In: Proceedings of the CVPR, pp. 14288–14297 (2022)
35. Ye, M., Lan, X., Li, J., Yuen, P.C.: Hierarchical discriminative learning for visible thermal person re-identification. In: Proceedings of the AAAI, pp. 7501–7508 (2018)
36. Ye, M., Ruan, W., Du, B., Shou, M.Z.: Channel augmented joint learning for visible-infrared recognition. In: Proceedings of the ICCV, pp. 13547–13556 (2021)
37. Ye, M., Shen, J., Crandall, D.J., Shao, L., Luo, J.: Dynamic dual-attentive aggregation learning for visible-infrared person re-identification. In: Proceedings of the ECCV, pp. 229–247 (2020)
38. Ye, M., Shen, J., Lin, G., Xiang, T., Shao, L., Hoi, S.C.H.: Deep learning for person re-identification: a survey and outlook. In: IEEE TPAMI, pp. 2872–2893 (2022)
39. Ye, M., Shen, J., Shao, L.: Visible-infrared person re-identification via homogeneous augmented tri-modal learning. In: IEEE TIFS, pp. 728–739 (2021)
40. Zhang, Y., Wang, H.: Diverse embedding expansion network and low-light cross-modality benchmark for visible-infrared person re-identification. In: Proceedings of the CVPR, pp. 2153–2162 (2023)
41. Zhang, Y., Yan, Y., Li, J., Wang, H.: MRCN: a novel modality restitution and compensation network for visible-infrared person re-identification. In: Proceedings of the AAAI, pp. 3498–3506 (2023)
42. Zhao, Z., Liu, B., Chu, Q., Lu, Y., Yu, N.: Joint color-irrelevant consistency learning and identity-aware modality adaptation for visible-infrared cross modality person re-identification. In: Proceedings of the AAAI, pp. 3520–3528 (2021)
43. Zhong, Z., Zheng, L., Kang, G., Li, S., Yang, Y.: Random erasing data augmentation. In: Proceedings of the AAAI, pp. 13001–13008 (2020)

Quadratic Polynomial Residual Network for No-Reference Image Quality Assessment

Xiaodong Fan[✉]

Faculty of Electrical and Control Engineering, Liaoning Technical University,
Huludao, Liaoning 125105, China
bhdxfxd@163.com

Abstract. Residual connection has become an essential structure of deep neural networks. In residual connection, shallow features are directly superimposed to deep features without any processing. In this paper, a quadratic polynomial residual module is designed to increase the nonlinear fitting ability of the network. As the name suggests, this module superimposes quadratic polynomials of shallow features onto deep features. In this way, the series of two modules has the fitting ability of a quartic polynomial. The fitting ability of the network increases exponentially with the number of layers. According to Taylor's theorem, it can be concluded that this module effectively improves the fitting ability of the network. Meanwhile, the image patches containing more information have greater contribution to image quality assessment. The patches are screened according to the two-dimensional information entropy, which reflects the information amount of patches. Based on the above two points, a quadratic polynomial residual network with entropy weighting and multi-receptive field structure is proposed for no-reference image quality assessment. The experimental results show that the proposed algorithm achieves high accuracy and more effectively fits the human visual system.

Keywords: Image quality assessment · Deep learning · Residual network

1 Introduction

As the eyes of intelligent equipment, computer vision is an essential part of the intelligent system, such as automatic driving, intelligent monitoring and face recognition. Image acquisition, transmission and storage inevitably cause image distortions before entering the system. The quality of the image directly determines the working effect of the system. Therefore, the first part of the vision system is to evaluate the quality of the input image and give a hint when the image

Supported by This work was supported by National Natural Science Foundation of China [No. 52177047, 62203197], Liaoning Province Education Administration [No. LJKZ1026, LJKZ1030].

quality does not meet the requirements. Image quality assessment (IQA) models lever intelligent techniques to evaluate image quality by simulating human visual characteristics. When the original undistorted image is available, the conventional method is to compare the distorted image with the original undistorted image to obtain a distortion measure. In this case, the undistorted image is called the reference image, and such methods are defined as full reference image quality assessment (FR-IQA). On the contrary, when the reference images are unaccessible, it is termed no-reference image quality assessment (NR-IQA), which is also named blind IQA (BIQA). Unfortunately, in practice, NR-IQA is the overwhelming majority with more practical value and wide application potential. In the past decade, convolutional neural networks (CNN) have gradually become the mainstream algorithm in the field of computer vision. Meanwhile, ResNet [1] is the most widely applied CNN model by far. The residual connection in the model makes the neural network deeper, which greatly improves the function fitting ability of the model. In fact, the residual connection is an identity map in parallel to the convolutional block. If the input feature map is denoted as x and the output of the convolutional block is denoted as $C(x)$, the output of the residual block is denoted as $C(x) + x$. At the same time, the activation function is generally ReLu in ResNet. It follows that the forward process can be regarded as the composition of linear functions. Therefore, the highest degree of the function formed by the forward process is one and the whole network can be regarded as a block linear function. In this paper, an attempt is made to replace the residual connection by a quadratic polynomial as showed in Fig. 1. It is obvious that this block degenerates to the residual block when $W_2 = 0, b = 0$ and W_1 is the identity matrix. It's a much broader model with much stronger generalization ability. At the same time, two blocks composition can fit a quartic polynomial exactly, and n blocks composition can fit a polynomial of degree 2^n. According to Taylor's theorem, the proposed model can efficiently fit any continuous function with infinite derivatives.

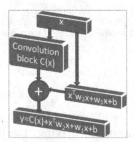

Fig. 1. Quadratic polynomial residual block

It has been confirmed that the image features extracted at different stages of deep convolutional neural networks correspond to different receptive fields, and deep features correspond to larger receptive fields. In the proposed network,

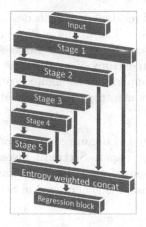

Fig. 2. Multi-receptive field structure

ResNet50 serves as the backbone network, and the features at different stages are extracted and concatenated into the regression block to achieve quality scoring as showed in Fig. 2. In this way, the quality characteristics of the image can be extracted comprehensively. The existing IQA algorithms usually segment images into patches for quality prediction. Patches at different locations carry different amounts of information, and obviously patches with more information are more noteworthy. Image entropy reflects the average amount of information in an image as a statistical feature of the image sharpness [2]. The larger the entropy, the richer of the information and the better the image quality will be [3]. Therefore, it is a reasonable and effective method to lever the image entropy as a measure of patch importance for weighting and screening patches. Meanwhile, in CNN, features extracted from different convolution kernels are variant image information. For example, some convolution kernels extract image edge, while others extract texture. The conventional networks treat the features as perfectly parallel and equally important. Obviously, the importance of the different features should vary for different visual tasks. Image entropy is an appropriate measure of the feature importance. This paper will apply it as a basis to determine the channel attention weight and filter patches.

Based on the above points, the proposed model quadratic polynomial residual network (QPRN) possesses the ability of high-order polynomial fitting with the multi-receptive field structure for NR-IQA. First, the distorted images are divided into patches and the two-dimensional information entropy of each patch is calculated. The patches with their entropies greater than a threshold are fed into the multi-receptive field QPRN for training. Whereafter, the features from different stages of the network are concatenated and sent to the regression block with their entropies as weight. Finally, the proposed model is validated and tested on the benchmark data sets. The experimental results show that the proposed algorithm is slightly superior to the existing NR-IQA algorithms. The main contributions of the proposed method can be summarized as follows:

1) The quadratic polynomial residual module is designed to enhance the nonlinear fitting ability of ResNet. The fitting ability of the whole network grows exponentially with the composition of the proposed modules.
2) The larger the information entropy of a patch, the richer the information will be. Therefore, a patch sampling and weighted concat strategy based on the two-dimensional information entropy was designed to screen out patches and features for IQA.

The remainder of the paper is organized as follows. Section 2 presents some works closely related to this paper. Section 3 investigates the structure and parameter setting of the proposed network. Experimental results are presented in Sect. 4. Finally, the conclusion is generalized in Sect. 5.

2   Related Work

2.1   IQA and Information Entropy

Information entropy is a measure of the information content in an image. Furthermore, the perceived quality is closely related to the information content. A method for evaluating digital radiographic images is presented based the information entropy [4]. The discrete cosine transform intensifies the differences of entropy. A blind IQA approach for natural images based on entropy differences in the discrete cosine transform domain is proposed [5]. Based on the local image entropy and its variance with additional color adjustment, a visual quality assessment for 3D-printed surfaces was proposed [6]. The perceptual quality of the image strongly depends on the non-maximum suppression (NMS). The image quality can be expressed by a sequence of entropy obtained with thresholded NMS [7]. In Reference [8], the reduced-reference features for IQA are computed by measuring the entropy of each discrete wavelet transform sub-band. Liu et al. [9]. calculated the one-dimensional entropy of the image patch in the spatial domain and frequency domain as the image features to fed the CNN. Since digital images are two-dimensional, the two-dimensional information entropy can better reflect the spatial distribution of pixels. Therefore, the two-dimensional information entropy is employed instead of one-dimensional information entropy in the designed residual CNN.

2.2   IQA and Deep Learning

With the rise of artificial intelligence, CNN has achieved remarkable success in many computer vision tasks. Kang et al. [10] design a CNN for NR-IQA with 3232 patches divided from the original image as the input. The average score of patches was adapted to predict the whole image quality, which made up for the lack of the training data. However, the network cannot distinguish between distortion types. Bosse et al. [11,12] proposed deeper networks that could predicted both distortion types and image quality. Deeper networks contain more parameters and require more training data. Kim et al. [13,14] proposed a CNN framework with the random clipping method to obtain more training data. On this basis, in order to obtain higher accuracy and better generalization ability, Yang et al. [15] constructed a NR-IQA algorithm based on the double-stream CNN structure, which appended gradient images as the auxiliary channel in addition to the original images, so that the network can extract more features from different distortions. Talebi and Milanfar [16] improved the CNN framework with image aesthetics. Li et al. [17] evaluated the importance of image semantics and designed a CNN based on semantic feature aggregation. However, the above IQA networks only focus on the numerical features of pixels, and ignore the spatial distribution of pixels. Li, Wang and Hou [18] proposed a two-channel convolutional neural network (CNN) based on shuffle unit for stereoscopic image quality assessment. Ma, Cai1 and Sun [19] proposed a no-reference image quality assessment method based on a multi-scale convolutional neural network, which integrates both global information and local information of an image. Madhusudana et al. [20] obtain image quality representations in a self-supervised manner with the prediction of distortion type and degree as an auxiliary task. Despite high correlation numbers on existing IQA datasets, CNN-based models may be easily falsified in the group maximum differentiation competition [21]. Zhang et al. [22] proposed a continual learning method for a stream of IQA datasets.

In summary, two-dimensional information entropy is closely related to image quality, and residual connection improves CNN's performance in IQA tasks. Next, the specific design of the proposed network will be developed around the above two centers.

3   Design of Network

In this section, the mechanism of patch sampling with the two-dimensional information entropy is explained. Finally, the structure and parameter setting of the proposed network model are clarified.

3.1   Two-Dimensional Information Entropy for Patch Sampling

The training for CNN needs abundant data. In order to make up for the shortage of samples, the original image is usually sampled with patches, and the final image quality score is based on the average prediction of the patches. In this way, each patch has the same quality score with the original image, the difference between patches were ignored. However, the differences between patches are inevitable. As shown in Fig. 3, the red box patch as the foreground of the image is the main factor of the image quality, while the blue box as the background is not the decisive factor. Experiments show that the entropy of the red patch with 6.547 is greater than the blue box with 3.862. Therefore, the information entropy can represent the difference between patches.

Information entropy is a statistical features of images, which reflects the average amount of information. It contains a large amount of structure information and can be measure the sharpness of images, areas with higher entropy are more

138   X. Fan

Fig. 3. Entropy differences between patches

likely to attract attention. A patch selecting strategy was proposed based on the
two-dimensional information entropy, which can effectively provide the spatial
information. The two-dimensional information entropy is defined as

$$ H = - \sum_{x_1=0}^{255} \sum_{x_2=0}^{255} p(x_1, x_2) \log_2 p(x_1, x_2) \tag{1} $$

where x_1 is the gray value of the current pixel, x_2 is the mean value of grays
in the neighborhood, and $p(x_1, x_2)$ is the joint probability density of x_1 and x_2.
Figure 4 shows the calculation of x_1 and x_2 in eight neighborhoods, where x_1
represents the gray value at pixel (i, j), x_2 is the mean value of grays in the
white area.

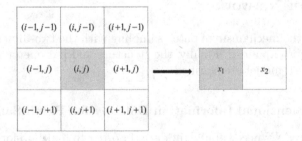

Fig. 4. Example of x_1 and x_2 for eight neighborhoods

The joint probability density $p(x_1, x_2)$ can be expressed as

$$ p(x_1, x_2) = \frac{f(x_1, x_2)}{MN} \tag{2} $$

where $f(x_1, x_2)$ is the frequency of the feature pairs (x_1, x_2), and M, N represents
the size of the image.

The following example illustrates the influence of two-dimensional information entropy on IQA. The image butterfly is split into patches. The two-
dimensional information entropy of all patches are extracted from the original

image and distorted image as shown in Fig. 5 (b, e). Through the proposed CNN, the predicted scores of the patches are obtained, and the error with the ground truth is shown in Fig. 5 (c, f). It can be seen that patches with low entropy have smooth texture, while patches with high entropy have prominent texture. For the distortion image, the similar conclusion can be obtained. The patch with smaller entropy corresponds to the higher prediction error, that is, the entropy is inversely proportional to the prediction error. Therefore, the patch with higher entropy should be chose to train the network, meanwhile the patch with smaller entropy can be ignored.

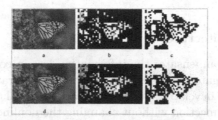

Fig. 5. Image, patch entropy and prediction errors. (a) (d) Original image and distortion image, (b) (e) Patchs entropy, and (c) (f) Prediction errors.

Fig. 6. Selected patches with different thresholds

Figure 6 shows the selected patches whose entropy are greater than some thresholds. Different shapes represent different thresholds. For example, the blue hexagon represents the sampling points with threshold 0. It was found that the sample were evenly distributed in the whole image for the threshold 0. In fact, the patch in the smooth area has little reference to IQA. When the threshold is increased to 5.5, the sampling patches locate on the regions with more complex structures. However, huge thresholds make the number of patches selected too few, the information loss of the image is serious. Therefore, we should adopt a moderate threshold. As shown in Fig. 6, when the threshold was around 5.0, the selecting patches are concentrated in the representative area. Therefore, it is appropriate that the threshold is set to 5.

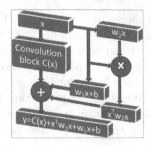

Fig. 7. Structure of quadratic polynomial residual

3.2   Network Architecture

Residual connection compensates the information loss in convolution and pooling operations. It mitigates gradient disappearance and gradient explosion, and enables the network to develop to depth. The early residual connection generally adopts linear jump connection without changing the nonlinear properties of the network. In order to enhance the nonlinear fitting ability of the network, the quadratic residual connection is designed, and its structure is shown in Fig. 7.

The quadratic residual connection consists of two linear layers in parallel, one of which is multiplied with the input features to form a quadratic term. The final jump connection forms a quadratic polynomial of the input feature and is therefore termed quadratic polynomial residual (QPR). Each residual block is composed of two parallel channels, one for convolution and the other for QPR. The data from the two channels are fused by addition in each residual block. It is not difficult to conclude from the figure that the series of two QPR blocks has the fitting ability of quartic polynomials. By analogy, the network's fitting ability increases exponentially with the depth.

Figure 8 shows the designed network contains five stages with QPR blocks as basic units. The input image is sampled repeatedly with randomly windows to form patches. The convolution blocks follow Resnet's basic convolution block structure. In fact, the receptive field of human vision usually changes in real time. Meanwhile, different stages of convolutional neural networks correspond to different scale receptive fields. The network adopts a multi-scale receptive field (MSRF) structure as shown in Fig. 8. The features extracted at each stage are entropy weighted and concated as inputs to the quality regression network.

4   Experiment Result

Pearsons Linear Correlation Coefficient (PLCC) and Spearman Rank Order Correlation Coefficient (SROCC) are selected to evaluate the performance of IQA networks. Comparative experiments were conducted to verify the availability of the proposed block for entropy selected patches (ESP), entropy weighted concat (EWC), multi-scale receptive field (MSRF) and quadratic polynomial residual (QPR). The experimental data are the average of ten-fold cross-validation.

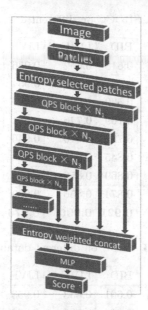

Fig. 8. Architecture of quadratic polynomial residual network

Table 1. Ablation experiment.

ESP	EWC	MSRF	QPR	SROCC	PLCC
				0.958	0.952
✓				0.961	0.957
✓	✓			0.967	0.959
✓	✓	✓		0.971	0.977
✓	✓	✓	✓	**0.976**	**0.987**

Table 1 shows the corresponding SROCC and PLCC in the ablation experiments. The experimental results show that the proposed network block effectively improves the accuracy of the network.

Table 2 and Table 3 show the SROCC and LCC of our method compared with state of the art IQA methods. The top performances were highlighted in bold. Experimental results show that our algorithm achieves the highest SROCC among all algorithms listed in Table 2. Relevant algorithms can be found in the references at the first column. Table 3 shows the PLCC obtained by all the algorithms involved in the experiment. The proposed algorithm achieves the best results.

Table 2. SROCC on different databases.

Methods	BID	LIVE-C	LIVE	CSIQ
Kang [10]	0.618	0.662	0.956	0.788
Bosse [12]	0.681	0.698	0.954	0.812
Kim [14]	0.539	0.595	0.961	0.815
Yan [15]	0.709	0.746	0.951	0.892
Xu [23]	0.721	0.640	0.946	0.741
Zhang [24]	0.845	0.851	0.968	0.946
Su [25]	0.869	0.859	0.962	0.923
Pan [26]	0.874	0.856	0.969	0.929
Proposed	**0.921**	**0.906**	**0.976**	**0.958**

Table 3. PLCC on different databases.

Methods	BID	LIVE-C	LIVE	CSIQ
Kang [10]	0.601	0.633	0.953	0.822
Bosse [12]	0.612	0.709	0.963	0.834
Kim [14]	0.576	0.613	0.962	0.803
Yan [15]	0.703	0.738	0.953	0.867
Xu [23]	0.736	0.678	0.947	0.823
Zhang [24]	0.859	0.678	0.971	0.959
Su [25]	0.878	0.882	0.966	0.942
Pan [26]	0.883	0.893	0.971	0.951
Proposed	**0.891**	**0.916**	**0.987**	**0.953**

5   Conclusion

Quadratic polynomial residual connections are designed in the network to strengthen the nonlinear fitting ability of the network. The concatenation of two designed modules has the same fitting ability as fourth-order polynomials. The fitting ability of the network increases exponentially with the number of layers. On the other hand, the information entropy is employed to measure the effectiveness of training data. In this way, the training data can be refined to improve training efficiency. Experimental results showed that the residual block improve the accuracy.

It can be seen from the experimental data that large-scale experiments on the model have not been carried out. Whether this model will exert its advantages in classification and recognition tasks is unknown. Therefore, Further experiments will be carried out to analyze the sensitivity of hyper-parameters on larger data sets. Meanwhile, the model should be applied to classification and recognition tasks to verify the fitting property. On the other hand, it will be a meaningful issue to analyze and prove the function fitting property of quadratic polynomial residuals from mathematical theory.

References

1. He, K., Zhang, X., Ren, S., Sun, J.: Deep residual learning for image recognition. In: 2016 IEEE Conference on Computer Vision and Pattern Recognition, pp. 770–778. IEEE, Las Vegas (2016)
2. Chen, X., Zhang, Q., Lin, M., Yang, G., He, C.: No-reference color image quality assessment: from entropy to perceptual quality. EURASIP J. Image Video Process. **77**(1), 1–12 (2019)
3. Deng, G.: An entropy interpretation of the logarithmic image processing model with application to contrast enhancement. IEEE Trans. Image Process. **18**(5), 1135–1140 (2009)
4. Tsai, D.Y., Lee, Y., Matsuyama, E.: Information entropy measure for evaluation of image quality. J. Digit. Imaging **21**, 338–347 (2008)
5. Yang, X., Li, F., Zhang, W., He, L.: Blind image quality assessment of natural scenes based on entropy differences in the DCT domain. Entropy **20**, 885–906 (2018)
6. Okarma, K., Fastowicz, J.: Improved quality assessment of color surfaces for additive manufacturing based on image entropy. Pattern Anal. Appl. **23**, 1035–1047 (2020)
7. Obuchowicz, R., Oszust, M., Bielecka, M., Bielecki, A., Pirkowski, A.: Magnetic resonance image quality assessment by using non-maximum suppression and entropy analysis. Entropy **22**(2), 220–236 (2020)
8. Golestaneh, S.A., Karam, L.: Reduced-reference quality assessment based on the entropy of DWT coefficients of locally weighted gradient magnitudes. IEEE Trans. Image Process. **25**(11), 5293–5303 (2016)
9. Liu, L., Liu, B., Huang, H., Bovik, A.C.: No-reference image quality assessment based on spatial and spectral entropies. Signal Process. Image Commun. **29**(8), 856–863 (2014)
10. Kang, L., Ye, P., Li, Y., Doermann, D.: Convolutional neural networks for no-reference image quality assessment. In: 2014 IEEE Conference on Computer Vision and Pattern Recognition, pp. 1733–1740. IEEE, Columbus (2014)
11. Bosse, S., Maniry, D., Wiegand, T., Samek, W.: A deep neural network for image quality assessment. In: IEEE International Conference on Image Processing, pp. 3773–3777. IEEE, Phoenix (2016)
12. Bosse, S., Maniry, D., Mller, K.R., Wiegand, T., Samek, W.: Deep neural networks for no-reference and full-reference image quality assessment. IEEE Trans. Image Process. **27**(1), 206–219 (2018)
13. Kim, J., Hui, Z., Ghadiyaram, D., Lee, S., Lei, Z., Bovik, A.C.: Deep convolutional neural models for picture-quality prediction: challenges and solutions to data-driven image quality assessment. IEEE Signal Process. Magaz. **34**(6), 130–141 (2017)
14. Kim, J., Lee, S.: Fully deep blind image quality predictor. IEEE J. Select. Topics Signal Process. **11**(1), 206–220 (2017)
15. Yan, Q., Gong, D., Zhang, Y.: Two-stream convolutional networks for blind image quality assessment. IEEE Trans. Image Process. **28**(5), 2200–2211 (2018)

16. Talebi, H., Milanfar, P.: NIMA: neural image assessment. IEEE Trans. Image Process. **27**(8), 3998–4011 (2018)
17. Li, D., Jiang, T., Lin, W., Ming, J.: Which has better visual quality: the clear blue sky or a blurry animal. IEEE Trans. Multim. **21**(5), 1221–1234 (2019)
18. Li, S., Wang, M., Hou, C.: No-reference stereoscopic image quality assessment based on shuffle-convolutional neural network. In: IEEE Visual Communications and Image Processing, pp. 1–4. IEEE, Sydney (2019)
19. Ma, Y., Cai, X., Sun, F.: Towards no-reference image quality assessment based on multi-scale convolutional neural network. Comput. Model. Eng. Sci. **123**(1), 201–216 (2020)
20. Madhusudana, P.C., Birkbeck, N., Wang, Y., Adsumilli, B., Bovik, A.C.: Image quality assessment using contrastive learning. IEEE Trans. Image Process. **31**, 4149–4161 (2022)
21. Wang, Z., Ma, K.: Active fine-tuning from GMAD examples improves blind image quality assessment. IEEE Trans. Pattern Anal. Mach. Intell. **44**(9), 4577–4590 (2022)
22. Zhang, W., Li, D., Ma, C., Zhai, G., Yang, X., Ma, K.: Continual learning for blind image quality assessment. IEEE Trans. Pattern Anal. Mach. Intell. **45**(3), 2864–2878 (2023)
23. Xu, J., Ye, P., Li, Q., Du, H., Liu, Y., Doermann, D.: Blind image quality assessment based on high order statistics aggregation. IEEE Trans. Image Process. **25**(9), 4444–4457 (2016)
24. Zhang, W., Ma, K., Yan, J., Deng, D., Wang, Z.: Blind image quality assessment using a deep bilinear convolutional neural network. IEEE Trans. Circuits Syst. Video Technol. **30**(1), 36–47 (2018)
25. Su, S., et al.: Blindly assess image quality in the wild guided by a self-adaptive hyper network. In: Proceedings of the IEEE/CVF Conference on Computer Vision and Pattern Recognition, pp. 3667–3676. IEEE, Seattle (2020)
26. Pan, Q., Guo, N., Qingge, L., Zhang, J., Yang, P.: PMT-IQA: progressive multi-task learning for blind image quality assessment. https://arxiv.org/abs/2301.01182. Accessed 3 Jan 2023

Interactive Learning for Interpretable Visual Recognition via Semantic-Aware Self-Teaching Framework

Hao Jiang, Haowei Li, Junhao Chen, Wentao Wan, and Keze Wang[✉]

Computer Science and Technology, Sun Yat-sen University, Guangzhou, China
{jiangh227,lihw59,wanwt3}@mail2.sysu.edu.cn, kezewang@gmail.com

Abstract. Aimed at tracing back the decision-making process of deep neural networks on the fine-grained image recognition task, various interpretable methods have been proposed and obtained promising results. However, the existing methods still have limited interpretability in aligning the abstract semantic concepts with the concrete image regions, due to the lack of human guidance during the model training. Attempting to address this issue and inspired by the machine teaching techniques, we formulate the training process of interpretable methods as an interactive learning manner by concisely simulating the human learning mechanism. Specifically, we propose a semantic-aware self-teaching framework to progressively improve the given neural network through an interactive teacher-student learning protocol. After initialing from the well-trained parameters of the given model, the teacher model focuses on minimally providing informative image regions to train the student model to generate interpretable predictions (i.e., semantic image regions) as good feedback. These feedback can encourage the teacher model to further refine the alignment of semantic concepts and image regions. Besides, our proposed framework is compatible with most of the existing network architectures. Extensive and comprehensive comparisons with the existing state-of-the-art interpretable approaches on the public benchmarks demonstrate that our interactive learning manner showcases an improved interpretability, a higher classification accuracy, and a greater degree of generality.

Keywords: Interpretable visual recognition · Interactive learning · Semantic-aware self-teaching

1 Introduction

To improve the interpretability of deep neural networks, many researchers have presented a range of approaches to shed light on the internal workings of these

Supplementary Information The online version contains supplementary material available at https://doi.org/10.1007/978-981-99-8546-3_12.

complex models [2,18,36]. Among these approaches, a straightforward strategy for addressing the interpretability challenge of deep neural networks is to visualize their feature representations [28]. However, there remains a gap between network visualization and semantic concept, which is crucial for improving interpretability. To bridge this gap, many works have been proposed to leverage the high-level features within deep neural networks to extract and grasp abstract semantics concepts [12,13], upon which auto-encoding [1] and prototype learning [4,26,31] methods are proposed. These methods attempt to establish a connection between image regions and abstract semantic concepts. Within these methods, a number of prototypes are utilized to focus "distracted" prototypes onto discriminative parts of the data, thereby improving the overall interpretability and accuracy [31]. However, the unidirectional approach adopted by recent studies [4,30,31] for image feature learning is incongruent with the interactive nature of human cognition and understanding. According to the current research about the human visual system, the process of recognizing salient image features and comprehending the image seems to be a complex and interactive one [33]. It involves both the self-generated feedback mechanism of attentional deployment, where selective processing determines which features are critical for encoding the integration of prior knowledge and social influence to inform the interpretation of the visual stimuli. Therefore, the interactive learning mechanism is essential for improving the robustness and explanation of interpretable neural networks.

Based on the attempt to get guidance in the training process and fit the cognition mechanism of humans, we focus on facilitating a teacher-student interactive learning manner for the interpretability of visual recognition models. The traditional teacher-student model [11,19] is characterized by a unidirectional flow of information, with limited interaction between the two parties. Machine teaching can be considered as an inverse problem to machine learning since its training objective is to find the smallest training set [38]. Naturally, this is compatible with the goal of interpretable visual recognition. We take the learner to be the human user, and the target semantics to be (a part of) the black-box recognition system that needs interpretation.

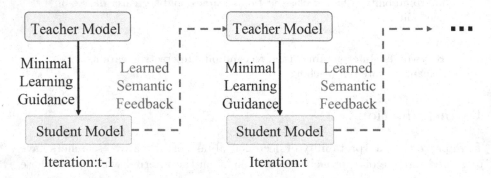

Fig. 1. The process of our interactive learning manner.

In this work, we propose an interactive semantic-aware self-teaching framework to progressively enable and improve the interpretability of the given neural network. As illustrated in Fig. 1, the teacher initially acquires well-trained parameters of a given model which performs well but lacks interpretability. Subsequently, the teacher model focuses on providing informative image regions to train the student model to generate interpretable predictions (i.e., semantic image regions) as good feedback. This implies that although the student model is initially a black box, the teacher model can obtain information about its parameters from the well-trained process, as well as representative image regions information. Hence, the teacher can create a cluster of compacted samples and patches that exhibit contextual relevance to the student models' selected patches feature. Consequently, with the guidance of the teacher, the black-box student model and reasoning process can gradually become more transparent. To demonstrate the compatibility and efficiency of our proposed framework, we evaluate it in various network structures, including CNN, ViT, and Swin-Transformer. We conduct experiments on three studies: bird species recognition (CUB), car model recognition (Cars), and dog species recognition (Dogs). A series of ablation studies are performed to investigate the influence of our proposed framework. Our framework shows outstanding performance with regard to generalization and accuracy, outperforming other interpretable approaches.

The **main** contribution of our work can be summarized as follows: i) We propose an interactive learning manner for enhancing network interpretability. Drawing on the bi-directional communication mechanism of human cognitive architecture, our interactive learning manner facilitates interpretable visual recognition; ii) We introduce a semantic-aware self-teaching framework to accurately capture features that exhibit varying degrees of granularity, ranging from coarse to fine; iii) Our proposed interactive learning framework can be readily compatible with diverse structural networks. Extensive and comprehensive experimental results on a variety of benchmarks demonstrate the effectiveness and superiority of our proposed framework.

2   Related Work

Machine Teaching involves identifying an optimal training set based on a given student model and target objective. Zhu et al. [37] proposed a comprehensive framework for this paradigm, which has been subsequently enriched by numerous works from various perspectives. In view of the iterative nature of the learning process, Liu et al. [20] introduced the concept of iterative machine teaching to facilitate rapid convergence of the learner model. Zhang et al. [34] further advanced the efficiency of machine teaching by introducing a more intelligent paradigm, referred to as one-shot machine teaching, which facilitates faster convergence using fewer examples. However, it should be noted that these machine teaching methods have only shown promising results on small-scale datasets, which means their practical applicability to real-world downstream tasks remains limited. As such, our work aims to address this gap by focusing on developing frameworks that are specifically designed to handle these downstream tasks. Considering the application, machine teaching has been applied in many situations, such as security [24], human-robot/computer interaction [23] and education [14]. However, there has been limited research on using machine teaching to enhance the interpretability of neural networks. Thus, our work contributes to this research gap by proposing a novel framework by extending machine teaching techniques in interpretable visual recognition.

Interpretable Representation Learning aims at generating feature representations that are interpretable and transparent, thereby enhancing the explainability and reliability of machine learning models. There are two primary directions in interpretable representation research: self-interpretable models and post-hoc analysis methods. A self-interpretable model is intentionally constructed to possess transparency and interpretability. Some related research has focused on using high-level convolutional layers to learn concrete parts and concepts of image regions. For example, Zhang Zhu et al. [35] constrain each filter response in the high-level convolutional layers to correspond to a specific part of the object, resulting in clearer semantic interpretations. Chen Zhu et al. [5] introduce a module called concept whitening in which the axes of the latent space are aligned with known common concepts. Post-hoc analysis methods aim to enhance the interpretability of well-trained deep neural networks through techniques such as feature visualization [3], saliency analysis [32], and gradient-based approaches [28]. In recent years, there has been a growing body of research in the field of prototype-based interpretable representation learning. For instance, Chen Zhu et al. [4] introduce ProtoPNet, a method that leverages category-specific prototypes based on convolutional neural networks (CNNs) to accurately perceive and recognize discriminative parts of objects. Additionally, Xue Zhu et al. [31] propose ProtoPFormer, which extends the ViT architecture with prototype-based methods to enable interpretable visual recognition. However, the existing methods still have limited interpretability in aligning the abstract semantic concepts with the concrete image regions, due to the lack of human guidance during the model training. Attempting to address this issue, we formulate the training process of interpretable methods as an interactive learning manner to concisely simulate the human learning mechanism.

3   Methodology

We propose an interactive learning manner for interpretable visual recognition via semantic-aware self-teaching. The proposed framework, shown in Fig. 2, comprises two key components: the student model, consisting of a backbone network $B(\cdot)$ and a patch selection module, and the self-teaching teacher model $T(\cdot)$. Initially, the input image is partitioned into patches of size 16×16 and subsequently forwarded to the student model and teacher model. The role of the teacher model at this stage is to initialize the category-concept bank. Relevant patch-feature pairs are identified via the patch selection module implemented in the student model. Consequently, the patch-feature pairs are transmitted to the category-concept bank for a query, whereby the bank promptly provides a patch-specific feature response. Based on the degree of similarity between the response and the selected patch feature, the identification of semantic patches can be achieved. Upon identification, the semantic patches are forwarded to the teacher model in order to update the category-concept bank, which can then be utilized to index the student model's patch pool as semantic guidance. Ultimately, by utilizing the semantic patch suggestions provided by the teacher model, the student model can acquire exceptional interpretability-enhancing abilities while maintaining optimal performance. Due to the limited page space, a more detailed description of our contrastive semantic loss can be found in the supplementary file.

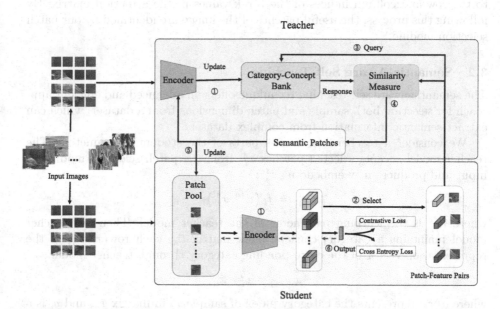

Fig. 2. The overview of our proposed interactive learning manner for interpretable visual recognition via semantic-aware self-teaching. The input image is divided into patches of size 16 × 16 and sent to the student model and teacher model. The student model has a patch selection module that identifies relevant patch-feature pairs. These pairs are then used to query a bank of category-concepts, which quickly provides patch-specific feature responses. By comparing the response to the selected patch feature, semantic patches can be identified. These semantic patches are then passed to the teacher model to update the category-concept bank. The updated bank can be used by the student model as semantic guidance for indexing its patch pool.

3.1   Patch Selection Strategy

Our framework is compatible with both isotropic and pyramid architectures. As an example, we will use the Swin Transformer [21], which is an implementation of the pyramid architecture. Swin Transformer adopts computing self-attention within local windows. The windows are arranged to evenly partition the image in a non-overlapping manner. The process of selecting image patches from Swin Transformer involves several steps. Firstly, the attention weights matrix (A) is obtained from the attention mechanism of the Swin Transformer:

$$x_i^{ps} = f_{ps}\left(w^{ps}, x_i\right), x_{ij}^{ps} \in Patches\left(x_i^{ps}\right), N = |Patches\left(x_i^{ps}\right)|$$

where x_i is an original input sample and w^{ps} represents parameters in Swin Transformer.

Secondly, the top-k$(1 \leq k \leq N)$ elements with the highest attention weights are identified. Lastly, patches are selected from the input image corresponding to the row and column indices of the top-k values in the attention matrix. By following this process, the useful patches of the image are identified by our patch selection module.

3.2   Semantic-Aware Self-Teaching

The semantic-aware self-teaching technique offers an advanced and efficient approach for selecting both sample and patch dimensions from a dataset, which can extract semantic information from complex datasets.

We consider a teacher who knows parts of the student's parameters. The teacher model accepts selected patches x_i^{sl} from the patch selection module as input and produces new embedding x_i^{te}:

$$x_i^{te} = f_t\left(w^{te}, x_i^{sl}\right)$$

where w^{te} is the learning parameter of the teacher model. Then our teacher model maintains a category-concept bank matrix $\bar{x}_c$, each row($\bar{x}_{c(x_i)}$) is the representation vector of the corresponding category through teacher model:

$$\bar{x}_{c(x_i)} = y_i^T \cdot \bar{x}_c$$

where $c\left(x_i\right)$ represents the category index of sample x_i in matrix $\bar{x}_c$, and y_i is a one-hot column vector represent the label of x_i. At first, we randomly initialize x_c, In each training epoch, We update the category-concept bank utilizing sample embedding selected by the teacher through the following two-step optimization.

After obtaining the category-concept bank, we do semantic patches optimization in order to obtain patches with more semantic concept information which would help improve model interpretability. And we also do sample optimization by selecting more important samples.

Semantic Patch Optimization. While optimizing semantic patches, we construct a matrix PS. It has b rows and N columns where b is the input batch size and N is the number of patches in each sample. PS_{ij} represents the similarity between the jth patch feature x_{ij}^{te} of the ith sample in the input batch and the corresponding category-concept patch feature $x_{c(x_i),j}$, we use cosine similarity:

$$PS_{ij} = sim\left(x_{ij}^{te},\ \bar{x}_{c(x_i),j}\right) = \frac{\langle x_{ij}^{te}, \bar{x}_{c(x_i),j}\rangle}{\left\|x_{ij}^{te}\right\| \cdot \left\|\bar{x}_{c(x_i),j}\right\|}. \tag{1}$$

Then top-k_p^{te} similar patch features in each sample are further selected as our semantic patches:

$$\hat{PS}_{ij} = \underset{j \in range(1,N+1)}{Sort} (PS_{ij})$$

$$A_{p,ij}^{te} = \begin{cases} \text{softmax} \left(\hat{PS}_{ij} \right) & j \leq k_p^{te}. \\ \text{softmax} \left(\hat{PS}_{ij} + (-100) \right) & j > k_p^{te}. \end{cases} \quad (2)$$

$$x_{p,i}^{te} = A_{p,i}^{te} \odot x_i^{ps}$$

where $x_{p,i}^{te}$ is the new sample after semantic patches optimization.

Sample Optimization. Procedures of patch optimization are similar to semantic patch optimization. We construct a sample similarity vector S. It is a b dim vector with each element representing the similarity between current sample feature x_i^{te} and corresponding category representation $\hat{x}_{c(x_i)}$ in our category-concept bank. Where b is the batch size of input samples. Then we select top-k_s^{te} samples, those we think are more useful.

After the above two-step optimization, the teacher model produces selected useful samples with semantic patches $x_{s,p}^{te}$. We use it to update our category-concept bank following the momentum mechanism:

$$\bar{x}_c = \alpha \bar{x}_c + (1 - \alpha) x_{s,p}^{te} \quad (3)$$

where α is a hyperparameter that balances the weights of the current sample and corresponding category-concept representation in the maintained category-concept bank while updating. Finally, We use them as input data for the student model to predict their labels:

$$y' = f_s \left(w^s, x_{s,p}^{te} \right)$$

where w^s is the learning parameter of the student model.

Algorithm 1 describes the training of our semantic-aware self-teaching framework. Note that the training process is a mode of interactive learning that involves patch-feature pairs feedback and semantic patches guidance.

4   Experiments

Experimental Settings: We perform a series of experiments on three datasets, namely CUB-200-2011 [29], Standford Cars [17], and Stanford Dogs [15]. Our proposed framework is compared against the state-of-the-art (SOTA) methods in terms of both accuracy and interpretability. The CUB dataset comprises 5994 images for training and 5794 for testing, representing over 200 avian species. The Cars dataset comprises 16,185 images of vehicles, representing 196 distinct car classes. The Dogs dataset is a fine-grained classification dataset, encompassing 120 individual dog breeds. All images are resized to 384 × 384 pixels.

Algorithm 1: The semantic-aware teacher

1: Initialize the teacher parameter w_{te}^0 with a well pre-trained student parameter w_s^0, randomly initialize the category-concept bank $\bar{x}_c^0$;
2: Set $t = 0$ and the maximal interactive number T, learning parameter $w^t = (w_{ps}^t, w_{te}^t, w_s^t)$;
3: **while** w^t has not converged or $t < T$ **do**
4:    Student accepts input batch x^t, selects patches, feeds into teacher, produces z^t;
5:    Teach semantic patches and important samples to Student:

$$z_p^t = A_p^{te} \odot z^t, z_{s,p}^t = (A_s^{te})^T \cdot z_p^t$$

6:    Use the selected semantic patches and samples to perform the update:

$$w^{t+1} = w^t - \eta_t \frac{\partial \mathcal{L}\left((w^t, x^t), y^t\right)}{\partial w^t}$$

7:    Update teacher's category-concept bank:

$$\bar{x}_{c(x_i^t)}^{t+1} = \alpha \bar{x}_{c(x_i^t)}^t + (1 - \alpha) z_{p,i}^t$$

8:    $t \leftarrow t + 1$
9: **end while**

Backbones: We explore three prominent network architectures, namely ResNet-50 [10], ViT-B/16 [8] and SwinT-L [21], as the backbones in our experimental setup. These networks are initialized with official pre-trained weights on the ImageNet dataset [6].

Metric: Due to the lack of a unified index to evaluate the interpretability of visual recognition using semantic alignment features, we describe the semantic alignment score. The semantic alignment score serves to measure the accuracy with which the selected interpretable regions by the proposed method correspond to the semantics of the original image. The semantic alignment score comprises two components, namely the Intersection over Union (IoU) and the semantic measure. The IoU represents the ratio of the intersection of the semantic patches selected by our framework and the ground truth semantic labels of the original image. We only employed the original dataset without any data augmentation during both the training and inference stages, and the ground truth with semantic labels was solely used for evaluating interpretability. The semantic measure refers to the vector similarity between the ground truth concept region and the semantic patch selected by our framework, which is then scaled up by a factor of 100. The formal expression is as follows:

$$\text{Semantic Alignment Score} = \text{IoU} \times (100 \times \text{Semantic Measure})$$

Here "Semantic Measure" is calculated according to the following formula:

$$T = \frac{1}{K} \sum_{i=1}^{K} \frac{\langle S_i^{gt}, S_i^{sl} \rangle}{\|S_i^{gt}\| \cdot \|S_i^{sl}\|}$$

where T represents the average semantic measure of K sub-concept semantic regions, S_i^{gt} represents the concatenated feature vector of the "i-th" ground truth sub-concept semantic region patches, S_i^{sl} represents the concatenated feature vector of our selected patches which have been divided into the "i-th" sub-concept semantic region. The feature vector of each patch is produced by our well-trained student model.

Implementation Details: The proposed framework is implemented in PyTorch. Pre-trained using ImageNet, our backbones f contain ResNet [10], Vision Transformer [8], and Swin Transformer [21]. The input images are standardized by resizing them to 384 × 384 pixels. We split the image into 16×16 patches for all backbones. The batch size is set to 16. All models are trained for 50 epochs with AdamW optimizer [22] and cosine LR schedule with warmup steps. The learning rate is initialized as 5e-4 for CUB-200-2011, 2e-4 for Stanford Dogs, and 2e-3 for Stanford Cars. Training and testing are conducted using four NVIDIA 3090 GPUs.

Table 1. The interpretability (Semantic Alignment Score) and accuracy comparison of CNN-based ViT-based and Swin-T-based methods.

Method	Venue	Semantic Alignment Score			Accuracy		
		CUB	Cars	Dogs	CUB	Cars	Dogs
ResNet50 Baseline [10]	CVPR2016	0	0	0	84.5	92.9	88.1
ProtoTree [26]	CVPR2021	67.4	72.8	66.9	82.2	86.6	85.3
Def. ProtoPNet [7]	CVPR2022	80.5	79.3	70.6	**85.2**	89.4	86.5
ProtoPool [27]	ECCV2022	83.4	78.5	69.7	81.1	88.9	87.2
ResNet50+Ours	–	**88.4**	**82.9**	**77.6**	84.9	**93.1**	**88.2**
ViT Baseline [8]	ICLR2021	0	0	0	90.2	93.5	91.2
TransFG [9]	AAAI2022	50.3	49.8	47.4	**90.9**	**94.1**	90.4
ViT-B+Ours	–	**80.5**	**79.3**	**76.4**	90.4	93.7	**91.3**
Swin-T-L Baseline [21]	ICCV2021	0	0	0	91.7	94.5	90.1
ViT-NeT [16]	ICML2023	72.4	74.5	70.3	91.7	**95.0**	90.3
Swin-T+Ours	–	**80.6**	**79.3**	**75.4**	**91.8**	94.8	90.3

Performance Comparisons: We collect all competitive methods that report their results on CUB, Cars, and Dogs. Table 1 illustrates the top-1 accuracy achieved by our interactive learning framework, alongside other competitive methods, across the involved datasets employing three different types of backbones, including ResNet-50 [10], ViT-B/16 [8], and SwinT-L [21]. For a fair comparison, all input images are uncropped.

i) We find that our proposed framework exhibits excellent compatibility and is adaptable to networks with different structures. As shown in Table 1, our framework can serve as a plugin and be integrated with ResNet, ViT, and Swin-T models, achieving consistent and promising performance across various architectures. Moreover, our proposed framework demonstrates interpretability in accommodating diverse network architectures.

*ii) We find that our proposed framework achieves an optimal balance between accuracy and interpretability.*It can be observed that the baseline approach exhibits relatively higher performance when compared to other self-interpretable baseline techniques [7,25,27] owing to the non-interpretable characteristics of the base method. But our framework demonstrates a substantially improved performance of accuracy as well as in semantic alignment score.

iii) We find that our proposed framework can achieve state-of-art interpretability and competitive accuracy performance. We compare our proposed interactive learning manner with ViT-NeT [16], a SOTA method of interpretable network. Our framework consistently achieves higher semantic alignment scores across all datasets, thereby exemplifying enhanced interpretability. As Table 1 shows, in all backbone structures, our framework can achieve a higher semantic alignment score, outperforming ProtoPool [27] by 11%, TransFG [9] by 60%, and Vit-NeT [16] by 8%. And our framework can achieve a competitive accuracy performance in all backbone structures, outperforming ProtoPool [27] by 4.2% most, TransFG [9] by 0.9% most.

Visualization Analyses: Figure 3 exemplifies the transparent reasoning process that underpins our semantic-aware self-teaching framework for the test evaluation. As illustrated in Fig. 3, our proposed method identifies the most representative semantic concept for the Brown Creeper (above) as their backs and for the Western Meadowlark(below) as their breasts, which aligns with reality. Based on our findings, we find that each class demonstrates a distinct semantic focus. As additional examples, the Brewer Blackbird is most commonly associated with its tail, while the Geococcyx is predominantly identified by its neck. Our framework effectively captures the connection between image patches and semantic features, resulting in accurate classification and localization of regions.

Table 2. Ablation studies of patch selection module and machine teaching module.

Method	ResNet-50			ViT-B/16			SwinT-L		
	CUB	Cars	Dogs	CUB	Cars	Dogs	CUB	Cars	Dogs
Ours w/o Self-teaching	85.5	93.5	88.7	91.0	94.0	91.7	92.3	95.3	90.8
Ours w/o Patch Selection	84.2	92.2	87.8	90.0	93.1	89.7	91.3	94.1	89.7
Ours	84.9	93.1	88.2	90.4	93.7	91.3	91.8	94.8	90.3

Ablation Studies: We perform an ablation analysis on the CUB, Cars, and Dogs datasets using three distinct structural backbones to evaluate the constituent elements of our proposed interactive learning paradigm. As shown in Table 2, our study considers the effect of the patch selection module and teacher model. The patch selection module can improve recognition accuracy performance at most 0.8% to ensure the object level interpretability. The teacher

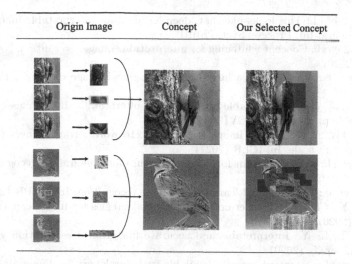

Fig. 3. The interpretable reasoning process of our proposed method. More illustration examples can be found in the supplementary file

model slightly decreases the recognition accuracy performance at most 0.3% while ensuring the concept level interpretability. This suggests that there may be a trade-off between semantic interpretability and performance, where an increase in one could lead to a decrease in the other. However, by employing an interactive learning manner, the concomitant benefits of enhanced semantic interpretability and performance can be effectively realized.

5   Conclusion

In this work, we proposed a semantic-aware self-teaching framework for interpretable visual recognition via semantic-aware self-teaching. Our proposed framework employs the teacher model to adaptively guide the student model to concentrate on more representative semantic patches according to its feedback. Moreover, our proposed framework is compatible with most of the existing network architectures. In the future, we will investigate how to apply our framework for explainable video analytics.

References

1. Alvarez-Melis, D. et al.: Towards robust interpretability with self-explaining neural networks. In: NIPS (2018)
2. Beckh, K., et al.: Explainable machine learning with prior knowledge: an overview. arXiv (2021)
3. Bychkov, D., et al.: Deep learning based tissue analysis predicts outcome in colorectal cancer. Sci. Rep. (2018)

4. Chen, C., et al.: This looks like that: deep learning for interpretable image recognition. Neural Inf. Process. Syst. (2019)
5. Chen, Z., et al.: Concept whitening for interpretable image recognition. Nat. Mach. Intell. (2020)
6. Deng, J., et al.: Imagenet: a large-scale hierarchical image database. In: CVPR (2009)
7. Donnelly, J., et al.: Deformable protopnet: an interpretable image classifier using deformable prototypes. In: CVPR (2022)
8. Dosovitskiy, A., et al.: An image is worth 16x16 words: transformers for image recognition at scale. In: ICLR (2021)
9. He, J., et al.: Transfg: a transformer architecture for fine-grained recognition. In: AAAI (2022)
10. He, K., et al.: Deep residual learning for image recognition. In: CVPR (2016)
11. Huang, Y., Chen, J.: Teacher-critical training strategies for image captioning. In: CVPR (2020)
12. Huang, Z., Li, Y.: Interpretable and accurate fine-grained recognition via region grouping. arXiv (2020)
13. Ji, R., et al.: Attention convolutional binary neural tree for fine-grained visual categorization. In: CVPR (2019)
14. Johns, E., et al.: Becoming the expert - interactive multi-class machine teaching. In: CVPR (2015)
15. Khosla, A., et al.: Novel dataset for fine-grained image categorization: stanford dogs. In: Proceedings of the CVPR Workshop on Fine-Grained Visual Categorization (FGVC) (2011)
16. Kim, S., et al.: Vit-net: interpretable vision transformers with neural tree decoder. In: ICML (2023)
17. Krause, J., et al.: 3d object representations for fine-grained categorization. In: ICCV (2013)
18. Linardatos, P., et al.: Explainable AI: a review of machine learning interpretability methods. Entropy (2020)
19. Liu, R., et al.: Teacher-student training for robust tacotron-based TTS. In: ICASSP (2020)
20. Liu, W., et al.: Iterative machine teaching. Mach. Learn. (2017)
21. Liu, Z., et al.: Swin transformer: hierarchical vision transformer using shifted windows. In: ICCV (2021)
22. Loshchilov, I., Hutter, F.: Decoupled weight decay regularization. Learning (2017)
23. Meek, C., et al.: Analysis of a design pattern for teaching with features and labels. arXiv (2016)
24. Mei, S., Zhu, X.: Using machine teaching to identify optimal training-set attacks on machine learners. In: AAAI (2015)
25. Nauta, et al.: Neural prototype trees for interpretable fine-grained image recognition. arXiv (2020)
26. Nauta, M., et al.: Neural prototype trees for interpretable fine-grained image recognition. In: CVPR (2020)
27. Rymarczyk, D., et al.: Interpretable image classification with differentiable prototypes assignment. In: ECCV (2022)
28. Selvaraju, R.R., et al.: Grad-cam: visual explanations from deep networks via gradient-based localization. In: ICCV (2016)
29. Wah, C., et al.: The caltech-ucsd birds-200-2011 dataset (2011)
30. Wang, J., et al.: Interpretable image recognition by constructing transparent embedding space. In: ICCV (2021)

31. Xue, M., et al.: Protopformer: concentrating on prototypical parts in vision transformers for interpretable image recognition. arXiv (2022)
32. Zech, J.R., et al.: Variable generalization performance of a deep learning model to detect pneumonia in chest radiographs: a cross-sectional study. PLoS Med. (2018)
33. Zeng, X., Sun, H.: Interactive image recognition of space target objects. IOP Conf. Ser. (2017)
34. Zhang, C., et al.: One-shot machine teaching: cost very few examples to converge faster. arXiv (2022)
35. Zhang, Q., et al.: Interpretable convolutional neural networks. In: CVPR (2017)
36. Zhang, X., et al.: Explainable machine learning in image classification models: an uncertainty quantification perspective. Knowl. Based Syst. (2022)
37. Zhu, X., et al.: Machine teaching: an inverse problem to machine learning and an approach toward optimal education. In: AAAI (2015)
38. Zhu, X., et al.: An overview of machine teaching. arXiv (2018)

Adaptive and Compact Graph Convolutional Network for Micro-expression Recognition

Renwei Ba[✉], Xiao Li, Ruimin Yang, Chunlei Li, and Zhoufeng Liu

School of Electrical and Information Engineering, Zhongyuan University of Technology,
Zhengzhou 450007, Henan, China
2021106196@zut.edu.cn

Abstract. Micro-expression recognition (MER) is a very challenging task since the motion of Micro-expressions (MEs) is subtle, transient and often occurs in tiny regions of face. To build discriminative representation from tiny regions, this paper proposes a novel two stream network, namely Adaptive and Compact Graph Convolutional Network (ACGCN). To be specific, we propose a novel Cheek Included Facial Graph to build more effective structural graph. Then, we propose the Tightly Connected Strategy to adaptively select structural graph to build compact and discriminative facial graph and adjacency matrix. We design the Small Region module to enlarge the interested feature in tiny regions and extract effective feature to build strong and effective node representations. We also adopt the spatial attention to make the network focus on the visual feature of salient regions. Experiments conducted on two micro-expressions datasets (CASME II, SAMM) show our approach outperforms the previous works.

Keywords: Micro-expressions Recognition · Graph convolutional Network · Adjacency Matrix · Tightly Connected Strategy

1 Introduction

Facial Micro-expressions (MEs) can reflect people's real emotion and provide effective information for psychotherapy [3], public security [16] and other important areas. Considering the huge contributions of MER in affective computing, MER is valuable to study and has received a lot of attention in recent years. Unlike the macro-expressions, as shown in Fig. 1 MEs are involuntary facial movements which are subtle, rapid (typically less than 200 ms) and low intensity. Thus, how to learn discriminative and robust feature included in whole MEs image can be very challenging task.

The previous hand-crafted approach such as 3D histograms of oriented gradients [20] (3DHOG), histogram of optical flow [14] (HOOF), and local binary pattern-three orthogonal planes [25] (LBP-TOP) achieved comparable performance micro-expressions recognition tasks. But the subtle and transient motions make the hand-crafted methods unable to generate robust representation between different MEs.

Recently, convolutional neural networks (CNN) have shown their powerful learning and representing ability and researchers have made superior performance on MER tasks, [22,24]. For example, Lei et al. [11] input the patches of eyebrows and mouth

Q. Liu et al. (Eds.): PRCV 2023, LNCS 14433, pp. 158–169, 2024.
https://doi.org/10.1007/978-981-99-8546-3_13

Fig. 1. The others and happiness categories of MEs of a person. The interest features are in tiny regions with very low intensity (green points). Cheek regions can also provide useful information (blue points). (Color figure online)

into the designed temporal convolutional network to extract robust graph representation of MEs. Meanwhile, Lei et al. [10] take the patches of eyebrows and mouth into the proposed node learning and edge learning block to learn effective graph representation of MEs. Lo et al. [15] employ the GCN on extracted AU feature to explore the dependency between AU node for MER. Wei et al. [26] obtain the low-order geometric information by adopt GCN on the landmark coordinates of eyebrows, mouth and nose. The aforementioned works have shown promising results on MER tasks. However, the manual selection causes a wastage of useful structure information. Thus, how to adaptively build the compact and discriminative structural information is still a problem in MER. Meanwhile, by observing the dataset samples, we found the cheek region which close to nose also provided useful structural information.

To alleviate these problems, we propose a novel two stream MER network for MER. In GCN stream, we propose a novel Cheek Included Facial Graph (CIFG) which additionally add the cropped patches of cheek region into facial graph to build a more effective facial graph. Then we propose the Tightly Connected Strategy (TCS) to automatically build compact facial graph. When solving the frequent Subgraph Mining problem, researchers search the frequent subgraph by evaluating the connect relationship of nodes [12]. The problem of building compact and discriminative facial graph is similar to frequent Subgraph Mining problem. Thus, the TCS is designed to directly select the tightly connected elements in adjacency matrix and facial graph to build compact and discriminative facial graph. The Adaptive Relationship Matrix (ARM) is designed to change the connect relationship between nodes and also makes the TCS become automatically. Then, we propose a Small Region (SR) module to better extract the local and global visual feature in the compact facial graph and project the feature maps into graph nodes. Then, the graph nodes are input into GCN to adaptively learn discriminative structural feature. In CNN stream, we take the ResNet-18 [5] as the backbone and assemble the spatial attention mechanism into the network to learn attentive feature. Finally, we fuse the attentive visual information and structural information for the classification. The main contributions of this paper can be summarized as follows:

(1) We propose a novel Cheek Included Facial Graph (CIFG) to offer discriminative structural information. Then, we propose the Tightly Connected Strategy (TCS) and Adaptive Relationship Matrix (ARM) to adaptively build the discriminative structure graph. This strategy can save the computation cost and offer more effective structural information to improve the MER accuracy.

(2) This paper proposes a novel Small Region (SR) module to better explore the visual feature in small activated regions. SR module can enlarge the size of interest feature and extract local and global visual feature to build discriminative node representation.

(3) We assemble the CNN stream and GCN stream to construct ACGCN, the network can learn discriminative and robust feature to recognize the MEs categories. The network outperforms the state-of-the-art works a large margin on CASME II dataset and achieve comparable results on SAMM dataset.

2   Related Work

The hand-crafted methods can be classified into three main directions, Optical Flow (OF) based method, Histograms of Oriented Gradients (HOG) based method and local binary pattern (LBP) based method [14,18,20,21,25]. The hand-crafted works [17,23] are concentrated on finding the key-points and build the hand-crafted discriminative feature around the key-points. The hand-crafted methods have shown promising performance on MER, while these methods still short of adaptability when comparing to Deep CNNs. Then, researchers start to employ the Deep Convolutional Neural Network (Deep CNNs) to build visual representations for MER on account of the powerful generalization and representation ability of Deep CNNs.

2.1   Structure Graph

The structure graph in MER can be divided into three kinds: the region of interests which corresponds to landmarks, the calculated distance and angle change of landmarks and the Action Units (AUs).

Zhou et al. [30] trained the network to extract the feature of AU under the guided of annotated label in datasets and took the extracted feature map as a graph of structure. In a similar way, Lo et al. [15] transformed the node representations of AU by one-hot encoder and the node representations are took as structural graph. Unlike Lo et al. and Xie et al., Kumar et al. [9] constructed the graph directly with the detected coordinates of landmarks. Lei et al. [10] adopted the cropped patches of mouth and eyebrows and also 9 AUs to build discriminative structure representation. Focusing on the effectiveness of structure feature, Wei et al. [26] selected 14 landmarks in a single frame as the spatial information and aggregated the temporal information by staking the selected landmarks of the Onsetframe, Apexframe and Offsetframe to construct the Geometric Movement Graph (GM-Graph). The experiment results of the above works have demonstrated that reducing the huge redundant information in structure information can build more effective and discriminative structural representation. The reduced structural information can not only save the computation cost but also improve the performance of model. While all the structure relation graph in the aforementioned works was obtained by manual selection, which causes the wastage of effective feature and lack of adaptability. To alleviated these problems, this paper proposes an automatically select method to automatically select the facial graph during the training without any artificial experience. We also added the cheek region of landmarks into the initial graph to offer additional discriminative structural information which overlooked by these works.

The Adjacency Matrix: Lei et al. [11] employed the temporal convolutional network (TCN) on the node feature vector, Kumar et al. [9] adopted the Graph Attention Networks [23] (GAN) on the feature of node locations and the feature constructed by two directional optical flow. Both the two networks can build the structure representations directly on the feature maps which means they didn't need to define the adjacency matrix. Zhou et al. [30] fed the extracted AU feature maps into GCN to build the structural representations. The adjacency matrix proposed by Lo et al. [15] is modeled by the conditional possibility of two co-occurred AUs. They calculated the conditional probability $P(U_i|U_j)$ to denote the connect weights in adjacency matrix, which is as follows:

$$A_{ij} = P(U_i|U_j) = \frac{N_{i \cap j}}{N_j} \tag{1}$$

$N_{i \cap j}$ is the amount occurrence of the U_i and U_j, N_j is the total occurrence j-th AU. In the same way, Lei et al. [10] also utilized the AUs information provided in the datasets to generate the adjacency matrix. Inspired by Wei et al. [26], we first randomly initialize the adjacency matrix. Then, we design an Adaptive Relationship Matrix (ARM) to learn the suitable connection relationships during the training and add the ARM with the randomly initialized adjacency matrix to update the relationship between nodes. In this way, the TCS can automatically select the most discriminative nodes to build compact structure graph.

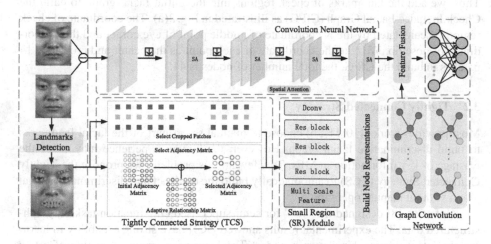

Fig. 2. The structure of the proposed ACGCN.

3   Method

In this section, the architecture of the proposed ACGCN is introduced. As illustrate in Fig. 2, the proposed ACGCN consists of two branches, the CNN branch extracts the visual feature on the difference between Onset and Apex frame. The GCN branch aims to automatically build compact and discriminative facial graph and extract strong and discriminative structure feature. Finally, the features of CNN and GCN are fused for MEs classification.

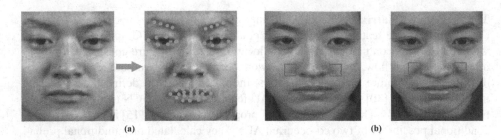

(a)   (b)

Fig. 3. (a) An illustration of the select 44 facial landmarks. (b) The left sample is disgust ME, the right sample is happiness ME, the difference of cheek region between two samples turns out the cheek region can also provide effective information.

3.1 Cheek Included Facial Graph

To build more discriminative facial map, we first select the landmarks of eyebrows, eyes, mouth and nose wing to build the initial facial graph. Since the dlib [6] detector predicts wrong location of eyebrow tail, the landmarks of the eyebrow tail don't be considered. In addition, by observing the MEs samples as shown in Fig. 3(b), we found the cheek regions which close to nose can also provide effective structural information. Thus, we add the landmarks of cheek regions into the initial facial graph to build the Cheek Included Facial Graph (CIFG) as illustrated in Fig. 3(a). The coordinate of cheek regions is calculated by the X coordinate of middle point of eye corner and the Y coordinate of nose tip. We take the coordinate of each point as the center and crop 14×14 patches to build the CIFG, the total number of nodes is 44.

3.2 Tightly Connected Strategy

Most previous works [10,11,26] choose to manual select part of the facial landmarks to build discriminative facial graph. However, the human experience may cause the wastage of key information in MEs. To alleviate this problem, we propose the Tightly Connected Strategy (TCS) to automatically build the compact and discriminative facial graph without manual operation.

Kipf et al. [8] designed the GCN for semi-supervised classification task on graph-structured data. The experimental results showed GCN have strong capability to learning the latent relationship between nodes. The layer-wise propagation rule of a two layer GCN is as follows:

$$H^{l+1} = \sigma(\tilde{D}^{-\frac{1}{2}} \tilde{A} \tilde{D}^{-\frac{1}{2}} H^l W^l) \tag{2}$$

The $\tilde{A} \in R^{N \times N}$ is the adjacency matrix which represents the node relationship in graph, $W^l \in R^{N \times D}$ is one layer weight matrix, $W^l \in R^{N \times D}$ is the input matrix of lth layer, $D_{ii} = \sum_j \tilde{A}_{ij}$ and σ is the activation function. The output of GCN is $H^{(l+1)} \in R^{N \times D_l}$, N is the number of nodes and D_l is the dimension of each nodes. Then the output is passed to softmax function to predict which class each node belongs to. Which means, one row ($1 \times H \times W$) in output ($N \times H \times W$) represent one node in graph and one node can be classified into D_l kinds of class. While the MER is one label classification, thus we take the output as the node representation rather than the possibility of class. The $a_{i,j} \in \tilde{A}$ is the edge weight connect the node and edge. We can simply sum all edges of one node with:

$$A_i = \sum_j a_{i,j} , A_i \in 1 \times N \qquad (3)$$

The obtained A_i can show how tightly one node connected to the others. We employ the torch.max() function to get the index of maximum in A_i and delete the maximum by index. The same operation is repeated for 36 times and all the index are appended into a list, then we have maximum index list $MAX = [i_1, i_2, ..., i_{36}]$. The compact and effective facial graph and adjacency matrix can be obtained with:

$$H^{new} = H[i] , A^{new} = A[i] \ if \ i \subset MAX \qquad (4)$$

Adaptive Relationship Matrix: The adjacency matrix is randomly initialized before the training. While during the process of TCS, the data in adjacency matrix is fixed which means the node relationship is defined randomly. TCS can selects the nodes according to the sum value of all edges of one node, this will cause the unreasonable selection. To alleviate this problem and also make the selection adaptively, we propose an Adaptive Relationship Matrix (ARM) as A_{ARM} to automatically change the connect weights during the training. A_{AD} is another randomly initialized adjacency matrix which have same size as A, and we adaptively learn the connect relationships on A_{AD} as follows:

$$A_{ARM} = \alpha(FC(AvgPool(A_{AD}))) \bigotimes A_{AD} \qquad (5)$$

A_{AMG} is the output, FC is fully Connected layer, $AvgPool$ is average pooling layer and α is the sigmoid function. By multiplying the output of the sigmoid function with the A_{AD}, the higher weight can reinforce the interested edges in A_{AD} and the lower weight can reduce the value of interference edges. Then, we add the A_{ARM} with the adjacency matrix A and obtain the new propagation rule of layer-wise GCN as:

$$Z = softmax((\tilde{A} + A_{ARM}) \, ReLU((\tilde{A} + A_{ARM})XW^0)W^1) \qquad (6)$$

By adding the adjacency matrix with ARM, the TCS can adaptively selects the valuable nodes in structure graph.

3.3 Small Region Module

To construct effective node representations, we design the Small Region (SR) module as shown in Fig. 4. The SR module is elaborate designed to extract discriminative visual feature in cropped patches which are small size.

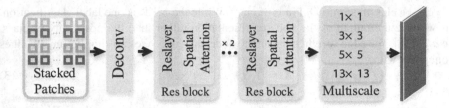

Fig. 4. Small Region block is constructed by deconvolution layer, Reslayer, spatial attention and multiscale block.

The cropped patches are first input to Deconv layer to enlarge the size of feature mpas. Then we employ 5 ResNet layers to extract the visual features in the enlarged feature maps. To further explore the visual feature, we build the multiscale block with 4 different convolutional branches. Since the large kernels focus on extracting global information and small kernels focus on extracting local information, the 4 branches are constructed with 4 different size of convolution kernels which are 1, 3, 5, 13 to pay attention to both local and global visual feature in feature maps. We also add the shortcut way to keep the interested feature and facilitate the training. The feature maps from the 4 branch and shortcut are concatenated to build the robust visual feature representation which size is (B, 576, H, W).

Then we employ two 1×1 convolution layer on the feature map, one layer aims to map the feature map into the reduced feature, which operation is to reduce the number of channel into 512 and resize the dimension into (B, 576, H*W). The other one aims to map the feature map into projecting matrix which dimension is (B, 36, H*W). Finally, we obtain the node representation by utilizing the torch.bmm() function on the reduced feature and projecting matrix.

3.4   Adaptive and Compact Graph Convolutional Network

In GCN stream, we employ a two-layer GCN on the obtained node representations and adjacency matrix to learn discriminative structural representation. GCN take the compact node representations as the input which dimension is 36×512. The dimension of first layer is 512×1024, the dimension of second layer is 1024×2048 and the size of output is 36×2048.

In CNN stream, we employ a ResNet-18 as the backbone to extract the local and global visual feature on the difference between Onset frame and Apex frame. The CNN consists of 4 ResNet layers, and reduce the feature resolution from 224×224 to 14×14. To make the CNN focus on activated regions in facial image, we adopt the spatial attention block and embed it into all layers to achieve spatial attention. By adopting the channel max pooling (CMP) strategy and channel average pooling (CAP) strategy, we obtained the max representation and average representation, both size of them are $1\times H\times W$. To adaptively learn the attention of the input feature, we first concatenate the two representations on the first dimension, and then map the $2\times H\times W$ representations into $1\times H\times W$ attention weight matrix by a convolution layer and a sigmoid activation.

The spatial attention is achieved by multiplying the input feature with the attention matrix. To facilitate the training process, an additional residual connection is added into the block either.

For the final classification, we first reduce the size of CNN stream feature form 14×14 to 12×12 to fit the feature size of the GCN stream. Then the features of GCN stream and CNN stream are fused and passed into fully connection layer to predict MEs categories. The standard Cross Entropy loss function is adopted to train the whole network.

4 Experiments

4.1 Datasets and Implementation Details

CASME II [19] include 26 participants from one ethnicity, select 255 videos from all raw videos and annotate seven MEs categories: surprise, happiness, disgust, sadness, repression, fear and others. In SAMM [2], all 32 subjects are from 13 different ethnicities, and the total number of samples is 159 with eight MEs categories based on self-report. Both two datasets are collected under laboratory environment with high-speed cameras at 200 fps. Follow the previous work [11], we select 5 classes from the two datasets to evaluate our proposed network. The Summary of the two dataset for 5 classes is shown as Table 1. During the training process, we adopt the AdamW optimizer with the initial 8e-3 learning rate with 0.987 exponential decrease rate in 45 epochs. The leave-one-subject-out (LOSO) strategy with accuracy(ACC) and F1-score evaluation matrices is adopted to evaluate the model performance in all experiments.

Table 1. Summary of the data distributions for CASME II and SAMM for 5 classes.

Expressions	CASMEII %	Expressions	SAMM %
Dis	63	Ang	57
Hap	32	Hap	26
Sur	25	Sur	15
Rep	27	Con	12
Other	99	Other	26

4.2 Quantitative Results

We compare our ACGCN network with several state-of-the-art methods. From Table 2, on SAMM, our network outperforms DSSN, AMAN, Graph-TCN and AUGCN by 17.45%/22.06%, 5.98%/1.68%, 2.78%/0.84% and 3.52%/0.24% with respect to Accuracy/F1-score. We can also see that ACGCN outperforms the higher results of previous methods on CASMEII. Specifically for Five-class MER tasks, our method outperforms DSSN, Graph-TCN, AUGCN and STA-GCN by 10.73%/7.54%, 7.53%/8.05%, 7.24%/10.04% and 5.43%/9.55% with respect to Accuracy/F1-score. The performance on SAMM is suboptimal due to the data imbalance problem. Data augmentation [10,11] is effective to alleviate this problem, while the previous works [7,27] also face the same problems.

Table 2. Comparisions with SORT works on CASME II and SAMM.

Methods	CASMEII %		SAMM %	
	Accuracy	F1-score	Accuracy	F1-score
LBP-TOP [28]	39.68	35.89	39.68	35.89
Bi-WOOF [13]	57.89	61.00	–	–
DSSN (2019) [7]	70.78	72.97	57.35	46.44
Graph-TCN (2020) [11]	73.98	72.46	75	69.85
AUGCN(2021) [10]	74.27	70.47	74.26	70.45
STA-GCN(2021) [29]	76.08	70.96	–	–
AMAN(2022) [27]	75.40	71.25	68.85	66.82
ours	**81.51**	**80.51**	**77.78**	**70.69**

Table 3. Evaluation of each stream.

Methods	Accuracy %	F1-score %
Baseline(ResNet-18)	**67.23**	**65.97**
CNNstream	**71.01**	**66.51**
GCNstream	**54.20**	35.49
ACGCN	**81.51**	**80.51**

4.3   Ablation Study

To further evaluate the effectiveness of the proposed CNN and GCN branch, we conduct the ablation experiments on CASME II dataset with 5 classes. We set the ResNet-18 as the baseline model, and keep only one branch in each experiment a time.

As shown in Table 3, the spatial attention can significantly improve the performance by 3.78%/0.54%. As shown in Fig. 5, the image in the second row is visualization results of MLAN and the MLAN-based CAM++ [1] result. The Action Units (AUs) [4] of happiness are mainly AU 12, AU 14 and AU 15, which are lip corners pulled obliquely/down. And the interested regions in attention maps of happiness are around the lip corners. The Action Units (AUs) of surprise are mainly AU 1 and AU 2 which are the inner/outer portion of eyebrows raised separately. And the interested regions in attention maps of surprise are around the inner portion of eyebrows. The Action Units (AUs) of others are mainly AU 4 which is eyebrow lowered and drawn together and the attention maps of others are around the whole eyebrows. The visualization results show the network can be guided to pay attention to the local salient regions. The added cheek region is also demonstrated to be effective according to the visualization result of the happiness and others categories of MEs. While, the performance is imperfect when there is no CNN branch, which demonstrate the visual feature is indispensable. Finally, when we assemble the CNN and GCN branches to build our ACGCN network, the whole network outperforms the baseline by 7.23%/10.94% on CASME II dataset respect to Accuracy/F1-score.

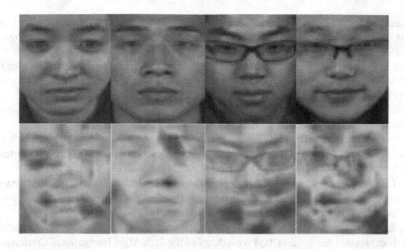

Fig. 5. The visulization of CASMEII dataset. From left to right represents 4 micro-expressions which are the happiness, surprise, others and others.

5   Conclusion

In this paper, we explored a method to build the compact structure graph without manual selection and designed a novel two branch network for MER task. Spatial Attention was adopted in CNN branch to focus on the visual feature in tiny regions in face. The TCS, ARM and SR module was proposed in GCN branch to automatically build compact and discriminative node representations on CIFG. A two layer GCN was adopted to learn structural features on the compact facial graph. Extensive experiments illustrated that our ACGCN network achieves much higher performance than state-of-the-arts on CASME II dataset and also on SAMM dataset. Though we have proposed the method to automatically build the facial graph, the whole network is still a two stage model since the dlib detector is adopted to extract the facial landmarks. In the future work, we will explore an end-to-end model to build strong structure representation for MER.

References

1. Chattopadhay, A., Sarkar, A., Howlader, P., Balasubramanian, V.N.: Grad-cam++: generalized gradient-based visual explanations for deep convolutional networks. In: 2018 IEEE Winter Conference on Applications of Computer Vision (WACV), pp. 839–847. IEEE (2018)
2. Davison, A.K., Lansley, C., Costen, N., Tan, K., Yap, M.H.: SAMM: a spontaneous micro-facial movement dataset. IEEE Trans. Affect. Comput. **9**(1), 116–129 (2016)
3. Ekman, P.: Lie catching and microexpressions. Phil. Decept. **1**(2), 5 (2009)
4. Ekman, P., Friesen, W.V.: Facial action coding system. Environ. Psychol. Nonverbal Behav. (1978)
5. He, K., Zhang, X., Ren, S., Sun, J.: Deep residual learning for image recognition. In: Proceedings of the IEEE Conference on Computer Vision and Pattern Recognition, pp. 770–778 (2016)

6. Kazemi, V., Sullivan, J.: One millisecond face alignment with an ensemble of regression trees. In: Proceedings of the IEEE Conference on Computer Vision and Pattern Recognition, pp. 1867–1874 (2014)
7. Khor, H.Q., See, J., Liong, S.T., Phan, R.C., Lin, W.: Dual-stream shallow networks for facial micro-expression recognition. In: 2019 IEEE International Conference on Image Processing (ICIP), pp. 36–40. IEEE (2019)
8. Kipf, T.N., Welling, M.: Semi-supervised classification with graph convolutional networks. arXiv preprint arXiv:1609.02907 (2016)
9. Kumar, A.J.R., Bhanu, B.: Three stream graph attention network using dynamic patch selection for the classification of micro-expressions. In: Proceedings of the IEEE/CVF Conference on Computer Vision and Pattern Recognition, pp. 2476–2485 (2022)
10. Lei, L., Chen, T., Li, S., Li, J.: Micro-expression recognition based on facial graph representation learning and facial action unit fusion. In: Proceedings of the IEEE/CVF Conference on Computer Vision and Pattern Recognition, pp. 1571–1580 (2021)
11. Lei, L., Li, J., Chen, T., Li, S.: A novel graph-tcn with a graph structured representation for micro-expression recognition. In: Proceedings of the 28th ACM International Conference on Multimedia, pp. 2237–2245 (2020)
12. Li, L., Ding, P., Chen, H., Wu, X.: Frequent pattern mining in big social graphs. IEEE Trans. Emerg. Topics Comput. Intell. $\mathbf{6}$(3), 638–648 (2021)
13. Liong, S.T., See, J., Wong, K., Phan, R.C.W.: Less is more: micro-expression recognition from video using apex frame. Signal Process. Image Commun. $\mathbf{62}$, 82–92 (2018)
14. Liu, Y.J., Zhang, J.K., Yan, W.J., Wang, S.J., Zhao, G., Fu, X.: A main directional mean optical flow feature for spontaneous micro-expression recognition. IEEE Trans. Affect. Comput. $\mathbf{7}$(4), 299–310 (2015)
15. Lo, L., Xie, H.X., Shuai, H.H., Cheng, W.H.: MER-GCN: micro-expression recognition based on relation modeling with graph convolutional networks. In: 2020 IEEE Conference on Multimedia Information Processing and Retrieval (MIPR), pp. 79–84. IEEE (2020)
16. O'sullivan, M., Frank, M.G., Hurley, C.M., Tiwana, J.: Police lie detection accuracy: the effect of lie scenario. Law Hum Behav. $\mathbf{33}$(6), 530 (2009)
17. Pan, H., Xie, L., Lv, Z., Li, J., Wang, Z.: Hierarchical support vector machine for facial micro-expression recognition. Multimedia Tools Appl. $\mathbf{79}$, 31451–31465 (2020)
18. Pfister, T., Li, X., Zhao, G., Pietikäinen, M.: Differentiating spontaneous from posed facial expressions within a generic facial expression recognition framework. In: 2011 IEEE International Conference on Computer Vision Workshops (ICCV Workshops), pp. 868–875. IEEE (2011)
19. Pfister, T., Li, X., Zhao, G., Pietikäinen, M.: Recognising spontaneous facial micro-expressions. In: 2011 International Conference on Computer Vision, pp. 1449–1456. IEEE (2011)
20. Polikovsky, S., Kameda, Y., Ohta, Y.: Facial micro-expressions recognition using high speed camera and 3d-gradient descriptor (2009)
21. Polikovsky, S., Kameda, Y., Ohta, Y.: Facial micro-expression detection in hi-speed video based on facial action coding system (FACS). IEICE Trans. Inf. Syst. $\mathbf{96}$(1), 81–92 (2013)
22. Van Quang, N., Chun, J., Tokuyama, T.: Capsulenet for micro-expression recognition. In: 2019 14th IEEE International Conference on Automatic Face & Gesture Recognition (FG 2019), pp. 1–7. IEEE (2019)
23. Velickovic, P., et al.: Graph attention networks. STAT $\mathbf{1050}$(20), 10–48550 (2017)
24. Verma, M., Vipparthi, S.K., Singh, G.: Affectivenet: affective-motion feature learning for microexpression recognition. IEEE Multimedia $\mathbf{28}$(1), 17–27 (2020)

25. Wang, Y., See, J., Phan, R.C.-W., Oh, Y.-H.: LBP with six intersection points: reducing redundant information in LBP-TOP for micro-expression recognition. In: Cremers, D., Reid, I., Saito, H., Yang, M.-H. (eds.) ACCV 2014. LNCS, vol. 9003, pp. 525–537. Springer, Cham (2015). https://doi.org/10.1007/978-3-319-16865-4_34

26. Wei, J., Peng, W., Lu, G., Li, Y., Yan, J., Zhao, G.: Geometric graph representation with learnable graph structure and adaptive au constraint for micro-expression recognition. arXiv preprint arXiv:2205.00380 (2022)

27. Wei, M., Zheng, W., Zong, Y., Jiang, X., Lu, C., Liu, J.: A novel micro-expression recognition approach using attention-based magnification-adaptive networks. In: ICASSP 2022-2022 IEEE International Conference on Acoustics, Speech and Signal Processing (ICASSP), pp. 2420–2424. IEEE (2022)

28. Zhao, G., Pietikainen, M.: Dynamic texture recognition using local binary patterns with an application to facial expressions. IEEE Trans. Pattern Anal. Mach. Intell. 29(6), 915–928 (2007)

29. Zhao, X., Ma, H., Wang, R.: STA-GCN: spatio-temporal AU graph convolution network for facial micro-expression recognition. In: Ma, H., et al. (eds.) PRCV 2021. LNCS, vol. 13019, pp. 80–91. Springer, Cham (2021). https://doi.org/10.1007/978-3-030-88004-0_7

30. Zhou, L., Mao, Q., Dong, M.: Objective class-based micro-expression recognition through simultaneous action unit detection and feature aggregation. arXiv preprint arXiv:2012.13148 (2020)

Consistency Guided Multiview Hypergraph Embedding Learning with Multiatlas-Based Functional Connectivity Networks Using Resting-State fMRI

Wei Wang[1] and Li Xiao[1,2](✉)

[1] Department of Electronic Engineering and Information Science,
University of Science and Technology of China, Hefei 230052, China
xiaoli11@ustc.edu.cn
[2] Institute of Artificial Intelligence, Hefei Comprehensive National Science Center,
Hefei 230088, China

Abstract. Recently, resting-state functional connectivity network (FCN) analysis via graph convolutional networks (GCNs) has greatly boosted diagnostic performance of brain diseases in a manner that can refine FCN embeddings by treating FCN as irregular graph-structured data. In this paper, we propose a **C**onsistency **G**uided **M**ultiview **HyperG**raph **E**mbedding **L**earning (CG-MHGEL) framework to integrate FCNs based on multiple brain atlases in multisite studies. First, we model brain network as a hypergraph and develop a multiview hypergraph convolutional network (HGCN) to extract a multiatlas-based FCN embedding for each subject. Here, we employ HGCN rather than GCN to capture more complex information in brain networks, due to the fact that a hypergraph can characterize higher-order relations among multiple vertexes than a widely used graph. Moreover, in order to preserve between-subject associations to promote optimal FCN embeddings, we impose a class-consistency regularization in the embedding space to minimize intra-class dissimilarities while maximizing inter-class dissimilarities for subjects, as well as a site-consistency regularization to further penalize the dissimilarities between intra-class but inter-site subjects. The learned multiatlas-based FCN embeddings are finally fed into fully connected layers followed by the soft-max classifier for brain disease diagnosis. Experimental results on the ABIDE demonstrate the effectiveness of our method for autism spectrum disorder (ASD) identification. Furthermore, the detected ASD-relevant brain regions can be easily traced back with biological interpretability.

Keywords: Autism · embedding · functional connectivity · graph convolution networks · hypergraph

Supplementary Information The online version contains supplementary material available at https://doi.org/10.1007/978-981-99-8546-3_14.

1 Introduction

In recent decades, resting-state functional magnetic resonance imaging (rs-fMRI)-derived functional connectivity network (FCN), characterized by functional connectivity (FC) between paired brain regions-of-interest (ROIs) [17], has been extensively studied as a promising tool to identify potential neuroimaging biomarkers for brain disease diagnosis [8,18]. Specifically, FCN is naturally delineated as a graph, for which vertexes represent the spatially distributed but functionally linked ROIs, and edges indicate the between-ROI FCs that are quantified as the dependence between blood oxygenation level-dependent (BOLD) time series of paired ROIs. Changes in functional connectivity at a brain-wide level are expected to be associated with brain diseases, thereby providing novel insights into the underlying pathophysiological mechanisms. We note that "FCN" and "FC" hereafter in this paper are referred to as rs-fMRI derived FCN and rs-fMRI derived FC, respectively, unless stated otherwise.

A large number of machine/deep learning models have attracted considerable attention in FCN-based automated diagnosis of brain diseases, such as support vector machine (SVM) for discrimination between Parkinson's disease and multiple system atrophy [1], logistic regression for Alzheimer's disease early detection [24], and deep neural networks (DNNs) for schizophrenia identification [3], among others. However, the FCN matrix consisting of all FCs is directly vectorized and fed into the aforementioned models, such that rich topological structure information among ROIs is ignored. In order to uncover discriminative patterns in brain networks, FCN should be regarded as irregular graph-structured data with vertexes being ROIs and fed into a diagnostic model as a whole, where the FC profile of each node serves as the node features.

More recently, due to the fact that graph convolutional networks (GCNs) [13] can well handle irregular graph structures with graph convolutions to propagate features of adjacent vertexes, they have been utilized to obtain FCN embeddings by accounting for both vertex features (i.e., FC profiles of ROIs) and topological properties among ROIs, and thus have greatly improved diagnostic performance of brain diseases. For example, a Siamese GCN was investigated to learn a graph similarity metric between FCNs for classifying autism spectrum disorder (ASD) patients [9]. A graph embedding learning (GEL) model was presented for diagnosis of major depressive disorder, in which an FCN embedding was first learned via GCN and input into the soft-max classifier [12]. Similarly, a multiview graph embedding learning (MGEL) model was introduced to leverage complementary information from FCNs based on multiple atlases, resulting in a multiatlas-based FCN embedding with better performance of subsequent ASD diagnosis than using single-atlas-based FCN [4].

Current GCNs, although effective, still suffer from two main deficiencies when applied to FCN analysis. First, in GCNs, vertex features are propagated along between-ROI connections on a simple graph. However, the brain network is likely more complex than can be fully defined by pairwise relations between ROIs, and there often exist high-order relations among multiple (more than two) ROIs [23]. Second, the FCN embedding for every individual subject is considered independently during the learning process of GCNs, failing to manage between-subject associations that may promote to learn optimal embeddings of FCNs. To address these issues, in this paper we propose a consistency guided multiview hypergraph embedding learning (CG-MHGEL) framework to accommodate multiatlas-based FCNs for brain disease diagnosis in multisite studies, as illustrated in Fig. 1 and briefly described in the following.

1) We first develop a multiview hypergraph convolutional network (HGCN) to integrate multiple FCNs constructed on different atlases for each subject. To be specific, for each subject, we learn an FCN embedding vector with respect to every atlas via an HGCN [7] followed by a combination of maximum and average poolings, and then a multiatlas-based FCN embedding is obtained as a weighted concatenation of all the learned FCN embeddings across atlases. We note here that as a generation of the traditional graph, a hypergraph [2] with vertexes being ROIs can represent high-order relations among multiple ROIs, in which a hyperedge is generated by linking an ROI and its multiple nearest neighbors. As such, we model brain network as a hypergraph for each atlas and utilize HGCN instead of GCN to capture more complex information in brain network.

2) Besides, we impose two consistency regularizations on the learned multiatlas-based FCN embeddings of subjects to deal with between-subject associations across classes and sites: a class-consistency regularization to simultaneously minimize intra-class dissimilarities and inter-class similarities of the embeddings, as well as a site-consistency regularization to further penalize the dissimilarities between intra-class but inter-site subjects. Basically, the class-consistency regularization aims to encourage compactness within each class and separation between classes to promote the learning of multiatlas-based FCN embeddings discriminative across classes, while the site-consistency regularization helps to compensate for between-site heterogeneity that arises from differences in neuroimaging scanners and/or scanning protocols at multiple sites and may lead to biasing FCN analysis.

3) Ultimately, the learned multiatlas-based FCN embeddings are fed into a few fully connected layers and the soft-max classifier for brain disease diagnosis.

To demonstrate the effectiveness of the proposed CG-MHGEL in this paper, we conduct extensive experiments for autism spectrum disorder (ASD) identification using the Automated Anatomical Labeling (AAL) [16] and Harvard-Oxford (HO) [5] atlases based FCNs from the top three sites with the largest sample sizes in the publicly available ABIDE dataset [6]. Experimental results show that our CG-MHGEL yields superior classification accuracy compared with several other competing methods. In addition, we exploit Gradient-weighted Class Activation Mapping (Grad-CAM) [14] in the CG-MHGEL to discover discriminative ROIs as potential neuroimaging biomarkers associated with ASD.

2 Methods

Notation: Throughout this paper, we use uppercase boldface, lowercase boldface, and normal italic letters to represent matrices, vectors, and scalars, respectively. Let I be the identity matrix. The element at the i-th row and j-th column of a matrix A is denoted as $A_{i,j}$. The superscript T denotes the vector/matrix transpose, $\text{tr}(\cdot)$ denotes the trace of a matrix, and $|\cdot|$ denotes the cardinality of a set.

2.1 Hypergraph and Hypergraph Construction with FCN

Given a weighted hypergraph $\mathcal{G} \triangleq (\mathcal{V}, \mathcal{E}, W)$, $\mathcal{V} = \{v_1, v_2, \cdots, v_{|\mathcal{V}|}\}$ is the vertex set, $\mathcal{E} = \{e_1, e_2, \cdots, e_{|\mathcal{E}|}\}$ is the hyperedge set with each hyperedge $e_i \in \mathcal{E}$ being assigned a weight $w(e_i)$ for $1 \le i \le |\mathcal{E}|$, and $W = \text{diag}(\{w(e_i)\}_{i=1}^{|\mathcal{E}|}) \in \mathbb{R}^{|\mathcal{E}| \times |\mathcal{E}|}$ is a diagonal matrix of hyperedge weights. Notice that each hyperedge is a subset of $\mathcal{V}$ and all hyperedges have $\bigcup_{i=1}^{|\mathcal{E}|} e_i = \mathcal{V}$. The hypergraph structure can be indicated by the incidence matrix $H \in \mathbb{R}^{|\mathcal{V}| \times |\mathcal{E}|}$ with elements $h(v_i, e_j) = 1$ (or a real value if H is weighted) when vertex $v_i \in e_j$, and 0 otherwise. The vertex degree and hyperedge degree are respectively defined by $d(v_i) = \sum_{e_j \in \mathcal{E}} w(e_j) h(v_i, e_j)$ for $1 \le i \le |\mathcal{V}|$ and $d(e_j) = \sum_{v_i \in \mathcal{V}} h(v_i, e_j)$ for $1 \le j \le |\mathcal{E}|$. Let $D_v \in \mathbb{R}^{|\mathcal{V}| \times |\mathcal{V}|}$ and $D_e \in \mathbb{R}^{|\mathcal{E}| \times |\mathcal{E}|}$ be respectively diagonal matrices of the vertex degrees and the hyperedge degrees, i.e., $D_v = \text{diag}(\{d(v_i)\}_{i=1}^{|\mathcal{V}|})$ and $D_e = \text{diag}(\{d(e_j)\}_{j=1}^{|\mathcal{E}|})$. More details about hypergraphs can be found in [2].

Let $X = [x_1, x_2, \cdots, x_N] \in \mathbb{R}^{N \times N}$ be an FCN matrix of a training subject, where $x_i \in \mathbb{R}^N$ is the FC profile of the i-th ROI for $1 \le i \le N$, and N is the number of ROIs as defined by a specific atlas. Let vertexes $\{v_i\}_{i=1}^N$ be the ROIs and x_i be vertex features of v_i for $1 \le i \le N$. According to the nearest-neighbor based hypergraph generation scheme [7], given an ROI (i.e., centroid) v_j, a hyperedge e_j is composed of v_j itself and its $K - 1$ nearest-neighboring vertexes, where the similarity between vertexes v_i and v_j is measured by

$$\text{Sim}(v_i, v_j) = e^{-\|x_i - x_j\|_2^2 / 2\sigma^2}, \tag{1}$$

and here σ is empirically set to the mean of the Euclidean distances of all vertex pairs. In this way, we obtain an FC-based hypergraph with a total of N hyperedges for the subject. We calculate the weighted incidence matrix $H \in \mathbb{R}^{N \times N}$ as, for $1 \le i, j \le N$,

$$h(v_i, e_j) = \begin{cases} \text{Sim}(v_i, v_j), & v_i \in e_j \\ 0, & v_i \notin e_j. \end{cases} \tag{2}$$

2.2 Proposed CG-MHGEL with Multiatlas-Based FCNs

In this paper, we propose a CG-MHGEL framework to accommodate multiatlas-based FCNs for distinguishing brain disease patients (PTs) and healthy controls (HCs) that are collected from multiple sites, as shown in Fig. 1.

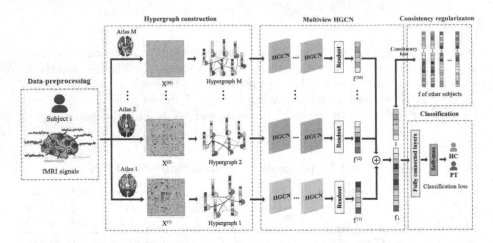

Fig. 1. Illustration of the proposed CG-MHGEL framework. We first model each brain network as a hypergraph and learn a multiatlas-based FCN embedding via a multiview HGCN for each subject. Moreover, in the embedding space, we impose class- and site-consistency regularizations to account for between-subject associations associated with classes and sites, respectively. The learned multiatlas-based FCN embeddings are finally fed into fully connected layers and the soft-max classifier for disease identification.

Multiview HGCN for Multiatlas-Based FCN Embedding. Suppose that each brain is parcellated based on M different atlases individually. Let $\boldsymbol{X}^{(m)} \in \mathbb{R}^{N_m \times N_m}$ denote the FCN matrix of a training subject for the m-th atlas, where N_m denotes the number of ROIs in the m-th atlas. With respect to every single-atlas-based FCN data, we independently apply an HGCN [7] with L layers to extract an FCN embedding for each training subject, i.e., the $(l+1)$-the layer is

$$\boldsymbol{X}_{(l+1)}^{(m)} = \phi\left((\boldsymbol{D}_v^{(m)})^{-\frac{1}{2}}\boldsymbol{H}^{(m)}\boldsymbol{W}^{(m)}(\boldsymbol{D}_e^{(m)})^{-1}(\boldsymbol{H}^{(m)})^T(\boldsymbol{D}_v^{(m)})^{-\frac{1}{2}}\boldsymbol{X}_{(l)}^{(m)}\boldsymbol{\Theta}_{(l)}^{(m)}\right),\tag{3}$$

where $\boldsymbol{X}_{(0)}^{(m)} := \boldsymbol{X}^{(m)}$ serves as input to the first layer, and each layer is accompanied with an HGCN weight matrix $\boldsymbol{\Theta}_{(l)}^{(m)}$ and a nonlinear activation function ϕ. The hyperedge weight matrix $\boldsymbol{W}^{(m)} \in \mathbb{R}^{N_m \times N_m}$ is learnable. To reflect class-specific interactions among ROIs, we generate the shared hypergraph structure in (3), i.e., the incidence matrix $\boldsymbol{H}^{(m)}$, by

$$\boldsymbol{H}^{(m)} = \boldsymbol{H}_{\text{HC}}^{(m)} \| \boldsymbol{H}_{\text{PT}}^{(m)},\tag{4}$$

instead of by averaging the FCN matrices of all training subjects irrespective of class, where $\|$ stands for the concatenation operation, and the incidence matrices $\boldsymbol{H}_{\text{HC}}^{(m)}$ and $\boldsymbol{H}_{\text{PT}}^{(m)}$ are calculated based on the average FCN matrices of healthy controls and brain disease patients in the training set, respectively.

After the HGCN, we perform both the row-wise maximum and average pooling operations on the output of the L-th layer in (3), i.e., $\boldsymbol{X}_{(L)}^{(m)} \in \mathbb{R}^{N_m \times d_m}$ with

d_m being the vertex feature dimension, and then concatenate the resulting maximum value vector and average value vector as the readout $\boldsymbol{f}^{(m)} \in \mathbb{R}^{2N_m}$. To integrate complementary information from M different brain atlases, we aggregate all the readouts $\{\boldsymbol{f}^{(m)}\}_{m=1}^M$ as the final feature vector, i.e.,

$$\boldsymbol{f} = \boldsymbol{f}^{(1)} \parallel \boldsymbol{f}^{(2)} \parallel \cdots \parallel \boldsymbol{f}^{(M)}, \tag{5}$$

which we call the multiatlas-based FCN embedding for the training subject.

Class- and Site-Consistency Regularizations for Subjects. One can see that the multiatlas-based FCN embeddings above fail to encode similarities or dissimilarities between subjects. From a discriminative point of view, the greater the intra-class similarities and inter-class dissimilarities for subjects are, the better the classification performance is. We thus impose a class-consistency regularization to render the learned multiatlas-based FCN embeddings of subjects with the same label as close as possible, and those of subjects with different labels as separate as possible. Specifically, assume that there are Q training subjects with labels $\{y_i\}_{i=1}^Q$. Let $\boldsymbol{f}_i \in \mathbb{R}^{2\sum_{m=1}^M N_m}$ for $1 \leq i \leq Q$ be the multiatlas-based FCN embeddings. The class-consistency regularization is so defined by

$$\mathcal{L}_{\text{cc}} = \underbrace{\sum_{y_i=y_j} \frac{1}{Q}\|\boldsymbol{f}_i - \boldsymbol{f}_j\|_2}_{\text{intra-class similarities}} - \underbrace{\sum_{y_i \neq y_j} \frac{1}{Q}\|\boldsymbol{f}_i - \boldsymbol{f}_j\|_2}_{\text{inter-class similarities}}. \tag{6}$$

In addition, to compensate for the inter-site data heterogeneity between subjects from different sites (e.g., S sites), we impose a site-consistency regularization on the learned multiatlas-based FCN embeddings to further penalize the dissimilarities between intra-class but inter-site subjects as

$$\mathcal{L}_{\text{sc}} = \sum_{\substack{y_i=y_j \\ s_i \neq s_j}} \frac{1}{Q}\|\boldsymbol{f}_i - \boldsymbol{f}_j\|_2, \tag{7}$$

where $s_i \in \{1, 2, \cdots, S\}$ denotes the site index of the i-th training subject.

Classifier and Loss Function. The learned multiatlas-based FCN embedding $\boldsymbol{f}$ for each training subject is finally connected to multiple fully connected layers before being fed into the soft-max classifier, and the loss function of the proposed CG-MHGEL is

$$\mathcal{L} = \mathcal{L}_{\text{ce}} + \alpha \cdot \mathcal{L}_{\text{cc}} + \beta \cdot \mathcal{L}_{\text{sc}}, \tag{8}$$

where $\mathcal{L}_{\text{ce}}$ is the cross-entropy loss for disease classification, and α, β are tunable hyperparameters that control the relative contribution of respective regularization terms.

Table 1. Demographic information of the subjects from the ABIDE in this study.

Site	Category (ASD/HC)	Sex (F/M)	Age (mean)
NYU	71/96	132/35	15.418
UM1	36/44	57/23	13.775
USM	38/22	60/0	23.870
Total	145/162	249/58	16.642

3 Experiment

3.1 Experimental Settings

Data and Preprocessing. The ABIDE dataset is a publicly available, multi-site imaging repository [6], which consists of structural and functional images of both ASD and HC subjects that were collected from 17 different sites. In this study, we used the rs-fMRI data of 145 ASD and 162 HC cases from the top three sites with the largest sample sizes, namely the New York University (NYU), the University of Michigan (UM), and the University of Utah School of Medicine (USM); see a brief summary of the selected subjects in Table 1. We downloaded the rs-fMRI data that had been previously preprocessed by a standard pipeline with the Data Processing Assistant for Resting-State fMRI (DPARSF) [21] from http://preprocessed-connectomes-project.org/abide, where the concrete image acquisition parameters and preprocessing procedures can be readily found as well. Each brain was partitioned into 116 and 111 ROIs based on the Automated Anatomical Labeling (AAL) [16] and Harvard-Oxford (HO) [5] atlases, respectively. As a consequence, we derive two FCN matrices for each subject, i.e., $X^{(1)} \in \mathbb{R}^{116 \times 116}$ with the AAL and $X^{(2)} \in \mathbb{R}^{111 \times 111}$ with the HO, by computing the Fisher's z-transformed Pearson's correlation coefficient between BOLD time series of each pair of ROIs.

Implementation Details. In all experiments, we conducted 10-fold intra-site cross validation five times. For 10-fold intra-site cross validation, the whole set of subjects was first split into 10 disjoint folds by keeping roughly the same proportions of sites and diagnostic groups across folds; then each fold was successively selected as the test set and the other 9 subsets were used the training set. The mean and standard deviation (sd) of classification results in all the test sets were reported in terms of accuracy, sensitivity, precision, and F1-score.

3.2 Experimental Results and Analysis

Comparison Results. For ASD classification, we compared the proposed CG-MHGEL with several other related baselines, i.e., SVM [1], DNN [22], GEL [12], and HGEL [7] that were single-view methods working on single-atlas-based FCN, and MGEL [4] that was used to fuse the two-atlas-based FCNs.

Table 2. ASD classification performance (mean (sd)) of different methods.

Method	Atlas	Accuracy	Sensitivity	Precision	F1-score
SVM	AAL	65.79 (1.49)	61.48 (1.56)	65.24 (2.02)	62.55 (1.77)
SVM	HO	67.17 (1.14)	62.66 (1.92)	66.80 (1.79)	63.77 (1.57)
DNN	AAL	67.16 (1.14)	61.01 (1.90)	66.31 (1.45)	62.74 (1.58)
DNN	HO	68.21 (1.72)	61.98 (2.16)	68.89 (2.04)	64.51 (2.13)
GEL	AAL	64.75 (1.45)	61.92 (3.17)	63.10 (2.06)	61.74 (2.47)
GEL	HO	68.08 (1.52)	65.92 (3.75)	67.41 (1.57)	65.66 (2.08)
HGEL	AAL	64.81 (2.16)	59.90 (3.02)	64.11 (2.44)	61.28 (2.35)
HGEL	HO	68.53 (2.05)	64.26 (2.63)	68.17 (3.20)	65.35 (2.14)
MGEL	AAL+HO	68.00 (1.47)	64.01 (0.97)	67.48 (2.88)	64.15 (1.28)
CG-MGEL	AAL+HO	70.14 (1.05)	67.14 (1.25)	69.65 (1.19)	67.53 (0.90)
CG-MHGEL	AAL+HO	**71.67 (0.62)**	**67.60 (2.01)**	**72.00 (1.07)**	**68.99 (1.10)**

The involved hyperparameters in the proposed CG-MHGEL were optimally set as follows. $K = 3$, $\alpha = 2 \times 10^{-3}$, $\beta = 5 \times 10^{-10}$. A two-layer HGCN with 256 and 64 neurons was trained for every single-atlas-based FCN, where the activation function σ was ReLU. The numbers of neurons in the three fully connected layers were respectively $128, 32, 2$. The number of epochs was 50, and the learning rate of the first 30 epochs was 10^{-3} and 10^{-4} of the remaining epochs. The batch size was 16 and the weight decay was 5×10^{-4}. For SVM and DNN, the lower triangular portion of the FCN matrix was vectorized as input. SVM used a linear kernel and a cost parameter $C = 0.1$. In DNN, the numbers of neurons in fully connected layers were respectively $1024, 128, 32, 2$ and the dropout rate was 0.2. For GEL, HGEL, MGEL and CG-MGEL the settings and hyperparameters were the same as those employed in the proposed CG-MHGEL except for the regularization parameters that were tuned to be optimal alone.

As shown in Table 2, the proposed CG-MHGEL outperformed the other five competing baseline methods for ASD vs. HC classification on all the four evaluation metrics. Specifically, compared with the single-view methods, the proposed CG-MHGEL improved the classification performance by 3.14% to 5.92% in terms of accuracy, which suggests that fusing FCNs based on multiple atlases can help boost the performance in comparison to using single-atlas-based FCN. Moreover, the proposed CG-MHGEL achieved 3.67% higher accuracy than MGEL and 1.53% higher accuracy than CG-MGEL. This implies that our method can learn better FCN embeddings than the graph-based method by means of using a hypergraph to characterize high-order relations among ROIs in brain network analysis and shows the effectiveness of our developed two consistency regularizations.

Ablation Studies. To further evaluate the effectiveness of the class-consistency (CC) regularization and the site-consistency (SC) regularization in our proposed CG-MHGEL, we executed three additional experiments for ablation

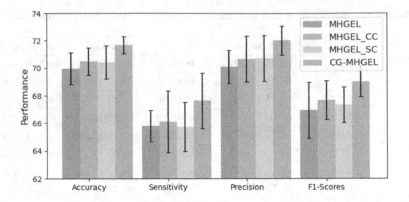

Fig. 2. Performance comparison for ablation studies of the two consistency regularizations in the proposed CG-MHGEL.

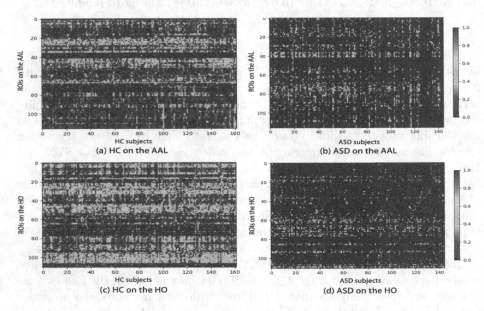

Fig. 3. Class-specific heat-maps generated by using Grad-CAM to the proposed CG-MHGEL.

studies as displayed in Fig. 2. The ablation methods, i.e., MHGEL, $MHGEL_{CC}$, and $MHGEL_{SC}$, represent the CG-MHGEL without these two consistency regularizations, without the SC regularization, and without the CC regularization, respectively. Through comparing the performance of $MHGEL_{CC}$ and $MHGEL_{SC}$ with that of MHGEL, it indicates that both the CC and SC regularizations were beneficial to promoting optimal FCN embeddings. The best performance of the CG-MHGEL demonstrates that simultaneously considering the two consistency regularizations can better aggregate associations between subjects.

Most Discriminative ROIs. To detect the most discriminative ASD-relevant ROIs, we quantified the relative contribution of each ROI to the classification performance of the proposed CG-MHGEL by adopting Grad-CAM [14]. Specifically, Grad-CAM exploited information from the last hypergraph convolution layer and the fully connected layers to obtain the class activation values for each node (i.e., ROI) per subject. For each atlas, the class-specific heat-maps were shown in Fig. 3. We then computed the average activation value of each ROI across all the subjects (irregardless of class) to reflect the regional contribution to this classification task. For ease of illustration, we visualized the top 10 most discriminative ASD-relevant ROIs for each atlas by BrainNet Viewer [20] in Fig. 4, most of which have already been reported in the previous studies. For example, precuneus cortex, supramarginal gyrus, and superior temporal have crucial functions for supporting the social cognition and interactions, which are proved to be deficient in ASD patients [10,11]. A decreased regional homogeneity was found in middle frontal gyrus and superior parietal lobule of patients with ASD [15]. The evidence shows that there was a reduced brain activation in ASD in right cuneus [19].

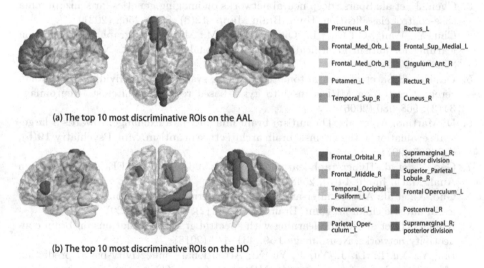

(a) The top 10 most discriminative ROIs on the AAL

(b) The top 10 most discriminative ROIs on the HO

Fig. 4. Visualization of the top 10 most discriminative ROIs for each atlas, whose full names can be found in Tables S1 and S2 in Supplementary material.

4   Conclusion

In this paper, we proposed a CG-MHGEL method to extract a multiatlas-based FCN embedding of each subject for multisite ASD studies, where HGCN instead of GCN was used to capture more complicated information in brain networks, and class- and site-consistency regularizations were imposed in the embedding space to account for between-subject associations associated with classes and sites, respectively. Experimental results on the ABIDE demonstrated the superiority of our method over other related ones for ASD classification. Furthermore, we identified ASD-relevant ROIs that can be validated with previous findings. In future work, we will design a more powerful contrastive regularization scheme to incorporate associations between subjects across classes and sites in the proposed method, and test on all sites of the ABIDE.

Acknowledgments. This work was supported in part by the National Natural Science Foundation of China under Grant 62202442, and in part by the Anhui Provincial Natural Science Foundation under Grant 2208085QF188.

References

1. Baggio, H.C., et al.: Cerebellar resting-state functional connectivity in Parkinson's disease and multiple system atrophy: characterization of abnormalities and potential for differential diagnosis at the single-patient level. NeuroImage Clin. **22**, 101720 (2019)
2. Berge, C.: Graphs and Hypergraphs. North-Holland, Amsterdam (1976)
3. Chen, J., et al.: Sparse deep neural networks on imaging genetics for schizophrenia case-control classification. Hum. Brain Mapp. **42**(8), 2556–2568 (2021)
4. Chu, Y., Wang, G., Cao, L., Qiao, L., Liu, M.: Multi-scale graph representation learning for autism identification with functional MRI. Front. Neuroinform. **15**, 802305 (2022)
5. Desikan, R.S., et al.: An automated labeling system for subdividing the human cerebral cortex on MRI scans into gyral based regions of interest. Neuroimage **31**(3), 968–980 (2006)
6. Di Martino, A., et al.: The autism brain imaging data exchange: towards a large-scale evaluation of the intrinsic brain architecture in autism. Mol. Psychiatry **19**(6), 659–667 (2014)
7. Gao, Y., et al.: Hypergraph learning: methods and practices. IEEE Trans. Pattern Anal. Mach. Intell. **44**(5), 2548–2566 (2022)
8. Guo, X., et al.: Altered inter-and intrahemispheric functional connectivity dynamics in autistic children. Hum. Brain Mapp. **41**(2), 419–428 (2020)
9. Ktena, S.I., et al.: Metric learning with spectral graph convolutions on brain connectivity networks. Neuroimage **169**, 431–442 (2018)
10. Liu, Y., Xu, L., Li, J., Yu, J., Yu, X.: Attentional connectivity-based prediction of autism using heterogeneous rs-fMRI data from CC200 atlas. Exp. Neurobiol. **29**(1), 27 (2020)
11. Monk, C.S., et al.: Abnormalities of intrinsic functional connectivity in autism spectrum disorders. Neuroimage **47**(2), 764–772 (2009)
12. Qin, K., et al.: Using graph convolutional network to characterize individuals with major depressive disorder across multiple imaging sites. EBioMedicine **78**, 103977 (2022)
13. Scarselli, F., Gori, M., Tsoi, A.C., Hagenbuchner, M., Monfardini, G.: The graph neural network model. IEEE Trans. Neural Netw. **20**(1), 61–80 (2008)
14. Selvaraju, R.R., et al.: Grad-CAM: visual explanations from deep networks via gradient-based localization. In: Proceedings of the IEEE International Conference on Computer Vision, pp. 618–626 (2017)

15. Shukla, D.K., Keehn, B., Müller, R.A.: Regional homogeneity of fMRI time series in autism spectrum disorders. Neurosci. Lett. **476**(1), 46–51 (2010)
16. Tzourio-Mazoyer, N., et al.: Automated anatomical labeling of activations in SPM using a macroscopic anatomical parcellation of the MNI MRI single-subject brain. Neuroimage **15**(1), 273–289 (2002)
17. Van Den Heuvel, M.P., Pol, H.E.H.: Exploring the brain network: a review on resting-state fMRI functional connectivity. Eur. Neuropsychopharmacol. **20**(8), 519–534 (2010)
18. Wang, M., Huang, J., Liu, M., Zhang, D.: Modeling dynamic characteristics of brain functional connectivity networks using resting-state functional MRI. Med. Image Anal. **71**, 102063 (2021)
19. Wichers, R.H., et al.: Modulation of brain activation during executive functioning in autism with citalopram. Transl. Psychiatry **9**(1), 286 (2019)
20. Xia, M., Wang, J., He, Y.: BrainNet viewer: a network visualization tool for human brain connectomics. PLoS ONE **8**(7), e68910 (2013)
21. Yan, C.G., Wang, X.D., Zuo, X.N., Zang, Y.F.: DPABI: data processing & analysis for (resting-state) brain imaging. Neuroinformatics **14**, 339–351 (2016)
22. Yang, X., Schrader, P.T., Zhang, N.: A deep neural network study of the ABIDE repository on autism spectrum classification. Int. J. Adv. Comput. Sci. Appl. **11**(4), 1–66 (2020)
23. Yu, S., et al.: Higher-order interactions characterized in cortical activity. J. Neurosci. **31**(48), 17514–17526 (2011)
24. Zhang, X., Hu, B., Ma, X., Xu, L.: Resting-state whole-brain functional connectivity networks for MCI classification using L2-regularized logistic regression. IEEE Trans. Nanobiosci. **14**(2), 237–247 (2015)

A Diffusion Simulation User Behavior Perception Attention Network for Information Diffusion Prediction

Yuanming Shao[1,2] (ID), Hui He[2(✉)] (ID), Yu Tai[2] (ID), Xinglong Wu[2] (ID), and Hongwei Yang[2] (ID)

[1] State Key Laboratory of Communication Content Cognition, People's Daily Online, Beijing 100733, China
`ymshao@stu.hit.edu.cn`
[2] Harbin Institute of Technology, Harbin, China
`{hehui,taiyu,yanghongwei}@hit.edu.cn`, `xlwu@stu.hit.edu.cn`

Abstract. Information diffusion prediction is an essential task in understanding the dissemination of information on social networks. Its objective is to predict the next user infected with a piece of information. While previous work focuses primarily on the analysis of diffusion sequences, recent work shifts towards examining social network connections between users. During the diffusion of information, users are expected to send and receive information. However, few works analyze the sending and receiving behavior of users during information diffusion. We design a **D**iffusion **S**imulation **U**ser **B**ehavior **P**erception **A**ttention **N**etwork (DSUBPAN). First, based on the social network graph, we construct a diffusion simulation heterogeneous network graph, which simulates diffusion, and obtain the sending and receiving behavior of users during information diffusion. Second, we utilize a user behavior fuse transformer to fuse different user behaviors. Then, we employ an attention network to perceive the time information and user sequence information in the information diffusion sequence. Finally, we utilize a dense layer and a softmax layer to predict the next infected user. Our model outperforms baseline methods on two real-world datasets, demonstrating its effectiveness.

Keywords: Information diffusion prediction · Social network · Graph convolutional network

1 Introduction

With the increase of social networks, a vast number of individuals actively engage with these platforms, where they can share their personal information and repost content from others. As a result, a substantial amount of information is spread among users on social networks, such as Twitter. Information diffusion prediction has been extensively studied so far. And models for information diffusion prediction tasks play a crucial role in areas such as product promotion [10], community detection [1], epidemiology [15], and other fields.

Q. Liu et al. (Eds.): PRCV 2023, LNCS 14433, pp. 182–194, 2024.
https://doi.org/10.1007/978-981-99-8546-3_15

The diffusion of a piece of information always shapes into an information cascade sequence (or simply cascade). Therefore, we also refer to information diffusion prediction as cascade prediction. A cascade is represented by the information of different users and their infection times. Taking Twitter as an example, for a tweet, we consider the user's retweeting behavior to be infected, and all users who retweet the tweet and their retweeting time can be seen as a cascaded sequence. There are different behaviors of users in a cascaded sequence. In the process of information diffusion, users will receive information and send information, and other users related to these two behaviors are not the same. However, previous works [17,21,26] rarely consider the user's sending behavior and receiving behavior. Most previous works focus on the diffusion sequence [12,20,24] and some works additionally consider the social network [23,25,26]. These methods simply embed different users differently and do not take into account the different behaviors that may occur during the diffusion of information, and the range of users infected by these behaviors may be different.

To address the issue of inadequate representation of diverse user sending and receiving behaviors in existing methods, we design a **D**iffusion **S**imulation **U**ser **B**ehavior **P**erception **A**ttention **N**etwork (DSUBPAN). First, we construct a diffusion simulation heterogeneous graph and use a multiple-layer graph convolutional network to simulate the information diffusion and obtain the sending embeddings and receiving embeddings of users. Second, we utilize a user behavior fusion transformer to fuse the two behaviors. Then, we employ an attention network to perceive the time information and user sequence information in the cascade. Finally, we utilize the dense layer and softmax layer to predict the next infected user. The main contributions of this paper can be summarized as follows:

- We build a heterogeneous graph, which not only utilizes the social network graph information but also simulates the process of information diffusion, to perceive the sending behaviors and receiving behaviors of users. Then we use a multiple-layer graph convolutional network to obtain the sending embeddings and receiving embeddings of users.
- We utilize a user behavior fusion transformer to fuse two representations. And we utilize an attention network to perceive the time information and user sequence information in the cascade.
- Our model outperforms the baseline methods on two real datasets, proving its effectiveness.

2   Related Work

We review some previous studies on information diffusion prediction from the diffusion path based methods and social graph based methods.

Diffusion Path Based Methods. This category of methods uses the given observed cascade sequence to predict the next infected user. In order to model the diffusion process, earlier studies presuppose a diffusion model. For example, the independent cascade (IC) model [8] assigns an independent diffusion probability between every two users. The performance of such methods [5,8,11] depends on the assumptions of the presupposed diffusion model. Later, researchers begin to use different deep learning models for information diffusion prediction tasks, such as recurrent neural networks [7,17] (RNN) and attention mechanisms [21,22]. These studies typically utilize representation learning techniques to learn representations of diffusion paths from cascade sequences instead of assuming a prior diffusion model. For example, DeepDiffuse [7] utilizes a long-short term memory (LSTM) to perceive cascade sequences and use the attention mechanism to handle infected timestamps. To perceive the users' long-term dependency, NDM [22] uses the convolutional neural network and attention mechanism. HiDAN [21] captures dependency-aware user representations in the cascade by constructing an attention network.

Social Graph Based Methods. In social networks, users follow each other when sharing interests and hobbies, increasing the likelihood of information diffusion through friend-to-friend transmission [25]. Previous studies [19,25] focus on improving prediction performance by considering social relations and utilizing the topology of social graphs to predict information diffusion direction more accurately. With the development of graph neural networks, some models [3,18,26,27] also incorporate graph convolutional networks to learn user representations on social graphs. For example, TopoLSTM [16] is based on the LSTM model, which shares the hidden state between users who have relationships in a cascade sequence, to introduce the social graph information. FOREST [23] aggregates the neighborhood features to obtain the user embedding and uses LSTM to perceive the correlation information of user representations in the cascade sequence to perform micro cascade prediction. DyHGCN [26] uses the cascade sequences and social graph to create a heterogeneous graph, which captures global diffusion relations. However, most existing methods only assign a learnable representation to each user and fail to consider their different behaviors during information diffusion. Our approach differs by not only considering the global diffusion of information based on the social graph but representing the sending and receiving behavior of users separately.

3   Methodology

3.1   Problem Definition

Suppose there is a set of users $\mathcal{V}$ and a set of cascades $\mathcal{C}$. Each cascade $c_i \in \mathcal{C}$ represents the sequence of infected users and their corresponding infection times. For each cascade in the set $\mathcal{C}$, $c_i = \{(u_1, t_1), (u_2, t_2), ..., (u_{|c_i|}, t_{|c_i|})\}$, with infected user $u_j \in \mathcal{V}$ arranged in ascending order according to their respective infection times. The length of cascade $|c_i|$ is denoted as n for convenience. Additionally, we let $N = |\mathcal{V}|$ denote the total number of users in $\mathcal{V}$.

A social graph $\mathcal{G} = (\mathcal{V}, \mathcal{E})$ is provided if information diffusion occurs on a social network. Here, the node set of graph $\mathcal{V}$ is the set of users, and the edges set $\mathcal{E} = \{\langle u_p, u_q \rangle | u_p \in \mathcal{V}, u_q \in \mathcal{V}, u_p \neq u_q, u_p \text{follows} u_q\}$ denotes relationships between them. $\mathcal{G}$ is a directed graph.

The information diffusion prediction task is formalized as: given an observed cascade $c_i = \{(u_1, t_1), (u_2, t_2), ..., (u_n, t_n)\}$, the model predicts the next infected user $u_{n+1} \in \mathcal{V}$ by calculating the infected probability $p(u_{n+1}|c_i)$ for each user in the set of users $\mathcal{V}$.

3.2   Model Framework

Figure 1 shows the architecture of DSUBPAN. Our model consists of four modules. First, the diffusion simulation user behavior embedding module constructs a diffusion simulation heterogeneous graph and applies the graph convolutional network to obtain two embeddings for each user in the cascade, i.e., sending embeddings and receiving embeddings. Second, the user behavior fusion transformer involves perceiving information diffusion among users in a cascade and fusing two embeddings of each user into one representation. Then the cascade perception attention network utilizes a time embedding layer and a multiple-head self-attention network to perceive user sequence information and time informa-

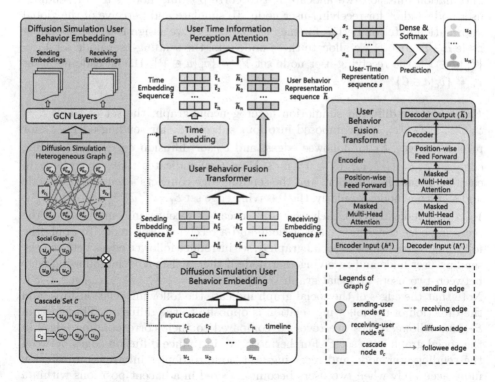

Fig. 1. The architecture of our proposed DSUBPAN.

tion within the cascade. Finally, we employ the dense layer and softmax layer to predict the probabilities of future infected users. Similar to [23,26], the prediction outcome is a sequence of users.

3.3 Diffusion Simulation User Behavior Embedding

To simulate the information diffusion path of the input cascade, we build a diffusion simulation heterogeneous graph to simulate the diffusion process, and then we use a user perception graph convolutional network to obtain the sending embedding and the receiving embedding of each user in the input cascade.

Diffusion Simulation Heterogeneous Graph. Given cascade set $\mathcal{C}$, user set $\mathcal{V}$, and the social graph $\mathcal{G} = (\mathcal{V}, \mathcal{E})$, to simulate information diffusion processes, we build our diffusion simulation heterogeneous graph. $\widetilde{\mathcal{G}} = (\widetilde{\mathcal{V}}, \widetilde{\mathcal{E}})$. Graph $\widetilde{\mathcal{G}}$ comprises three types of nodes and four types of edges.

Nodes. For the node set $\widetilde{\mathcal{V}}$ in our diffusion simulation heterogeneous graph, $\widetilde{\mathcal{V}} = \widetilde{\mathcal{V}}_s \cup \widetilde{\mathcal{V}}_r \cup \widetilde{\mathcal{V}}_c$, where $\widetilde{\mathcal{V}}_s$ represents the sending-user node set, $\widetilde{\mathcal{V}}_r$ represents the receiving-user node set, and $\widetilde{\mathcal{V}}_c$ represents the cascade node set. For each user $u \in \mathcal{V}$, to simulate a user's sending behavior and receiving behavior during information diffusion, we allocate it two corresponding nodes, $\widetilde{v}_u^s$ for sending-user node and $\widetilde{v}_u^r$ for receiving-user node. To simulate and perceive of the global information diffusion better, for each cascade $c \in \mathcal{C}$, we assign it a corresponding node $\widetilde{v}_c$. Therefore, we allocate three node sets: the sending-user node set $\widetilde{\mathcal{V}}_s = \{\widetilde{v}_u^s | u \in \mathcal{V}\}$; the receiving-user node set $\widetilde{\mathcal{V}}_r = \{\widetilde{v}_u^r | u \in \mathcal{V}\}$; the cascade node set $\widetilde{\mathcal{V}}_c = \{\widetilde{v}_c | c \in \mathcal{C}\}$.

Edges. In our diffusion simulation heterogeneous graph, the set of edges $\widetilde{\mathcal{E}} = \widetilde{\mathcal{E}}_s \cup \widetilde{\mathcal{E}}_r \cup \widetilde{\mathcal{E}}_f \cup \widetilde{\mathcal{E}}_d$ is decomposed into four subsets: $\widetilde{\mathcal{E}}_s$ for sending edges, $\widetilde{\mathcal{E}}_r$ for receiving edges, $\widetilde{\mathcal{E}}_f$ for followee edges, and $\widetilde{\mathcal{E}}_d$ for diffusion edges. The sending edge set $\widetilde{\mathcal{E}}_s = \{\langle \widetilde{v}_u^s, \widetilde{v}_c \rangle | \widetilde{v}_u^s \in \widetilde{\mathcal{V}}_s, \widetilde{v}_c \in \widetilde{\mathcal{V}}_c, u \in c\}$ indicates that each user u in cascade c sends information, and the sending edge connects a sending-user node and a cascade node. Similarly, the receiving edge set $\widetilde{\mathcal{E}}_r = \{\langle \widetilde{v}_c, \widetilde{v}_u^r \rangle | \widetilde{v}_c \in \widetilde{\mathcal{V}}_c, \widetilde{v}_u^r \in \widetilde{\mathcal{V}}_r, u \in c\}$ indicates that each user u in cascade c receives information, and the receiving edge connects a cascade node and a receiving-user node. To apply the following relationships in social graph $\mathcal{G}$, we introduce followee edges between the sending-user node and receiving-user node if there exists a social following edge between two users in social graph $\mathcal{G}$. That is, $\widetilde{\mathcal{E}}_f = \{\langle \widetilde{v}_{u_q}^s, \widetilde{v}_{u_p}^r \rangle | \langle u_p, u_q \rangle \in \mathcal{E}\}$. Note that the edges of the social graph indicate the following relationships, and the direction of the following relation is opposite to the diffusion. Therefore, in $\widetilde{\mathcal{E}}_f$, our followee edges are reversed compared to their corresponding edges in the social graph edge set $\mathcal{E}$. Furthermore, we introduce diffusion edges between the sending-user node and receiving-user node to depict information diffusion more accurately when two users become infected in adjacent positions within a cascade. Specifically, for each cascade c_i, we construct a subset of the diffusion edges $\widetilde{\mathcal{E}}_d^i = \{\langle \widetilde{v}_{u_j}^s, \widetilde{v}_{u_{j+1}}^r \rangle | c_i = \{(u_1, t_1), ..., (u_n, t_n)\}, j < n, j \in \mathbb{N}^+\}$, and set of diffusion edges is the union of these subsets: $\widetilde{\mathcal{E}}_d = \bigcup_i \widetilde{\mathcal{E}}_d^i$.

User Behavior Perception Graph Convolutional Network. After constructing the heterogeneous graph $\widetilde{\mathcal{G}} = (\widetilde{\mathcal{V}}, \widetilde{\mathcal{E}})$, to extract node features, we employ a multiple-layer graph convolutional network (GCN) [9]. Specifically, we first assign each node $\widetilde{v} \in \widetilde{\mathcal{V}}$ a corresponding learnable embedding vector $z_{\widetilde{v}}^0 \in \mathbb{R}^d$, where d denotes the embedding dimension. Then, in the lth layer GCN, for each node $\widetilde{v} \in \widetilde{\mathcal{V}}$, we calculate the embedding $z_{\widetilde{v}}^l \in \mathbb{R}^d$ by:

$$z_{\widetilde{v}}^l = \sigma(W_g^l \sum_{v \in \mathcal{N}(\widetilde{v})} \frac{z_v^l}{|\mathcal{N}(\widetilde{v})|}), \tag{1}$$

where $W_g^l \in \mathbb{R}^{d \times d}$ represents the weight matrix of GCN in the lth layer. $\sigma(\cdot)$ denotes a sequence of functions including ReLU activation, dropout, and layer normalization operations. $\mathcal{N}(\cdot)$ is a function that finds all neighbors of the input node. $|\mathcal{N}(\cdot)|$ represents the total number of neighbors.

We denote the total number of GCN layers as L. For a given user u_j in the input cascade c, we first identify two nodes, the sending-user node $\widetilde{v}_{u_j}^s$ and the receiving-user $\widetilde{v}_{u_j}^r$ node in the heterogeneous graph. Then, we utilize their embeddings in the Lth layer GCN to obtain the sending embedding h_j^s and receiving embedding h_j^r:

$$v_s = \widetilde{v}_{u_j}^s \in \widetilde{\mathcal{V}}_s, \quad v_r = \widetilde{v}_{u_j}^r \in \widetilde{\mathcal{V}}_r,$$
$$h_j^s = z_{v_s}^L, \quad h_j^r = z_{v_r}^L. \tag{2}$$

Thus, for an input cascade $c_i = \{(u_1, t_1), ..., (u_n, t_n)\}$, we derive two embedding sequences: the sending embedding sequence $h^s = [h_1^s, ..., h_n^s]$ and the sending embedding sequence $h^r = [h_1^r, ..., h_n^r]$.

3.4 User Behavior Fusion Transformer

To perceive the direction of information diffusion in input cascades and fuse user behavior embeddings, we design a user behavior fusion transformer based on the transformer model [14]. This module enables us to gain further insight into how information diffuses among users and fuses their sending and receiving embeddings into a single representation of their behavior. As a result, we generate a user behavior representation sequence from the sending embeddings sequence and receiving embeddings sequence.

Our user behavior fusion transformer comprises an encoder and a decoder. The encoder's input is the sending embedding sequence $h^s = [h_1^s, ..., h_n^s]$ and it produces the encoding sequence of the sending embedding as output. The decoder, on the other hand, takes in both the output sequence of the encoder and the receiving embedding sequence $h^r = [h_1^s, ..., h_n^s]$ as input, producing a user behavior sequence $\tilde{h} = [\tilde{h}_1, ..., \tilde{h}_n]$ that serves as our user behavior fusion transformer's final output.

The encoder module consists of a masked multi-head attention network and a position-wise feedforward network. We use a masked multi-head attention network to encode the input embedding sequence h^s, and then transform its semantics with the position-wise feedforward network. The attention mask prevents future information leakage, and the query, key, and value of the attention network are all h^s.

The decoder has an additional masked multi-head attention network compared to the encoder. It uses one attention network to encode h^r, with the query, key, and value all being h^r. Then another attention network integrates the encoder output as the query and the previous attention network output as the key and value. Also, a position-wise feedforward network is utilized to transform the semantics of the second layer attention's output sequence, and we finally obtain a user behavior sequence $\tilde{h} = [\tilde{h}_1, ..., \tilde{h}_n]$.

3.5   Cascade Perception Attention Network

Previous studies [2,4,11] demonstrate that the time of infection for previous users can significantly impact information diffusion. To perceive the time information, we convert the timestamps into time embeddings. First, we extract all timestamps from cascade set $\mathcal{C}$, and obtain a set of timestamps $\mathcal{T}$. Second, we find the maximal timestamps t_{max} and minimum timestamps t_{min} in set $\mathcal{T}$. Then, we uniformly divide the timestamps into T intervals with equal length Δt:

$$\Delta t = (t_{max} - t_{min})/T. \tag{3}$$

Next, we partition all timestamps into T intervals, i.e., $\{[t_{min}, t_{min}+\Delta t), [t_{min}+\Delta t, t_{min} + 2\Delta t), ..., [t_{min} + (T-1)\Delta t, t_{max}]\}$. For q^{th} interval, we will allocate a learnable vector $t_q \in \mathbb{R}^d$. Consequently, every timestamp t_j in the input cascade is transformed into a vector denoted as $\tilde{t}_j$ and we obtain an embedding sequence of time $\tilde{t} = [\tilde{t}_1, ..., \tilde{t}_n]$ by:

$$q = \min\{\lceil (t_j - t_{min})/\Delta t \rceil, T\}, \quad \tilde{t}_j = t_q. \tag{4}$$

With the time embedding sequence $\tilde{t} = [\tilde{t}_1, ..., \tilde{t}_n]$ and the user behavior representation sequence $\tilde{h} = [\tilde{h}_1, ..., \tilde{h}_n]$ at hand, we employ a user time information perception attention network to perceive user sequence information and time information, yielding a user-time representation sequence $s = [s_1, ..., s_n]$:

$$\tilde{q}^u_{j,h} = \tilde{W}^u_{q,h}\tilde{h}_j + \tilde{b}^u_{q,h}, \quad \tilde{q}^t_{j,h} = \tilde{W}^t_{q,h}\tilde{t}_j + \tilde{b}^t_{q,h},$$

$$\tilde{k}_{j,h} = \tilde{W}_{k,h}\tilde{h}_j + \tilde{b}_{k,h}, \quad \tilde{v}_{j,h} = \tilde{W}_{v,h}\tilde{h}_j + \tilde{b}_{v,h},$$

$$\tilde{\alpha}^h_{j,k} = \frac{(\tilde{q}^u_{j,h})\tilde{k}_{k,h} + (\tilde{q}^t_{j,h})^T\tilde{k}_{k,h}}{2\sqrt{\tilde{d}_h}} + \tilde{m}_{j,k}, \quad \tilde{w}^h_{j,k} = \frac{\exp(\tilde{\alpha}^h_{j,k})}{\sum_{l=1}^n \exp(\tilde{\alpha}^h_{j,l})}, \tag{5}$$

$$\tilde{s}^T_{j,h} = \sum_{k=1}^n \tilde{w}^h_{j,k}\tilde{v}^T_{j,h}, \quad s_j = \mathbf{LayerNorm}(\tilde{h}_j + (\tilde{W}_s[\tilde{s}^T_{j,1}; ...; \tilde{s}^T_{j,\tilde{H}}]^T + \tilde{b}_s)),$$

where h represents the head number, and $\tilde{H}$ represents the total number of heads. $\tilde{\boldsymbol{W}}_{q,h}^u, \tilde{\boldsymbol{W}}_{q,h}^t, \tilde{\boldsymbol{W}}_{k,h}, \tilde{\boldsymbol{W}}_{v,h} \subset \mathbb{R}^{d \times \tilde{d}_h}$ and $\tilde{\boldsymbol{W}}_s \subset \mathbb{R}^{(H \cdot \tilde{d}_h) \times d}$ represent weight matrices. $\tilde{\boldsymbol{b}}_{q,h}^u, \tilde{\boldsymbol{b}}_{q,h}^t, \tilde{\boldsymbol{b}}_{k,h}, \tilde{\boldsymbol{b}}_{v,h} \in \mathbb{R}^{\tilde{d}_h}$ and $\tilde{\boldsymbol{b}}_s \in \mathbb{R}^d$ represent the bias vectors. We denote the head dimension as $\tilde{d}_h$. The symbol $[;]$ denotes vector concatenation. The **LayerNorm**$(\cdot)$ refers to the function of layer normalization. $\tilde{m}_{j,k}$ is the mask to prevent the leakage of future users, and it is calculated by:

$$\tilde{m}_{j,k} = \begin{cases} 0, & j \le k, \\ -\infty, & j > k. \end{cases} \tag{6}$$

3.6 Diffusion Prediction

From the user-time representation sequence $s = [s_1, ..., s_n]$, for information diffusion prediction, we utilize a dense layer to change each user-time representation into users' infected probabilities $\tilde{\boldsymbol{y}} = [\tilde{\boldsymbol{y}}_1, ..., \tilde{\boldsymbol{y}}_n]$:

$$\tilde{\boldsymbol{y}} = \boldsymbol{W}s + \boldsymbol{b}, \tag{7}$$

where $\boldsymbol{W} \in \mathbb{R}^{d \times N}$ represents the weight matrix, and $\boldsymbol{b} \in \mathbb{R}^N$ represents the bias vector. Next, we utilize a softmax layer to normalize the infected probabilities $\tilde{\boldsymbol{y}}$ resulting in $\hat{\boldsymbol{y}} = [\hat{\boldsymbol{y}}_1, ..., \hat{\boldsymbol{y}}_n] \in [0, 1]$. For each element $\tilde{\boldsymbol{y}}_j = [\tilde{y}_{j,1}, ..., \tilde{y}_{j,N}]^T \in \tilde{\boldsymbol{y}}$, we compute its normalized values by:

$$\hat{y}_{j,k} = \frac{\exp(\tilde{y}_{j,k})}{\sum_{l=1}^N \exp(\tilde{y}_{j,l})}. \tag{8}$$

This allows us to achieve the prediction goal of determining the probability of the next infected user, where the user with the largest probability value is the next infected user.

In addition, we utilize the cross-entropy loss function to train the parameters:

$$\mathcal{L} = -\sum_{j=2}^n \sum_{k=1}^N y_{j,k} \log \hat{y}_{j,k}, \tag{9}$$

where $y_{j,k}$ indicates whether user k is infected or not at position j of the input cascade. If user k is infected on position j, $y_{j,k} = 1$; otherwise, $y_{j,k} = 0$. We update the parameters of the model by employing the Adam optimizer.

4 Experiment

Datasets. To evaluate our model, we utilize the Twitter dataset [6] and Douban dataset [28], which are two widely adopted public real-world datasets. In previous works [23, 26], these datasets are partitioned into training, validation, and testing sets based on cascades with an 8:1:1 ratio. Table 1 presents detailed statistics of the two datasets.

Table 1. Statistics of Twitter and Douban.

Datasets	#Cascades	#Users	#Edges	Avg. Length
Twitter	3,442	12,627	309,631	32.60
Douban	10,602	23,123	348,280	27.14

Baselines. We compared DSUBPAN with several information diffusion prediction models to assess its effectiveness. Some of these baselines are based on the diffusion path, i.e. DeepDiffuse [7], NDM [22], while others consider the social graph, i.e., TopoLSTM [16], Forest [23], DyHGCN [26], and MS-HGAT [13].

Metrics. We always view information diffusion prediction as a ranking task. We compute the infection probabilities of users and rank them. Therefore, we use HITS scores on top k (hits@k) and the mean average precision on top k (mAP@k). A model performs better if it gets higher values on these metrics.

Parameter Settings. For hyperparameter settings, we set the graph embedding dimension d to 64 and set the number of GCN layers to 2. In our user behavior fusion transformer and cascade perception attention network, we use multiple head attention networks with a head number of 4, and a head dimension of 32. We set the number of time intervals T to 50. In addition, the dropout rate in our model is set to 0.3. We set the batch size to 16 and the learning rate to 0.001 for the Adam optimizer.

4.1 Results

Table 2. Comparative experiments results and the improvement on two datasets (%). The performance scores of the leading baseline are underlined. The best performance scores among all compared models are indicated in bold.

Datasets	Twitter						Douban					
Metrics	hits			mAP			hits			mAP		
Models	@10	@50	@100	@10	@50	@100	@10	@50	@100	@10	@50	@100
DeepDiffuse	4.47	14.45	21.42	1.50	1.89	2.00	7.29	15.26	21.03	2.87	3.21	3.29
NDM	21.08	31.27	38.22	14.32	14.78	14.88	10.42	18.32	23.29	5.67	6.03	6.10
TopoLSTM	11.44	27.82	36.35	3.84	4.58	4.70	9.82	20.66	27.48	3.71	4.22	4.32
FOREST	25.41	39.26	47.53	16.90	17.52	17.64	14.77	26.08	33.01	8.14	8.65	8.75
DyHGCN	25.80	43.05	53.44	15.67	16.45	16.60	_16.32_	_28.55_	_35.70_	_9.18_	_9.73_	_9.83_
MS-HGAT	**30.49**	_47.00_	_57.13_	_18.62_	_19.37_	_19.51_	14.92	26.43	33.29	8.06	8.58	8.68
DSUBPAN	30.06	**47.74**	**57.51**	**18.66**	**19.47**	**19.61**	**17.69**	**30.82**	**38.31**	**9.77**	**10.36**	**10.47**
Improvement	-1.41	1.57	0.67	0.21	0.52	0.51	8.45	7.94	7.30	6.38	6.50	6.48

Comparative Experiments. Table 2 shows the results of the comparative experiments. Except hits@10, DSUBPAN performs better on other evaluation metrics on two real-world datasets compared with the baselines, which proves DSUBPAN's effectiveness. Specifically, we have the following observations:

(1) DSUBPAN improves by more than 6% on all metrics on the Douban dataset. On the Twitter dataset, DSUBPAN gets the lower score on hits@10 than the recent baseline MS-HGAT, but our model improves by more than 15% compared with MS-HGAT on the Douban dataset. It is speculated that MS-HGAT is more suitable for situations with fewer cascades sequences.

(2) Among all baseline methods, the top two performers, MS-HGAT and DyHGCN, are both recent social graph based methods. This indicates that the information diffusion prediction model considering the relationships between users performs better. DSUBPAN is also designed based on this method.

(3) Compared with DyHGCN, which considers both the social graph and diffusion graph, we further consider the sending behavior and receiving behavior of users based on utilizing these two types of graphs to simulate information diffusion on heterogeneous graphs. Experimental results show that DSUBPAN which considers different user behaviors has better performance.

Table 3. Ablation study results on two datasets (%).

Datasets	Twitter						Douban					
Metrics	hits			mAP			hits			mAP		
Models	@10	@50	@100	@10	@50	@100	@10	@50	@100	@10	@50	@100
DSUBPAN	**30.06**	**47.74**	**57.51**	**18.66**	**19.47**	**19.61**	**17.69**	**30.82**	**38.31**	**9.77**	**10.36**	**10.47**
w/o Graph	24.10	37.10	45.02	16.09	16.67	16.78	14.86	25.62	32.28	8.36	8.85	8.95
w/o Trans	28.90	45.58	54.96	18.09	18.85	18.98	15.91	27.52	34.38	8.81	9.33	9.43
w/o Time	29.84	47.59	57.39	18.55	19.36	19.50	15.83	28.53	35.98	8.48	9.06	9.16

Ablation Study. We compare the three variants obtained after changing or omitting different modules of DSUBPAN to investigate the relative importance of each module in DSUBPAN. Specifically, instead of obtaining the sending embedding and the receiving embedding by using diffusion simulation heterogeneous graph and GCN layers, we directly assign two learnable embedding representations by an embedding layer to each user as the sending embedding and the receiving embedding (i.e., w/o Graph). Instead of using the user behavior fusion transformer, we fuse the sending embedding and receiving embedding by naive feature concat and a dense layer (i.e., w/o Trans). We omit the additional time embedding query vector in user time information perception and only use the user behavior representation as the query vector, at which point the user time information perception module changes into an ordinary multi-head self-attention (i.e., w/o Time). Table 3 shows the results of the ablation experiments.

The results of the ablation experiments indicate that The effect of the three variant models is decreased compared with DSUBPAN in all indicators. And The drop in performance is most pronounced when we replace the diffusion simulation user behavior embedding module in DSUBPAN, which indicates that this module is the most important.

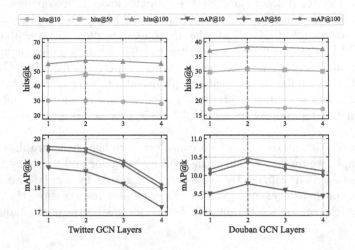

Fig. 2. Parameter analysis on the two datasets.

Parameter Analysis. We further analyze one of the significant hyperparameters of DSUBPAN, the number of GCN layers, on the Twitter dataset and Douban dataset. The results are shown in Fig. 2. When the number of GCN layers is greater than 1, due to overfitting, all evaluation metrics gradually decrease as the number of layers grows on both datasets. Compared to the 1-layer GCN, except for the reduction of mAP@k on the Twitter dataset, the 2-layer GCN performs better on the other five metrics on the two datasets. Therefore, 2-layer GCN achieves better performance.

5 Conclusion

In this paper, we propose a novel DSUBPAN model for information diffusion prediction. In order to better predict the direction of diffusion, we consider both the social graph and the sending behavior and receiving behavior of users. We simulate information diffusion by constructing a novel heterogeneous graph, learn the sending and receiving representations of users, and utilize a transformer to fuse the two representations. We design experiments on two real-world datasets, Twitter and Douban. DSUBPAN shows over 5% improvement in hits@k and mAP@k compared with the baseline methods, demonstrating its effectiveness.

Acknowledgment. This work was supported in part by the Joint Funds of the National Key Research and Development Program of China (2020YFB1406902) and the Fundamental Research Funds for the State Key Laboratory of Communication Content Cognition (A12003).

References

1. Barbieri, N., Bonchi, F., Manco, G.: Cascade-based community detection. In: WSDM, pp. 33–42 (2013)
2. Cao, Q., Shen, H., Cen, K., Ouyang, W., Cheng, X.: Deephawkes: bridging the gap between prediction and understanding of information cascades. In: CIKM, pp. 1149–1158 (2017)
3. Cao, Z., Han, K., Zhu, J.: Information diffusion prediction via dynamic graph neural networks. In: CSCWD, pp. 1099–1104. IEEE (2021)
4. Chen, X., Zhang, K., Zhou, F., Trajcevski, G., Zhong, T., Zhang, F.: Information cascades modeling via deep multi-task learning. In: SIGIR, pp. 885–888 (2019)
5. Gomez Rodriguez, M., Leskovec, J., Schölkopf, B.: Structure and dynamics of information pathways in online media. In: WSDM, pp. 23–32 (2013)
6. Hodas, N.O., Lerman, K.: The simple rules of social contagion. Sci. Rep. **4**(1), 4343 (2014)
7. Islam, M.R., Muthiah, S., Adhikari, B., Prakash, B.A., Ramakrishnan, N.: Deep-diffuse: predicting the 'who' and 'when' in cascades. In: ICDM, pp. 1055–1060. IEEE (2018)
8. Kempe, D., Kleinberg, J., Tardos, É.: Maximizing the spread of influence through a social network. In: SIGKDD, pp. 137–146 (2003)
9. Kipf, T.N., Welling, M.: Semi-supervised classification with graph convolutional networks. arXiv preprint arXiv:1609.02907 (2016)
10. Leskovec, J., Adamic, L.A., Huberman, B.A.: The dynamics of viral marketing. TWEB **1**(1), 5-es (2007)
11. Rodriguez, M.G., Leskovec, J., Balduzzi, D., Schölkopf, B.: Uncovering the structure and temporal dynamics of information propagation. Netw. Sci. **2**(1), 26–65 (2014)
12. Sankar, A., Zhang, X., Krishnan, A., Han, J.: Inf-vae: a variational autoencoder framework to integrate homophily and influence in diffusion prediction. In: WSDM, pp. 510–518 (2020)
13. Sun, L., Rao, Y., Zhang, X., Lan, Y., Yu, S.: MS-HGAT: memory-enhanced sequential hypergraph attention network for information diffusion prediction. In: AAAI, vol. 36, pp. 4156–4164 (2022)
14. Vaswani, A., et al.: Attention is all you need. Adv. Neural Inf. Process. Syst. **30** (2017)
15. Wallinga, J., Teunis, P.: Different epidemic curves for severe acute respiratory syndrome reveal similar impacts of control measures. Am. J. Epidemiol. **160**(6), 509–516 (2004)
16. Wang, J., Zheng, V.W., Liu, Z., Chang, K.C.C.: Topological recurrent neural network for diffusion prediction. In: ICDM, pp. 475–484. IEEE (2017)
17. Wang, Y., Shen, H., Liu, S., Gao, J., Cheng, X.: Cascade dynamics modeling with attention-based recurrent neural network. In: IJCAI, vol. 17, pp. 2985–2991 (2017)
18. Wang, Z., Chen, C., Li, W.: Attention network for information diffusion prediction. In: WWW, pp. 65–66 (2018)

19. Wang, Z., Chen, C., Li, W.: A sequential neural information diffusion model with structure attention. In: CIKM, pp. 1795–1798 (2018)
20. Wang, Z., Chen, C., Li, W.: Joint learning of user representation with diffusion sequence and network structure. TKDE **34**(3), 1275–1287 (2020)
21. Wang, Z., Li, W.: Hierarchical diffusion attention network. In: IJCAI, pp. 3828–3834 (2019)
22. Yang, C., Sun, M., Liu, H., Han, S., Liu, Z., Luan, H.: Neural diffusion model for microscopic cascade prediction. arXiv preprint arXiv:1812.08933 (2018)
23. Yang, C., Tang, J., Sun, M., Cui, G., Liu, Z.: Multi-scale information diffusion prediction with reinforced recurrent networks. In: IJCAI, pp. 4033–4039 (2019)
24. Yang, J., Leskovec, J.: Modeling information diffusion in implicit networks. In: ICDM, pp. 599–608. IEEE (2010)
25. Yang, Y., et al.: Rain: social role-aware information diffusion. In: AAAI, vol. 29 (2015)
26. Yuan, C., Li, J., Zhou, W., Lu, Y., Zhang, X., Hu, S.: DyHGCN: a dynamic heterogeneous graph convolutional network to learn users' dynamic preferences for information diffusion prediction. In: Hutter, F., Kersting, K., Lijffijt, J., Valera, I. (eds.) ECML PKDD 2020. LNCS (LNAI), vol. 12459, pp. 347–363. Springer, Cham (2021). https://doi.org/10.1007/978-3-030-67664-3_21
27. Zhang, Y., Lyu, T., Zhang, Y.: Cosine: community-preserving social network embedding from information diffusion cascades. In: AAAI, vol. 32 (2018)
28. Zhong, E., Fan, W., Wang, J., Xiao, L., Li, Y.: Comsoc: adaptive transfer of user behaviors over composite social network. In: SIGKDD, pp. 696–704 (2012)

A Representation Learning Link Prediction Approach Using Line Graph Neural Networks

Yu Tai[1], Hongwei Yang[1], Hui He[1]([✉]), Xinglong Wu[1],
and Weizhe Zhang[1,2]

[1] Harbin Institute of Technology, Harbin 150001, China
{taiyu,yanghongwei,hehui,wzzhang}@hit.edu.cn, xlwu@stu.hit.edu.cn
[2] Pengcheng Laboratory, Shenzhen 518055, China

Abstract. Link prediction problem aims to infer the potential future links between two nodes in the network. Most of the existing methods exhibit limited universality and are only effective in specific scenarios, while also neglecting the issue of information loss during model training. To address such issues, we propose a link prediction method based on the line graph neural network (NLG-GNN). Firstly, we employ Node2Vector to learn the latent feature representation vector of each node in the network. Secondly, we extract the local subgraphs surrounding the target link and transform them into the corresponding line graphs. Then, we design a Graph Convolutional Network (GCN) to learn the structural feature representation vector of the node through the line graph. Finally, we combine the latent and structural features through the output layer to predict the target links. We execute extensive experiments on 17 diverse datasets, demonstrating the superior performance and faster convergence of our NLG-GNN method over all baseline methods.

Keywords: Node embedding · Graph neural networks · Line graph · Link prediction

1 Introduction

Network structure pervades our daily life, encompassing social networks, protein-protein interaction networks, and transportation networks. Diverse interactions among distinct entities form the links in the network. Accurately predicting potential links significantly enhance the performance of various applications ranging from knowledge graph completion [23] and recommender systems [30] to biological networks [16].

Due to the valuableness of predicting links for network structure, many studies have been conducted and can be categorized into three types: (1) Topological similarity-based approaches [15] suppose that the greater the similarity between two nodes in a network, the higher the likelihood of a link between them. Such necessitate manual analysis for designing evaluation indicators with strong

Q. Liu et al. (Eds.): PRCV 2023, LNCS 14433, pp. 195–207, 2024.
https://doi.org/10.1007/978-981-99-8546-3_16

assumptions. (2) Maximum likelihood-based approaches [7] assume the network has its own formation mode and integrate existing link information to calculate a likelihood value. Such have high complexity and long time consumption, making them unsuitable for large-scale networks. (3) Representation learning-based approaches [2,25,35] enables automatic acquisition of network information that is advantageous for link prediction.

However, in some relevant methods [10,33], the utilization of learned feature information for pooling operations may result in information loss and adversely impact prediction performance. To address this issue, we propose NLG-GNN to directly represent features of the target link rather than characterizing features from the entire enclosing subgraph. Specifically, to ensure minimal information loss, we construct the line graph node feature by concatenating the two connected node features, which is more effective than mean or weighted sum pooling in preserving information. In addition, most existing methods are limited to specific scenarios and lack the universality and versatility required for adaptation in different situations. Therefore, there are two major issues for the link prediction methods. So the first issue (**CH1**) is: *How to better circumvent information loss during model training?* The second issue (**CH2**) is: *How to ensure the universality of model?*

To address the above two issues, in this paper, we propose a new end-to-end model NLG-GNN to efficiently explore the effective information in link prediction. The characteristics and main contributions of our NLG-GNN are as follows:

(1) Targeting **CH1**, we design a novel representation learning method based on line graph neural networks [34], which incurs less information loss compared to traditional methods.
(2) Targeting **CH2**, we collect and process 17 representative datasets, including dense and sparse graphs with varying network properties to validate the universality of our model.
(3) We carry out experiments with 17 real-world datasets to predict possible links, indicating that NLG-GNN markedly surpasses the current state-of-the-art link prediction approaches.

2   Related Work

In this section, we review the studies on link prediction from the following three categories: Topological similarity-based approaches, Maximum likelihood-based approaches, and Representation learning-based approaches.

Topological Similarity-based Approaches. Topological similarity-based approaches suppose assume that greater similarity between two nodes in a network increases the likelihood of a link between them. These approaches can be specified by local structure [1,4] or global structure [13,15]. For the local structure, the well-known common neighbor (CN) index [4], Salton index, and Jaccard index focus on the directly connected common neighbors. Adamic-Adar (AA) index [1] and Resource Allocation (RA) index center on the degree of common neighbors. For the global structure, it focuses on the path length between two nodes to capture more topological information. Katz index [13] considers the paths in a network for all lengths. It assigns similarity weight to each path. The longer the path, the lower the weight. LHN-II index [15] supposes that if the neighbor nodes of two nodes are similar, then there is a probability that a link between these two nodes. Topological similarity-based approaches have low computational complexity, but their prediction performance are restricted by their limitations, such as strong assumption, poor versatility, and information access restrictions.

Maximum Likelihood-Based Approaches. Maximum likelihood-based approaches assume that the network possesses its own organizational or formation mode, and then integrates existing link information to calculate a likelihood value for the network. The likelihood value increases as the network links approach their true distribution. The classical likelihood-based approaches comprise three methods: Hierarchical Structure Model (HSM) [7], Stochastic Block Model (SBM) [11], and Loop Model (LM) [20]. LM model outperforms both HSM and SBM in terms of prediction accuracy. The advantage of maximum likelihood-based approaches is its ability to comprehend network structure. However, it may struggle with large networks due to the inherent complexity in its own theory.

Representation Learning-Based Approaches. Representation Learning-based approaches utilize vector operations to compute node similarity, demonstrating a high level of expertise in link prediction task. In the initial stage, various embedding techniques such as Deepwalk [21], LINE [26], Node2Vec [10], and SDNE [28] were proposed to acquire a low-dimensional vector representation for each node in the graph structure. These embedding methods only capture the topology of network structure and fail to effectively integrate information from individual nodes. To this end, umerous graph-based methods have emerged. SEAL [33], VGAE [14], VGNAE [3], Neo-GNNS [32], PS2 [25], NNESF [9], and Meta-iKG [35] all focus on learning the structural feature of the network. However, TDGNN [22], TGAT [31], CNN-LSTM [24] and DNLP [6] take time series information into account to predict links that change over time. Representation learning-based approaches automatically extract valuable information from network structures, providing deeper insights and further improvements for link prediction.

3 Methodology

Problem Formalization. Given an undirected graph $\mathcal{G}(V, E, \mathbf{M}, \mathbf{A})$, $V = \{v_1, v_2, \cdots, v_N\}$ denotes the set of nodes referring to specific entities, and $E = \{e_1, e_2, \cdots, e_M\}, E \in \mathbb{R}^{N \times N}$ represents the set of edges carrying the relationships/interactions of nodes. $\mathbf{M} \in \mathbb{R}^{N \times N}$ is the node information matrix for

V. $\mathbf{A}$ is the adjacency matrix of graph $\mathcal{G}$. Then $a_{i,j}$ equals 1 if a link exists between v_i and v_j; otherwise, it is set to 0. The target of our link prediction problem is to predict the relationship between v_i and v_j and map an optical function f to train as Eq. (1).

$$f\left(v_i, v_j\right) = \begin{cases} 1, e_{i,j} \in E, \\ 0, \text{ Otherwise} \end{cases}. \tag{1}$$

3.1   The NLG-GNN Framework

As shown in Fig. 1, we propose an deep learning-based method, called NLG-GNN. It includes three parts and we will explain in the following subsections.

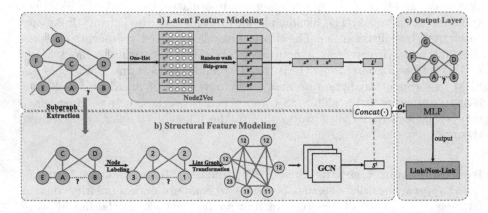

Fig. 1. Illustration of our NLG-GNN: a) Latent features modeling is devised by Node2Vec. b) Structural feature modeling is further divided into four steps: 1) A subgraph surrounding two target nodes is extracted. 2) Node labeling function assigns labels to each node to characterize the structural weight on the target link. 3) Line graph transformation converts the subgraph into its line graph, and 4) Structural representation learning is constructed by GCN. c) Output layer is to predict the probability of the target link.

a) Latent Feature Learning. The latent features of nodes refer to the unobservable characteristics of nodes in a graph, which need to be inferred through representation learning methods. We utilize Node2Vec to establish the latent features. Node2Vector utilizes a combination of depth-first search and breadth-first search to generate a node sequence via biased random walk, ultimately obtaining the vector representation of the node through Skip-Gram [17] model. We generate the latent feature vector of the link to be predicted by concatenating the latent feature vectors of node x and node y:

$$\mathbf{Emb} = \text{Node2Vec}\left(\mathbf{A}, l, n\right), \tag{2}$$

$$\mathbf{L}_{(x,y)} = \text{Concat}\left(\mathbf{Emb}_x, \mathbf{Emb}_y\right), \tag{3}$$

where l and n represent the length of the random work sequence, and the number of sequences. $Concat(\cdot)$ is a concatenation operation.

b) Structural Feature Modeling. Structural feature modeling is divided into four steps: first, we extract the surrounding closed subgraph of the target link. Then, we design a node labeling function that assigns labels to each node to characterize the structural weight on the target link. Next, we convert the extracted closed subgraph into its corresponding line graph. Finally, we construct GCN to learn the structural feature from this line graph, generating a feature vector S^i for the target link.

Subgraph Extraction and Node Labeling. We extract the subgraph from node a and node b and their neighbors within h hops, which is defined as $\mathcal{G}^h_{(a,b)}$, where (a,b) is the target link to be predicted. In this model, we extract the two-hop closed subgraph $\mathcal{G}^2_{(a,b)}$ for the target link from the network $\mathcal{G}$. Next, we assign a label to each node as its node information, and apply Eq. (4) to label each individual node i:

$$l(i) = 1 + \min(d_a, d_b) + (d/2)[(d/2) + (d\%2) - 1], \qquad (4)$$

where d_a and d_b denote the distances from node i to node a and b in the graph, respectively, with d being the sum of d_a and d_b. The label value represents the relative distance from each node to the target link, serving as a means of conveying structural information within the closed subgraph.

Line Graph Transformation. A line graph for a original graph $\mathcal{G} = \{V, E\}$ is described as $I(\mathcal{G}) = \{\tilde{V}, \tilde{E}\}$. In $I(\mathcal{G})$, the node of $I(\mathcal{G})$ matches an edge of $\mathcal{G}$. An edge exists between two nodes when the corresponding two edges in $\mathcal{G}$ have a common vertex. Firstly, the edge set E_{sub} of the closed subgraph $\mathcal{G}^h_{(a,b)}$ is taken as the node set $\tilde{V}$ of the line graph $I(\mathcal{G})$. The labels of corresponding nodes on $\mathcal{G}^h_{(a,b)}$ are concatenated to represent the new label $Label'_e$ for each node e in $\tilde{V}$. Then, a connection operation is performed for nodes in $\tilde{V}$ that share common vertices on graph $\mathcal{G}^h_{(a,b)}$ and their resulting edges are added to form edge set $\tilde{E}$. These steps complete the transformation from closed subgraph to line graph.

Structural Representation Learning. We employ GCN to learn the line graph and obtain the structural feature vector for the node of the line graph, extracting feature information from all neighbor nodes of each node. A node in the line graph refers to a link in the initial closed subgraph, so the embedding of the node is utilized as the structural feature vector S^i of the target link.

We input the adjacency matrix $\tilde{\mathbf{A}}$ of $I(\mathcal{G})$ and the node information matrix $\tilde{\mathbf{M}}$ to the GCN layer. Each row of $\tilde{\mathbf{M}}$ expresses a feature vector representing a node in the line graph, which includes the label $\mathbf{Label}'_{(a,b)}$ and original feature vectors of corresponding two nodes:

$$\mathbf{M}_{(a,b)} = Concat(\mathbf{Label}'_{(a,b)}, \mathbf{R}_a, \mathbf{R}_b), \qquad (5)$$

where $\mathbf{R}_a$ and $\mathbf{R}_b$ are the original feature vectors of node a and node b.

For a line graph $I(\mathcal{G})$, the node embedding of (a, b) at layer k in a GCN is represented by $\mathbf{h}_{(a,b)}^{(k)}$. Then the embedding at layer $k + 1$ is:

$$\mathbf{h}_{(a,b)}^{(k+1)} = \left(\mathbf{h}_{(a,b)}^{(k)} + \gamma \sum_{d \in \mathcal{P}_{(a,b)}} \mathbf{h}_d^{(k)} \right) \mathbf{R}^{(k)}, \tag{6}$$

where $\mathcal{P}_{(a,b)}$ is the set of neighbors of (a, b). γ is a normalized coefficient. $\mathbf{R}^{(k)}$ is the weight matrix in the k-th layer. The first layer in the GCN is initialized as $\mathbf{h}_{(a,b)}^{(0)} = \mathbf{M}_{(a,b)}$.

c) **Output Layer.** Based on latent feature learning module and structural feature learning module, we derive the potential feature vector L^i and structural feature vector S^i of the target link respectively. Then we concatenate them to obtain the final representation O^i for the target link (a, b). We employ MLP to map the representation vector O^i to the probability of the existence of the target link:

$$\mathbf{O}_{(a,b)} = Concat(\mathbf{S}_{(a,b)}, \mathbf{L}_{(a,b)}), \tag{7}$$

$$\mathbf{c}_{(a,b)} = ReLU(\mathbf{O}_{(a,b)}\mathbf{W}^1 + \mathbf{b}^1), \tag{8}$$

$$q_{(a,b)} = Softmax(\mathbf{s}_{(a,b)}\mathbf{W}^2 + \mathbf{b}^2), \tag{9}$$

where $\mathbf{s}_{(a,b)}$ represents the output of the first layer of MLP. $q_{(a,b)}$ denotes the final prediction result obtained from the second layer of MLP, which is equipped with ReLU and Softmax activation functions. $\mathbf{W}^1$ and $\mathbf{W}^2$ are weight matrices while $\mathbf{b}^1$ and $\mathbf{b}^2$ denote bias terms.

Our NLG-GNN employs binary cross-entropy minimization as the objective function and trains the model by minimizing the cross-entropy loss of target links in the training set:

$$\mathcal{L} = -\sum_{l \in L_t} (z_l \log (q_l) + (1 - z_l) \log (1 - q_l)), \tag{10}$$

where L_t represents the collection of objective links that are to be predicted. q_l denotes the probability of link l's existence and $z_l \in (0, 1)$ is a binary label, 1 and 0 indicate its existence or absence, respectively. When updating the model parameters, NLG-GNN model employs the Adam optimization method [8] for enhanced the performance.

4   Experimental Setup

Datasets. We evaluate the performance of our NLG-GNN on 17 publicly available network datasets from diverse domains, including YST, SMG, USAir[1],

---

[1] http://vlado.fmf.uni-lj.si/pub/networks/data/.

HPD, NS [19], KHN, GRQ, PB, NSC, LDG, Yeast [27], Power [29], ZWL, EML, C.ele [29], BUP, and ADV [18]. The effectiveness of the NLG-GNN model is fully validated through experiments. Specific statistics for these datasets are presented in Table 1.

Table 1. The statistics of datasets

Datasets	Area	#Nodes	#Links	Degree
LDG	Co-authorship	8,324	41,532	9.98
HPD	Biology	8,756	32,331	7.38
YST	Biology	2,284	6,646	5.82
C.ele	Biology	279	2,148	14.46
SMG	Co-authorship	1,024	4,916	9.60
USAir	Transportation	332	2,126	12.81
BUP	Political Blogs	105	441	8.40
PB	Political Blogs	1,222	16,714	27.36
ADV	Social Network	5,155	39,285	15.24
EML	Emails	1,133	5,451	9.62
ZWL	Co-authorship	6,651	54,182	16.29
NSC	Co-authorship	1,461	2,742	3.75
KHN	Co-authorship	3,772	12,718	6.74
NS	Co-authorship	1,589	2,742	3.45
GRQ	Co-authorship	5,241	14,484	5.53
Power	Power Network	4,941	6,594	2.67
Yeast	Biology	2,375	11,693	9.85

Baselines. In order to assess the predictive performance of our NLG-GNN model, we select 10 of the most representative link prediction methods for comprehensive comparative experiments, which are across three categories: (1) Topological similarity-based approaches: CN [4], Katz [13], PageRank (PR) [5], and SimRank (SR) [12], (2) Maximum Likelihood-based approaches: SBM [11], and (3) Representation Learning-based approaches: Node2Vec (N2V) [10], GAE [14], VGAE [14], VGNAE [3], SEAL [33], and NNESF [9]. Moreover, our models: NLG-GNN^{-N2V}, NLG-GNN^{-Lab}, and NLG-GNN^{-Lin}.

Implementation Details. For our NLG-GNN model, the number (n) and length (l) of sequences are set to 10 and 20, respectively. The dimension of embedding d_e is 16. Hops for the subgraph extraction is set to 2. For baseline models, the damping factor of Katz, PageRank, and SimRank are set to 0.001, 0.85 and 0.8, respectively. As for SEAL, the number (n) and length (l) of sequences are set to 10 and 80, respectively. The dimension of embedding d is

128. For GAE, VGAE, VGNAE, and SEAL models, we adopt the default parameter settings provided by their respective works. We utilize Average Precision (AP) and Area Under the Curve (AUC) as metrics for assessing the performance of link prediction problem.

4.1 Experimental Results

Performance Analysis. The comparative experiments between our NLG-GNN and 10 baseline methods are conducted on 17 datasets. The links are partitioned randomly into training and test sets, with an equal number of non-existent links generated as negative example. We select 80% of the links for training, and the results are shown in Tables 2. It is evident that similarity-based approaches such as CN, Katz, PR, and SR fail to perform well on all datasets due to their rigid assumptions and inability to handle diverse data types. The SBM method based on maximum likelihood exhibits the poorest overall performance while representation Learning-based approaches that can automatically learn effective network information generally outperform all baselines. The performance of Node2Vec exhibits relative stability across different datasets. However, it falls short of optimal efficacy due to its reliance only on latent feature. This limitation highlights the contribution of structural feature to our NLG-GNN model. GAE and VGAE are methods for analyzing network structural data, with VGAE performing better than GAE by applying KL divergence to prevent overfitting. Additionally, VGNAE incorporates L_2 normalization to address the

Table 2. Comparison of NLG-GNN with baselines for AUC and AP(%). The top two performances are distinguished by bold and underline, respectively.

Model	USAir		NS		PB		Yeast		Power		BUP		EML		SMG		NSC	
	AUC	AP	AUC	AP	AUC	AP	AUC	AP	AUC	AP	AUC	AP	AUC	AP	AUC	AP	AUC	AP
CN	93.12	92.64	93.74	93.19	91.82	90.34	90.31	89.28	55.61	58.09	90.00	86.90	79.50	78.07	78.13	77.39	97.49	97.45
Katz	92.21	93.61	94.03	94.37	92.66	92.17	91.84	94.03	60.39	73.55	87.26	87.31	88.78	90.00	85.97	88.48	97.42	98.32
PR	93.64	94.70	93.88	94.71	93.67	92.20	92.14	94.49	59.86	74.79	91.04	89.55	89.69	91.43	89.07	91.37	97.55	98.24
SR	79.16	69.96	94.07	93.75	76.52	64.94	91.44	92.98	71.36	70.05	86.34	82.35	87.03	87.19	78.15	71.06	95.69	95.74
SBM	70.60	71.15	51.79	51.42	58.85	57.52	51.20	51.19	50.22	50.21	58.48	58.48	51.03	50.87	56.48	54.98	55.40	55.71
N2V	86.20	81.88	90.42	94.24	85.53	85.03	93.14	94.49	70.07	75.62	80.20	82.70	84.15	82.36	77.83	77.45	96.33	96.01
GAE	89.90	92.39	91.82	94.81	91.45	92.26	93.03	95.01	66.28	70.69	89.82	90.08	87.12	87.29	79.40	84.27	96.79	97.40
VGAE	91.20	93.16	92.02	95.04	91.98	92.32	93.33	95.43	68.73	73.07	91.95	91.06	88.82	87.82	82.98	86.39	97.28	97.95
SEAL	<u>95.62</u>	95.13	97.33	97.17	<u>94.27</u>	<u>93.99</u>	96.82	<u>97.36</u>	<u>81.34</u>	<u>83.64</u>	92.69	<u>91.29</u>	<u>91.33</u>	<u>91.67</u>	89.59	89.27	<u>99.50</u>	<u>99.48</u>
VGNAE	91.26	94.08	92.00	96.73	91.98	92.86	95.39	96.45	74.58	80.77	<u>93.63</u>	91.22	88.82	89.85	84.67	90.35	98.76	98.84
NNESF	95.47	<u>95.73</u>	<u>97.40</u>	<u>97.70</u>	93.97	93.73	95.30	96.31	72.14	74.66	88.97	87.85	88.78	90.27	<u>90.17</u>	<u>90.47</u>	99.46	99.45
NLG-GNN	**97.52**	**96.83**	**98.32**	**98.65**	**94.87**	**94.74**	**97.39**	**97.90**	**82.37**	**85.02**	**95.55**	**92.10**	**91.75**	**92.04**	**91.72**	**92.36**	**99.85**	**99.57**
IMP	1.99	1.15	0.94	0.97	0.64	0.80	0.59	0.55	1.27	1.65	2.05	0.89	0.46	0.40	1.72	2.09	0.35	0.09

Model	KHN		GRQ		HPD		ADV		ZWL		LDG		YST		C.ele	
	AUC	AP	AUC	AP	AUC	AP	AUC	AP	AUC	AP	AUC	AP	AUC	AP	AUC	AP
CN	74.74	74.12	88.27	88.24	69.42	69.27	86.53	86.14	90.23	90.01	83.85	83.52	65.89	65.77	84.53	81.19
Katz	83.95	87.98	89.85	92.62	85.44	88.11	92.06	93.58	95.33	96.36	92.64	94.73	80.06	85.89	85.34	84.80
PR	88.88	91.75	89.96	93.33	87.43	91.1	92.29	94.11	97.19	97.30	93.76	96.17	80.95	85.72	89.56	87.68
SR	80.19	78.06	89.64	93.17	81.27	83.76	87.05	82.04	94.58	95.46	90.12	88.64	73.82	76.99	74.65	65.39
SBM	53.73	52.44	48.48	49.00	50.85	50.69	50.31	50.36	50.78	50.49	51.38	50.90	51.91	51.28	60.86	60.19
N2V	82.63	83.28	91.63	93.79	80.21	81.43	77.15	77.98	94.17	94.31	91.48	91.83	77.24	80.36	80.12	78.39
GAE	82.83	88.25	90.35	93.35	82.57	87.19	87.64	90.59	95.82	96.01	93.28	94.92	76.14	82.44	80.59	81.88
VGAE	83.03	88.73	90.8	94.52	82.79	87.91	87.9	91.55	95.85	96.28	93.94	95.43	80.32	84.77	82.43	82.06
SEAL	<u>92.53</u>	<u>93.08</u>	<u>97.47</u>	<u>97.94</u>	<u>92.24</u>	<u>93.51</u>	95.00	95.13	<u>97.44</u>	<u>97.52</u>	<u>96.45</u>	<u>96.61</u>	<u>90.81</u>	<u>91.80</u>	<u>87.46</u>	<u>86.09</u>
VGNAE	85.08	92.03	90.8	95.87	84.86	90.17	89.9	92.76	96.31	96.46	95.21	96.22	80.32	87.70	86.41	83.95
NNESF	91.15	92.50	96.79	97.46	90.01	91.21	**95.10**	**95.36**	95.53	96.39	94.63	95.60	89.69	90.28	85.25	84.73
NLG-GNN	**93.72**	**94.07**	**97.73**	**98.10**	**92.73**	**93.63**	<u>95.00</u>	<u>95.27</u>	**97.8**	**97.73**	**97.23**	**96.98**	**92.02**	**92.84**	**90.06**	**88.13**
IMP	1.29	1.06	0.27	0.16	0.53	0.13	-0.11	-0.09	0.37	0.22	0.81	0.38	1.33	1.13	2.97	2.37

issue of indistinguishable isolated node embeddings in VGAE, resulting in superior outcomes compared to VGAE. SEAL automatically learns the distribution of links from the dataset and achieves the optical performance among the baseline models. However, SEAL exhibits poor performance on two datasets BUP and SMG, indicating its limited capability in handling sparse graphs. Our NLG-GNN model outperforms NNESF on 16 datasets (including dense and sparse graphs), except for ADV, as it leverages both the latent and structural features within the network while aggregating link information from two nodes into a line graph node to prevent loss of effective features during learning. Consequently, NLG-GNN can learn more effective feature vectors for representing links to be predicted, resulting in superior performance stability compared to other baseline methods.

Table 3. Comparison of NLG-GNN with its variants for ACU and AP(%).

Model	USAir		NS		PB		Yeast		Power		C.ele	
	AUC	AP	AUC	AP	AUC	AP	AUC	AP	AUC	AP	AUC	AP
NLG-GNN-N2V	95.18	95.05	98.41	98.30	94.83	93.48	97.10	97.33	81.65	97.33	88.33	86.20
NLG-GNN-Lab	89.17	89.50	56.86	58.69	86.41	81.81	80.68	76.67	51.27	76.67	72.56	76.32
NLG-GNN-Lin	95.39	95.01	98.30	98.52	94.12	92.10	97.01	97.00	81.06	97.00	87.39	86.60
NLG-GNN	**97.52**	**96.83**	**98.52**	**98.55**	**94.87**	**93.74**	**97.39**	**97.90**	**82.37**	**97.90**	**90.06**	**88.13**

For all datasets, 10%–90% of the links are sampled as positive training sets, while the remaining are positive test sets. An equal number of negative links are sampled for both training and test sets. Four representation learning-based methods (i.e., NLG-GNN, SEAL, VGNAE and VGAE) are compared by AUC on six datasets to evaluate their performance. Experimental results are presented

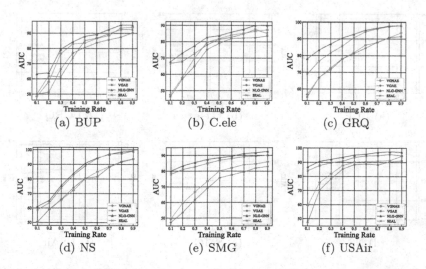

(a) BUP	(b) C.ele	(c) GRQ
(d) NS	(e) SMG	(f) USAir

Fig. 2. AUC evaluation with varying proportions of the training dataset on six datasets.

in Fig. 2. From the results, our proposed NLG-GNN surpasses three state-of-the-art baseline methods with varying amounts of training data, demonstrating the insensitivity of NLG-GNN to experimental data quantity.

Ablation Study. We perform the ablation study of NLG-GNN on six datasets. From the results in Table 3, NLG-GNN is markedly superior to other variants. Specifically, when we remove the node labeling module (i.e., NLG-GNN^{-Lab}), AUC and AP exhibit a significant decline, which validates that effective label information is crucial for characterizing the structural feature of closed subgraphs during feature extraction. In addition, when we eliminate the latent feature learning and line graph transformation of NLG-GNN, (i.e., NLG-GNN^{-N2V} and NLG-GNN^{-Lin}, respectively), AUC and AP decrease to a varying degree, indicating the effectiveness of these two modules.

Parameter Analysis. We conducted experiments to select the optimal parameters for the length of random walk sequences and dimension of feature vectors

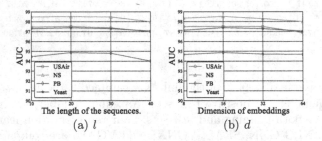

(a) l (b) d

Fig. 3. Parameter analysis on USAir, NS, PB and Yeast datasets

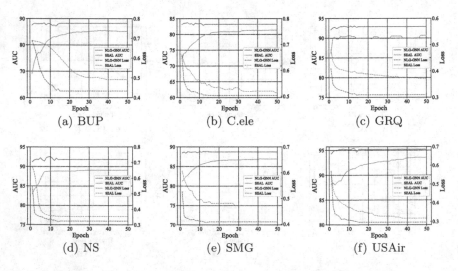

(a) BUP   (b) C.ele   (c) GRQ

(d) NS   (e) SMG   (f) USAir

Fig. 4. Comparison of model loss and AUC between our NLG-GNN and SEAL method on six datasets.

on USAir, NS, PB and Yeast datasets. Figure 3 indicates that the best AUC was achieved when l and d are 20 and 16, respectively.

Convergence Speed Analysis. To analyze the convergence speed of the model, we trained NLG-GNN and SEAL on different datasets with 50% training data. We record loss and AUC values for each epoch during the training process, as depicted in Fig. 4. Our NLG-GNN exhibits faster convergence than SEAL, requiring only 10 to 15 epochs for optimal performance, while the latter necessitates 50 epochs of training to converge. Therefore, NLG-GNN possesses the capability to decrease training duration and enhance productivity.

5 Conclusion

We propose a novel representation learning-based method, called NLG-GNN, which fuses the representation of latent and structural features. First, based on Node2Vector, we learn the latent feature about the network. Next, based on line graph neural networks, we consolidate dispersed features from two nodes into a single node to minimize information loss. Finally, we combine the latent and structural feature vector to predict positive links via a two-layer MLP. Extensive experiments on 17 datasets demonstrate that the performance of our proposed NLG-GNN surpasses that of other baseline methods. Moreover, NLG-GNN exhibits robustness to variations in dataset and maintains optimal performance across different proportions of training data. In addition, our model has fewer parameters and faster convergence than other baseline methods. For our future research, we will focus on exploring the effective representation of links in heterogeneous networks and take the interpretability of model into account.

Acknowledgements. This work was supported in part by the Joint Funds of the National Natural Science Foundation of China (Grant No. U22A2036), the National Key Research and Development Program of China (2020YFB1406902), the Key-Area Research and Development Program of Guangdong Province (2020B0101360001), and the GHfund C (20220203, ghfund202202033706).

References

1. Adamic, L.A., Adar, E.: Friends and neighbors on the web. Soc. Netw. **25**(3), 211–230 (2003)
2. Agibetov, A.: Neural graph embeddings as explicit low-rank matrix factorization for link prediction. Pattern Recogn. **133**, 108977 (2023)
3. Ahn, S.J., Kim, M.: Variational graph normalized autoencoders. In: CIKM 2021, pp. 2827–2831 (2021)
4. Barabási, A.L., Albert, R.: Emergence of scaling in random networks. Science **286**(5439), 509–512 (1999)
5. Brin, S., Page, L.: The anatomy of a large-scale hypertextual web search engine. Comput. Netw. **30**(1–7), 107–117 (1998)

6. Chen, J., Zhang, J., Chen, Z., Du, M., Xuan, Q.: Time-aware gradient attack on dynamic network link prediction. IEEE Trans. Knowl. Data Eng. **35**(02), 2091–2102 (2023)
7. Clauset, A., Moore, C., Newman, M.E.: Hierarchical structure and the prediction of missing links in networks. Nature **453**(7191), 98–101 (2008)
8. Duchi, J., Hazan, E., Singer, Y.: Adaptive subgradient methods for online learning and stochastic optimization. J. Mach. Learn. Res. **12**(7), 2121–2159 (2011)
9. Fang, Z., Tan, S., Wang, Y., Lu, J.: Elementary subgraph features for link prediction with neural networks. IEEE Trans. Knowl. Data Eng. **35**(04), 3822–3831 (2023)
10. Grover, A., Leskovec, J.: node2vec: scalable feature learning for networks. In: SIGKDD, pp. 855–864 (2016)
11. Guimerà, R., Sales-Pardo, M.: Missing and spurious interactions and the reconstruction of complex networks. Proc. Natl. Acad. Sci. U.S.A. **106**(52), 22073–22078 (2009)
12. Jeh, G., Widom, J.: Simrank: a measure of structural-context similarity. In: SIGKDD, pp. 538–543 (2002)
13. Katz, L.: A new status index derived from sociometric analysis. Psychometrika **18**(1), 39–43 (1953)
14. Kipf, T.N., Welling, M.: Variational graph auto-encoders. arXiv preprint arXiv:1611.07308 (2016)
15. Leicht, E.A., Holme, P., Newman, M.E.: Vertex similarity in networks. Phys. Rev. E **73**(2), 026120 (2006)
16. Li, M., et al.: Heterologous expression of arabidopsis thaliana rty gene in strawberry (fragaria× ananassa duch.) improves drought tolerance. BMC Plant Biol. **21**(1), 1–20 (2021)
17. Mikolov, T., Chen, K., Corrado, G., Dean, J.: Efficient estimation of word representations in vector space. arXiv preprint arXiv:1301.3781 (2013)
18. Newman, M.E.: The structure of scientific collaboration networks. Proc. Natl. Acad. Sci. U.S.A. **98**(2), 404–409 (2001)
19. Newman, M.E.: Finding community structure in networks using the eigenvectors of matrices. Phys. Rev. E **74**(3), 036104 (2006)
20. Pan, L., Zhou, T., Lü, L., Hu, C.K.: Predicting missing links and identifying spurious links via likelihood analysis. Sci. Rep. **6**(1), 22955 (2016)
21. Perozzi, B., Al-Rfou, R., Skiena, S.: Deepwalk: online learning of social representations. In: SIGKDD, pp. 701–710 (2014)
22. Qu, L., Zhu, H., Duan, Q., Shi, Y.: Continuous-time link prediction via temporal dependent graph neural network. In: WWW 2020, pp. 3026–3032 (2020)
23. Ren, X., Bai, L., Xiao, Q., Meng, X.: Hierarchical self-attention embedding for temporal knowledge graph completion. In: WWW 2023, pp. 2539–2547 (2023)
24. Selvarajah, K., Ragunathan, K., Kobti, Z., Kargar, M.: Dynamic network link prediction by learning effective subgraphs using CNN-LSTM. In: IJCNN 2020, pp. 1–8. IEEE (2020)
25. Tan, Q., et al.: Bring your own view: graph neural networks for link prediction with personalized subgraph selection. In: WSDM 2023, pp. 625–633 (2023)
26. Tang, J., Qu, M., Wang, M., Zhang, M., Yan, J., Mei, Q.: Line: large-scale information network embedding. In: WWW 2015, pp. 1067–1077 (2015)
27. Von Mering, C., et al.: Comparative assessment of large-scale data sets of protein-protein interactions. Nature **417**(6887), 399–403 (2002)
28. Wang, D., Cui, P., Zhu, W.: Structural deep network embedding. In: SIGKDD, pp. 1225–1234 (2016)

29. Watts, D.J., Strogatz, S.H.: Collective dynamics of small-world networks. Nature **393**(6684), 440–442 (1998)
30. Wu, X., He, H., Yang, H., Tai, Y., Wang, Z., Zhang, W.: PDA-GNN: propagation-depth-aware graph neural networks for recommendation. In: World Wide Web, pp. 1–22 (2023)
31. Xu, D., Ruan, C., Korpeoglu, E., Kumar, S., Achan, K.: Inductive representation learning on temporal graphs. arXiv preprint arXiv:2002.07962 (2020)
32. Yun, S., Kim, S., Lee, J., Kang, J., Kim, H.J.: NEO-GNNS: neighborhood overlap-aware graph neural networks for link prediction. Proc. Adv. Neural Inf. **34**, 13683–13694 (2021)
33. Zhang, M., Chen, Y.: Link prediction based on graph neural networks. Proc. Adv. Neural Inf. **31**, 1–11 (2018)
34. Zhang, Z., Sun, S., Ma, G., Zhong, C.: Line graph contrastive learning for link prediction. Pattern Recogn. **140**, 109537 (2023)
35. Zheng, S., Mai, S., Sun, Y., Hu, H., Yang, Y.: Subgraph-aware few-shot inductive link prediction via meta-learning. IEEE Trans. Knowl. Data Eng. **35**(6), 6512–6517 (2023)

Event Sparse Net: Sparse Dynamic Graph Multi-representation Learning with Temporal Attention for Event-Based Data

Dan Li[1], Teng Huang[1(✉)], Jie Hong[1], Yile Hong[1], Jiaqi Wang[1], Zhen Wang[2], and Xi Zhang[3,4(✉)]

[1] Institute of Artificial Intelligence and Blockchain Guangzhou University, Guangzhou, China
huangteng1220@gzhu.edu.cn
[2] Zhejiang Lab Kechuang Avenue, Zhongtai Sub-District, Yuhang District, Hangzhou, Zhejiang, China
[3] School of Arts, Sun Yat-sen University, Guangzhou, China
[4] College of Music, University of Colorado Boulder, Boulder, USA
xizhangpiano@gmail.com

Abstract. Graph structure data has seen widespread utilization in modeling and learning representations, with dynamic graph neural networks being a popular choice. However, existing approaches to dynamic representation learning suffer from either discrete learning, leading to the loss of temporal information, or continuous learning, which entails significant computational burdens. Regarding these issues, we propose an innovative dynamic graph neural network called Event Sparse Net (ESN). By encoding time information adaptively as snapshots and there is an identical amount of temporal structure in each snapshot, our approach achieves continuous and precise time encoding while avoiding potential information loss in snapshot-based methods. Additionally, we introduce a lightweight module, namely Global Temporal Attention, for computing node representations based on temporal dynamics and structural neighborhoods. By simplifying the fully-connected attention fusion, our approach significantly reduces computational costs compared to the currently best-performing methods. We assess our methodology on four continuous/discrete graph datasets for link prediction to assess its effectiveness. In comparison experiments with top-notch baseline models, ESN achieves competitive performance with faster inference speed.

Keywords: dynamic graph representations · self-attention mechanism · light sparse temporal model · link prediction

Supported by the National Natural Science Foundation of China under Grant 62002074 and 62072452; Supported by the Shenzhen Science and Technology Program JCYJ20200109115627045, in part by the Regional Joint Fund of Guangdong under Grant 2021B1515120011.

1   Introduction

The capability of dynamic graph neural networks(DGNNs) to effectively grasp intricate temporal connections and interactions has led to significant interest. By introducing an additional temporal dimension to accumulate variations in embeddings or representations, DGNNs have become powerful tools in diverse domains such as dynamic social networks [1], bioinformatics [2], neuroscience [3], interconnected protein networks [4], smart recommendation engines [5,6], visual information recognition [7–9], remote sensing [10–12], cybersecurity [13,14].

To address the challenges posed by time-varying graphs, preprocessing raw dynamic graph representations is crucial. These representations capture the continuous evolution of the graph, including events such as node emergence or disappearance and link addition/deletion [15–19]. Previous studies [20–29] have categorized these raw representations into two primary segments: dynamic continuous graphs and dynamic discrete graphs. Dynamic continuous projects of the original data form onto a 2D graph with temporal dimension, maximizing the information about graph evolution. However, these continuous networks are computationally complex [19]. On the other hand, sampling snapshots of graphs Periodically allows discrete graphs to capture structural representations, resulting in simpler evolving networks but with a loss of temporal information [15]. Our objective is to discover an efficient method for encoding raw dynamic graph representations, reducing the loss of temporal information, and simplifying the representation learning process.

Extracting and analyzing temporal graph learning patterns is the main objective of DGNNs on refined dynamic temporal graphs. An efficient and powerful network is crucial for this task. Some studies [21,25,26] analyze features in dynamic graph sequences by recurrent neural networks (RNNs). Nevertheless, DGNNs that utilize RNNs as the underlying framework are time-intensive and obtain terrible results with large time steps. Transformer-based approaches [27,28,30] provide adaptive and interpretable attention mechanisms over past embeddings, surpassing RNN-based DGNNs in long-term analysis. Nevertheless, the standard transformer [31] with fully-connected attention requires high computational resources for time-dependent sequences [32]. Our goal is to simplify the conveyance of temporal information along the time dimension while achieving satisfactory performance in link prediction tasks. To accomplish this, we propose a lightweight module named Global Temporal Attention (GTA). The GTA module efficiently computes temporal information using simplified sparse attention conjunctions for both graph tasks.

The general configuration of ESN's architecture is depicted in Fig. 1, which comprises two key components: Local Self Attention (LSA) for analyzing local structural patterns within snapshots, and GTA for capturing global temporal patterns.

The novel model we propose was assessed by conducting research on dual continuous datasets, with a specific emphasis on inductive link prediction tasks. The experimental results provide evidence that ESN exhibits better performance when applied to continuous graph data compared to current networks. Additionally, we introduced a simplified version of ESN that utilizes only LSA and GTA. This abbreviated version was employed to learn representations of discrete datasets, tackling not only the inductive link prediction task but also the transductive link prediction task. Further experiments with this simplified version demonstrate that it not only achieves superior accuracy and efficiency but also exhibits accelerated performance involving four separate datasets that specifically pertain to dynamic graphs. Summarizing the contributions of this paper:

- The establishment of the ESN framework is aimed at achieving an optimal effect balance between efficiency and accuracy when working with dynamic continuous/discrete datasets.
- We propose GTA, a lightweight temporal self-attention module. ESN, based on GTA, significantly reduces computational requirements compared to the solutions based on RNN and transformer when applied to continuous dynamic graph datasets.
- The simplified ESN model, featuring only LSA and GTA, additionally has the potential to be applied to discrete dynamic graph datasets. Our experiments indicate that ESN outperforms SOTA approaches in inductive/transductive link prediction tasks.

To unveil our findings regarding dynamic graph datasets and networks, this paper has been structured into several distinct sections. Firstly, Sect. 2 offers an exhaustive analysis of the latest research that has been carried out in this domain. Subsequently, in Sect. 3, we present an intricate mathematical elucidation of ESN. Moving on to Sect. 4, we provide a comprehensive breakdown of our experimental design and results. Finally, readers can find our concluding remarks in Sect. 5.

2   Related Work

2.1   Graph Representations

Graph representations encompass both static and dynamic aspects, with the latter incorporating time as a critical parameter. In the realm of dynamic graphs, the raw representation captures essential information, including node interactions [33] and timestamped edges [15], reflecting events like the creation and removal of nodes and links. Research efforts in graph representation learning highlight two specific areas: dynamic continuous and dynamic discrete graphs [15–17,19]. Dynamic continuous graphs project raw representations onto a 2D graph of time, providing a comprehensive view of the graph's evolution. In contrast, dynamic discrete graphs group graph embeddings based on temporal granularity [16,18], offering snapshots of the graph across distinct temporal intervals [34]. However, dynamic continuous graphs tend to be computationally complex, while dynamic discrete graphs may suffer from information loss and imbalanced temporal information distribution. Thus, finding an optimal balance between computational efficiency and preserving temporal information remains a key challenge in representation learning for dynamic graphs.

2.2 Dynamic Graph Neural Network

RNN-based DGNNs [22,35] summarize temporal information but are computationally expensive and scale poorly on long sequences [27]. Transformer-based DGNNs [27,28] handle temporal information along the time dimension but suffer from high computational costs. In our approach, ESN, we introduce a lightweight module called GTA. GTA reduces inference time by exchanging information only among 1-hop neighbors and the relay node. Experimental results show that ESN outperforms existing approaches in both continuous and discrete representations. Our aim is to achieve an optimal trade-off between accuracy and efficiency in dynamic graph representation learning.

3 Methods

ESN comprises two distinct components: LSA and GTA in Fig. 1. Given a dynamic graph composed of continuous time periods, LSA is used to extract the local features of the dynamic graph for further capturing the correlation between different graphs at different time states. This is achieved by assembling self-attention mechanisms on each time block. Since dynamic graphs contain information along the temporal dimension, a specially designed lightweight GTA architecture is employed to capture unique features along the temporal dimension and refine global semantics.

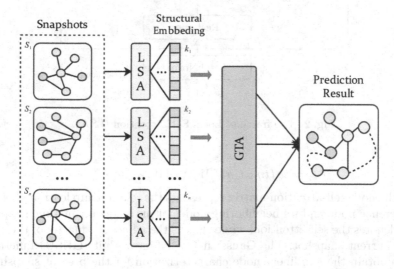

Fig. 1. The overall architecture of ESN consists of two main parts: Local Self Attention (LSA) and Global Temporal Attention (GTA).

3.1   Local Self Attention

By encoding the original input f_{in}, N time-dependent snapshots are obtained and each snapshot is subjected to local feature extraction using LSA in Fig. 2. The snapshot is reshaped to obtain node representations X_m and X_n for the reference node m and its neighboring nodes n respectively, which serve as input data for LSA. Node representations are multiplied by the corresponding snapshot weight matrix W_s and an exponential linear unit (ELU) activation function σ is applied to speed up training and then processed by self-attention calculation, as shown in Eq. 1.

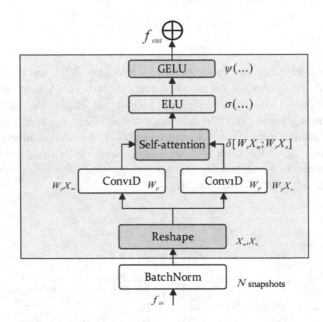

Fig. 2. The frame of Local Self Attention (LSA).

$$\beta_{mn} = softmax(\sigma(\delta[W_s X_m; W_s X_n] \cdot A_{mn})) \tag{1}$$

where the node self-attention matrix represents the relationship learning between the reference node and its neighboring nodes at an independent time snapshot. The δ denotes the self-attention, while A_{mn} is the adjacency matrix of the graph for the current snapshot. The Gaussian Error Linear Unit (GELU) function is used to obtain the non-linear node characterization for the present snapshot in Eq. 2.

$$SE = \psi \left(\sum_{n \subseteq N_m} \beta_{mn} \cdot W_s X_n \right) \tag{2}$$

where SE is structural embeddings of current snapshot. ψ indicates the GELU function for the output representations of LSA.

LSA adds position embedding to each snapshot. As a dynamic graph composed of a continuous period of time, there is a certain temporal connection between each node in the graph. However, since local semantics are difficult to obtain global information, to facilitate the acquisition of global semantic information in the next step, the paper chooses to add temporal positional information to LSA.

3.2 Global Temporal Attention

In order to obtain global structural changes of each snapshot over time, further global features are obtained based on local features. The transformer-based computing mechanism is proposed to integrate them using fully connected attention, which can achieve good accuracy on different dynamics (discrete or continuous). Yet, this computational approach utilizes a large amount of processing capabilities, which limits its widespread application. This article proposes a lightweight architecture in Fig. 3, GTA, which significantly reduces the computation time required for the model and effectively handles global features in the temporal dimension.

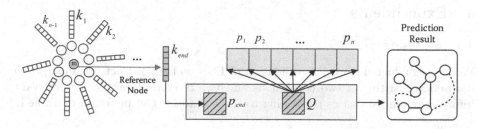

Fig. 3. The frame of Global Temporal Attention (GTA).

The GTA is composed of a series of snapshots and separate relay snapshots on time dimensions. Only a portion of nodes in each snapshot is used for testing, while others are used for training and updating the model. The relay snapshot and other snapshots are connected through the self-attention mechanism, and all snapshots are aggregated and distributed, which also determines the ability of the relay snapshot to learn global representation. Therefore, our proposed lightweight architecture, GTA, can reduce frequent access to each snapshot in spatial terms, thereby achieving the goal of integrating global information.

From Fig. 3, it can be observed that the reference node is represented as a series of inputs to the model across all time blocks. Additionally, the values across all time blocks are averaged to obtain Q, which is used to initialize the relay snapshot. To update the contextual information of p_i, the following information is required as shown in Eq. 3.

$$c_i^t = [z_{i-1}^{t-1}; z_i^{t-1}; z_{i+1}^{t-1}; p_i; m^{t-1}] \tag{3}$$

where c_i^t refers to the contextual information of the i-th snapshot, z_{i-1}^{t-1} and z_{i+1}^{t-1} refer to the update information of the adjacent neighbors of the i-th snapshot, p_i refers to its own embedding, z_i^{t-1} represents the final time step's self-state, and m^{t-1} denotes the update information of the relay snapshot.

The current snapshot's temporal state z_i^t and the relay snapshot's state m_t are processed through the self-attention mechanism in the model. Subsequently, both states undergo a softmax operation defined by Eq. 4 and Eq. 5. This results in a nonlinear data representation of the current snapshot.

$$\eta_i^t = softmax \left(\frac{z_i^{t-1} W_q \cdot (c_i^t W_k)^T}{\sqrt{d}} \right) \tag{4}$$

$$\theta^t = softmax \left(\frac{r^{t-1} W_q' \cdot ([r^{t-1}; Z^t] W_k')^T}{\sqrt{d'}} \right) \tag{5}$$

where η_i^t and θ^t are the self-attention coefficients of the current snapshot and relay snapshot, $W_q, W_k, W_i, W_q', W_k', W_i'$ are learning parameters, and the feature dimension are d, d'. Z^t is temporal snapshot, r^t is relay snapshot.

4   Experiments

4.1   Datasets

We have conducted empirical validation of ESN on four distinct dynamic graph datasets, consisting of two continuous and two discrete datasets. The relevant information and statistics pertaining to these datasets can be found in Table 1.

Table 1. Four distinct dynamic graph datasets.

Dataset	Link	Node	Time Consumption(s)/Time Steps
Reddit_{Con}	672447	10984	2678390 s
Wikipedia_{Con}	157474	9227	2678373 s
ML-10M_{Dis}	43760	20537	13
Enron_{Dis}	2347	143	16

Reddit and Wikipedia. The datasets presented in [25] showcase the activity of users and their editorial contributions to Reddit and Wikipedia within a single month. The data is represented through continuously evolving graphs, with dynamic labels reflecting the users' current state on their respective platforms. It is worth mentioning that the dataset of Reddit includes 10,984 nodes combined with a staggering 672,447 links. In contrast, Wikipedia's dataset is composed of 9,227 nodes and a modest 157,474 links.

Enron and ML-10M. The pair of discrete graph datasets presented herein pertain to the realm of network communications. Enron, for instance, encompasses a total of 143 nodes corresponding to individual employees, with a staggering 2,347 links denoting email interactions between them. On the other hand, ML-10M is comprised of a significantly larger dataset featuring a grand sum of 20,537 nodes that correspond to users holding distinct tags, with 43,760 interconnective links representing their respective interactions. [36]

4.2 Setup

In the beginning stage of the training, [37] is employed to instigate initialization of the learnable parameters W for each layer as a preventive measure against sudden vanishing or exploding of the gradient. GELU culminates in generating the conclusive non-linear representations that stem from GSA. ELU is an activation function for both temporal snapshots and relays within GTA. By utilizing binary cross-entropy loss, a proficient dissemination of probabilities is attained for anticipated link predictions, spanning inference, and transductive projects. The count of adaptable time-related snapshots varies, contingent on the specific dataset and task at hand, with Reddit and Wikipedia featuring 16 and 12 of them, respectively. We also use dropout in the range of 0.3 to 0.7 to avoid over-fitting.

4.3 Continuous Data Inductive Learning

We conducted inductive learning on four models - Const-TGAT, TGAT, ESN*, and ESN - using two dynamic continuous graph datasets, Reddit and Wikipedia in Table 2. Evidently, ESN* obtained a high accuracy of 87.55% and demonstrated the shortest inference time of 20.36 s on the Reddit dataset. On the Wikipedia dataset, it achieved an even lower inference time of 11.35 s. The TGAT model utilized a time attention coefficient matrix to aggregate time representations, which resulted in significantly higher accuracy in prediction tasks compared to Const-TGAT, as well as substantial reductions in inference time on both the Reddit and Wikipedia datasets. Therefore, our results further demonstrate that our GTA module can greatly improve model inference speed while maintaining accuracy and efficiently implementing downstream tasks.

Table 2. The link prediction results (accuracy%) of inductive learning task on dynamic continuous graph datasets. Left: Accuracy of inductive learning); Right: Inference time(s). ESN* combines GTA and the time encoding component of TGAT.

Dataset	Const-TGAT	TGAT	ESN*	ESN	TGAT	ESN*	ESN
Reddit	78.26	85.63	**88.74**	87.22	37.45 s	25.62 s	**20.36 s**
Wikipedia	73.19	**81.77**	80.23	75.31	19.22 s	18.11 s	**11.35 s**

By introducing the GTA module to modify TGAT and represent it using ESN*, we presented the results in Table 2. Since TGAT can perform functional time encoding and standard conversion, the GTA module replaces the converter function after modification. As shown in the data from Table 2, the modified ESN* achieved a high prediction accuracy of 88.74% and 80.23% on the Reddit and Wikipedia datasets, respectively, while also reducing inference time by 11.83 s and 1.11 s, further demonstrating the validity of the GTA module. Compared to ESN, TGAT had a higher prediction accuracy of 6.46% on the Wikipedia dataset, but the ESN model was faster by 7.87 s in terms of inference speed. Therefore, our ESN model is more popular in practical scenarios.

4.4   Discrete Data Inductive Learning

Table 3 employs two discrete datasets, Enron and ML-10M, to conduct inductive learning experiments on models such as GAT, DynGEM, DynAERNN, and ESN. Obviously, the ESN model achieves optimal experimental results on both datasets, with a link prediction accuracy of up to 90.26% on Inductive learning, exceeding that of the DynAERNN model by a maximum of 22.41%. The experimental results on the ML-10M dataset are generally better than those on the Enron dataset, as the former has significantly more links and nodes. On both datasets, the Inductive experimental results of the DynGEM model are respectively 19.43% and 10.94% lower than those of our ESN model. As opposed to ESN, which leverages the transform's fully-connected attention conjunction architecture to extract temporal patterns over time, DynGEM employs dense layers and recurrent layers in its analysis of time graphs' evolution. Not only is the model lighter in magnitude, but its accuracy is also higher. Therefore, the GTA structure of ESN has certain effectiveness and competitiveness in inductive learning on discrete datasets.

Table 3. The link prediction results (accuracy %) of inductive/transductive learning task on discrete graph datasets.

Dataset	DynAERNN	DynGEM	GAT	DySAT	Event Sparse Net
Enron_{Ind}	57.43	60.41	66.54	76.24	**79.84**
ML-10M_{Ind}	86.72	79.32	79.63	88.53	**90.26**
Enron_{Tran}	70.12	65.42	73.61	78.25	**84.06**
ML-10M_{Tran}	86.29	81.70	84.93	93.10	**94.28**

4.5   Discrete Data Transductive Learning

We also conducted link prediction tasks under transductive learning, using the same discrete dataset and baselines. As shown in Table 3, with the ML-10M dataset, the ESN model and DySAT model based on the Transformer achieved better link prediction accuracy than the model based on RNN through transductive learning. Among them, the accuracy of the Transformer-based model reached 94.28%, while the prediction accuracy of the RNN-based model was 81.70%, which is 12.58% lower than the highest accuracy. Therefore, this indicates that the self-attention mechanism has a relatively good performance in transductive learning.

5   Conclusion

Throughout this article, we offer a thorough analysis of ESN, a novel framework designed to handle link prediction tasks with high accuracy and efficiency through inductive and transductive learning. The model comprises two modules, LSA and GTA, with the former responsible for learning local structural representations of each encoding snapshot in relation to the temporal dimension, while the lightweight GTA extracts global features over time. By leveraging the advantages of a lightweight architecture, our approach based on GTA significantly reduces computational complexity when processing continuous dynamic graph datasets compared to RNN and Transformer-based solutions, while maintaining accuracy. To assess the effectiveness of the solution we propose, we conducted experiments on four dynamic continuous/discrete graph datasets. Results demonstrate that ESN achieves competitive performance in both inference speed and accuracy.

References

1. Zhu, T., Li, J., Hu, X., Xiong, P., Zhou, W.: The dynamic privacy-preserving mechanisms for online dynamic social networks. IEEE Trans. Knowl. Data Eng. **34**(06), 2962–2974 (2022)
2. Grover, A., Leskovec, J.: node2vec: scalable feature learning for networks. In: Proceedings of the 22nd ACM SIGKDD International Conference on Knowledge Discovery and Data Mining, pp. 855–864 (2016)
3. Goering, S., Klein, E.: Fostering neuroethics integration with neuroscience in the brain initiative: comments on the nih neuroethics roadmap. AJOB Neurosci. **11**(3), 184–188 (2020)
4. Fout, A.M.: Protein interface prediction using graph convolutional networks. Ph.D. dissertation, Colorado State University (2017)
5. Jiang, L., Cheng, Y., Yang, L., Li, J., Yan, H., Wang, X.: A trust-based collaborative filtering algorithm for e-commerce recommendation system. J. Ambient. Intell. Humaniz. Comput. **10**(8), 3023–3034 (2019)
6. Ying, R., He, R., Chen, K., Eksombatchai, P., Hamilton, W.L., Leskovec, J.: Graph convolutional neural networks for web-scale recommender systems. In: Proceedings of the 24th ACM SIGKDD International Conference on Knowledge Discovery & Data Mining, pp. 974–983 (2018)
7. Chen, C., Huang, T.: Camdar-adv: generating adversarial patches on 3d object. Int. J. Intell. Syst. **36**(3), 1441–1453 (2021)

8. Shi, Z., Chang, C., Chen, H., Du, X., Zhang, H.: PR-NET: progressively-refined neural network for image manipulation localization. Int. J. Intell. Syst. **37**(5), 3166–3188 (2022)
9. Yan, H., Chen, M., Hu, L., Jia, C.: Secure video retrieval using image query on an untrusted cloud. Appl. Soft Comput. **97**, 106782 (2020)
10. Ai, S., Koe, A.S.V., Huang, T.: Adversarial perturbation in remote sensing image recognition. Appl. Soft Comput. **105**, 107252 (2021)
11. Wang, X., Li, J., Li, J., Yan, H.: Multilevel similarity model for high-resolution remote sensing image registration. Inf. Sci. **505**, 294–305 (2019)
12. Wang, X., Li, J., Yan, H.: An improved anti-quantum mst3 public key encryption scheme for remote sensing images. Enterp. Inf. Syst. **15**(4), 530–544 (2021)
13. Li, J., et al.: Efficient and secure outsourcing of differentially private data publishing with multiple evaluators. IEEE Trans. Depend. Secure Comput. **19**(01), 67–76 (2022)
14. Yan, H., Hu, L., Xiang, X., Liu, Z., Yuan, X.: PPCL: privacy-preserving collaborative learning for mitigating indirect information leakage. Inf. Sci. **548**, 423–437 (2021)
15. Barros, C.D., Mendonça, M.R., Vieira, A.B., Ziviani, A.: A survey on embedding dynamic graphs. ACM Comput. Surv. (CSUR) **55**(1), 1–37 (2021)
16. Cai, H., Zheng, V.W., Chang, K.C.-C.: A comprehensive survey of graph embedding: problems, techniques, and applications. IEEE Trans. Knowl. Data Eng. **30**(9), 1616–1637 (2018)
17. Cui, P., Wang, X., Pei, J., Zhu, W.: A survey on network embedding. IEEE Trans. Knowl. Data Eng. **31**(5), 833–852 (2018)
18. Kazemi, S.M., et al.: Representation learning for dynamic graphs: a survey. J. Mach. Learn. Res. **21**(70), 1–73 (2020)
19. Skarding, J., Gabrys, B., Musial, K.: "Foundations and modeling of dynamic networks using dynamic graph neural networks: a survey. IEEE Access **9**, 79143–79168 (2021)
20. Pang, Y., et al.: SPARSE-DYN: sparse dynamic graph multirepresentation learning via event-based sparse temporal attention network. Int. J. Intell. Syst. **37**(11), 8770–8789 (2022)
21. Chen, J., Xu, X., Wu, Y., Zheng, H.: GC-LSTM: graph convolution embedded LSTM for dynamic link prediction. arXiv preprint arXiv:1812.04206 (2018)
22. Goyal, P., Chhetri, S.R., Canedo, A.: dyngraph2vec: capturing network dynamics using dynamic graph representation learning. Knowl.-Based Syst. **187**, 104816 (2020)
23. Pang, Y., et al.: Graph decipher: a transparent dual-attention graph neural network to understand the message-passing mechanism for the node classification. Int. J. Intell. Syst. **37**(11), 8747–8769 (2022)
24. Jiang, N., Jie, W., Li, J., Liu, X., Jin, D.: GATRUST: a multi-aspect graph attention network model for trust assessment in OSNS. IEEE Trans. Knowl. Data Eng. **01**, 1–1 (2022)
25. Kumar, S., Zhang, X., Leskovec, J.: Predicting dynamic embedding trajectory in temporal interaction networks. In: Proceedings of the 25th ACM SIGKDD International Conference on Knowledge Discovery & Data Mining, pp. 1269–1278 (2019)
26. Ma, Y., Guo, Z., Ren, Z., Tang, J., Yin, D.: Streaming graph neural networks. In: Proceedings of the 43rd International ACM SIGIR Conference on Research and Development in Information Retrieval, pp. 719–728 (2020)

27. Sankar, A., Wu, Y., Gou, L., Zhang, W., Yang, H.: DYSAT: deep neural representation learning on dynamic graphs via self-attention networks. In: Proceedings of the 13th International Conference on Web Search and Data Mining, pp. 519–527 (2020)

28. Xu, D., Ruan, C., Korpeoglu, E., Kumar, S., Achan, K.: Inductive representation learning on temporal graphs. arXiv preprint arXiv:2002.07962 (2020)

29. Zhou, L., Yang, Y., Ren, X., Wu, F., Zhuang, Y.: Dynamic network embedding by modeling triadic closure process. In: Proceedings of the AAAI Conference on Artificial Intelligence, vol. 32, no. 1 (2018)

30. Chen, X., Zhang, F., Zhou, F., Bonsangue, M.: Multi-scale graph capsule with influence attention for information cascades prediction. Int. J. Intell. Syst. **37**(3), 2584–2611 (2022)

31. Vaswani, A., et al.: Attention is all you need. Adv. Neural Inf. Process. Syst. **30**, 5998–6008 (2017)

32. Guo, Q., Qiu, X., Liu, P., Shao, Y., Xue, X., Zhang, Z.: Star-transformer. arXiv preprint arXiv:1902.09113 (2019)

33. Latapy, M., Viard, T., Magnien, C.: Stream graphs and link streams for the modeling of interactions over time. Soc. Netw. Anal. Min. **8**(1), 1–29 (2018)

34. Taheri, A., Gimpel, K., Berger-Wolf, T.: Learning to represent the evolution of dynamic graphs with recurrent models. In: Companion Proceedings of The 2019 World Wide Web Conference, pp. 301–307 (2019)

35. Hajiramezanali, E., Hasanzadeh, A., Duffield, N., Narayanan, K.R., Zhou, M., Qian, X.: Variational graph recurrent neural networks. arXiv preprint arXiv:1908.09710 (2019)

36. Klimt, B., Yang, Y.: The enron corpus: a new dataset for email classification research. In: Boulicaut, J.-F., Esposito, F., Giannotti, F., Pedreschi, D. (eds.) ECML 2004. LNCS (LNAI), vol. 3201, pp. 217–226. Springer, Heidelberg (2004). https://doi.org/10.1007/978-3-540-30115-8_22

37. Glorot, X., Bengio, Y.: Understanding the difficulty of training deep feedforward neural networks. In: Proceedings of the Thirteenth International Conference on Artificial Intelligence and Statistics, pp. 249–256. JMLR Workshop and Conference Proceedings (2010)

Federated Learning Based on Diffusion Model to Cope with Non-IID Data

Zhuang Zhao[1], Feng Yang[1,2,3(✉)], and Guirong Liang[1]

[1] School of Computer, Electronics and Information, Guangxi University,
Nanning 530004, Guangxi, China
yf@gxu.edu.cn
[2] Guangxi Key Laboratory of Multimedia Communications Network Technology,
Guangxi University, Nanning 530004, Guangxi, China
[3] Key Laboratory of Parallel and Distributed Computing in Guangxi Colleges and
Universities, Guangxi University, Nanning, China

Abstract. Federated learning is a distributed machine learning paradigm that allows model training without centralizing sensitive data in a single place. However, non independent and identical distribution (non-IID) data can lead to degraded learning performance in federated learning. Data augmentation schemes have been proposed to address this issue, but they often require sharing clients' original data, which poses privacy risks. To address these challenges, we propose FedDDA, a data augmentation-based federated learning architecture that uses diffusion models to generate data conforming to the global class distribution and alleviate the non-IID data problem. In FedDDA, a diffusion model is trained through federated learning and then used for data augmentation, thus mitigating the degree of non-IID data without disclosing clients' original data. Our experiments on non-IID settings with various configurations show that FedDDA significantly outperforms FedAvg, with up to 43.04% improvement on the Cifar10 dataset and up to 20.05% improvement on the Fashion-MNIST dataset. Additionally, we find that relatively low-quality generated samples that conform to the global class distribution still improve federated learning performance considerably.

Keywords: Federated learning · Non-IID data · Diffusion model · Data augmentation

1 Introduction

In recent years, federated learning [1] has gained widespread adoption in various scenarios, such as intelligent medicine, financial risk control, and the Internet of Things. As a distributed machine learning method, federated learning faces various challenges, one of which is the non independent and identical distribution (non-IID) data problem [2]. Non-IID data can significantly impact the performance of federated learning. Figure 1 illustrates the performance of Federated

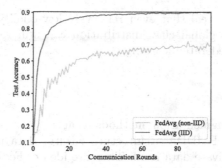

Fig. 1. The performance of FedAvg in non-IID and IID environments.

Averaging (FedAvg) [1] on the Cifar10 dataset in IID and non-IID data settings, showing a substantial performance gap between the two.

Many scholars have studied the non-IID data problem [2,3] in federated learning and proposed various solutions [4–8]. Some scholars use algorithmic approaches for optimization, such as adding regular terms to the objective function [7], taking into account differences in client local steps [6], and utilizing client gradient similarity knowledge [4]. In addition, some scholars have proposed schemes based on data augmentation [9–12]. Hangyu et al. [2] pointed out that data augmentation schemes can significantly improve the performance of federated learning. However, most data augmentation-based solutions require clients to share their original data in some form, which poses a risk of privacy data disclosure. To the best of our knowledge, this issue has not been satisfactorily addressed in the literature.

Non-IID data leads to deviations between clients' local optimization goals and global optimization goals, resulting in a decline in federated learning performance. Since this gap is mainly caused by data distribution, we hope to solve this problem from the perspective of data. How to alleviate data heterogeneity (non-IID) without disclosing clients' original data has aroused our thinking. Recently, the diffusion model [13–15] has inspired us. Diffusion models are a class of probability generation models that can generate high-quality samples. Moreover, the training process of diffusion models does not rely on adversarial training [16], which may cause instability and mode collapse [15–17]. By combining the diffusion model, we propose a three-stage federated learning architecture. The diffusion model is used to generate data that conforms to the global class distribution to alleviate the problem of non-IID data.

Our contributions are summarized as follows:

1. We propose a novel federated learning architecture, FedDDA, which utilizes the diffusion model of federated training for data augmentation and mitigates the challenges posed by non-IID data, all while preserving client raw data privacy.
2. The effectiveness of FedDDA is demonstrated by comparing its experimental performance with state-of-the-art methods in a large number of non-IID environments.

3. The experiments reveal that even relatively low-quality generated samples conforming to the global class distribution can still result in considerable improvements in federated learning.

2   Method

Unlike other data augmentation methods that rely on sharing data among clients, the proposed FedDDA architecture ensures data privacy and security by keeping clients' private data local. The core idea of FedDDA is to use diffusion models to generate samples that alleviate the problem of degraded learning performance in federated learning caused by non-IID data. An overview of Fed-DDA is shown in Fig. 2. Below we describe the three stages of FedDDA.

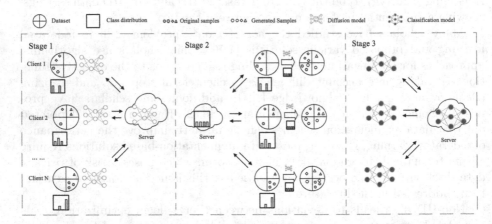

Fig. 2. Overview of proposed FedDDA Architecture. The core idea of the proposed FedDDA is to use a diffusion model to generate samples to alleviate the problems caused by non-IID data. FedDDA includes three stages: 1) The clients train a diffusion model through federated learning, which will be used for the second stage. 2) The clients upload their own class distribution information to the server, which aggregates this information to obtain a global class distribution. Then, each client generates samples using the diffusion model under the guidance of the global class distribution, and obtains a new dataset; 3) The clients use the new datasets to train a classification model through federated learning.

2.1   The First Stage

In this stage, we aim to train a diffusion model using federated learning. Once training is complete, the resulting diffusion model will be used for data augmentation in the second stage. Next, we will introduce the diffusion model constructed in this stage, as well as the process of federally training the diffusion model.

Diffusion Model. Diffusion models are a class of probabilistic generative models used in deep learning that allow for the generation of high-dimensional and complex data distributions. The purpose of diffusion models [13,14] is to train a deep neural network for denoising, using this deep neural network to gradually denoise from randomly sampled Gaussian noise to obtain high-quality samples. The training of diffusion model includes a forward noise adding process and a reverse noise removing process. In the forward process, the Gaussian noise ϵ of the T step is added to a sample x, producing a sequence of noisy samples $x^1, \ldots, x^T$. In the reverse process, a deep neural network θ is used to predict the Gaussian noise added in the forward process. The optimization objective function of the diffusion model is simplified as follows:

$$L_{DM}(\theta; x, t) = \mathbb{E}_{\mathbf{x}, \epsilon \sim \mathcal{N}(0,1), t}\left[\|\epsilon - \epsilon_\theta(x^t, t)\|^2\right], \tag{1}$$

with t uniformly sampled from $\{1, \ldots, T\}$. Later, we need to introduce conditions [17,18] into the diffusion model, so that conditions can be used to guide the process of sample generation. Specifically, attention mechanism [19] is introduced in neural networks to achieve a guided generation process through conditions. Attention mechanism is used to attend to both the intermediate representation $\varphi_\theta(x^t)$ of the input and the intermediate representation $\tau_\theta(y)$ of the label information. The definition of the query Q, key K, and value V matrices is as follows:

$$Q = W_Q \cdot \varphi_\theta(x^t), \quad K = W_K \cdot \tau_\theta(y), \quad V = W_V \cdot \tau_\theta(y), \tag{2}$$

where W_Q, W_K, W_V are learnable weight matrices. The output of a cross-attention layer is as follows:

$$\text{Attention}(Q, K, V) = \text{softmax}\left(\frac{QK^T}{\sqrt{d}}\right)V, \tag{3}$$

where d is the dimension of the attention matrix. This produces a new representation that combines information from both the input and the label information. Therefore, conditions [17] are introduced in L_{DM}:

$$L_{DM}(\theta; x, y, t) = \mathbb{E}_{\mathbf{x}, y, \epsilon \sim \mathcal{N}(0,1), t}\left[\|\epsilon - \epsilon_\theta(x^t, y, t)\|^2\right]. \tag{4}$$

Federated Training Diffusion Model. Federated learning is a distributed machine learning paradigm that focuses on protecting private data while training models. Federated Averaging (FedAvg) [1] is one of the most widely used algorithms in federated learning. In FedAvg, the model training process is conducted on the client side, and the parameter updates of the model are implemented through communication with a central server. Finally, a global model is federally trained using the data of all clients, and no private client data is uploaded during the entire training process. In the first stage of FedDDA, the optimization objective of the federal training diffusion model is:

$$\arg\min L(\theta) := \sum_{n=1}^{N} \frac{D_n}{D} L_n(\theta), \text{where } L_n(\theta) = L_{DM}(\theta; x, y, t), \tag{5}$$

where L_n is the local empirical loss of client n, N represents the number of all clients, D_n represents the number of samples from the n-th client's private dataset $\mathcal{D}_n$, D represents the number of samples in the dataset $\mathcal{D} = \{\mathcal{D}_1, \mathcal{D}_2, \cdots, \mathcal{D}_N\}$. The meaning of this optimization goal is to train a global model to minimize the total empirical loss on the entire dataset. The process of federated training diffusion models is shown in Algorithm 1.

Algorithm 1. Federated Training Diffusion Model

1: **procedure** SERVER-SIDE OPTIMIZATION
2:      Initialization: global model θ_0
3:      **for** each round $r = 1$ to R **do**
4:          S = (random set of clients)
5:          **for** each client $n \in S$ in paralled **do**
6:              $\theta_{t+1}^n \leftarrow$ ClientLocalUpdate(n, θ_t)
7:              $\theta_{t+1} \leftarrow \sum_{n=1}^{N} \frac{D_n}{D} \theta_{t+1}^n$

8: **procedure** CLIENTLOCALUPDATE(n, θ)
9:      **for** each local epoch $e = 1$ to E **do**
10:         **for** mini-batch $(\xi_x, \xi_y) \subseteq \mathbb{D}_n$ **do**
11:             # Sample a mini-batch t from the uniform distribution
12:             $\xi_t \sim \text{Uniform}(\{1, \ldots, T\})$
13:             # Sample a mini-batch ϵ from the Gaussian distribution
14:             $\xi_\epsilon \sim \mathcal{N}(\mathbf{0}, \mathbf{I}))$
15:             $\theta \leftarrow \theta - \eta \nabla_\theta L_{DM}(\theta; \xi_x, \xi_y, \xi_t)$
16:     Send θ to server

2.2   The Second Stage

The main task of this stage is for each client to use a diffusion model to generate samples that conform to the global class distribution, thereby reducing data heterogeneity (non-IID). This stage can be described as the following steps:

1. Each client computes their own class count vector, denoted by $V_n = \{v_1, v_2, \cdots, v_c\}$, where n is the client index, c is the number of classes, and v_i is the number of samples for class i.
2. Each client sends their class count vector V_n to the server.
3. The server aggregates all of the class count vectors to obtain the global class distribution vector, denoted by $G = \{g_1, g_2, \cdots, g_c\}$. The global class distribution vector is computed as follows:

$$g_i = \frac{\sum_{n=1}^{N} V_n[i]}{\sum_{n=1}^{N} \sum_{j=1}^{c} V_n[j]}. \tag{6}$$

4. The server sends the global class distribution vector G back to each client.

5. The client calculates the total number of samples S_n to be generated according to the parameter ω, that is, $S_n = \omega \times D_n$, and then samples S_n labels from the global class distribution G to obtain a label set $\mathcal{L}$. $\mathcal{L}$ can be expressed as:

$$\mathcal{L} = \{y_1, \cdots, y_{S_n}\}, \text{where } y_i \sim G, i = 1, 2, \cdots, S_n. \tag{7}$$

6. The client uses a diffusion model to generate samples under the guidance of labels in $\mathcal{L}$.
7. The client adds the generated samples to their own dataset.

The process of the second stage is shown in Algorithm 2.

Algorithm 2. Data Augmentation

1: **procedure** SERVER-SIDE
2:      Receive class count vector from clients
3:      Calculate global class distribution G, via (6)
4:      Send global class distribution G to all clients
5: **procedure** CLIENT-SIDE
6:      Calculate its own class count vector V_n
7:      Send the class count vector V_n to the server
8:      Wait for receiving global class distribution G
9:      Initialize a dataset $\mathcal{D}'$
10:     $S_n = \omega \times D_n$
11:     Generate label set $\mathcal{L} = \{y_1, \cdots, y_{S_n}\}$, via (7)
12:     **for** each label $y_i \in \mathcal{L}$ **do**
13:          Generate sample x_i from label y_i using diffusion model
14:          Add (x_i, y_i) to $\mathcal{D}'$
15:     $\mathcal{D}_n = \mathcal{D}_n \cup \mathcal{D}'$

2.3 The Third Stage

The goal of the third stage of FedDDA is to use the new datasets obtained in the second stage to train a task-specific model through federated learning. Taking a classification task as an example, clients need to federally train a neural network w for classification tasks. At this stage, the optimization objective of federated learning is:

$$\arg\min L(w) := \sum_{n=1}^{N} \frac{D_n}{D} L_n(w), \text{where } L_n(w) = \frac{1}{D_n} \sum_{i=1}^{D_n} L_{CE}(w; x_i, y_i). \tag{8}$$

L_{CE} is the cross-entropy loss, which measures the difference between the predicted values and the ground truth labels. The process of federated training classification model refers to Appendix A.

3   Experiments

3.1   Setup

Datasets. Our experiments aim to evaluate the performance of our proposed architecture and other federated learning methods for image classification tasks on two widely used benchmark datasets, Cifar10 [20] and Fashion-MNIST [21]. On each dataset, we conducted two types of non-IID experiments to analyze the impact of different data distributions on federated learning methods. The data distribution diagram of clients is shown in Appendix B.

In the first type of non-IID experiment, we refer to the non-IID data partitioning strategy of McMahan et al. in [1], in which the data was sorted by labels, divided into fixed-size fragments, and then randomly sampled s fragments for each participant. In this data distribution, we obtain two different degrees of data partitioning by adjusting s (i.e., s = 2 and s = 4).

In the second type of non-IID experiment, we sample data for each participant from Dirichlet distribution $Dir(\gamma)$ [22] to control the degree of data imbalance, where smaller γ values correspond to more imbalanced data distributions. We considered two different γ values, i.e., $\gamma = 0.1$ and $\gamma = 1.0$.

Baseline. To demonstrate the effectiveness of our proposed training architecture, we compared it with several classic federated learning algorithms, including FedAvg, FedProx, FedNova, and Scaffold. For the experiments, we used U-net [23] as the backbone network for the diffusion model and ResNet18 [24] as the backbone network for the classification task. When training the ResNet18 classification network, we empirically set the learning rate to 0.001 and the batch size to 128. When training the diffusion model, we empirically set the learning rate to 0.0002, the batch size to 64, step T to 1000 and performed 400 rounds of federated training.

Implementation. The implementation of the code is based on the PyTorch framework and the simulation is performed on a single machine with an RTX 3090 GPU having 24 GB of VRAM and an Intel(R) Xeon(R) Platinum 8350C CPU @ 2.60 GHz. The complete code will be available.

3.2   Performance Comparison

The following paragraphs describe a comparison of experimental results between the proposed FedDDA architecture and classic federated learning algorithms on the Cifar10 and Fashion-MNIST datasets. As the proposed architecture requires additional training of a diffusion model, to ensure fairness in the comparison, the local epoch of other methods is set to 5 while that of FedDDA is set to 2 during the training of the classification network.

Table 1 and Table 2 show the highest testing accuracy performances of several federated learning methods in the two non-IID settings. In the extreme

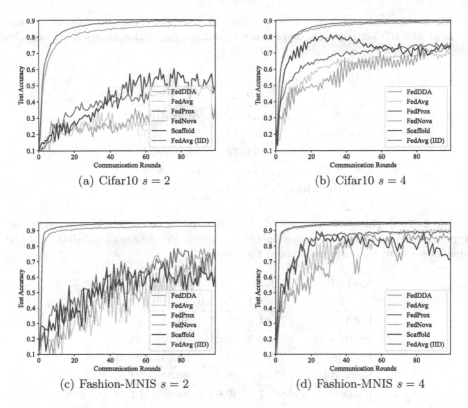

(a) Cifar10 $s = 2$ (b) Cifar10 $s = 4$

(c) Fashion-MNIS $s = 2$ (d) Fashion-MNIS $s = 4$

Fig. 3. The performance comparison of FedDDA, FedAvg, FedProx and Scaffold in the first type of non-IID setting, and FedAvg(IID) indicates the performance of FedAvg in IID setting.

non-IID experiment (when $s = 2$ in the first type of non-IID settings), Fed-DDA significantly outperforms other methods. On the Cifar10 dataset, FedDDA improves accuracy by 43.04 compared to the FedAvg and by 26.73 compared to the second-ranked Scaffold. On the Fashion-MNIST dataset, FedDDA achieves the highest improvement of 20.05 compared to FedAvg. In both types of non-IID experiments, as the degree of non-IID increases, FedDDA's advantages over other federated learning methods become more apparent. In the non-IID setting, there is a difference between the clients' optimization goals and the global optimization goals, which will lead to the decline in the performance of federated learning. By adding generated samples conforming to the global distribution, the difference in optimization direction is alleviated, so the experimental performance of proposed FedDDA is very close to the performance in the IID settings. Figure 3 shows the test accuracy variation curves of these federated learning methods in the first non-IID setting. Due to space limitations, the second non-IID case is in Appendix C.

Table 1. The highest test accuracy (%) for 100 rounds of communication in the first type of non-IID settings. The parameter s controls the degree of non-IID, and the smaller s means the more unbalanced the distribution.

Method	Cifar10		Fashion-MNIST	
	$s=2$	$s=4$	$s=2$	$s=4$
FedAvg [1]	44.04	72.3	72.81	91.07
FedProx [7]	49.87	76.29	78.43	89.74
FedNova [6]	40.94	76.17	78.5	90.49
Sacffold [8]	60.32	81.23	69.53	89.51
FedDDA (our)	87.05	89.34	92.86	94.43

Table 2. The highest test accuracy (%) for 100 rounds of communication in the second type of non-IID settings. The parameter γ controls the degree of non-IID, and the smaller γ means the more unbalanced the distribution.

Method	Cifar10		Fashion-MNIST	
	$\gamma = 0.1$	$\gamma = 1.0$	$\gamma = 0.1$	$\gamma = 1.0$
FedAvg [1]	75.81	90.17	86.54	94.66
FedProx [7]	77.56	89.23	88.36	94.45
FedNova [6]	74.8	89.87	90.55	94.39
Sacffold [8]	77.54	90.35	90.31	94.36
FedDDA (our)	88.17	89.6	93.89	94.64

3.3   Experimental Factors Analysis

Next, we aim to analyze the impact of generated sample quality and quantity on FedDDA. The experiments are conducted under the first type of non-IID setting with $s = 4$.

Sample Quality. In this experiment, we aim to analyze the impact of generated sample quality on FedDDA. The parameter ω that controls the number of samples generated is set to 2.0. To measure the quality of generated samples, we use Fréchet Inception Distance (FID) [25] as the evaluation metric. Smaller FID values indicate higher quality of generated samples. We control the quality of generated samples by adjusting the sampling step number T in the diffusion model. The results of the experiment are consistent with our subjective expectations, where higher quality generated images result in improved test accuracy. It is worth noting that even low-quality generated samples (generated with only 10 sampling steps) are still able to improve the performance of federated learning. Table 3 shows the FID values under different sampling steps and the corresponding test accuracy of FedDDA.

Sample Quantity. Considering that it is not practical to generate too many samples in real-world scenarios, We set the maximum value of ω to 4. Firstly, the

Table 3. The corresponding relationship between the quality (FID) of generated samples and the test accuracy (%), the quality of generated samples is controlled by the sampling step number T. A smaller FID value indicates a higher quality of generated samples. The column of $T = 0$ indicates that no generated samples is added for comparison.

T	0	10	20	50	100	1000
FID	–	48.41	35.49	26.25	21.86	19.56
Cifar10	72.3	84.76	85.89	86.66	87.32	89.34
FID	–	20.27	15.70	10.48	7.55	3.06
Fashion-MNIST	91.07	93.15	93.38	93.77	93.95	94.43

sampling step number T is set to 1000 to obtain relatively high-quality generated samples. Experiments show that even using a small number of relatively high-quality generated samples ($\omega = 0.01$) can significantly improve the performance of federated learning (test accuracy on Cifar10 improved from 72.3 to 80.23, on Fashion-MNIST test accuracy improved from 91.07 to 92.88 compared to FedAvg without generated samples). Table 4 shows the test accuracy of FedDDA under different quantities of relatively high-quality generated samples.

Table 4. The correspondence between sample quantity (controlled by ω) and test accuracy (%) in the case of relatively high-quality generated samples.

ω	0.0	0.01	0.05	0.1	0.5	1.0	2.0	4.0
Cifar10	72.3	80.23	82.65	85.71	87.75	88.48	89.34	89.49
Fashion-MNIST	91.07	92.88	93.11	93.52	94.21	94.29	94.43	94.53

Next, we analyze the relationship between the quantity of low-quality generated samples and the performance of FedDDA. The sampling step number T is set to 10, resulting in FID values of 48.41 and 20.27 for the generated samples on Cifar10 and Fashion-MNIST, respectively. Appendix D shows the partially generated samples at this time, and many generated samples in Cifar10 are difficult to distinguish which class they belong to, even for humans. The experiments show that even a small number of low-quality generated samples can still improve

Table 5. The correspondence between sample quantity (controlled by ω) and test accuracy (%) in the case of low quality generated samples.

ω	0.0	0.01	0.05	0.1	0.5	1.0	2.0	4.0
Cifar10	72.3	73.07	78.59	77.99	83.43	84.54	85.31	85.56
Fashion-MNIST	91.07	92.68	92.59	92.46	92.83	93.51	93.38	93.53

the performance of FedDDA. When the quantity of low-quality generated samples increases, FedDDA can still achieve decent performance. Table 5 provides the test accuracy of FedDDA under different quantities of low-quality generated samples.

4   Conclusion

In this work, we propose FedDDA, a federated learning architecture utilizing diffusion model for data augmentation, to deal with the non-IID data problem in practical applications. Through comparisons with several classic federated learning methods on Cifar10 and Fashion-MNIST datasets, we demonstrated the effectiveness of our proposed architecture. Finally, by controlling the quality and quantity of generated samples, we find that: 1) when the quality of generated samples is high, a small number of generated samples that conform to the global class distribution can significantly improve the performance of federated learning; 2) by increasing the number of low quality generated samples that conform to the global class distribution, federated learning performance can still be improved.

References

1. McMahan, B., Moore, E., Ramage, D., Hampson, S., y Arcas, B.A.: Communication-efficient learning of deep networks from decentralized data. In: Artificial Intelligence and Statistics, pp. 1273–1282. PMLR (2017)
2. Zhu, H., Jinjin, X., Liu, S., Jin, Y.: Federated learning on non-IID data: a survey. Neurocomputing **465**, 371–390 (2021)
3. Li, Q., Diao, Y., Chen, Q., He, B.: Federated learning on non-IID data silos: an experimental study. In: 2022 IEEE 38th International Conference on Data Engineering (ICDE), pp. 965–978. IEEE (2022)
4. Palihawadana, C., Wiratunga, N., Wijekoon, A., Kalutarage, H.: FedSim: similarity guided model aggregation for federated learning. Neurocomputing **483**, 432–445 (2022)
5. Xin, B., et al.: Federated synthetic data generation with differential privacy. Neurocomputing **468**, 1–10 (2022)
6. Wang, J., Liu, Q., Liang, H., Joshi, G., Poor, H.V.: Tackling the objective inconsistency problem in heterogeneous federated optimization. In: Advances in Neural Information Processing Systems, vol. 33, pp. 7611–7623 (2020)
7. Li, T., Sahu, A.K., Zaheer, M., Sanjabi, M., Talwalkar, A., Smith, V.: Federated optimization in heterogeneous networks. In: Proceedings of Machine Learning and Systems, vol. 2, pp. 429–450 (2020)
8. Karimireddy, S.P., Kale, S., Mohri, M., Reddi, S., Stich, S., Suresh, A.T.: SCAFFOLD: stochastic controlled averaging for federated learning. In: International Conference on Machine Learning, pp. 5132–5143. PMLR (2020)
9. Jeong, E., Oh, S., Kim, H., Park, J., Bennis, M., Kim, S.L.: Communication-efficient on-device machine learning: Federated distillation and augmentation under non-IID private data. arXiv preprint arXiv:1811.11479 (2018)

10. Qiong, W., Chen, X., Zhou, Z., Zhang, J.: FedHome: cloud-edge based personalized federated learning for in-home health monitoring. IEEE Trans. Mob. Comput. **21**(8), 2818–2832 (2020)
11. Duan, M., Liu, D., Chen, X., Liu, R., Tan, Y., Liang, L.: Self-balancing federated learning with global imbalanced data in mobile systems. IEEE Trans. Parallel Distrib. Syst. **32**(1), 59–71 (2020)
12. Zhao, Y., Li, M., Lai, L., Suda, N., Civin, D., Chandra, V.: Federated learning with non-IID data. arXiv preprint arXiv:1806.00582 (2018)
13. Ho, J., Jain, A., Abbeel, P.: Denoising diffusion probabilistic models. In: Advances in Neural Information Processing Systems, vol. 33, pp. 6840–6851 (2020)
14. van den Berg, R., Bottou, L., Kohler, J., Gans, A.: Denoising diffusion implicit models. In: International Conference on Learning Representations (2021)
15. Dhariwal, P., Nichol, A.: Diffusion models beat GANs on image synthesis. In: Advances in Neural Information Processing Systems, vol. 34, pp. 8780–8794 (2021)
16. Goodfellow, I., et al.: Generative adversarial networks. Commun. ACM **63**(11), 139–144 (2020)
17. Rombach, R., Blattmann, A., Lorenz, D., Esser, P., Ommer, B.: High-resolution image synthesis with latent diffusion models. In: Proceedings of the IEEE/CVF Conference on Computer Vision and Pattern Recognition, pp. 10684–10695 (2022)
18. Mirza, M., Osindero, S.: Conditional generative adversarial nets. Comput. Vis. Image Underst. **150**, 145–153 (2014)
19. Vaswani, A., et al.: Attention is all you need. In: Advances in Neural Information Processing Systems, vol. 30 (2017)
20. Krizhevsky, A., Hinton, G.: Learning multiple layers of features from tiny images (2009)
21. Xiao, H., Rasul, K., Vollgraf, R.: Fashion-MNIST: a novel image dataset for benchmarking machine learning algorithms. arXiv preprint arXiv:1708.07747 (2017)
22. Yurochkin, M., Agarwal, M., Ghosh, S., Greenewald, K., Hoang, N., Khazaeni, Y.: Bayesian nonparametric federated learning of neural networks. In: International Conference on Machine Learning, pp. 7252–7261. PMLR (2019)
23. Ronneberger, O., Fischer, P., Brox, T.: U-Net: convolutional networks for biomedical image segmentation. In: Navab, N., Hornegger, J., Wells, W.M., Frangi, A.F. (eds.) MICCAI 2015. LNCS, vol. 9351, pp. 234–241. Springer, Cham (2015). https://doi.org/10.1007/978-3-319-24574-4_28
24. He, K., Zhang, X., Ren, S., Sun, J.: Deep residual learning for image recognition. In: Proceedings of the IEEE Conference on Computer Vision and Pattern Recognition, pages 770–778 (2016)
25. Szegedy, C., Ioffe, S., Vanhoucke, V., Alemi, A.: Inception-v4, inception-ResNet and the impact of residual connections on learning. In: Proceedings of the AAAI Conference on Artificial Intelligence, vol. 31 (2017)

SFRSwin: A Shallow Significant Feature Retention Swin Transformer for Fine-Grained Image Classification of Wildlife Species

Shuai Wang[1,2], Yubing Han[1,2(✉)], Shouliang Song[1,2], Honglei Zhu[1,2], Li Zhang[1,2], Anming Dong[1,2], and Jiguo Yu[3]

[1] Key Laboratory of Computing Power Network and Information Security, Ministry of Education, Shandong Computer Science Center (National Supercomputer Center in Jinan), Qilu University of Technology (Shandong Academy of Sciences), Jinan 250353, China
cauhanbing@163.com
[2] School of Computer Science and Technology, Qilu University of Technology (Shandong Academy of Sciences), Jinan 250353, China
[3] Big Data Institute, Qilu University of Technology (Shandong Academy of Sciences), Jinan 250353, China

Abstract. Fine-grained image classification of wildlife species is a task of practical value and has an important role to play in the fields of endangered animal conservation, environmental protection and ecological conservation. However, the small differences between different subclasses of wildlife and the large differences within the same subclasses pose a great challenge to the classification of wildlife species. In addition, the feature extraction capability of existing methods is insufficient, ignoring the role of shallow effective features and failing to identify subtle differences between images well. To solve the above problems, this paper proposes an improved Swin Transformer architecture, called SFRSwin. Specifically, a shallow feature retention mechanism is proposed, where the mechanism consists of a branch that extracts significant features from shallow features, is used to retain important features in the shallow layers of the image, and forms a dual-stream structure with the original network. SFRSwin was trained and tested on the communal dataset

This work was supported in part by the National Natural Science Foundation of China (NSFC) under Grant 62272256, the Shandong Provincial Natural Science Foundation under Grants ZR2021MF026 and ZR2023MF040, the Innovation Capability Enhancement Program for Small and Medium-sized Technological Enterprises of Shandong Province under Grants 2022TSGC2180 and 2022TSGC2123, the Innovation Team Cultivating Program of Jinan under Grant 202228093, and the Piloting Fundamental Research Program for the Integration of Scientific Research, Education and Industry of Qilu University of Technology (Shandong Academy of Sciences) under Grants 2021JC02014 and 2022XD001, the Talent Cultivation Promotion Program of Computer Science and Technology in Qilu University of Technology (Shandong Academy of Sciences) under Grants 2021PY05001 and 2023PY059.

Q. Liu et al. (Eds.): PRCV 2023, LNCS 14433, pp. 232–243, 2024.
https://doi.org/10.1007/978-981-99-8546-3_19

Stanford Dogs and the small-scale dataset Shark species, and achieved an accuracy of 93.8% and 84.3% on the validation set, an improvement of 0.1% and 0.3% respectively over the pre-improvement period. In terms of complexity, the FLOPs only increased by 2.7% and the number of parameters only increased by 0.15%.

Keywords: Fine-grained image classification · Wildlife species classification · Transformer

1 Introduction

In taxonomy, organisms are classified according to domains, kingdoms, phyla, orders, families, genera and species. At the species level, there are subspecies, which are phenotypically similar clusters of a species that are distributed in subgeographic areas within the range of the species and are taxonomically distinct from other populations of the species. In the case of the tiger, for example, traditional taxonomy considers that there are eight subspecies of tiger, including the Bengal, Caspian, Northeastern, Javan, South China, Balinese, Sumatran and Indo-Chinese tigers. Regrettably, however, some subspecies of tigers have become endangered due to severe damage to their range and habitat ecosystems, chemical pollution, climate change and random killing by humans.

Indeed, advanced technology [6,8,15] can help improve the efficiency and effectiveness of wildlife conservation efforts. We know that for wildlife conservation work, identifying (distinguishing) specific endangered subspecies is a prerequisite for all work. Therefore, identifying subspecies of wildlife through images collected in the field can assist in the conservation of such subspecies to a certain extent. For example, by recording and counting the numbers and distribution of a subspecies, and by detecting anomalies in time.

Image classification of wildlife species belongs to the category of fine-grained image classification [12,16]. In the natural world, the same species has variable body shape, rich texture and similar appearance. Compared with coarse-grained image classification, image classification of wildlife species has the characteristics of small inter-class differences and large intra-class differences, making it significantly more difficult to classify. In practice, the captured images are affected by uncertainties such as illumination, occlusion and lens distortion, further increasing the difficulty of classifying wildlife species. In terms of algorithms, for some existing machine learning methods, their ability to distinguish subtle differences between images is insufficient.

The recent Transformer [5] and its variant model [13] have shown better learning ability than traditional CNNs in fine-grained classification, providing a breakthrough learning architecture and self-focusing mechanism for wildlife fine-grained image classification tasks. Therefore, this paper proposes a network shallow significant feature retention Swin Transformer [9,10] network (SFRSwin) for fine-grained image classification of wildlife species. Specifically, this paper designs a structure based on the Swin Transformer that contains a branch from the shallow layer of the network. SFRSwin is able to capture significant features from the shallow layers of the image that will help identify subtle differences in fine-grained images of wildlife species.

The main contributions of this paper are threefold:

A. The swin transformer, which has the advantage of low spatio-temporal overhead, was used as the backbone and improved upon to form a new network designed to identify fine-grained categories of wildlife.
B. A branch is designed to retain the shallow significant features, using convolution and maximum pooling to retain the shallow significant features of the network and add them element by element with the output of the swin transformer blocks before the final output, where the added significant features will help to identify subtle differences in the images.
C. Random data enhancement is introduced, including random scaling and random level flipping, to solve the problem of insufficient training data for small sample datasets, while improving model robustness.

The paper is organized as follows. Section 2 briefly describes the available classic fine-grained image classification methods. Our network architecture is described in Sect. 3. Model evaluation and analysis are presented in Sect. 4, and concluding remarks are presented in Sect. 5.

2   Related Works

2.1   Convolutional Neural Network

CNNs have shown excellent performance in classification tasks, and researchers can transfer the knowledge learned from general-purpose classification tasks to the field of fine-grained image classification [11]. Classification methods based on local detection, end-to-end and attention mechanisms are three stages in the development of convolutional neural networks. Innovative ideas for these classification methods include the following four main directions: learning more discriminative feature representations through powerful deep models, using pose alignment operations to eliminate effects such as pose bias and shooting angle, constructing local region feature representations, and segmentation-based target object localization.

Classification Algorithms Based on Local Detection. The main process of local detection-based classification algorithms is to improve the accuracy of fine-grained classification by first detecting and localizing local areas in an image and then extracting features from these local areas. This two-stage approach usually uses a multi-stage alternating or cascading structure, but this structure makes the training process complex. In addition, to reduce the interference of background noise, the algorithms are often filtered using annotation information, but the high cost of obtaining annotation information limits the practicality of the algorithms. For example, the Part RCNN algorithm proposed by Zhang et al. [3] improves the classification accuracy of the CUB200-2011 dataset to 73.89% using geometric constraints and boundary constraints to optimize the detection results, with the geometric constraints improving the classification accuracy by about 1%.

Bilinear Feature Learning Algorithms. To address the complexity of the training process, researchers have proposed a series of bilinear feature learning algorithms that do not require additional annotation information and can learn more discriminative features directly, thus avoiding the problem of insufficient annotation information during the training process. Although such algorithms can simplify the training process, they can suffer from the dimensional disaster problem due to the high dimensionality of the features they produce. To alleviate this problem, methods such as aggregating low-dimensional features or designing loss functions can be used. For different features of an image, common operations include concatenation, summation and pooling, but human vision processing relies mainly on two neural pathways, for object recognition and object localization respectively. Therefore, Lin et al. [7] proposed a bilinear convolutional neural network model, which uses two different convolutional neural networks to collaboratively complete category detection and feature extraction of fine-grained images, and obtains the final bilinear features as a basis for classification through bilinear operations and feature normalization processing. The method achieved an accuracy of 84.1% on the CUB200-2011 dataset. The bilinear operation achieves the fusion of different features through the operation of outer product multiplication, but the bilinear feature vector obtained by this process is a high-dimensional quadratic extension feature, making it impossible to be applied to large-scale real scenes. In addition, convolutional neural networks use the same convolutional kernel scale for feature extraction, leading to a homogenization of convolutional features, which is not conducive to the capture of information from locally minute regions in fine-grained classification, and the feature extraction capability is slightly inadequate.

Attention Mechanism. In the image domain, the attention mechanism is mainly implemented by masking operations, where new weights are used to identify key regions of an image, and the trained deep learning model can be used to learn the regions to be attended to in the input new image. Thus, classification algorithms based on the attention mechanism are able to learn and understand the underlying information in the image autonomously. Furthermore, it is difficult to describe the differences between subordinate classes in a single convolutional neural network model, which makes fine-grained classification difficult. To address this problem, the researchers attempted to combine object-level convolutional networks and part-level convolutional networks for multi-scale feature learning, and set constraints to filter representative features. The nature of the attentional mechanism is similar to that of humans observing things in the outside world. Humans usually observe things by quickly scanning the panorama and then quickly targeting local areas to focus on based on brain signals, eventually forming the focus of attention. In fine-grained classification tasks, the application of attentional mechanisms can detect multiple local regions that are evenly distributed, and their corresponding features can be used to distinguish between different classes of target objects. For example, the recurrent attentional convolutional neural network proposed by Fu et al. [4] learns discriminable regional attention and local region feature representations at multiple scales. This method enables the network to progressively locate discriminable regions to produce higher confidence predictions. The method achieves accuracies of 85.3%, 87.3% and 92.5% on the CUB200-2011, Stanford Dogs and Stanford Cars datasets respectively.

2.2   Vision Transformer

The ViT model was proposed by Dosovitskiy et al. [2] 2020. The model captures long-range dependencies through a self-attentive module to extract global features of an image, which effectively improves classification accuracy. However, the model can only capture the correlation between pixels within a single image sample, resulting in insufficient output feature extraction and a large number of parameters. In addition, the model uses the class patch output from the final Transformer layer as the final feature representation, which has a large amount of redundancy and results in poor discriminative feature representation. Although ViT breaks the disadvantage that convolutional neural networks cannot capture long-range dependencies, it is weak in capturing multi-size features because the token size seen by the Transformer block of each layer is sixteen times the down-sampling rate, and although the global modeling capability is obtained through global self-attentiveness, the features processed are single-size and low resolution. In addition, global self-attention leads to a computational complexity that is a multiple of the square of the image size. Although the Transformer model has some disadvantages in the processing of visual problems, such as being computationally intensive, it still has the best value for improvement.

3   Methodology

The proposed network is based on the Swin Transformer with a new shallow feature retention branch to form a dual stream structure, as shown in Fig. 1. Within the shallow feature retention branch, significant features are retained through maximum pooling and point-wise addition is performed with the output of Swin Transformer blocks to achieve feature element-by-element summation.

3.1   Self-attentive Mechanism Based on Shifted Windows

This paper uses the Swin Transformer, an improved Vision Transformer, as the global feature extraction backbone. its core mechanism, W-MSA, is to perform the transformer's operations within a small window. Because Swin Transformer

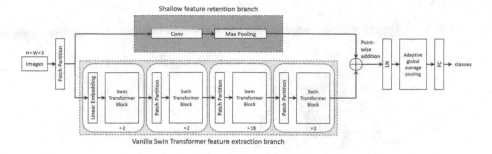

Fig. 1. Diagram of the proposed network.

	downsp. rate (output size)	Swin-T	Swin-S	Swin-B	Swin-L
stage 1	4× (56×56)	concat 4×4, 96-d, LN	concat 4×4, 96-d, LN	concat 4×4, 128-d, LN	concat 4×4, 192-d, LN
		win. sz. 7×7, dim 96, head 3 × 2	win. sz. 7×7, dim 96, head 3 × 2	win. sz. 7×7, dim 128, head 4 × 2	win. sz. 7×7, dim 192, head 6 × 2
stage 2	8× (28×28)	concat 2×2, 192-d, LN	concat 2×2, 192-d, LN	concat 2×2, 256-d, LN	concat 2×2, 384-d, LN
		win. sz. 7×7, dim 192, head 6 × 2	win. sz. 7×7, dim 192, head 6 × 2	win. sz. 7×7, dim 256, head 8 × 2	win. sz. 7×7, dim 384, head 12 × 2
stage 3	16× (14×14)	concat 2×2, 384-d, LN	concat 2×2, 384-d, LN	concat 2×2, 512-d, LN	concat 2×2, 768-d, LN
		win. sz. 7×7, dim 384, head 12 × 6	win. sz. 7×7, dim 384, head 12 × 18	win. sz. 7×7, dim 512, head 16 × 18	win. sz. 7×7, dim 768, head 24 × 18
stage 4	32× (7×7)	concat 2×2, 768-d, LN	concat 2×2, 768-d, LN	concat 2×2, 1024-d, LN	concat 2×2, 1536-d, LN
		win. sz. 7×7, dim 768, head 24 × 2	win. sz. 7×7, dim 768, head 24 × 2	win. sz. 7×7, dim 1024, head 32 × 2	win. sz. 7×7, dim 1536, head 48 × 2

Fig. 2. Detailed architectures of the Swin Transformer.

has two advantages: (1) it solves the problem of too much computational overhead per layer in ViT; (2) it uses shifted windows to achieve cross-window connectivity, enabling the model to take into account the contents of other adjacent windows. As a result, the proposed model has a more significant advantage over traditional CNN-based approaches in establishing global remote dependencies of body parts.

In practice, this paper strictly follows the hierarchical architecture of the Swin Transformer, which consists of four different structures of the network, see Fig. 2. It consists of four encoding stages, each containing a certain number of Swin Transformer blocks, including window-based multi-headed self-attention and shift-window-based multi-headed self-attention. This hierarchical architecture provides the flexibility to model at a variety of scales and has linear computational complexity relative to image size. In this paper, we use Swin-Base (Swin-B) as a baseline on which to base our subsequent research.

Retention of Shallow Significant Features. In this paper, a shallow significant feature retention mechanism is designed, which forms a dual-stream structure with the original Swin Transformer through a shallow feature extraction branch. The added branch is designed to retain significant information in the shallow features from the input, helping the network to distinguish subtle

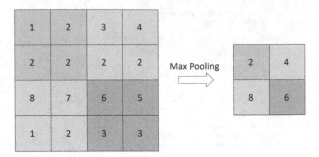

Fig. 3. Maximum pooling operation.

differences from similar subjects in the image. This mechanism allows shallow, fine-grained features to be fused with deeper, abstract features, while providing a multi-scale feature representation that allows the network to better capture features at different levels.

Inspired by the existing feature fusion method [14], point-wise addition is chosen as the feature fusion method for SFRSwin. Point-wise addition is an operation of element-by-element addition between two tensors with the same shape. Point-wise addition is a common means of cross-channel or cross-layer feature fusion of models, which is simple to operate and has good scalability and efficiency. the specific operation of point-wise addition is formulated as:

$$v = v_1 + v_2, v = \{x_i \mid x_i = v_1[i] + v_2[i], i = 1, \ldots, n\}, \tag{1}$$

where the eigenvectors $v_1 \in R^n$ and $v_2 \in R^n$.

Since point-wise addition requires both tensors to have the same shape, SFRSwin uses convolution and pooling operations within the shallow feature retention branch to initially extract shallow features while changing their shape to meet the point-wise addition requirement.

Maximum Pooling. The pooling layer is one of the common components in current convolutional neural networks. The pooling layer mimics the human visual system to reduce the dimensionality of the data and represent the image with higher-level features. The pooling layer dramatically reduces the network model parameters and computational cost, and also reduces the risk of overfitting the network to a certain extent. In summary, pooling layers have the following main effects: i. increase the network perceptual field; ii. suppress noise and reduce information redundancy; iii. reduce the amount of model computation, reduce the difficulty of network optimization and prevent network overfitting; and iv. make the model more robust to changes in the location of features in the input image.

Maximum pooling reduces the bias in the estimated mean due to parameter errors in the convolutional layer and preserves more texture information [1].

Maximum pooling can be expressed as:

$$s_{ij} = \max_{i=1,j=1}^{c} (F_{ij}) + b_2. \tag{2}$$

In the forward process, the maximum value in the image region is selected as the value after pooling in that region, as in Fig. 3.

As can be seen, the mechanism of maximum pooling is a good fit for the needs of fine-grained image classification tasks. The use of maximum pooling within the shallow feature extraction branch allows for the preservation of as many significant features as possible while meeting the dimensionality reduction requirements.

3.2   Random Data Enhancement

In practical applications, it is often the case that there is not enough training data. For small data sets, it is important to design stochastic data augmentation as it can increase the diversity of the data. Data augmentation methods essentially build on the limited amount of data available by generating incremental data according to rules to achieve the equivalent of a larger amount of data, without actually collecting more data. The aim of data augmentation methods is not only to increase the number of data samples, but more importantly to enhance the characteristics of the data itself. Data augmentation techniques can impose constraints on the data as required, adding a priori knowledge of the antecedent processes, such as removing or completing some information, to reduce the negative impact on the performance of the model for processing image tasks. The networks trained using random data augmented data theoretically have better classification performance. Therefore, in this paper, random scaling and random level flipping are added to the training process of SFRSwin to solve the problem of insufficient training data for small sample datasets and to improve the robustness of the model.

4   Evaluation

4.1   Datasets and Implementation Details

Stanford Dogs Dataset. In the field of fine-grained image classification, commonly used publicly available datasets include Caltech-UCSD Birds-200-2011, Stanford Dogs, Stanford Cars, 102 Category Flower Dataset, FGVC-Aircraft Benchmark and so on. Based on the application scenarios of the classification networks described in this paper, and the need for a comprehensive cross-sectional comparison with other classification models, the widely used Stanford Dogs dataset was chosen for the performance testing of the classification models. The dataset contains a total of 20,580 images of 120 dog breeds from around the world. The dataset was constructed using images and annotations from ImageNet to perform fine-grained image classification tasks.

We construct and train the SFRSwin with the PyTorch framework on a 64 bit Window 10 system configured with INTEL XEON Silver 4210R CPU @2.40 GHz, NVIDIA RTX 3090 GPU, and 128 GB memory. During the experiments, the original Swin Transformer (Swin-B) network was first trained and validated, followed by SFRSwin. The accuracy and loss curves of the validation set were compared between the two. This process fixed the superparameter for the purpose of controlling the variables.

Table 1. Comparative results on Stanford Dogs dataset.

Method	Backbone	Accuracy(%)
RA-CNN	VGG-19	87.3
SEP	ResNet-50	88.8
Cross-X	ResNet-50	88.9
ViT	ViT-B16	91.2
ViT-NeT	DeiT-III-B	93.6
Swin	Swin-Base	93.7
SFRSwin(ours)	Swin-Base	**93.8**

After a pre-experimental demonstration, the parameters for training and validation were determined. The dataset was divided into a training part and a validation part according to 6:4; the network used AdamW as the optimizer and the cross-entropy loss function to evaluate the model loss; the learning rate of the training process was set to 0.00001 and the batch size to 32, and the pre-training weights of Swin Transformer on ImageNet 1K were loaded for migration learning training.

Through experiments, the validation set accuracy of SFRSwin is 93.8%, an improvement of 0.1% compared to the original Swin-B (93.7%). At the same time, the loss extrema is smaller compared to the original network.

The classification performance of SFRSwin was compared with that of some mainstream classification networks on the Stanford Dogs dataset, and the results of the comparison are shown in Table 1.

Experimental results on the Stanford Dogs dataset show that the SFRSwin network has improved performance over the original Swin-B network, as well as improving performance over existing mainstream classification networks.

Shark Species Dataset. The network described in this paper is designed for a wildlife species classification scenario. Therefore, it is particularly important to use small-scale datasets to simulate the problem of insufficient data volume in real-world application scenarios. To validate the classification performance of the SFRSwin network on small-scale datasets, the Shark species dataset from Kaggle (https://www.kaggle.com/datasets/larusso94/shark-species) was chosen for this paper. This dataset contains 14 species of sharks with a total of 1523 images, which can simulate the actual collected shark images. The choice of this dataset can bring the experimental results closer to the real situation. Therefore, this dataset is well suited for validating the network performance in this paper.

The experimental conditions are set as follows: the dataset was divided into a training part and a validation part according to 6:4; the network used AdamW as the optimizer and the cross-entropy loss function was used to evaluate the model loss; the learning rate of the training process was set to 0.00001 and the batch size was 32. The experimental results are shown in Table 2. Through the experiment, the best result of SFRSwin has an accuracy of 84.3%, which is an improvement of 0.3% compared to the original Swin-B (83.9%). Meanwhile, the loss extremes are smaller compared to the original network. The confusion matrix results are shown in Fig. 4.

Table 2. Comparative results on Shark species dataset.

Method	Backbone	Accuracy(%)	Loss
Swin	Swin-Base	83.9	0.5926
SFRSwin(ours)	Swin-Base	**84.3**	**0.5804**

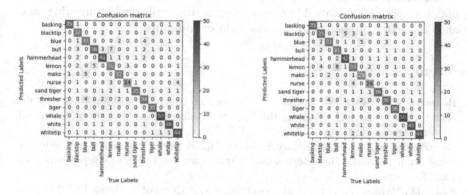

Fig. 4. Confusion matrix for the two models on the Shark species dataset, left: Swin-B, right: SFRSwin.

Table 3. Comparison of the complexity of SFRSwin and Swin-B.

Method	Backbone	FLOPs	#param.
Swin	Swin-Base	15.16892G	86.80M
SFRSwin(ours)	Swin-Base	15.57997G	86.93M

4.2 Model Complexity Analysis

In order to evaluate the complexity of SFRSwin, the number of parameters and the amount of computation were calculated on the same computing platform with a single NVIDIA RTX 3090 GPU with the help of the Thop library, and the results were compared with the pre-model, the results of which are shown in Table 3. The FLOPs were calculated to increase by 2.7% and the number of parameters by 0.15%.

5 Conclusion

This paper proposes a solution to the challenge of wildlife species classification with the help of fine-grained image classification. The paper improves on the Swin Transformer network, proposes a shallow feature retention mechanism and conducts related experiments.

After experiments on two datasets, the SFRSwin network with Swin-B as the backbone proved to have better fine-grained classification capability for wildlife species images. Notably, SFRSwin improves the performance of wildlife species classification without a significant increase in spatio-temporal overhead. More importantly, this performance improvement is not limited to large-scale datasets, but is also applicable to smaller datasets, making it more valuable for practical applications. The introduction of this technology provides more reliable technical support for applications in wildlife conservation and ecological monitoring, and expands the scope of applications in the field of computer vision.

References

1. Boureau, Y.L., Bach, F., LeCun, Y., Ponce, J.: Learning mid-level features for recognition. In: 2010 IEEE Computer Society Conference on Computer Vision and Pattern Recognition, pp. 2559–2566. IEEE (2010)
2. Dosovitskiy, A., et al.: An image is worth 16x16 words: transformers for image recognition at scale. In: International Conference on Learning Representations (2021)
3. Fleet, D., Pajdla, T., Schiele, B., Tuytelaars, T.: Computer vision–ECCV 2014–13th European Conference, Zurich, Switzerland, 6–12 September 2014, Proceedings, Part III. LNCS, vol. 8694. Springer, Cham (2014)
4. Fu, J., Zheng, H., Tao, M.: Look closer to see better: recurrent attention convolutional neural network for fine-grained image recognition. In: IEEE Conference on Computer Vision & Pattern Recognition (2017)
5. Han, K., et al.: A survey on vision transformer. IEEE Trans. Pattern Anal. Mach. Intell. **45**(1), 87–110 (2023). https://doi.org/10.1109/TPAMI.2022.3152247
6. Hodgson, J.C., Baylis, S.M., Mott, R., Herrod, A., Clarke, R.H.: Precision wildlife monitoring using unmanned aerial vehicles. Sci. Rep. **6**(1), 1–7 (2016)
7. Lin, T.Y., RoyChowdhury, A., Maji, S.: Bilinear CNN models for fine-grained visual recognition. In: Proceedings of the IEEE International Conference on Computer Vision, pp. 1449–1457 (2015)

8. Liu, H., et al.: TransiFC: invariant cues-aware feature concentration learning for efficient fine-grained bird image classification. In: IEEE Transactions on Multimedia, pp. 1–14 (2023). https://doi.org/10.1109/TMM.2023.3238548
9. Liu, Z., et al.: Swin Transformer V2: scaling up capacity and resolution. In: Proceedings of the IEEE/CVF Conference on Computer Vision and Pattern Recognition, pp. 12009–12019 (2022)
10. Liu, Z., et al.: Swin Transformer: hierarchical vision transformer using shifted windows. In: Proceedings of the IEEE/CVF International Conference on Computer Vision, pp. 10012–10022 (2021)
11. Qiu, C., Zhou, W.: A survey of recent advances in CNN-based fine-grained visual categorization. In: 2020 IEEE 20th International Conference on Communication Technology (ICCT), pp. 1377–1384. IEEE (2020)
12. Shen, Z., Mu, L., Gao, J., Shi, Y., Liu, Z.: Review of fine-grained image categorization. J. Comput. Appl. **43**(1), 51 (2023)
13. Su, T., Ye, S., Song, C., Cheng, J.: Mask-Vit: an object mask embedding in vision transformer for fine-grained visual classification. In: 2022 IEEE International Conference on Image Processing (ICIP), pp. 1626–1630. IEEE (2022)
14. Vaswani, A., et al.: Attention is all you need. In: Advances in Neural Information Processing Systems, vol. 30 (2017)
15. Wu, Z., et al.: Deep learning enables satellite-based monitoring of large populations of terrestrial mammals across heterogeneous landscape. Nat. Commun. **14**(1), 3072 (2023)
16. Zheng, M., et al.: A survey of fine-grained image categorization. In: 2018 14th IEEE International Conference on Signal Processing (ICSP), pp. 533–538 (2018). https://doi.org/10.1109/ICSP.2018.8652307

A Robust and High Accurate Method for Hand Kinematics Decoding from Neural Populations

Chinan Wang, Ming Yin, F. Liang, and X. Wang[✉]

The Key Laboratory of Biomedical Engineering of Hainan Province,
School of Biomedical Engineering, Hainan University, Haikou 570228, China
wangx@hainanu.edu.cn

Abstract. Offline decoding of movement trajectories from invasive brain-machine interface (iBMI) is a crucial issue of achieving cortical movement control. Scientists are dedicated to improving decoding speed and accuracy to assist patients in better controlling neuroprosthetics. However, previous studies treated channels as normal sequential inputs, merely considering time as a dimension representing channel information quantity. So, this inevitably leads underutilization of temporal information. Herein, a QRNN network integrated with a temporal attention module was proposed to decode movement kinematics from neural populations. It improves the performance by 3.45% compared to the state-of-the-art (SOTA) method. Moreover, this approach only incurs a increase of parameter less than 0.1% compared to the QRNN with same hyperparameter configuration. An information-theoretic analysis was performed to discuss the efficacy of the temporal attention module in neural decoding performance.

Keywords: iBMI · neural decoding · QRNN · attention mechanism

1 Introduction

With significant advancements in materials science, electronics, and computer technology, iBMI, as an interdisciplinary field, have been greatly benefited. large-scale wireless neural recording systems have emerged, providing a crucial technology for the next generation of brain-machine interfaces with high spatial resolution. Currently, iBMI has made groundbreaking advancements in restoring movement, sensation, and communication in paralyzed patients. IBMI allowed patients to control prosthetic arms and generat attempted handwriting, enabling communication at comparable speeds to non-disabled individuals [12].

The iBMI normally requires large-scale wireless neural recording systems and recording systems' thousands of channels significantly increase power consumption [15,25]. If the complete signal were transmitted through the system,under

Acknowledgment and supplementary materials can be viewed at the link: https://github.com/Charlyww/TQRNN-ibmi.git.

current wireless technology, the limit may be under several hundred channels. Thus, only the multi-unitactivity(MUA) features were sent to the data processing end [8]. Scientists have delicated themselves to find a high-performance MUA neural decoder to help patients control neural prosthetics with high speed and accuracy. Kalman Filter (KF), Multilayer Perceptron (MLP), Long Short-Term Memory (LSTM), and Quasi-Recurrent Neural Network (QRNN) have been used for neural decoding [1–3]. However, these methods do not fully exploit the temporal features of sequences. Herein, we focused on MUA neural decoding, and a novel QRNN decoder with added temporal attention is proposed to significantly enhance the offline decoding performance. There are mainly three benefits of improving the decoding performance. First, it increases the possibility of subjects leaning to cortical control. Second, a decoder that accurately predicts natural movements significantly reduces the cognitive load on subjects during adaptation, which is particularly important for human subjects. Third, considering the limited learning capacity of neurons in both animals and humans, using a decoder that aligns well with the subject's natural neural reflexes can lead to better performance in subsequent tasks.

Overall, our contributions are summarized as follows:

1. The concept of temporal attention was first introduced in the neural decoding of hand movement tasks using MUA in this work.
2. A stable and highly accurate decoder, Temporal-attention Quasi-recurrent neural network (TQRNN), was designed. It outperformed the current state-of-the-art (QRNN) by 3.45% in MUA decoding tasks.
3. Based on the experimental process, a theoretical analysis was conducted from an information theory perspective to explain the performance improvement.

The remaining chapters of this paper will be presented as follows: Related Works, Method, Experiments and Conclusion.

2   Related Works

2.1   iBMI Cortical Control Decoding Algorithm

The most widely used and classic algorithm in MUA cortical control is the Kalman filter, which was proposed in 1960 [10]. Compared to other complex decoders, its lower computational requirements provide low latency and easy deployment. Many excellent works have been conducted using the Kalman filter [19,27]. However, in recent years, with the rapid growth of integrated circuits, the latency of deep learning methods has also met the requirements for real-time control. As a result, many deep learning-based approaches have emerged in the field of MUA decoding for cortical control [6,22]. Currently, the most two effective methods in MUA motion decoding are LSTM and QRNN.

2.2   Attention Module

In previous studies, BAKS with QRNN can adequately captures data features and channel dimensions [1]. However, The temporal features was not fully utilized. Temporal information is only processed within the QRNN module by providing the previous time step input with a forget factor. While several temporal attention modules have been proposed in the past few years [9], they were not specifically tailored for iBMI decoding. Consequently, inspired from squeeze-and-excitation net (SENet) and efficient channel attention net (ECA-Net) [5,21], we have incorporated their ideas into QRNN to improve its utilization of temporal features. The proposed temporal attention module exhibited distinct differences compared to the previous. The effectiveness of the unique aspects of the proposed attention module was verified through subsequent ablative experiments.

3   Method

In this section, we present the decoding process, covering six aspects: neural recording system and behavioral task, experimental procedure of the cortical control, preprocessing with task data, extracting firing rate, temporal-attention QRNN, and evaluation metrics. Due to space limitations, we included the sections "Preprocessing with task data" and "Extracting Firing rate" in the supplementary materials.

3.1   Neural Recording System and Behavioral Task

The dataset used in this paper is from the public dataset [16]. As shown in Fig. 1, it involves a male monkey that underwent implantation of a 96-channel Utah electrode in primary motor cortex (M1) area and performed a random reach task. Random targets were presented on the screen, and the subject touched the target with his finger and held it for 450 ms. kinematics and neural data were recorded simultaneously. The recorded data includes the finger's position mapped on the screen and MUA from the 96 channels. For more details in experimental design, please refer to [13]. The data used in this study spans a time period of 10 months, from April 7, 2016, to January 31, 2017 (a total of 18 data segments).

3.2   Experimental Procedure of the Cortical Control

As shown in Fig. 1, the experimental procedure of the cortical control can be divided into two steps [6,22,23]: the offline training of the neural decoder and then the online executing. The first step involves the acquisition of neural signals, such as MUA and local field potentials (LFP) [7], from the motor cortex of subjects (e.g., macaques, humans [11]) using an invasive neural recording device. Concurrently, the corresponding hand movement of the subject is also recorded. These signals are then transmitted and packaged by neurophysiology recording system [14]. Subsequently, the collected data is used to train a neural decoder. The parameters of trained decoder are then transferred to a lower-latency processing device, typically a field-programmable gate array (FPGA) [26], where the decoder is deployed for real-time decoding of cursor or robotic arm movements based on the recorded neural signals [20]. As indicated by the pentagon star in the Fig. 1, the work presented in this paper was conducted during the offline decoding phase.

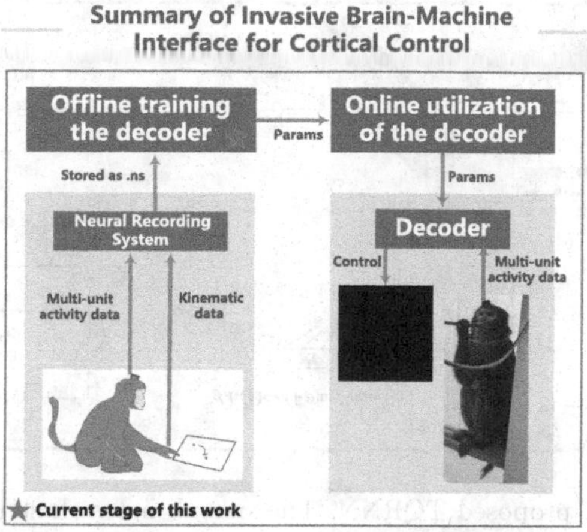

Fig. 1. Implementation Process of Invasive Brain-Machine Interface for cortical control.

3.3 Temporal-Attention QRNN

The overall architecture of the proposed TQRNN is illustrated in Fig. 2. The input is composed as $\mathbb{R}_{-}\{B, T, C\}$, B represents the batch size, T represents the timestep, and C represents the channel. The specific value of the batch size is determined through optimization iterations. The interval between every two consecutive time steps is 0.12 s. The optimal number of time steps determined by the optimizer is 5. C represents the number of channels. Although there are 96 channels in the electrodes, in practice, some channels may not record neural information or may have a low neural firing rate. These channels were not regarded as an entire neuron. Therefore, channels with a firing rate lower than 0.5 Hz are excluded. The remaining channels in different days range from 85 to 95.

The gray boxes in the figure represent the dimension information during data transmission. After the data input, it undergoes a time-attention module to extract temporal features. The module proposed in ours work was inspired by SENet and ECA-Net, but we made improvements based on the characteristics of iBMI. In comparison to SENet, we employed a convolutional module instead of fully connected layers. Compared to ECA-Net, we used two convolutional layers with a different kernel calculation logic. The calculation logic of ours is as follows:

$$\text{Kernel\_size} = \log_2 \left(\ln \left(\alpha + \exp(\text{Channels}) \right) \right) / \beta \qquad (1)$$

The 'channels' in the Temporal-Attention module represents T (number of time steps). Parameter α limits the kernel size for lower dimensions, while parameter β and the logarithmic function limit the kernel size for higher dimensions. This formula differs from the kernel size calculation formula of ECA in that it does not require adjusting parameters α and β when the number of channels changes. Due to space constraints, we have provided a detailed explanation of this in the supplementary materials.

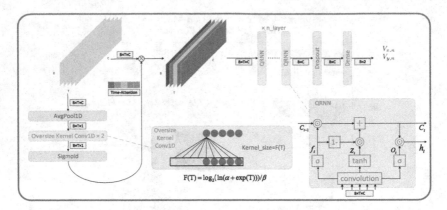

Fig. 2. Illustration of proposed TQRNN. The overall network structure consists of a Temporal attention module, QRNN layers, a Dropout layer, and a Dense layer. The Time Attention module comprises an average pooling layer, two convolutional layers, a sigmoid activation function, and a multiplier.

We have devised an oversize convolutional kernel, which provides the advantage of allowing each time step to contribute to the convolutional operation with information from all other time steps. Additionally, the parameters of the convolutional kernel are allowed to vary for each time step. For instance, in the case of the first and last time steps, only one parameter is shared. This approach ensures that each time step receives contributions from the entire temporal context while allowing flexibility in the parameterization of the convolutional kernel for different time steps.

After passing through the oversize convolutional kernel, we obtain the attention weights. These weights are then multiplied element-wise with the original inputs and transferred to the subsequent QRNN network. The visual depiction of the calculation flow for input and output is provided in the Fig. 2, and we present it again here:

$$
\begin{aligned}
\mathbf{z}_t &= \tanh\left(\mathbf{W}_z * \mathbf{X} + \mathbf{b}_z\right) \\
\mathbf{f}_t &= \sigma\left(\mathbf{W}_f * \mathbf{X} + \mathbf{b}_f\right) \\
\mathbf{o}_t &= \sigma\left(\mathbf{W}_o * \mathbf{X} + \mathbf{b}_o\right) \\
\mathbf{c}_t &= \mathbf{f}_t \odot \mathbf{c}_{t-1} + (1 - \mathbf{f}_t) \odot \mathbf{z}_t \\
\mathbf{h}_t &= \mathbf{o}_t \odot \mathbf{c}_t,
\end{aligned} \tag{2}
$$

The number of layers in the QRNN module is also determined through optimizer training. Finally, the predicted labels, which represent the direction of the fingers on the x and y axes, are outputted through a dropout module and a dense layer.

3.4   Evaluation Metrics

Two widely used evaluation metrics(Root Mean Square Error (RMSE) and Pearson's Correlation Coefficient (CC)) were employed to quantify the decoding performance. RMSE provides a direct measure of the improvement in decoding performance, while CC compensates for the inability of RMSE to capture errors in relation to the overall performance. The specific formulas are as follows:

$$\text{RMSE} = \sqrt{\sum_{i=1}^{N} (\hat{y}_i - y_i)^2 / N}$$

$$\text{CC} = \frac{\sum_{t=1}^{N} (y_t - \bar{y})(\hat{y}_t - \bar{\hat{y}}_t)}{\sqrt{\sum_{t=1}^{N} (y_t - \bar{y})^2} \sqrt{\sum_{t=1}^{N} (\hat{y}_t - \bar{\hat{y}}_t)^2}}$$

(3)

Where y_i represents the true labels, $\hat{y}$ represents the predicted labels, $\bar{y}$ is the mean of the true labels, $\bar{\hat{y}}$ is the mean of the predicted labels, and N is the total number of samples. A smaller RMSE indicates a smaller error between the predicted and true values. CC ranges from –1 to 1, and values closer to ±1 indicate a stronger linear correlation between the predicted and true values.

4 Experiments

4.1 Implementation Details

To optimize the decoders used in our experimental section, we applied Optuna, a Bayesian optimization framework [4]. In order to demonstrate the effectiveness and robustness of the proposed architecture, we used the same set of hyperparameters as the QRNN for all the ablation experiments. These hyperparameters were optimized for the original QRNN architecture and not specifically for the proposed TQRNN architecture. Kalman Filter decoder does not require optimization and only requires linear regression to estimate its matrix parameters. The detailed hyperparameters are listed in Table 1.

To evaluate the performance of the proposed method, we employed a k-fold growing-window forward validation scheme [17]. In this scheme, the minimum size of the training data within each session was set to 50% of the entire dataset and k was set to 5. This scheme is more suitable for evaluating the decoding performance of MUA as it considers the sequential information.

Table 1. Hyperparameter configuration

Hyperparameter	MLP	LSTM	QRNN/TQRNN
Time steps	1	5	5
n_layers	2	1	1
units	350	250	600
epoch	6	11	14
Batch size	64	32	96
Dropout	0.3	0.3	0.5
Learning rate	0.0035	0.0101	0.0072
Optimiser	RMSProp	RMSProp	RMSProp

4.2 Comparison of Decoding Results

As shown in Sect. 3, we proposed a Temporal-attention QRNN structure. And in this subsection, we have validated the high decoding performance and robustness of our method from four different perspectives. In Table 2, we present a comparison of its performance with other methods. Table 3 provides results from ablation experiments. Figure 3 provides snippet examples comparison of four deep learning methods. Figure 4 demonstrates the decoding performance comparison for long time scales, specifically for 18 d within a year after electrode implantation. The data used in Table 2, Table 3, Fig. 3 is from the 41st day after electrode implantation, while the data used for Fig. 4 spans 18 days within a year after electrode implantation.

Table 2. Comparison of RMSE, CC and Parameters for Different Methods

Methods	Fold-1	Fold-2	Fold-3	Fold-4	Fold-5	Avg. (std)	CC	Params
KF [24]	34.44	44.56	42.87	41.34	37.82	40.21 (4.07)	0.723	None
MLP [1]	26.86	34.90	34.38	37.13	30.30	32.71 (1.83)	0.817	156102
LSTM [18]	28.34	34.96	32.59	34.83	28.94	31.93 (1.41)	0.806	343502
QRNN (SOTA) [1]	23.82	34.54	32.30	35.02	28.14	30.77 (2.12)	0.826	334202
TQRNN (ours)	23.43	33.71	31.05	33.05	27.29	29.71 (1.93)	0.840	334256

In Table 2, KF is the most widely used method due to its real-time capabilities and minimal training data requirement. QRNN is the baseline of our proposed structure and the state-of-the-art (SOTA) method. It can be observed that in the cross-validation, our method outperforms all other methods in each fold. The average RMSE is reduced by 1.06 compared to the current SOTA, resulting in a performance improvement of 3.45%. The standard deviation (std) also slightly decreases, indicating improved decoding stability of the new structure. Most importantly, our approach achieves comprehensive performance improvements over QRNN (SOTA) while only introducing an additional 54 parameters, which is nearly negligible.

Figure 3 demonstratses a comparison of decoding performance in two degrees of freedom: finger motion along the x-axis and y-axis. In the x-axis decoding task (c,d), it can be observed that the proposed TQRNN outperforms the QRNN in decoding sudden velocity changes at 3.8 s and exhibits a smaller decoding error at 4.7 s. In the y-axis decoding task(g, h), compared to the QRNN, TQRNN demonstrates more precise decoding of three extremum at 5 s.

In Fig. 4, we present a comparison of RMSE and CC for the four methods in long-term decoding. It can be observed that our method consistently achieves the best performance across all experiments, with improvements in RMSE ranging from 0.1 to 1.06 compared to QRNN.

In Table 3, we conducted ablation experiments. Except for QRNN, all other methods are proposed for the first time in this study for MUA motion decoding.

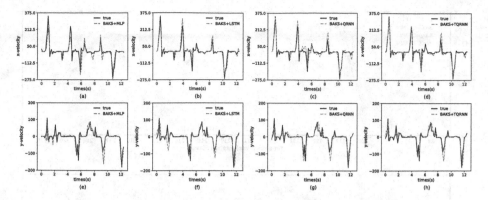

Fig. 3. Snippet examples of decoding results. From left to right, the colors represent MLP (red, RMSE = 32.71), LSTM (blue, RMSE = 31.93), QRNN (green, SOTA, RMSE = 30.77), and TQRNN (purple, proposed method, RMSE = 29.71). (Color figure online)

From top to bottom, they are: QRNN decoder with ECA-Net as the time attention module, QRNN decoder with SENet as the time attention module, TQRNN network with only one convolutional layer, and TQRNN network without the use of oversize convolutional kernels. It can be observed that although the method adopted in our study was not the best in all folds, the average RMSE achieved the best result of 29.71 in the ablation experiments.

4.3 Discussion

In the previous subsection, we have demonstrated the efficacy of our method by an experimental perspective. Herein, we discuss iBMI neural decoding from an information theory standpoint.

In Fig. 5, we depicts a Markov chain modeling for invasive brain-machine interface decoding task. Within this framework, we first define our objective as follows:

$$\min_{NRS,W_{1,2,3}} I(Y;N) - I(\hat{Y};N) \tag{4}$$

By considering only the part influenced by deep learning models, our problem can be reformulated as follows:

$$\min_{W_{1,2,3}} I(\hat{Y};X), \text{subject to } I(\hat{Y};Y) = I(X;Y) \tag{5}$$

However, the decoding of temporal information in the model is not fully detailed, as it is only processed within the QRNN module by providing the previous time step input with a forget factor. This approach assumes the distribution of information over time in advance and then performs fine-tuning. Consequently, this leads to significant information loss, as expressed in the following equation:

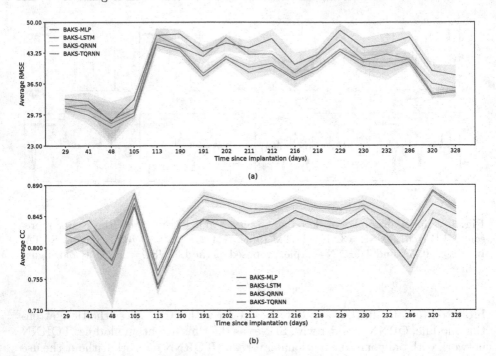

Fig. 4. Long-term decoding performance comparison across four methods. The light colors represent the range of std.

Table 3. Ablation study

Methods	Fold-1	Fold-2	Fold-3	Fold-4	Fold-5	Avg. (std)	CC
QRNN (SOTA)	23.82	34.54	32.30	35.02	28.14	30.77 (2.12)	0.826
ECA+QRNN	23.19	35.39	31.43	33.81	27.33	30.23 (2.22)	0.834
SENet+QRNN	23.92	35.01	30.92	32.84	27.55	30.05 (1.96)	0.838
TQRNN-1	23.22	34.74	31.47	33.02	27.10	29.91 (2.10)	0.839
TQRNN-k	23.44	33.82	30.94	32.96	28.94	30.02 (1.85)	0.839
TQRNN (Proposed)	23.43	33.71	31.05	33.05	27.29	29.71 (1.93)	0.840

$$L_{qrnn}(W_2) = I((T_1, C_1); Y) - I((T_2, C_2); Y) \qquad (6)$$

To address this problem, we first need to understand the basic theorem of two Markov chains. Firstly, in a Markov chain from X to Z, the mutual information at each layer decreases gradually, which can be expressed as follows:

$$I(X; Y) \geq I(X; Z) \text{ subject to Markov } X \rightarrow Y \rightarrow Z \qquad (7)$$

So, how can we reduce the information loss in Eq. 6 in this case? Our idea is to preprocess the time dimension before the QRNN module, making its features more prominent. This allows the QRNN module to more accurately interpret the information in the time dimension. Therefore, we can reformulate our problem

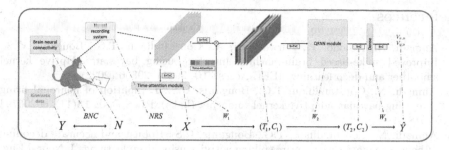

Fig. 5. A Markov chain modeling for invasive brain-machine interface motion decoding task

as follows:

$$\min_{W_1} L_{qrnn}(W_2), \text{subject to } I((T_1, C_1); Y) = I(X; Y) \tag{8}$$

Furthermore, in information theory, reversible transformations do not change the mutual information. This can be stated as the following theorem:

$$I(X; Y) = I(\psi(X); \phi(Y)), \text{where } \psi \text{ and } \phi \text{ are reversible transformations} \tag{9}$$

And time attention is a reversible transformation. So, we introduce the temporal attention module, and the experimental results are indeed satisfactory.

5 Conclusion

In this paper, we proposed a TQRNN neural decoder. It achieved a 3.45% improvement in performance compared to the current state-of-the-art on MUA decoding for hand movement tasks with the cost of parameters' increase of less than 0.1%. The Snippet examples also demonstrated that our proposed approach enables more accurate hand movement decoding. Ablation studies validated the rationale behind our improvements to QRNN. The decoding experiments on long-term data demonstrated the generality of our method. Additionally, this work introduced the concept of temporal attention for the first time in the neural decoding of hand movement tasks, and we have demonstrated, from an information-theoretic perspective, why temporal attention is effective in neural decoding.

Our next steps of work include: Firstly, integrating the decoder into real-time control and optimizing its online performance. Secondly, exploring the construction of a versatile channel attention module that can synergistically enhance decoding performance in conjunction with the temporal attention module. Thirdly, investigating more effective methods to utilize spike data for cortical control.

References

1. Ahmadi, N., Adiono, T., Purwarianti, A., Constandinou, T.G., Bouganis, C.S.: Improved spike-based brain-machine interface using bayesian adaptive kernel smoother and deep learning. IEEE Access **10**, 29341–29356 (2022)
2. Ahmadi, N., Constandinou, T.G., Bouganis, C.S.: Estimation of neuronal firing rate using bayesian adaptive kernel smoother (baks). PLoS ONE **13**(11), e0206794 (2018)
3. Ahmadi, N., Constandinou, T.G., Bouganis, C.S.: Robust and accurate decoding of hand kinematics from entire spiking activity using deep learning. J. Neural Eng. **18**(2), 026011 (2021)
4. Akiba, T., Sano, S., Yanase, T., Ohta, T., Koyama, M.: Optuna: a next-generation hyperparameter optimization framework. In: Proceedings of the 25th ACM SIGKDD International Conference on Knowledge Discovery & Data Mining, pp. 2623–2631 (2019)
5. An, G., Zhou, W., Wu, Y., Zheng, Z., Liu, Y.: Squeeze-and-excitation on spatial and temporal deep feature space for action recognition. In: 2018 14th IEEE International Conference on Signal Processing (ICSP), pp. 648–653. IEEE (2018)
6. Anumanchipalli, G.K., Chartier, J., Chang, E.F.: Speech synthesis from neural decoding of spoken sentences. Nature **568**(7753), 493–498 (2019)
7. Biasiucci, A., Franceschiello, B., Murray, M.M.: Electroencephalography. Curr. Biol. **29**(3), R80–R85 (2019)
8. Even-Chen, N., et al.: Power-saving design opportunities for wireless intracortical brain-computer interfaces. Nat. Biomed. Eng. **4**(10), 984–996 (2020)
9. Guo, M.H., et al.: Attention mechanisms in computer vision: a survey. Comput. Visual Media **8**(3), 331–368 (2022)
10. Kalman, R.E.: A new approach to linear filtering and prediction problems (1960)
11. Lansdell, B., Milovanovic, I., Mellema, C., Fetz, E.E., Fairhall, A.L., Moritz, C.T.: Reconfiguring motor circuits for a joint manual and BCI task. IEEE Trans. Neural Syst. Rehabil. Eng. **28**(1), 248–257 (2019)
12. Liang, F., et al.: Non-human primate models and systems for gait and neurophysiological analysis. Front. Neurosci. **17**, 1141567 (2023)
13. Makin, J.G., O'Doherty, J.E., Cardoso, M.M., Sabes, P.N.: Superior arm-movement decoding from cortex with a new, unsupervised-learning algorithm. J. Neural Eng. **15**(2), 026010 (2018)
14. Mitz, A.R., Bartolo, R., Saunders, R.C., Browning, P.G., Talbot, T., Averbeck, B.B.: High channel count single-unit recordings from nonhuman primate frontal cortex. J. Neurosci. Methods **289**, 39–47 (2017)
15. Nurmikko, A.: Challenges for large-scale cortical interfaces. Neuron **108**(2), 259–269 (2020)
16. O'Doherty, J.E., Cardoso, M.M., Makin, J.G., Sabes, P.N.: Nonhuman primate reaching with multichannel sensorimotor cortex electrophysiology. Zenodo (2017). https://doi.org/10.5281/zenodo.583331
17. Schnaubelt, M.: A comparison of machine learning model validation schemes for non-stationary time series data. Technical report, FAU Discussion Papers in Economics (2019)
18. Tseng, P.H., Urpi, N.A., Lebedev, M., Nicolelis, M.: Decoding movements from cortical ensemble activity using a long short-term memory recurrent network. Neural Comput. **31**(6), 1085–1113 (2019)

19. Vaskov, A.K., et al.: Cortical decoding of individual finger group motions using refit kalman filter. Front. Neurosci. **12**, 751 (2018)
20. Vilela, M., Hochberg, L.R.: Applications of brain-computer interfaces to the control of robotic and prosthetic arms. Handb. Clin. Neurol. **168**, 87–99 (2020)
21. Wang, Q., Wu, B., Zhu, P., Li, P., Zuo, W., Hu, Q.: ECA-NET: efficient channel attention for deep convolutional neural networks. In: Proceedings of the IEEE/CVF Conference on Computer Vision and Pattern Recognition, pp. 11534–11542 (2020)
22. Willett, F.R., Avansino, D.T., Hochberg, L.R., Henderson, J.M., Shenoy, K.V.: High-performance brain-to-text communication via handwriting. Nature **593**(7858), 249–254 (2021)
23. Willsey, M.S., et al.: Real-time brain-machine interface achieves high-velocity prosthetic finger movements using a biologically-inspired neural network decoder. In: bioRxiv, pp. 2021–08 (2021)
24. Wu, W., et al.: Neural decoding of cursor motion using a kalman filter. Adv. Neural. Inf. Process. Syst. **15**, 1–8 (2002)
25. Yin, M., et al.: Wireless neurosensor for full-spectrum electrophysiology recordings during free behavior. Neuron **84**(6), 1170–1182 (2014)
26. Zhang, X., et al.: The combination of brain-computer interfaces and artificial intelligence: applications and challenges. Ann. Transl. Med. **8**(11) (2020)
27. Zheng, Q., Zhang, Y., Wan, Z., Malik, W.Q., Chen, W., Zhang, S.: Orthogonalizing the activity of two neural units for 2d cursor movement control. In: 2020 42nd Annual International Conference of the IEEE Engineering in Medicine & Biology Society (EMBC), pp. 3046–3049. IEEE (2020)

Multi-head Attention Induced Dynamic Hypergraph Convolutional Networks

Xu Peng, Wei Lin, and Taisong Jin[✉]

Key Laboratory of Multimedia Trusted Perception and Efficient Computing, Ministry of Education of China, School of Informatics, Xiamen University, Xiamen, China
{penglingxiao,23020211153949}@stu.xmu.edu.cn, jintaisong@xmu.edu.cn

Abstract. Hypergraph neural networks (HGNNs) have recently attracted much attention from researchers due to the powerful modeling ability. Existing HGNNs usually derive the data representation by capturing the high-order adjacent relations in a hypergraph. However, incomplete exploration and exploitation of hypergraph structure result in the deficiency of high-order relations among the samples. To this end, we propose a novel hypergraph convolutional networks (M-HGCN) to capture the latent structured properties in a hypergraph. Specifically, two novelty designs are proposed to enhance the expressive capability of HGNNs. (1) The CNN-like spatial graph convolution and self-adaptive hypergraph incidence matrix are employed to capture both the local and global structural properties in a hypergraph. (2) The dual-attention scheme is applied to hyperedges, which can model the interactions across multiple hyperedges to form hypergraph-aware features. The experimental results on the benchmark citation network datasets demonstrate the superior performance of the proposed method over the existing strong baselines.

Keywords: Hypergraph Neural Network · Hypergraph Structure · Attention Mechanism

1 Introduction

Ubiquitous data with Non-Euclidean structures can be found in various real-world applications, which can be effectively modeled by graphs. Graph Neural Networks (GNNs) have been widely utilized to extract meaningful data representations from graph data, enabling them to be employed in diverse learning tasks such as action recognition, knowledge graph extraction, and recommendation system construction. However, graph is often only suitable to model the objects with pair-wise relations. For the more complex graph data, the learning performance of GNNs model is not promising as would be expected. Different from graph, hypergraph is composed of a vertex set and a hyperedge set, where each hyperedge can connect the arbitrary number of vertices. (A simple graph can be considered as a special hypergraph when a hyperedge connects only two

Q. Liu et al. (Eds.): PRCV 2023, LNCS 14433, pp. 256–268, 2024.
https://doi.org/10.1007/978-981-99-8546-3_21

vertices). Due to the flexibility and natural property of hyperedges, hypergraph has the superior modeling capabilities on multi-modal data and even more complex data (see Fig. 1). HGNN [5] is the first hypergraph neural network, which forms a vertex-hyperedge-vertex hypergraph convolution scheme. DHGNN [8] leverages dynamically changing hypergraph structure to propagate the vertex features. HCHA [1] proves that graph convolution is a special case of hypergraph convolution when a hyperedge connects only two vertices and introduces an attention mechanism between the vertex and the hyperedge.

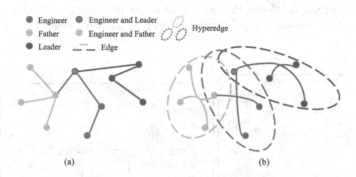

Fig. 1. Example of modeling people's social identities via graph and hypergraph. (a) Graph-based representation. Vertices represent people's identities and edges connect two people with the same identity. (b) Hypergraph-based representation. Hyperedge represents a group of people with the same identity.

Although the aforementioned HGNNs achieve promising performance on some benchmark graph datasets, the existing HGNNs generally focus on the propagation of vertex messages and ignore the utilization of the latent structured information within or across different hyperedges. Specifically, the advantages of modeling the high-order data correlations in each hyperedge of a hypergraph remain under explored. To further enhance the performance of HGNNs in real-world applications, it is crucial to design an innovative hypergraph-based deep learning architecture.

Inspired by the recent advances in GNNs, we propose a novel hypergraph convolutional networks method, termed M-HGCN, to derive structured data representation extracted from the hypergraph. Figure 2 shows the framework of the proposed method. For *vertex convolution*, ranking and max pooling operations are employed to transform the vertex data within a hyperedge into grid-like data, on which CNN-like spatial graph convolution and self-adaptive hypergraph incidence matrix are performed to derive the hyperedge features. For *hyperedge convolution*, self-attention mechanism and multi-head attention mechanism are leveraged on the hyperedge features to re-aggregate hyperedge features to vertex features. The contributions of our work are summarized as follows:

- We bridge the gap between graph convolution and hypergraph convolution to capture both the local neighborhood information within each hyperedge and the global structured property across different hyperedges.
- Each hyperedge is treated as a fully connected graph and the CNN-like spatial graph convolution is employed within each hyperedge to aggregate local neighboring structure. Furthermore, the self-adaptive hypergraph incidence matrix is proposed to make the vertex features reflect the global property cross a hyperedge.
- A hyperedge convolution with dual-attention mechanism is proposed to interact the hypergraph information among multiple hyperedges, which can further enhance the expressive capability of hypergraph neural networks.

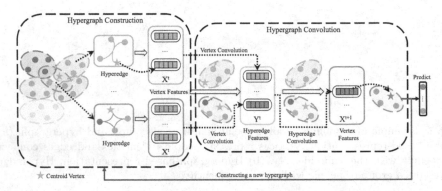

Fig. 2. The framework of the proposed method. First, hyperedges (red and green dashed lines) associated with a centroid vertex are generated from hypergraph construction. Second, vertex features within each hyperedge are propagated using *vertex convolution* to obtain hyperedge features. Third, hyperedge features are propagated to the respective vertices using *hyperedge convolution*. By stacking layers of *vertex convolution* and *hyperedge convolution* the data representation for the downstream learning task (in our case, node classification) is obtained. (Color figure online)

2   Related Work

2.1   Neural Networks on Graph

Graph is commonly employed to model intricate interactions between samples [6]. To encode graph-structured data, GNNs have been introduced, which are typically categorized into two groups: spatial-based approaches and spectral-based approaches.

From the spectral-based perspective, graph convolution operation is to design the trainable shared parameters of graph neural networks. The first graph convolutional network [2] is a local convolution based on spectral graph theory.

ChebNet [3] is designed to avoid the decomposition of the Laplacian matrix, and the shared parameters can be represented as Chebyshev polynomials. GCN [9] uses truncated Chebyshev polynomials as shared parameters to simplify the operation process.

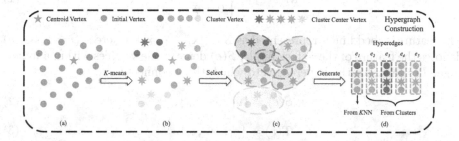

Fig. 3. Hypergraph construction. (a) Data samples. (b) K-means clustering result. (c) K-nearest neighbors (KNN) of centroid, including the centroid itself, are selected to form the main hyperedge. Adjacent hyperedges are created by selecting cluster centers closest to the centroid. (d) Derived hyperedges.

From the spatial-based perspective, graph convolution is used to aggregate neighbors feature. With the advent of GAT [11], the trainable aggregation operation is extended to the general graph structured data, where the attention mechanism is designed to aggregate the adjacent vertices feature. P-GNN [12] not only utilizes the vertex features but also takes into account the vertex positional information when performing the aggregation operation.

2.2   Neural Networks on Hypergraph

The modeling of pairwise relations via graphs, however, is limited in effectively capturing high-order relations among multiple samples. Hypergraph learning, on the other hand, has shown promising performance in learning data representation [14]. The concept of hypergraph learning was first introduced to model high-order data correlations, and it has gained popularity in subsequent studies [10,13]. Recently, HGNNs have been proposed to extend the convolution operation or attention mechanism from graphs to hypergraphs. To further enhance the performance of hypergraph-based representation learning, various HGNN models, such as DHSL [15] and HGNN$^+$ [7], have been developed to flexibly learn both the structures and parameters of hypergraph neural networks.

3   Methodology

3.1   Definitions and Notations

Hypergraph $\mathcal{G}$ is defined as $\mathcal{G} = (\mathcal{V}, \mathcal{E})$, where $\mathcal{V} = \{v_1, \cdots, v_s\}$ represents the vertex set. v denotes the centroid vertex. s is the total number of vertices. $\mathcal{E} =$

$\{e_1, \cdots, e_t\}$ is the hyperedge set. e denotes the hyperedge. $e_i = (v_1^{(i)}, \cdots, v_n^{(i)})$, n is the number of vertices in a hyperedge. We use an $|\mathcal{V}| \times |\mathcal{E}|$ incidence matrix $\mathbf{M}$ to denote hypergraph $\mathcal{G}$, whose entries are defined as

$$\mathbf{M}(v, e) = \begin{cases} 1, v \in e \\ 0, v \notin e, \end{cases} \tag{1}$$

The feature embedding is represented as $\mathbf{X} = [\mathbf{x}_1, \cdots, \mathbf{x}_q]$, where $\mathbf{x}_i (i = 1, \cdots, q)$ denotes the feature of the i-th example. $S(e)$ denotes the vertices within a hyperedge. $T(v)$ denotes the hyperedges to which the v connects.

$$S(e) = \{v_1, \cdots, v_n\}, \tag{2}$$

$$T(v) = \{e_1, \cdots, e_l\}, \tag{3}$$

where l is the number of hyperedges connecting v. $\mathbf{x}_v \in \mathbb{R}^{1 \times d}$ denotes a vertex feature embedding, d denotes feature dimension. When stacking all vertex features in $S(e)$, $\mathbf{X}_v \in \mathbb{R}^{n \times d}$ is derive. After vertex convolution, the hyperedge feature $\mathbf{x}_e \in \mathbb{R}^{1 \times d}$ is generated. When stacking all the hyperedge features in $T(v)$, $\mathbf{X}_e \in \mathbb{R}^{l \times d}$ is obtained.

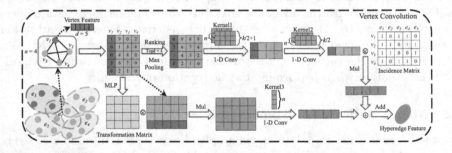

Fig. 4. Vertex convolution. (1) For the upper half, the vertex features are transformed into grid-like data. The red 1s indicate that randomly selected vertices and hyperedges to be connected. (2) For the bottom half, the vertex features are multiplied by a transformation matrix and aggregated using 1-D convolution. The resulting hyperedge feature is a combination of these features. (Color figure online)

3.2   Hypergraph Construction

Figure 3 shows the procedure of hypergraph construction. For each vertex, it's $n - 1$ nearest neighbors and itself are chosen to form $S(e)$. Besides, the k-means algorithm is performed on the whole feature map according to Euclidean distance, and $T(v)$ is determined by selecting the nearest $l - 1$ clusters for each vertex.

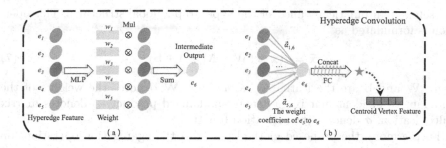

Fig. 5. Hyperedge convolution. (a) The hyperedge features are first aggregated via self-attention mechanism. (b) Multi-head attention mechanism is then applied to determine the importance of different hyperedge features in relation to the aggregated feature.

3.3   Vertex Convolution

In spatial graph convolution on a hypergraph, inconsistent neighbors pose a challenge due to varying receptive field sizes. To tackle this, the proposed method introduces a novel vertex convolution scheme. This scheme treats each hyperedge as a fully connected graph, where the 1-order neighborhood of each vertex in the hyperedge is identical. Furthermore, the feature of each vertex needs to be ranked, which is formulated as

$$\tilde{\mathbf{X}}_v = p(\mathbf{X}_v^\top, k),$$ (4)

where $p(\cdot)$ denotes the ranking of vertex feature in $\mathbf{X}_v$ and then perform max pooling to derive k-largest values in each vertex feature. $\cdot^\top$ represents matrix transposition.

After applying the function $p(\cdot)$ to the features of vertices within a hyperedge, $\tilde{\mathbf{X}}_v \in \mathbb{R}^{k \times n}$ can be interpreted as a one-dimensional (1-D) grid-like structure. As a result, the conventional convolutional operation can be expressed as

$$\mathbf{x} = c(\tilde{\mathbf{X}}_v),$$ (5)

where $c(\cdot)$ denotes 1-D convolution.

A self-adaptive hypergraph incidence matrix $\mathbf{M}_{adp} \in \mathbb{R}^{n \times d}$ is proposed to capture the local structure feature of the hypergraph while considering its global properties. The initial self-adaptive hypergraph incidence matrix is defined as

$$\mathbf{M}_{adp} = \Omega(\mathbf{M}),$$ (6)

where $\mathbf{M}$ denotes a hypergraph incidence matrix. $\Omega(\cdot)$ means that some regions in the matrix with value 0s are set to 1s.

The proposed vertex convolution scheme multiplies the extracted feature with the initial self-adaptive incidence matrix. As such, the proposed convolution scheme can capture the hidden spatial dependencies in the dynamic network by itself and automatically uncover the hypergraph global structures. This operation is formulated as

$$\tilde{\mathbf{x}} = \sigma(\mathbf{x} \otimes (\mathbf{W} \odot \mathbf{M}_{adp}) + \mathbf{b}), \tag{7}$$

where $\mathbf{W}$ and $\mathbf{b}$ are the learnable parameters. $\mathbf{W}$ denotes the weights of the self-adaptive incident matrix. $\odot$ denotes hadamard product. $\otimes$ denotes matrix multiplication. σ denotes the activation function.

Furthermore, the proposed method performs the aggregation operations on vertex features, defined as

$$\mathbf{H} = MLP(\mathbf{X}_v), \tag{8}$$

$$\hat{\mathbf{x}} = c(\mathbf{H} \otimes MLP(\mathbf{X}_v)), \tag{9}$$

where $\mathbf{H} \in \mathbb{R}^{n \times n}$ is the transformation matrix from Multi-layer Perception (MLP) for the feature permutating and weighting, $c(\cdot)$ denotes 1-D convolution, $\otimes$ denotes the matrix multiplication.

Hyperedge feature is generated by combining $\tilde{\mathbf{x}}$ and $\hat{\mathbf{x}}$

$$\mathbf{x}_e = \tilde{\mathbf{x}} + \hat{\mathbf{x}}, \tag{10}$$

The vertex convolution framework is shown in Fig. 4.

3.4   Hyperedge Convolution

The hyperedge convolution aims to re-aggregate hyperedge features to the vertex. To enhance the expressive capability of hyperedge features, we propose a novel hyperedge convolutional scheme based on a dual-attention mechanism. The proposed hyperedge convolution scheme leverages the self-attention mechanism to aggregate the hyperedge features and then adopts a multi-head attention mechanism on the aggregated hyperedge to interact with different hyperedges.

For self-attention mechanism, MLP is employed to generate the weights of hyperedges. Furthermore, the hyperedge features are aggregated by computing the weighted sum of hyperedge features, defined as

$$\mathbf{W} = softmax(MLP(\mathbf{X}_e)), \tag{11}$$

$$\mathbf{x}_{agg} = \mathbf{W} \odot \mathbf{X}_e, \tag{12}$$

where $\mathbf{W} = [\mathbf{w}_1, \cdots, \mathbf{w}_l]$ denotes each hyperedge weight. $\odot$ denotes the hadamard product. $\mathbf{x}_{agg}$ denotes the aggregated hyperedge feature.

Multi-head attention is defined as

$$\lambda_{j,agg} = \mathbf{a}[\mathbf{W}\mathbf{x}_j || \mathbf{W}\mathbf{x}_{agg}], \tag{13}$$

$$\alpha_{j,agg} = \frac{\exp(\lambda_{j,agg})}{\sum_{j \in T(u)} \exp(\lambda_{j,agg})}, \tag{14}$$

$$\mathbf{x}_v = \sigma \left[fc \left(\overset{M}{\underset{m=1}{\big\|}} \sum_{j \in T(u)} \alpha_{j,agg}^m \mathbf{W}^m \mathbf{x}_j \right) \right], \tag{15}$$

where $[\cdot \| \cdot]$ denotes the concatenation, $\mathbf{a}$ is a learnable weight vector that aims to map the concatenated feature to a real number. $\mathbf{W}$ is a shared weight matrix. $\lambda_{j,agg}$ denotes the importance of hyperedge $\mathbf{x}_j$'s feature to aggregated feature. $\alpha_{j,agg}$ represents the weight coefficient after the $softmax$ operation. M indicates the number of independent attention heads. fc denotes fully connected layer, $\sigma(\cdot)$ denotes activation function.

The hyperedge convolution framework is shown in Fig. 5.

Algorithm 1. M-HGCN

   Input: vertex features embedding $\mathbf{X}$; empty list $\mathbf{xlist}, \mathbf{ylist}$.
   Output: the node classification prediction results $\boldsymbol{f}$.
1 **for** $layers\ l = 1, \cdots, N$ **do**
2     $S(e), T(v), \mathbf{M}_{adp} \leftarrow HyperCons(\mathbf{X})$ **for** v **in** $\mathbf{X}$ **do**
3   **for** e **in** $T(v)$ **do**
4          $\mathbf{X}_v \leftarrow stack(S(e))$ $\mathbf{x}_e \leftarrow VertConv(\mathbf{X}_v)$ by Eqs. (4)~(10)
         $\mathbf{xlist}.insert(\mathbf{x}_e)$
5   **end**
6        $\mathbf{X}_e \leftarrow stack(\mathbf{xlist})$ $\mathbf{y} \leftarrow HyperConv(\mathbf{X}_e)$ by Eqs. (11)~(15)
       $\mathbf{ylist}.insert(\mathbf{y})$ $\mathbf{xlist}.clear()$
7   **end**
8     $\mathbf{X} \leftarrow stack(\mathbf{ylist})$ $\mathbf{ylist}.clear()$
9 **end**
10 $\boldsymbol{f} \leftarrow Predict(\mathbf{X})$

3.5 The Proposed Algorithm

The main procedure of the proposed M-HGCN method is listed in Algorithm 1. For each vertex, multiple $S(e)$ and $T(v)$ associated with it are generated after hypergraph construction ($HyperCons(\cdot)$). To update each vertex feature, vertex convolution ($VertConv(\cdot)$) and hyperedge convolution ($HyperConv(\cdot)$) are performed in $S(e)$ and $T(v)$, respectively. The new hypergraph structure is dynamically constructed based on the obtained vertex features. $stack(\cdot)$ represents stacking features operation.

4 Experiments

4.1 Datasets

We leverage the benchmark citation network datasets including Cora, Citeseer, Pubmed to conduct node classification experiments, where the details of different datasets are listed in Table 1.

Table 1. The statistic of benchmark citation network datasets.

Dataset	Cora	Citeseer	Pubmed
nodes	2,708	3,327	19,717
Feature	1,433	3,703	500
Training nodes	140	120	60
Validation nodes	500	500	500
Test nodes	1,000	1,000	1,000
Classes	7	6	3

For the Cora dataset, we follow an experimental setup in [8]. To prevent the influence of data distribution, the standard split and randomly selected different proportions of dataset are chosen as the training set. The standard split uses the fixed training samples with 5.2% of the dataset. The proportions of the randomly selected training set are chosen to be 2%, 5.2%, 10%, 20%, 30%, and 44%, respectively.

For the Citeseer and Pubmed datasets, we follow the experimental setup in [9]. The standard split for the Citeseer dataset uses 3.6% of the data for training, while the Pubmed dataset uses 0.3% for training.

4.2 Experimental Settings

The number of cluster centers is set as 400 and the nearest cluster is selected as the adjacent hyperedge. Each hyperedge consists of 128 vertices. A two-layer M-HGCN is employed in our experiment. Dropout layers are applied to avoid overfitting with rate of 0.5. Moreover, 8 largest values are chosen to perform max pooling operation during the vertex convolution. The randomly selected connection rate of the initial incident matrix is set to 50%. 8 attention heads are applied to the multi-head attention calculation. We minimize the cross-entropy loss function using Adam optimizer with learning rate of 0.001. The model is trained for a maximum of 1,000 epochs.

4.3 Results and Discussion

We repeat all experiments 30 times and report the average results of different classification accuracies. The experimental results are listed in Tables 2 and 3.

As shown in Table 2, for the standard split of three datasets, the proposed M-HGCN significantly and consistently outperforms the baseline methods, which achieves the best classification performance. Specifically, M-HGCN achieves better performance compared with LGCN, GAT and other graph-based methods. For example, M-HGCN obtains gains of 2.2%, 3.4%, and 1.9% compared with GCN. Compared with other hypergraph-based methods, *i.e.*, HGNN and DHGNN, M-HGCN can also yields better performance. In particular, M-HGCN outperforms HCHA by 1.0%, 2.5%, and 2.5%.

Table 2. Node classification accuracies (%) on the citation network datasets.

Method	Cora	Citeseer	Pubmed
GCN [9]	81.5	70.3	79.0
GAT [11]	83.0	72.5	79.0
HGNN [5]	81.6	71.9	80.1
LGCN [6]	83.3	73.0	79.5
DHGNN [8]	82.5	70.0	79.9
HNHN [4]	-	70.5	78.3
HCHA [1]	82.7	71.2	78.4
HGNN$^+$ [7]	80.9	70.4	-
M-HGCN (ours)	**83.7**	**73.7**	**80.9**

Table 3. Node classification accuracies on the Cora dataset with different splits. "#Train" stands for the number of training samples.

Label Rate(%)	#Train	GCN(%)	HGNN(%)	GAT(%)	DHGNN(%)	M-HGCN(%)
2	54	69.6	75.4	74.8	76.9	**80.5**
5.2	140	77.8	79.7	79.4	80.2	**81.2**
10	270	79.9	80.0	81.5	81.6	**83.3**
20	540	81.4	80.1	83.5	83.6	**85.1**
30	812	81.9	82.0	84.5	85.0	**85.5**
44	1,200	82.0	81.9	85.2	85.6	**86.3**

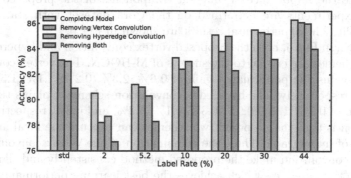

Fig. 6. Ablation studies on hypergraph convolution. "std" stands for standard split of Cora dataset. For the ablated model, MLP is applied to transform the vertex features or hyperedge features.

As shown in Table 3, for the Cora dataset with random split, M-HGCN achieves superior performance over DHGNN by a margin of 3.6%, 1.0%, 1.7%, 1.5%, 0.5%, 0.7% when 2%, 5.2%, 10%, 20%, 30%, 44% randomly sampled data is used as a training set, respectively. As the size of the training set increases, the performance of the compared methods increases accordingly. Compared to the baseline methods, the proposed method demonstrates the superior performance.

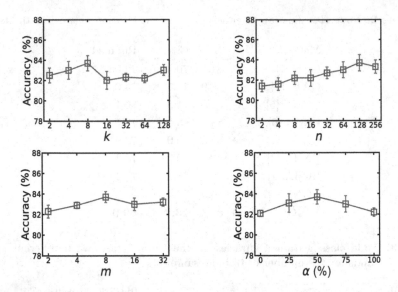

Fig. 7. Classification accuracy varies with changes in k, n, m, and α on the Cora standard split dataset.

4.4 Ablation Studies

To evaluate the importance of different components of the proposed method, ablation experiments are performed on the Cora dataset with standard and random split. The experimental results are shown in Fig. 6.

As shown in Fig. 6, both the proposed vertex convolution and hyperedge convolution schemes are crucial to the success of M-HGCN. The vertex convolution scheme explains the performance gain of 0.6%, 2.3%, 0.2%, 1.1%, 1.3%, 0.2%, and 0.2%, respectively. The hyperedge convolution scheme explains the performance gain of 0.7%, 1.8%, 0.9%, 0.3%, 0.1%, 0.3%, and 0.3%, respectively. The main reason is that the proposed two schemes can capture the local and global properties of hypergraph. Thus, the combination of the vertex convolution and hyperedge convolution make the proposed method consistently and significantly superior to the baselines, which achieves the best learning performance.

4.5 Hyperparameter Sensitivity

The main hyperparameters of the M-HGCN are k, n, m, and α. k represents the k-largest values in each vertex feature. n denotes the number of vertices in a hypergraph. m denotes the number of independent attention heads. α indicates the sparsity of the parts of the incident matrix where the vertices are not connected to the hyperedges (initial values in the incident matrix are denoted as 0). Figure 7 shows the impact of these parameters on node classification accuracy.

As shown in Fig. 7, the performance is stable when the parameters are varied within a large range. Specifically, the performance improved ($\alpha > 0$) after some connections are added between the vertices and hyperedges, which demonstrates that the utilization of the hypergraph structure facilitates the hypergraph-based data representation, and this is ensured by the fact that the performance increases as n grows.

On the other hand, the performance is degraded slightly when k is larger ($k > 16$). The main reason is that the features within a hyperedge are repeatedly extracted during the pooling process, causing redundant information to appear. As a result, the subsequent attention mechanism fails to capture the important information of the data. Moreover, the performance gradually stabilizes as m continues to increase.

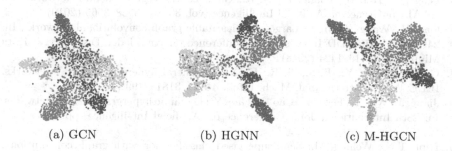

(a) GCN   (b) HGNN   (c) M-HGCN

Fig. 8. A t-SNE visualization of the computed feature representations of the GCN, HGNN and the M-HGCN model on the Cora dataset. Vertex colors denote classes.

4.6 Visualization

To demonstrate the superiority of our method intuitively, we select the Cora dataset for a visualization comparison. The output of the last layer convolution is visualized using the t-SNE method, and the result is shown in Fig. 8.

As shown in Fig. 8, compared to GCN and HGNN, the proposed method produces more identifiable clusters, which validates the effectiveness of our method.

5 Conclusion

In this paper, we have proposed the M-HGCN to discover the hypergraph structure, where the method in the spatial domain is introduced to capture the hypergraph structure representation. The experimental results demonstrate that our method yields consistently better performance than state-of-the-art HGNNs methods on the benchmark datasets. For future work, we will focus on hypergraph structure learning. How to make vertex messages propagate efficiently and precisely on the hypergraph is still an urgent problem to be solved.

Acknowledgments. This work was supported by National Key R&D Program of China (No. 2022ZD0118202) and the National Natural Science Foundation of China (No. 62376101, No. 62072386).

References

1. Bai, S., Zhang, F., Torr, P.H.: Hypergraph convolution and hypergraph attention. Pattern Recogn. **110**, 107637 (2021)
2. Bruna, J., Zaremba, W., Szlam, A., LeCun, Y.: Spectral networks and locally connected networks on graphs. arXiv preprint arXiv:1312.6203 (2013)
3. Defferrard, M., Bresson, X., Vandergheynst, P.: Convolutional neural networks on graphs with fast localized spectral filtering. In: Advances in Neural Information Processing Systems, vol. 29 (2016)
4. Dong, Y., Sawin, W., Bengio, Y.: HNHN: hypergraph networks with hyperedge neurons. arXiv preprin arXiv:2006.12278 (2020)
5. Feng, Y., You, H., Zhang, Z., Ji, R., Gao, Y.: Hypergraph neural networks. In: AAAI Conference on Artificial Intelligence, vol. 33, pp. 3558–3565 (2019)
6. Gao, H., Wang, Z., Ji, S.: Large-scale learnable graph convolutional networks. In: 24th ACM SIGKDD International Conference on Knowledge Discovery & Data Mining, pp. 1416–1424 (2018)
7. Gao, Y., Feng, Y., Ji, S., Ji, R.: HGNN+: general hypergraph neural networks. IEEE Trans. Pattern Anal. Mach. Intell. **45**(3), 3181–3199 (2022)
8. Jiang, J., Wei, Y., Feng, Y., Cao, J., Gao, Y.: Dynamic hypergraph neural networks. In: 28th International Joint Conference on Artificial Intelligence, pp. 2635–2641 (2019)
9. Kipf, T.N., Welling, M.: Semi-supervised classification with graph convolutional networks. arXiv preprint arXiv:1609.02907 (2016)
10. Tu, K., Cui, P., Wang, X., Wang, F., Zhu, W.: Structural deep embedding for hypernetworks. In: AAAI Conference on Artificial Intelligence, vol. 32, no. 1 (2018)
11. Veličković, P., Cucurull, G., Casanova, A., Romero, A., Lio, P., Bengio, Y.: Graph attention networks. arXiv preprint arXiv:1710.10903 (2017)
12. You, J., Ying, R., Leskovec, J.: Position-aware graph neural networks. In: International Conference on Machine Learning, pp. 7134–7143 (2019)
13. Zhang, M., Cui, Z., Jiang, S., Chen, Y.: Beyond link prediction: predicting hyperlinks in adjacency space. In: AAAI Conference on Artificial Intelligence, vol. 32, no. 1 (2018)
14. Zhang, R., Zou, Y., Ma, J.: Hyper-SAGNN: a self-attention based graph neural network for hypergraphs. arXiv preprint arXiv:1911.02613 (2019)
15. Zhang, Z., Feng, Y., Ying, S., Gao, Y.: Deep hypergraph structure learning. arXiv preprint arXiv:2208.12547 (2022)

Self Supervised Temporal Ultrasound Reconstruction for Muscle Atrophy Evaluation

Yue Zhang[1], Getao Du[2], Yonghua Zhan[2], Kaitai Guo[1], Yang Zheng[1], Jianzhong Guo[3], Xiaoping Chen[4], and Jimin Liang[1(✉)]

[1] School of Electronic Engineering, Xidian University, Xi'an 710071, Shaanxi, China
20021110123@stu.xidian.edu.cn, {ktguo,zhengy}@xidian.edu.cn,
jimleung@mail.xidian.edu.cn
[2] School of Life Science and Technology, Xidian University, Xi'an 710071, Shaanxi,
China
yhzhan@xidian.edu.cn
[3] School of Physics and Information Technology, Shanxi Normal University, Xi'an
710071, Shaanxi, China
guojz@snnu.edu.cn
[4] National Key Laboratory of Human Factors Engineering, China Astronaut
Research and Training Center, Beijing, China

Abstract. Muscle atrophy is a widespread disease that can reduce quality of life and increase morbidity and mortality. The development of non-invasive method to evaluate muscle atrophy is of great practical value. However, obtaining accurate criteria for the evaluation of muscle atrophy under non-invasive conditions is extremely difficult. This paper proposes a self-supervised temporal ultrasound reconstruction method based on masked autoencoder to explore the dynamic process of muscle atrophy. A score-position embedding is designed to realize the quantitative evaluation of muscle atrophy. Ultrasound images of the hind limb muscle of six macaque monkeys were acquired consecutively during 38 days of head-down bed rest experiments. Given an ultrasound image sequence, an asymmetric encoder-decoder structure is used to reconstruct the randomly masked images for the purpose of modelling the dynamic muscle atrophy process. We demonstrate the feasibility of using the position indicator as muscle atrophy score, which can be used to predict the degree of muscle atrophy. This study achieves the quantitative evaluation of muscle atrophy in the absence of accurate evaluation criteria for muscle atrophy.

Keywords: Masked autoencoder · Muscle atrophy quantization · Position embedding · Ultrasound image

This work is supported by the National Natural Science Foundation of China (U19B2030, 61976167, 62101416, 11727813) and the Natural Science Basic Research Program of Shaanxi (2022JQ-708).

1   Introduction

Muscle atrophy is a widespread illness occurring with weightlessness, aging, and various diseases, including neuromuscular disorders, diabetes, and cancer [1,2]. The loss of muscle mass and function can reduce quality of life and increase morbidity and mortality. It is of great practical value to develop accurate muscle evaluation models for diagnosing muscle atrophy and assessing the effectiveness of preventive and therapeutic approaches. Previous works have studied muscle atrophy by measuring muscle thickness [3–5], cross-sectional area [6,7] and volume [8], but the reliability of muscle atrophy evaluation by muscle size measurement still needs further study [5,7]. This paper aims to develop a more accurate model for muscle atrophy evaluation using the advanced artificial intelligence (AI) techniques.

In the field of medical image analysis adopting AI techniques, various diseases are usually evaluated using classification models trained on medical images with pathological grading. Many studies have demonstrated the feasibility of this approach with encouraging results [9–11]. In experiments of animal models with muscle atrophy induced through well-established hind limb unloading or head-down bed rest, similar to other disease evaluation models, an intuitive approach is to develop muscle atrophy evaluation models based on the time labels of hind limb unloading or bed rest. However, muscle atrophy is a complex physiological process, which is characterized by slow to fast fiber-type transition [12,13]. Therefore, there is no strong correlation between the time labels and the degree of muscle atrophy. In this paper, we attempt to develop a muscle atrophy evaluation model without using time labels as strong label.

Although it is difficult to obtain accurate evaluation criteria for evaluating muscle atrophy under non-invasive conditions, it is relatively easy to study the dynamics of muscle atrophy using self-supervised methods. Muscle atrophy is a continuous, steadily changing physiological process, and the state of the muscles changes dynamically during the process of muscle atrophy [12,13]. Therefore, the muscle status of adjacent sampling points should be most similar throughout the hind limb unloading or bed rest. Given enough sampling points, the dynamic process of muscle atrophy can be modelled by mutual prediction of muscle states under different sampling points. Motivated by this observation, this study extends the masked autoencoder to the task of predicting muscle state during dynamic muscle atrophy. We randomly mask the muscle state under continuous sampling points and learn an autoencoder to predict the muscle state. It is similar to the Masked Autoencoder [14] in the field of computer vision. The difference is that we will focus more on how to embed position information into the self-attention mechanism. Due to the continuity of muscle atrophy, the best performance in predicting muscle state can only be obtained by positional embedding that best matches the dynamic change process of muscle atrophy. This study compares the performance of muscle state prediction under various positional embedding and designs a score-position embedding to represent the dynamic change process of muscle atrophy.

2 Related Work

2.1 Muscle Atrophy Evaluation

Many approaches are available to evaluate muscle atrophy, including electromyo-graphy (EMG), bioelectrical impedance analysis (BIA) [15], ultrasonography [8], computed tomography (CT) [16] and magnetic resonance imaging (MRI) [17]. They are used to diagnose muscle atrophy by measuring muscle shrink-age strength, muscle thickness and cross-sectional area. However, quantitative evaluation criteria for muscle atrophy is difficult to obtain. This makes the quantitative results obtained by these methods lack comparable indicators. With the intensive research on intelligent medicine, more and more work is dedicated to developing intelligent models for evaluating diseases. Radiomics-based and deep learning-based approaches have become the focus of many studies [9,18]. Generally speaking, it is relatively easy to develop an evaluation model of a disease based on medical images with precise pathological grading. However, accurate evaluation criteria for muscle atrophy are difficult to obtain, especially for non-invasive methods, which severely increases the difficulty of developing muscle atrophy evaluation models.

Skeletal muscles are heterogeneous mixture of myofibers with different con-tractile, endurance and metabolic properties. Their response to the same stimu-lus is often enormous [19]. Previous studies have shown that the degree of muscle atrophy is related to the type of myofibers. The slow and fast myofibers show different effects. Slow type I myofibers are more severely atrophied than fast type II myofibers and a shift from type I to type II myofibers occurs [20,21]. This suggests that the degree of muscle atrophy is not strongly correlated with the time, which also increases the difficulty of developing muscle atrophy evaluation models.

Our study avoids using hind limb unloading or bed rest time as strong label to develop muscle atrophy evaluation model. The quantitative evaluation of muscle atrophy is achieved through a self-supervised method.

2.2 Masked Autoencoder

The idea of masked autoencoder, a form of more general denoising autoencoders, is widely used in natural language processing [22] and computer vision [14]. Masking as a type of noise dates back to at least a decade ago [23]. One of its most successful developments is BERT [22], which is conceptually masked autoencoding on language tokens. Denoising/masked autoencoding methods for computer vision have been making continuous progress [14,24]. A series of recent methods are based on Transformer architectures [25] and are towards a uni-fied solution between vision and language. MAE [14] introduced an asymmetric encoder-decoder architecture for masked image modeling. Masked visual mod-eling has been proposed to learn effective visual representations based on the simple pipeline of masking and reconstruction. Our study follows this line of research.

Because muscle atrophy is a dynamic process in which changes of muscle status are continuous and steadily variable during the process of muscle atrophy, it allows us to learn the dynamic process of muscle atrophy by reconstructing the signal that is masked under continuous sampling. This study proposes a method of temporal ultrasound reconstruction based on the masked autoencoder to explore the dynamic process of muscle atrophy.

3   Method

3.1   Animal Model and Ultrasonography

The head-down bed rest experiment (HD) was used to induce muscle atrophy of rhesus macaques under weightless conditions. Six macaques were immobilized in the recumbent position on experimental beds inclined at $-6°$ for 38 days. All activities of the macaques were carried out on the experimental bed throughout the experiment. Using the self-control method, data obtained from 6 macaques during the initial phase of bed rest were used as the normal control group (D0 CTRL) and data obtained during bed rest were used as the atrophy experimental group (D1-D38 HD).

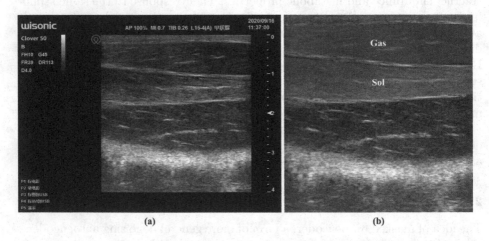

(a)   (b)

Fig. 1. Exemplary images of ultrasonography and muscle annotation. (a) Exemplary image obtained by the Clover 50 system. (b) Manual annotation of Soleus (Sol) and Gastrocnemius (Gas) muscles.

The Gastrocnemius (Gas) and Soleus (Sol) muscles of the macaques' left hind limb were selected as the main subjects. Ultrasound images of the macaques' left hind limb muscles were collected once each day while the macaques were in bed rest. The Clover 50 system was operated at frequency of 10 MHz and 40 mm depth for B-Mode acquisition. A video of 783 frames was obtained for each scan. Two hundred frames of muscle ultrasound images were randomly extracted from the video. The contours of the Gas and Sol muscles were manually delineated by one graduate student trained by a sonographer. An exemplar image is shown in Fig. 1.

3.2 Temporal Ultrasound Masked Autoencoder

Our approach is an extension of MAE [14] on temporal ultrasound data, as shown in Fig. 2. The following describes the proposed temporal ultrasound image reconstruction method.

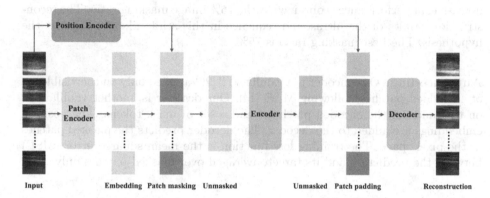

Fig. 2. Masked autoencoder on temporal ultrasound. The input patches are embedded by the patch encoder. Position embeddings from position encoder are added to the embedded patches. The encoder operates on the set of visible patch embeddings. The decoder then processes the full set of encoded patch embeddings and mask tokens to reconstruct the input.

Patch Encoder. Given an ultrasound image sequence with the length of 39 (D0-D38) of the same subject, each image is considered as a separate patch. A feature map of 768 channels is generated for each patch by three convolutional layers, and then the average value of each channel is calculated. The number of channels is consistent with the embedding dimension of each token in the encoder.

Position Encoder. We design a score-position embedding to quantitatively evaluate muscle atrophy. Each patch goes through three convolutional layers and one linear projection layer to generate the position embedding with the same dimension as the patch embedding. In order to avoid the position embedding learns random features that favor image reconstruction, which is contrary to the idea of position embedding, the last convolutional layer has only one channel and its output is averaged to a number (termed as position indicator hereafter) before being input to the linear projection layer. Finally, the linear projection layer maps the position indicator to the patch embedding dimension. In the following experiments, we verify the consistency of the position indicator obtained at multiple mask ratios and demonstrate that the position indicator can be used as the score of muscle atrophy.

Patch Masking. Patches are randomly masked from the patch embedding set according to a certain percentage. The masking strategy is referenced to the sampling strategy in MAE [14]. For MAE, it is assumed that the optimal masking ratio is related to the information redundancy of the data, and the optimal downstream performance is obtained at the 75% image masking ratio. The reconstruction results of the ultrasound sequence in this study likewise support the hypothesis. The best masking ratio is 75%.

Autoencoding. Our encoder is a vanilla ViT [26] applied only on the visible set of embedded patches, following MAE [14]. Our decoder is another vanilla ViT on the union of the encoded patch set and a set of mask tokens. The position embeddings are added to the decoder. The decoder predicts the masked patches in the pixel space. The training loss function is the mean squared error (MSE) between the prediction and its target, averaged over masked patches only.

4   Results

4.1   Implementation Details

In order to compare the different muscles during the dynamic changes of muscle atrophy, this study analyzed the Gas, Sol and the complete ultrasound images separately. All network models were implemented with PyTorch 1.9 on a single NVIDIA TITAN V graphics card. The learning ratio was initially set as 0.00015 and was decreased by cosine decay for 400 epochs in total. All models used adaptive moment estimation and weight decay (AdamW) optimizer with a batch size of 1. The leave-one-macaque-out cross-validation method was used for model evaluation.

4.2   Muscle Atrophy Evaluation by Masked Autoencoder

Masking Ratio. The influence of different masking ratios on the reconstruction performance of different muscles is shown in Fig. 3. Overall, for different muscles, the reconstruction performance increases steadily with the masking ratio until the sweet point. The optimal masking ratio for the ultrasound image reconstruction of different muscles is 75%.

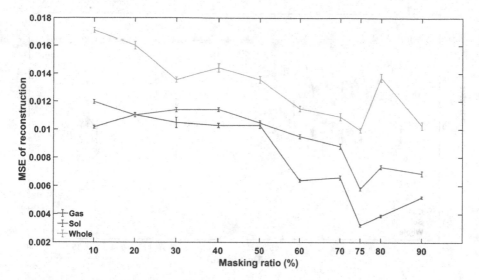

Fig. 3. Reconstruction performance for different muscles with different masking ratio. The y-axis is the mean squared error (MSE) between the prediction and its target, averaged over masked patches only.

Position Embedding. Figure 4 shows the position indicators for different muscles obtained at the optimal reconstruction performance. It can be noticed that during muscle atrophy, the variation trend of position indicators of different muscles is different. For the Gas muscle, the position indicator decreases overall with the duration of bed rest and remains largely stable from D10-D30.

For the Sol muscle, the position indicator shows a complex trend. It declines rapidly in the first few days of bed rest, rises gradually after D5, reaches the same level as Gas at D16, remains basically stable thereafter, and declines rapidly again after D30. The results may be attributed to the more complex transformation of the fiber types in the Sol muscle during muscle atrophy. In the skeletal muscles, multiple fiber types are generally intermingled within a single muscle group, and different muscle groups have varying proportions of fiber types [27]. The Gas is considered to be a fast-twitch fiber [21], while the Sol is considered to be a mixture of slow and fast-twitch fiber [21]. During muscle atrophy, the slow-twitch fibers in the Sol will transform into fast-twitch fibers, resulting in changes in the proportion and degree of atrophy of the various muscle fiber components in the Sol. In contrast, similar muscle fiber type transformation does not occur in the already fast-twitch Gas. This causes the position indicator of the Sol muscle to drift towards that of the Gas muscle.

For the complete ultrasound images, the position indicator shows a concave shape. More redundant information may have contributed to the inconsistency of this trend.

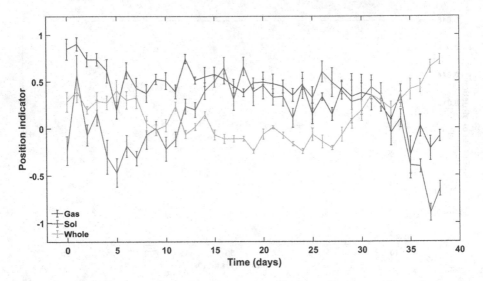

Fig. 4. Plot of position indicators at different bed rest time. The error bar represents the variation range of the position indicator for the six macaques at the same sampling point, showing a good consistency among different macaques.

4.3 Adequacy of Position Indicator

Due to the continuity of muscle atrophy, the position indicator can be used as the score of muscle atrophy only when the position embedding learns the relative positions of different muscle states in the ultrasound sequence. To demonstrate that the position embedding indeed learn the relative position rather than random features under the task of ultrasound image reconstruction, we conduct the following two experiments.

Correlation of Muscle Status. Due to the continuity of muscle atrophy, muscle ultrasound images from adjacent sampling points during bed rest have some degree of correlation. During model training, the network model can only perform optimal reconstruction performance once it has learned such correlation. To demonstrate that the task of temporal ultrasound image reconstruction drives the network model to learn such correlation, we visualize the self-attention matrix from the last layer of the decoder [26]. Because of the multi-head attention mechanism, multiple attention matrices would exist. Here all the attention matrices are averaged and normalized to 0-1. For the complete ultrasound images, the correlation around the diagonal is higher, as shown in Fig. 5(a). It suggests that the reconstruction of ultrasound images depended significantly on the correlation of muscle states at adjacent sampling points. For the Gas and the Sol, the distributions of the strong correlations in the attention matrix are scattered, but both spread from the diagonal to surrounding, as shown in Fig. 5(b) and (c).

The results shows that the task of temporal ultrasound image reconstruction motivated the network to learn the dynamic change process of muscle atrophy.

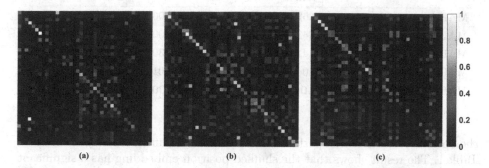

Fig. 5. Self attention matrix for (a) the complete ultrasound images, (b) the Gas and (c) the Sol. The result of multiplying and normalizing the Q and K matrices is considered as the correlation matrix, where Q and K are the Query and Key in the self-attention mechanism. Multiple correlation matrices can be obtained from the multi-headed self-attention, and they are averaged and normalized to 0-1.

Uniqueness of Position Indicator. We calculated the consistency of the position indicators obtained at different masking ratios with that obtained at a masking ratio of 75%, by Pearson correlation coefficients, as shown in Fig. 6. The results show excellent consistency of the position indicators, with correlation coefficients above 0.95. It suggests that the score-position embedding designed in this paper can learn the unique position embedding for different muscle states in the ultrasound sequence. Therefore, the position indicator can be used as the score of muscle atrophy.

4.4   Necessity of Position Embedding

After ensuring that the position indicator can be used as the score of muscle atrophy, we also want to investigate the contribution of the embedding of relative position information to the image reconstruction. In this paper, we perform two experiments to answer two questions: (1) Will shuffling the position embedding have a significant impact on reconstruction performance? (2) Do different position embedding methods affect the reconstruction performance?

Reconstruction of Shuffled Position Embedding. Since the task of temporal ultrasound image reconstruction motivates the position embedding to learn the relative position information in the ultrasound sequence, the performance of image reconstruction should be compromised if we shuffle the position embedding. We randomly shuffled the position embedding 1000 times and compared

Table 1. Image reconstruction performance under different position embedding.

Method	MSE, *mean ± SD*		
	Whole	Gas	Sol
MAE-No-Position	0.526 ± 0.0195	0.324 ± 0.0003	0.367 ± 0.0027
MAE-Sine-Cosine-Position	0.425 ± 0.0383	0.156 ± 0.0008	0.102 ± 0.0279
MAE-Learnable-Position	0.420 ± 0.0380	0.158 ± 0.0303	0.094 ± 0.0019
MAE-Shuffled-Position	0.040 ± 0.0018	0.029 ± 0.0011	0.024 ± 0.0017
MAE-Score-Position	**0.009 ± 0.0001**	**0.003 ± 0.00007**	**0.005 ± 0.00008**

the optimal reconstruction performance with the unshuffled results, as shown in
Table 1. The result shows that the shuffled position embedding has a significant
impact on the reconstruction performance. It demonstrates that the position
embedding is unique and the relative position are also unique in the ultrasound
sequence.

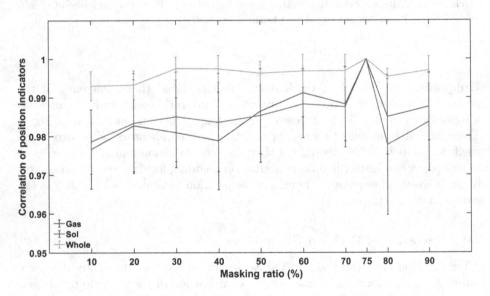

Fig. 6. Plot of correlation of the position indicator. The y-axis represents the Pearson
correlation coefficient between the position indicator obtained at different masking
ratios and that obtained at 75% masking ratio.

Comparison of Different Position Embedding. As previously hypothe-
sized, only the position embedding that best matched the dynamic change process of muscle atrophy would achieve the optimal muscle state prediction performance. How significant is the improvement in reconstruction performance of the

score-position embedding? We compared the reconstruction performance under various position embedding methods, as shown in Table 1. Without position embedding ('MAE-No-Position'), the performance of the reconstruction would suffer greatly. The reconstruction performance is also far from optimal when using a sine-cosine position embedding ('MAE-Sine-Cosine-Position') or a learnable position embedding ('MAE-Learnable-Position'). It suggests that the score-position embedding method designed in this paper ('MAE-Score-Position') is more capable of motivating the network to learn the dynamic change process for muscle atrophy.

5 Conclusion

This paper designs a self-supervised task of temporal ultrasound image reconstruction based on MAE for learning the dynamic change process of muscle atrophy. A score-position embedding is proposed to achieve quantitative evaluation of muscle atrophy without accurate labels. We demonstrated the feasibility of using the position indicator as the score of muscle atrophy by consistency testing, and verified the contribution of the position embedding by comparing the influence of different position embedding methods on the reconstruction performance. To our knowledge, this paper is the first to propose a quantitative evaluation method for muscle atrophy in the absence of accurate criteria.

References

1. Schiaffino, S.: Mechanisms regulating skeletal muscle growth and atrophy. FEBS J. **280**, 4294–4314 (2013)
2. Cohen, S.: Muscle wasting in disease: molecular mechanisms and promising therapies. Nat. Rev. Drug Discov. **14**(1), 58–74 (2014)
3. Parry, S.M.: Ultrasonography in the intensive care setting can be used to detect changes in the quality and quantity of muscle and is related to muscle strength and function. J. Crit. Care **30**(5), 1151–1159 (2015)
4. Yaeshima, K.: Mechanical and neural changes in plantar-flexor muscles after spinal cord injury in humans. Spinal Cord **53**(7), 526–533 (2015)
5. Kretić, D.: Reliability of ultrasound measurement of muscle thickness in patients with supraspinatus tendon pathology. Acta Clin. Croat. **57**(2), 335–341 (2018)
6. Puthucheary, Z.A.: Acute skeletal muscle wasting in critical illness. JAMA **310**(15), 1591–1600 (2013)
7. Ahtiainen, J.P.: Panoramic ultrasonography is a valid method to measure changes in skeletal muscle cross-sectional area. Eur. J. Appl. Physiol. **108**(2), 273–279 (2010)
8. Mele, A.: In vivo longitudinal study of rodent skeletal muscle atrophy using ultrasonography. Sci. Rep. **6**(1), 1–11 (2016)
9. Aerts, H.J.: Decoding tumour phenotype by noninvasive imaging using a quantitative radiomics approach. Nat. Commun. **5**(1), 1–9 (2014)
10. Vallières, M.: A radiomics model from joint FDG-PET and MRI texture features for the prediction of lung metastases in soft-tissue sarcomas of the extremities. Phys. Med. Biol. **60**(14), 5471–5480 (2015)

11. Huang, Y.Q.: Development and validation of a radiomics nomogram for preoperative prediction of lymph node metastasis in colorectal cancer. J. Clin. Oncol. **34**(18), 2157–2164 (2016)

12. Bodine, S.C.: Disuse-induced muscle wasting. Int. J. Biochem. Cell B. **45**(10), 2200–2208 (2013)

13. Globus, R.K.: Hindlimb unloading: rodent analog for microgravity. J. Appl. Physiol. **120**(10), 1196–1206 (2016)

14. He, K.: Masked autoencoders are scalable vision learners. In: CVPR (2022)

15. Looijaard, W. G.: Measuring and monitoring lean body mass in critical illness. Curr. Opin. Crit. Care. **24**(4), 241 (2018)

16. Cruz-Jentoft, A.J.: Sarcopenia: European consensus on definition and diagnosis-report of the european working group on sarcopenia in older people. Age Ageing **39**(4), 412–423 (2010)

17. Shah, P.K.: Lower-extremity muscle cross-sectional area after incomplete spinal cord injury. Arch. Phys. Med. Rehab. **87**(6), 772–778 (2006)

18. Xing, Y.: Classification of sMRI Images for Alzheimer's disease by using neural networks. In: PRCV, pp. 54–66 (2022)

19. Stein, T.: Metabolic consequences of muscle disuse atrophy. J. Nutr. **135**(7), 1824–1828 (2005)

20. Schneider, V.S.: Musculoskeletal adaptation to space flight, 4th edn. Springer, New York (2016)

21. Wang, J.: PGC-1α over-expression suppresses the skeletal muscle atrophy and myofiber-type composition during hindlimb unloading. Biosci. Biotech. Bioch. **81**(3), 500–513 (2017)

22. Devlin, J.: Bert: pre-training of deep bidirectional transformers for language understanding. In North American Chapter of the Association for Computational Linguistics (2019)

23. Vincent, P.: Stacked denoising autoencoders: learning useful representations in a deep network with a local denoising criterion. J. Mach. Learn. Res. (2010)

24. Chen, M. C.: Generative pretraining from pixels. In: ICML (2020)

25. Vaswani, A.: Attention is all you need. In: NIPS (2017)

26. Dosovitskiy, A.: An Image is Worth 16x16 Words: Transformers for Image Recognition at Scale. arXiv (2020)

27. Talbot, J.: Skeletal muscle fiber type: using insights from muscle developmental biology to dissect targets for susceptibility and resistance to muscle disease. WIREs Dev. Biol. **5**(4), 518–534 (2016)

Salient Object Detection Using Reciprocal Learning

Junjie Wu[1], Changqun Xia[2], Tianshu Yu[1], Zhentao He[1], and Jia Li[1,2(✉)]

[1] State Key Laboratory of Virtual Reality Technology and Systems, School of Computer Science and Engineering, Beihang University, Beijing 100191, China
jiali@buaa.edu.cn
[2] Peng Cheng Laboratory, Shenzhen 518000, China

Abstract. Typically, Objects with the same semantics may not always stand out in images with diverse backgrounds. Therefore, accurate salient object detection depends on considering both the foreground and background. Motivated by this observation, we proposed a novel reciprocal mechanism that considers the mutual relationships between background and foreground for salient object detection. First, we design a patulous U-shape framework comprising a shared encoder branch and two parallel decoder branches for extracting the foreground and background responses, respectively. Second, we propose a novel reciprocal feature interaction (RFI) module for the two decoder branches, allowing them to learn necessary information from each other adaptively. The RFI module primarily consists of a reciprocal transformer (RT) block that utilizes modulated window-based multi-head cross-attention (MW-MCA) to capture mutual dependencies between elements of the foreground and background features within the current two windows. Through the RFI module, the two decoder branches can mutually benefit each other and generate more discriminative foreground and background features. Additionally, we introduce a cooperative loss (CL) to guide the learning of foreground and background branches, which encourages our network to obtain more accurate predictions with clear boundaries and less uncertain areas. Finally, a simple but effective fusion strategy is utilized to produce the final saliency map. Extensive experiments on five benchmark datasets demonstrate the significant superiority of our method over the state-of-the-art approaches.

Keywords: Salient object detection · Reciprocal feature interaction · Window-based cross attention · Cooperative loss

1 Introduction

Salient object detection (SOD) usually aims to detect the most salient objects in a scene and segment their entire extent accurately. Essentially, it is to segment foreground objects that attract more attention compared to the background. The utilization of saliency detection can enhance various fields in computer vision and image processing, such as content-aware image editing [6], visual tracking [1], person re-identification [30] and image retrieval [5].

© The Author(s), under exclusive license to Springer Nature Singapore Pte Ltd. 2024
Q. Liu et al. (Eds.): PRCV 2023, LNCS 14433, pp. 281–293, 2024.
https://doi.org/10.1007/978-981-99-8546-3_23

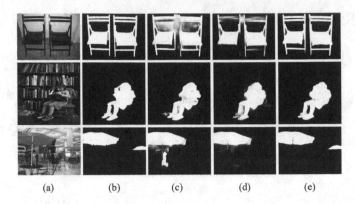

(a)   (b)   (c)   (d)   (e)

Fig. 1. The objects with same semantics are not always prominent in images with different backgrounds. (a) Images, (b) ground-truth, (c) and (d) results of [29] and [10], (e) results of our approach. As the background changes, the salient object also changes.

With the rapid development of convolutional neural networks (CNNs), which intelligently learn effective feature representation, numerous remarkable SOD works such as [4,25,37] have achieved promising results on the benchmarks. Specifically, [3] propose reverse attention to guide side-out residual learning for saliency refinement. [26] gather contextual information for refining the convolutional features iteratively with a recurrent mechanism from the global and local view. [27] devise a pyramid attention module and a salient edge detection module to enable the network to concentrate more on salient regions and refine object boundaries. However, these methods mainly focus on how to better integrate high-level and low-level features or multi-scale contextual information of the foreground in various ways. The dependency relationship between foreground and background is not fully explored and utilized.

In fact, objects with the same semantics show different degrees of visual attention in images with diverse backgrounds. As shown in Fig. 1, in the first row, the object 'chair' is easily noticeable and stands out as salient. However, in the second row, the saliency of the 'chair' diminishes due to changes in the background. In contrast, the 'person' becomes the salient object attracting more attention. Similarly, in the third row, the salient object shifts to the 'parasol', overshadowing the 'person'. Through these observations, we find that the saliency of objects varies along with the background. Thus, Considering the relationships between foreground and background elements can help better distinguish and identify salient objects and lead to more effective and reliable saliency detection algorithms in diverse scenes.

Inspired by this finding, we rethink the saliency detection task from the perspective of the foreground and background interaction and propose a novel reciprocal learning mechanism for salient object detection. As shown in Fig. 2, the network is a U-shape framework with a shared encoder branch with four stages and two parallel decoder branches. First, the network extracts common visual features from each encoder side-out and delivers them into the corresponding adaptive feature bridging (AFB) module. Then the progressive feature fusion (PFF) module integrates the AFB modules' outputs

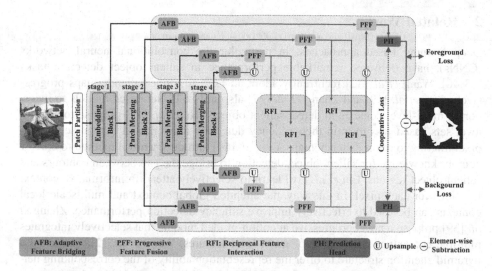

Fig. 2. Framework of the proposed approach. Block1/2/3/4: swin transformer block; AFB: adaptive feature bridging; PFF: progressive feature fusion; RFI: reciprocal feature interaction; PH: prediction head.

at the last two stages or the outputs of the AFB module and reciprocal feature interaction (RFI) module in two adjacent stages. The RFI module is designed to explore the reciprocal relationship between foreground and background and capture the long-range contextual dependencies between them, thus making the foreground and background features contribute to mutual improvement. Finally, the prediction head (PH) module is used to generate the binary foreground and background saliency maps, which are supervised by the proposed cooperative loss. The final saliency map is obtained by a very simple but effective background subtraction strategy in a cooperative supervision way. Experimental results on five public benchmark datasets show that our approach outperforms 13 state-of-the-art SOD models.

The main contributions of this paper are summarized as follows: 1) We revisit the problem of SOD from the new perspective of the reciprocal interplay between foreground and background. Compared with previous works, this scheme will pay more attention to exploring the inter-correlation between salient and non-salient elements, which is a more simple and direct solution. 2) We propose a novel reciprocal feature interaction (RFI) module with a reciprocal transformer (RT) block to deal with the reciprocal relationship between foreground and background and learn self-needed information from each other. In this way, the representation and discrimination of the foreground and background features would be enhanced. 3) According to the characteristics of foreground and background, we design cooperative loss to encourage our network to generate more accurate predictions with clear boundaries and less uncertain areas. 4) We conduct comprehensive experiments on five challenging datasets and achieve superior performance over the state-of-the-art approaches on all these datasets.

2 Related Work

Deep learning based approaches, in particular the convolutional neural networks (CNNs), have delivered remarkable performance in salient object detection tasks [13,40]. Wang *et al.* [12] recurrently adopt an FCN to refine saliency maps progressively. Luo *et al.* [19] and Zhang *et al.* [37] also utilize U-Net based models to incorporate multi-level contexts to detect salient objects. Hu *et al.* [11] propose to adopt a level sets based loss to train their saliency detection network and use guided superpixel filtering to refine saliency maps. Li *et al.* [14] introduce a novel method to borrow contour knowledge for salient object detection to bridge the gap between contours and salient object regions. Liu *et al.* [16] learn to selectively attend to informative context locations for each pixel. In this way, the attended global context and multiscale local contexts can be used to effectively improve saliency detection performance. Zhang *et al.* [39] propose a novel progressive attention-guided module that selectively integrates multiple contextual information of multi-level features. Wang *et al.* [27] proposed a pyramid attention structure to offer the representation ability of the corresponding network layer with an enlarged receptive field. Lately, with the remarkable advancements of transformers in the domain of Natural Language Processing (NLP), methods based on transformers have begun to emerge. Liu *et al.* [17] develop a pure transformer to propagate global contexts among image patches and use multi-level token fusion to get high-resolution results. Zhang *et al.* [36] take a step further by proposing a novel generative vision transformer with latent variables following an informative energy-based prior for salient object detection. In general, these methods are mainly designed to adaptively and effectively incorporate multi-level or multi-scale convolutional features, but ignore the full utilization of background information.

3 Approach

In this section, we give the details of the proposed Reciprocal Learning Network (RLNet) for salient object detection.

3.1 Base Network

We employ a pre-trained Swin Transformer [18] as the backbone to extract the visual features. From the four encoder stages, we obtain four outputs $E_i, i \in \{1,2,3,4\}$ (i is the stage index). Connecting the encoder and decoder branches, the AFB module enhances spatial representations of different-level features and boosts feature decoding. Here, we use the common convolution, dilated convolution [2], and asymmetric convolution [7], three kinds of convolutional blocks with different kernel sizes and shapes to enhance the spacial context and obtain the bridging features. For foreground and background branches, the bridging features are denoted as B_f^i and $B_b^i, i \in \{1,2,3,4\}$ respectively. To integrate and enhance features with different levels, the PFF module is utilized and calculated as follows:

$$Y_m = C_{3\times3}^2(CAT(X_1, X_2)) + C_{1\times1}(X_2),$$
$$Y = Y_m + C_{3\times3}^2(Y_m), \tag{1}$$

where X_1, X_2 are the inputs, Y is the output, Y_m is intermediate calculation results, CAT is the concatenation in the channel dimension, $C_{3\times3}^2$ is twice 3×3 convolution operations and $C_{1\times1}$ is the 1×1 convolution. As to the PH, we use several convolution layers to generate binary predictions.

3.2   Reciprocal Feature Interaction Module

To describe the reciprocal relationship between the foreground and background, we propose an RFI module with an RT block that can capture long-range contextual dependencies and useful local context, thus enhancing original feature representation. Next, we elaborate on the process of adaptively aggregating mutual contextual information.

As shown in Fig. 2, assume that for the foreground and background branches, the outputs of the PFF modules connected by the bridging features $B_f^i, B_b^i, i \in \{1, 2, 3\}$ are respectively $F_f^i, F_b^i, i \in \{1, 2, 3\}$ and the outputs of the RFI modules are denoted as $R_f^i, R_b^i, i \in \{2, 3\}$. The RFI module is calculated as follows:

$$R_f^i = \mathcal{F}_{RFI}(F_f^i, F_b^i) = C_{3\times3}(CAT(\mathcal{F}_{RT}(F_f^i, F_b^i), F_f^i)), i \in \{2, 3\},$$
$$R_b^i = \mathcal{F}_{RFI}(F_b^i, F_f^i) = C_{3\times3}(CAT(\mathcal{F}_{RT}(F_b^i, F_f^i), F_b^i)), i \in \{2, 3\}, \quad (2)$$

where $\mathcal{F}_{RFI}$ and $\mathcal{F}_{RT}$ are the RFI module and RT block, respectively. CAT is the concatenation in the channel dimension. $C_{3\times3}$ is a 3×3 convolution.

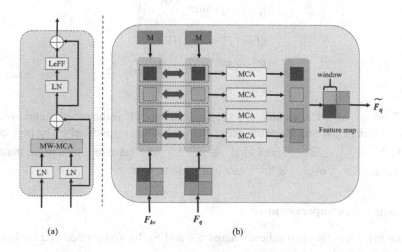

Fig. 3. Diagram of the reciprocal transformer block (a) and its main component MW-MCA (b).

Reciprocal Transformer. As shown in Fig. 3, the RT block consists of a MW-MCA module, followed by a Locally-enhanced Feed-Forward Network (LeFF). A LayerNorm (LN) layer is applied before each MW-MCA and LeFF. The LeFF is used to capture useful local context after the attention part [28]. The MW-MCA is used to capture the mutual long-range dependencies between two inputs, which divides two input features into non-overlapping windows, and every two windows at the same location form a window pair. Then a M-MCA is applied to the features in the window pair.

For the foreground branch, the computation of the RT block is described as follows:

$$\mathcal{F}_{RT}(F_f^i, F_b^i) = \mathcal{F}_{Le}(\widetilde{F_f^i}) + \widetilde{F_f^i}, \tag{3}$$

$$\widetilde{F_f^i} = \mathcal{F}_{MW}(LN(F_f^i), LN(F_b^i)) + F_f^i, \tag{4}$$

where $i \in \{2, 3\}$, $\mathcal{F}_{Le}$ and $\mathcal{F}_{MW}$ are the LeFF and MW-MCA, respectively. LN is the LayerNorm operation. $\widetilde{F_f^i}$ is the output of the residual connection after MW-MCA module. Next, we describe the details of the MW-MCA module.

Given two input feature maps $F_q, F_{kv} \in \mathbb{R}^{C \times H \times W}$ (C is the channel dimension, H and W are the height and width), we split F_q, F_{kv} into non-overlapping windows with size $M \times M$. Then we obtain the flattened and transposed features $F_q^j, F_{kv}^j \in \mathbb{R}^{M^2 \times C}, j \in \{1, 2, \ldots, N_w\}$, where $N_w = \frac{H}{M} \times \frac{W}{M}$ is the number of windows. After that, we perform modulated multi-head cross-attention (M-MCA) on each pair of windows and combine the results of all window pairs to obtain the output of the MW-MCA module. Suppose the head number is h and the dimension of each head is $d_h = C/h$. For the j-th window pair, the M-MCA module can be formulated as follows:

$$
\begin{aligned}
\mathcal{F}_M(F_q^j, F_{kv}^j) &= CAT(A_1^j, A_2^j, \ldots, A_h^j)W^{O^j}, \\
A_l^j &= softmax(\frac{QK^T}{\sqrt{d_h}})V, \\
Q &= (F_q^j + m_q)W_l^{Q^j}, \\
K &= (F_{kv}^j + m_{kv})W_l^{K^j}, \\
V &= (F_{kv}^j + m_{kv})W_l^{V^j},
\end{aligned}
\tag{5}
$$

where $\mathcal{F}_M$ is the M-MCA module, $l \in \{1, 2, \ldots, h\}$ and A_l^j is the cross-attention output of the l-th head. Q, K and V are the query, key and value, respectively. The projections $W_l^{Q^j}, W_l^{K^j}, W_l^{V^j} \in \mathbb{R}^{C \times d_h}$ and $W^{O^j} \in \mathbb{R}^{hd_h \times C}$ are learnable parameter matrices. The modulators $m_q, m_{kv} \in \mathbb{R}^{M^2 \times C}$ are learnable tensors.

3.3 Cooperative Supervision

After the PH, we obtain two saliency maps $S_\mathcal{F}$ and $S_\mathcal{B}$ for foreground and background by minimizing the loss

$$L_{CI} = D_c(Sig(S_\mathcal{B}), G_\mathcal{B}) + D_c(Sig(S_\mathcal{F}), G_\mathcal{F}), \tag{6}$$

where $Sig(\cdot)$ is the sigmoid function, $D_c(\cdot)$ means the cross-entropy loss and IoU loss, $G_\mathcal{F}$ and $G_\mathcal{B}$ are foreground and background groundtruth maps.

Figure 4 shows some prediction results and corresponding uncertainty maps calculated by their entropy. We can see that by using L_{CI}, the foreground and background

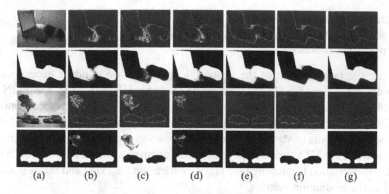

(a)   (b)   (c)   (d)   (e)   (f)   (g)

Fig. 4. Exploring the influence of cooperative loss. (a) Images and groundtruth; (b)–(g) the uncertainty maps and prediction results of foreground, background and final output without the cooperative loss (b)–(d) and with the cooperative loss (e)–(g); In the uncertainty map, the warmer the color, the greater the uncertainty of the corresponding prediction result.

maps can basically depict salient objects and distractors. However, there always exist some uncertain areas and unsatisfactory boundaries (see the uncertainty maps in Fig. 4 (b)–(d)) or leaving some areas mistakenly predicted. To reduce such error, we propose cooperative loss by constraining the relationship between foreground and background output. In this way, the network is encouraged to obtain more confident and accurate predictions and reduce uncertain areas. Here, we use Smooth L1 loss to implement the cooperative loss:

$$L_{\mathcal{SL}} = D_s(Sig(S_{\mathcal{F}}) + Sig(S_{\mathcal{B}}), \mathcal{I}) + D_s(Sig(S_{\mathcal{F}}) - Sig(S_{\mathcal{B}}), \mathcal{I}), \qquad (7)$$

where $D_s(\cdot)$ is the Smooth L1 loss function whose hyperparameter β is set 0.01 in the paper and $\mathcal{I}$ is the all-one matrix with the same size as $S_{\mathcal{F}}$ and $S_{\mathcal{B}}$. The overall loss $L_{\mathcal{CL}} = L_{\mathcal{CI}} + \lambda L_{\mathcal{SL}}$, here $\lambda = 0.2$ is a hyper-parameter to balance the two losses. Finally, the overall loss is used to train the proposed method.

3.4 Training and Inference

We run our experiments on a single NVIDIA GTX 3090 GPU with 22 GB memory. During training, the input image size is set to 384×384 and the path size is set to 8. The learning rate of the encoder layers is set to 0.05 with a weight decay of 0.0005 and momentum of 0.9. The learning rate of the rest layers in the proposed network is set to 10 times larger. The warm-up and linear decay strategy is used to adjust the learning rates. Data augmentation adopts random horizontal flipping and cropping. During testing, the proposed network removes all the losses, and each image is resized to 384×384. After the feed-forward process, the output of the network is composed of a foreground map and a background map. We use a very simple but effective fusion strategy based on background subtraction operation [22,38], i.e.,

$$Sal = relu(Sig(S_{\mathcal{F}}) - Sig(S_{\mathcal{B}})) \qquad (8)$$

where $relu(\cdot)$ means rectified linear unit function.

4   Experiments and Results

4.1   Experimental Setup

To evaluate the performance of the proposed approach, we conduct experiments on five popular datasets: ECSSD [32], PASCAL-S [15], HKU-IS [13], OMRON [33], DUTS [24]. DUTS is a large-scale dataset containing 10533 training images (denoted as DUTS-TR) and 5019 test images(denoted as DUTS-TE), and the images are challenging with salient objects of varied locations and scales as well as complex backgrounds. Here, we use DUTS-TR as our training set. In the comparisons, we adopt mean absolute error (MAE), weighted F-measure (F_β^w) [20], S-measure (S_m) [8] and mean E-measure (E_m) [9] as the evaluation metrics.

4.2   Comparisons with the State-of-the-Art

We compare our approach **RLNet** with 13 state-of-the-art methods, including F3Net [29], ITSD [41], MINet [21], CPD [31], BAS-NET [23], BAS-NET [23], AFNet [10], PICA-Net [16], R2Net [34], NIP [42], ACFFNet [35], VST [17] and LGVT [36].

Table 1. Performance of 13 SOTA and the proposed method on five benchmark datasets. Smaller MAE and larger F_β^w, S_m, E_m are better. The best results are in bold fonts.

Models	ECSSD				PASCAL-S				HKU-IS				DUTS-TE				OMRON			
	MAE↓	F_β^w↑	S_m↑	E_m↑	MAE↓	F_β^w↑	S_m↑	E_m↑	MAE↓	F_β^w↑	S_m↑	E_m↑	MAE↓	F_β^w↑	S_m↑	E_m↑	MAE↓	F_β^w↑	S_m↑	E_m↑
F3Net [29]	0.033	0.912	0.924	0.927	0.064	0.816	0.855	0.859	0.035	0.835	0.888	0.902	0.028	0.9	0.917	0.953	0.053	0.747	0.838	0.864
ITSD [41]	0.035	0.91	0.925	0.959	0.071	0.812	0.861	0.889	0.041	0.823	0.885	0.929	0.031	0.894	0.917	0.96	0.061	0.75	0.829	0.84
MINet [21]	0.034	0.905	0.925	0.957	0.071	0.808	0.856	0.869	0.037	0.813	0.884	0.926	0.029	0.889	0.919	0.96	0.056	0.718	0.833	0.869
CPD [31]	0.037	0.898	0.018	0.925	0.072	0.794	0.842	0.849	0.043	9795	0.869	0.886	0.034	0.875	0.905	0.944	0.056	0.719	**0.866**	0.87
BAS-NET [23]	0.037	0.904	0.918	0.921	0.076	0.793	0.832	0.847	0.048	0.803	0.866	0.884	0.032	0.889	0.909	0.946	0.056	0.751	0.865	0.869
AFNet [10]	0.042	0.886	0.913	0.918	0.07	0.797	0.844	0.846	0.046	0.785	0.867	0.879	0.036	0.869	0.905	0.942	0.057	0.717	0.826	0.853
PICA-Net [16]	0.053	0.867	0.894	0.915	0.083	0.772	0.835	0.083	0.053	0.754	0.874	0.856	0.04	0.84	0.905	0.936	0.062	0.695	0.832	0.833
R2Net [34]	0.029	0.925	0.929	-	0.061	0.841	0.863	-	0.031	0.864	0.9	-	0.025	0.914	0.925	-	0.047	0.777	0.85	-
NIP [42]	0.03	0.921	-	0.954	0.057	0.836	-	**0.93**	0.032	0.855	-	0.934	0.025	0.915	-	0.96	0.052	0.774	-	0.882
ACFFNet [35]	0.032	0.927	0.919	0.953	0.059	0.845	0.853	0.906	0.033	0.853	0.885	0.93	0.026	0.915	0.915	0.958	0.054	0.769	0.831	0.86
VST [17]	0.033	0.91	0.932	0.951	0.061	0.816	0.872	0.902	0.037	0.828	0.896	0.919	0.029	0.897	0.928	0.952	0.058	0.755	0.85	0.871
LGViT [36]	0.026	0.935	0.935	0.962	0.054	**0.855**	0.877	0.855	0.029	0.875	0.908	0.942	0.023	0.922	0.93	0.964	0.051	0.797	0.858	0.892
RLNet	**0.023**	**0.936**	**0.939**	**0.966**	**0.047**	0.852	**0.881**	0.922	**0.025**	**0.882**	**0.912**	**0.95**	**0.021**	**0.927**	**0.934**	**0.969**	**0.043**	**0.797**	0.865	**0.895**

The quantitative comparison results are shown in Table 1. From Table 1, we can see that our approach RLNet consistently outperforms the other approaches on all the datasets in terms of almost all evaluation metrics. It is worth noting that the MAE score of our method is significantly improved compared with the second-best results on the five datasets, e.g., 0.023 against 0.026 on ECSSD, 0.047 against 0.054 on PASCAL-S, 0.025 against 0.029 on HKU-IS, 0.043 against 0.051 on DUT-OMRON.

Figure 5 presents example saliency maps generated by our approach as well as other nine state-of-the-art methods. We can see that our method successfully highlights salient objects as coherent entities with well-defined boundaries. In contrast, many other methods fail to detect salient objects with largely altered appearances, as depicted in rows 1 to 3. These observations underscore the importance of reciprocal learning in addressing

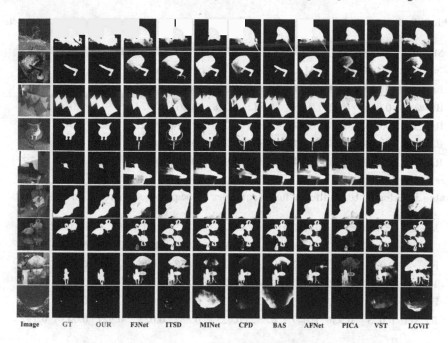

| Image | GT | OUR | F3Net | ITSD | MINet | CPD | BAS | AFNet | PICA | VST | LGViT |

Fig. 5. Qualitative comparisons of the state-of-the-art algorithms and our approach. GT means ground-truth masks of salient objects. The images are selected from five datasets for testing.

the challenges of object localization, object integrity, and boundary clarity in salient object detection (SOD). Furthermore, when salient objects share the same attributes (such as locations) with the background, many methods tend to mistakenly detect the background itself, as shown in rows 4 to 6. In our approach, the relative relationship between background and foreground is guaranteed by the novel reciprocal feature interaction module, which can capture the long-range contextual dependencies. Moreover, three extra examples of more difficult scenes are shown in the last three rows of Fig. 5. Even in these cases, our methods still obtain impressive results with accurate salient object localization. Overall, our method outperforms state-of-the-art methods in both quantitative and qualitative comparisons, demonstrating its superior performance.

4.3 Ablation Analysis

To validate the effectiveness of different components of the proposed method, we conduct experiments on the five datasets to compare the performance variations of our methods with different experimental settings.

Effectiveness of RFI and Cooperative Loss. To investigate the efficacy of the proposed RFI module, we conduct ablation experiments by introducing five different settings. As shown in Table 2, the first setting 'Bs' means the baseline model, i.e., the encoder branch and two decoder branches; the second setting 'Bs w/o B' means deleting the background decoder branch in the baseline model; the third setting 'Bs+CoL'

Table 2. Performance of different settings of the proposed approach on five benchmark datasets.

Models	ECSSD		PASCAL-S		HKU-IS		DUTS-TE		OMRON	
	MAE↓	F_β^w ↑	MAE↓	F_β^w ↑	MAE↓	F_β^w ↑	MAE↓	F_β^w ↑	MAE↓	F_β^w ↑
Bs	0.026	0.929	0.05	0.847	0.027	0.869	0.023	0.921	0.045	0.784
Bs w/o B	0.026	0.928	0.051	0.842	0.028	0.867	0.024	0.917	0.046	0.783
Bs+CoL	0.025	0.932	0.049	0.848	0.026	0.874	0.022	0.923	0.042	0.794
Bs+RFI	0.024	0.935	0.048	0.851	0.026	0.879	0.021	0.925	**0.041**	0.795
RLNet	**0.023**	**0.936**	**0.047**	**0.852**	**0.025**	**0.882**	**0.021**	**0.927**	0.043	**0.797**

Table 3. Performance of different fusion strategies of the proposed approach on five datasets.

Models	ECSSD		PASCAL-S		HKU-IS		DUTS-TE		OMRON	
	MAE↓	F_β^w ↑	MAE↓	F_β^w ↑	MAE↓	F_β^w ↑	MAE↓	F_β^w ↑	MAE↓	F_β^w ↑
F	0.028	0.919	0.052	0.835	0.03	0.858	0.026	0.906	0.047	0.778
1-B	0.026	0.926	0.05	0.842	0.028	0.862	0.024	0.913	0.045	0.781
F+1-B	0.034	0.906	0.056	0.824	0.034	0.837	0.031	0.89	0.051	0.761
RLNet	**0.023**	**0.936**	**0.047**	**0.852**	**0.025**	**0.882**	**0.021**	**0.927**	**0.043**	**0.797**

means the cooperative loss is adopted based on the baseline model; the fourth setting 'Bs+RFI' means the RFI module is added between the two decoder branches based on the baseline model; the last setting 'RLNet' means the RFI module and cooperative loss are both adopted based on the first setting.

From Table 2, we can see the performance of the 'Bs w/o B' model is reduced after removing background cues compared to the baseline model. It indicates that background cues provide a cooperative effect in SOD. In the third and fourth settings, The performance is improved to different degrees compared to the baseline. It indicates the effectiveness of the proposed RFI and cooperative loss. The RFI can capture the mutual relationship between foreground and background and obtain better context features. The cooperative loss can improve the discrimination ability of prediction results for uncertain areas like boundaries and obtain more confident and accurate predictions (see Fig. 4(e)–(g)). In the last setting, the performance is further improved compared to the third and fourth settings. This suggests that the RFI module and cooperative loss can work together to improve the performance of the network from different perspectives.

Effectiveness of Fusion Strategy. In this work, we use the background subtraction operation to fuse the foreground and background. We also analyze the impact of different fusion strategies on performance by designing other three different schemes. The first directly uses the result of foreground prediction as the final result. The second utilizes the inversion of the background map as the result. The third method is to average the results of the first and second schemes. The comparison of these three schemes and our RecNet is listed in Table 3. From Table 3, we find that the best performance is achieved by adopting the background subtraction way. This subtraction strategy not only increases pixel-level discrimination but also captures context contrast information.

5   Conclusion

In this paper, we revisit the problem of SOD from the perspective of reciprocal learning of background and foreground. Compared with previous work, this scheme will be more consistent with the essence of saliency detection, which may help develop new models. To solve this problem, we propose a novel reciprocal feature interaction module with a reciprocal transformer block to deal with mutual relationships between foreground and background. In this way, the feature discriminative power is enhanced to help perceive and localize foreground objects and background distractors. Moreover, we also propose cooperative loss to encourage our network to obtain more accurate and complementary predictions with clear boundaries and reduce uncertain areas. Extensive experiments on five benchmark datasets have validated the effectiveness of the proposed approach.

Acknowledgments. This work is partially supported by the National Natural Science Foundation of China under Grant 62132002 and Grant 62102206 and the Major Key Project of PCL (No. PCL2023AS7-1).

References

1. Borji, A., Frintrop, S., Sihite, D.N., Itti, L.: Adaptive object tracking by learning background context. In: 2012 IEEE Computer Society Conference on Computer Vision and Pattern Recognition Workshops, pp. 23–30. IEEE (2012)
2. Chen, L.C., Papandreou, G., Kokkinos, I., Murphy, K., Yuille, A.L.: DeepLab: semantic image segmentation with deep convolutional nets, atrous convolution, and fully connected CRFs. IEEE Trans. Pattern Anal. Mach. Intell. **40**(4), 834–848 (2017)
3. Chen, S., Tan, X., Wang, B., Hu, X.: Reverse attention for salient object detection. In: Proceedings of the European Conference on Computer Vision (ECCV), pp. 234–250 (2018)
4. Chen, X., Zheng, A., Li, J., Lu, F.: Look, perceive and segment: finding the salient objects in images via two-stream fixation-semantic CNNs. In: Proceedings of the IEEE International Conference on Computer Vision, pp. 1050–1058 (2017)
5. Cheng, M.M., Hou, Q.B., Zhang, S.H., Rosin, P.L.: Intelligent visual media processing: when graphics meets vision. J. Comput. Sci. Technol. **32**(1), 110–121 (2017)
6. Cheng, M.M., Zhang, F.L., Mitra, N.J., Huang, X., Hu, S.M.: Repfinder: finding approximately repeated scene elements for image editing. ACM Trans. Graph. (TOG) **29**(4), 1–8 (2010)
7. Ding, X., Guo, Y., Ding, G., Han, J.: ACNet: strengthening the kernel skeletons for powerful CNN via asymmetric convolution blocks. In: Proceedings of the IEEE/CVF International Conference on Computer Vision, pp. 1911–1920 (2019)
8. Fan, D.P., Cheng, M.M., Liu, Y., Li, T., Borji, A.: Structure-measure: a new way to evaluate foreground maps. In: Proceedings of the IEEE International Conference on Computer Vision, pp. 4548–4557 (2017)
9. Fan, D.P., Ji, G.P., Qin, X., Cheng, M.M.: Cognitive vision inspired object segmentation metric and loss function. Scientia Sinica Informationis **6**(6) (2021)

10. Feng, M., Lu, H., Ding, E.: Attentive feedback network for boundary-aware salient object detection. In: Proceedings of the IEEE/CVF Conference on Computer Vision and Pattern Recognition, pp. 1623–1632 (2019)

11. Hu, P., Shuai, B., Liu, J., Wang, G.: Deep level sets for salient object detection. In: Proceedings of the IEEE Conference on Computer Vision and Pattern Recognition, pp. 2300–2309 (2017)

12. Kuen, J., Wang, Z., Wang, G.: Recurrent attentional networks for saliency detection. In: Proceedings of the IEEE Conference on Computer Vision and Pattern Recognition, pp. 3668–3677 (2016)

13. Li, G., Yu, Y.: Visual saliency based on multiscale deep features. In: Proceedings of the IEEE Conference on Computer Vision and Pattern Recognition, pp. 5455–5463 (2015)

14. Li, X., Yang, F., Cheng, H., Liu, W., Shen, D.: Contour knowledge transfer for salient object detection. In: ECCV, pp. 355–370 (2018)

15. Li, Y., Hou, X., Koch, C., Rehg, J.M., Yuille, A.L.: The secrets of salient object segmentation. In: CVPR, pp. 280–287 (2014)

16. Liu, N., Han, J., Yang, M.H.: PiCANet: learning pixel-wise contextual attention for saliency detection. In: Proceedings of the IEEE Conference on Computer Vision and Pattern Recognition, pp. 3089–3098 (2018)

17. Liu, N., Zhang, N., Wan, K., Shao, L., Han, J.: Visual saliency transformer. In: Proceedings of the IEEE/CVF International Conference on Computer Vision, pp. 4722–4732 (2021)

18. Liu, Z., et al.: Swin transformer: hierarchical vision transformer using shifted windows. In: Proceedings of the IEEE/CVF International Conference on Computer Vision (ICCV) (2021)

19. Luo, Z., Mishra, A.K., Achkar, A., Eichel, J.A., Li, S., Jodoin, P.M.: Non-local deep features for salient object detection. In: CVPR (2017)

20. Margolin, R., Zelnik-Manor, L., Tal, A.: How to evaluate foreground maps? In: CVPR, pp. 248–255 (2014)

21. Pang, Y., Zhao, X., Zhang, L., Lu, H.: Multi-scale interactive network for salient object detection. In: Proceedings of the IEEE/CVF Conference on Computer Vision and Pattern Recognition, pp. 9413–9422 (2020)

22. Piccardi, M.: Background subtraction techniques: a review. In: 2004 IEEE International Conference on Systems, Man and Cybernetics (IEEE Cat. No. 04CH37583), vol. 4, pp. 3099–3104. IEEE (2004)

23. Qin, X., Zhang, Z., Huang, C., Gao, C., Dehghan, M., Jagersand, M.: Basnet: boundary-aware salient object detection. In: Proceedings of the IEEE/CVF Conference on Computer Vision and Pattern Recognition, pp. 7479–7489 (2019)

24. Wang, L., et al.: Learning to detect salient objects with image-level supervision. In: CVPR (2017)

25. Wang, T., Borji, A., Zhang, L., Zhang, P., Lu, H.: A stagewise refinement model for detecting salient objects in images. In: ICCV, pp. 4019–4028 (2017)

26. Wang, T., et al.: Detect globally, refine locally: a novel approach to saliency detection. In: CVPR, pp. 3127–3135 (2018)

27. Wang, W., Zhao, S., Shen, J., Hoi, S.C., Borji, A.: Salient object detection with pyramid attention and salient edges. In: Proceedings of the IEEE Conference on Computer Vision and Pattern Recognition, pp. 1448–1457 (2019)

28. Wang, Z., Cun, X., Bao, J., Zhou, W., Liu, J., Li, H.: Uformer: a general U-shaped transformer for image restoration. In: Proceedings of the IEEE/CVF Conference on Computer Vision and Pattern Recognition, pp. 17683–17693 (2022)

29. Wei, J., Wang, S., Huang, Q.: F^3net: fusion, feedback and focus for salient object detection. In: Proceedings of the AAAI Conference on Artificial Intelligence, vol. 34, pp. 12321–12328 (2020)

30. Wu, L., Wang, Y., Gao, J., Li, X.: Deep adaptive feature embedding with local sample distributions for person re-identification. Pattern Recogn. **73**, 275–288 (2018)
31. Wu, Z., Su, L., Huang, Q.: Cascaded partial decoder for fast and accurate salient object detection. In: Proceedings of the IEEE/CVF Conference on Computer Vision and Pattern Recognition, pp. 3907–3916 (2019)
32. Yan, Q., Xu, L., Shi, J., Jia, J.: Hierarchical saliency detection. In: CVPR, pp. 1155–1162 (2013)
33. Yang, C., Zhang, L., Lu, H., Ruan, X., Yang, M.H.: Saliency detection via graph-based manifold ranking. In: CVPR, pp. 3166–3173 (2013)
34. Zhang, J., Liang, Q., Guo, Q., Yang, J., Zhang, Q., Shi, Y.: R2Net: residual refinement network for salient object detection. Image Vis. Comput. **120**, 104423 (2022)
35. Zhang, J., Shi, Y., Zhang, Q., Cui, L., Chen, Y., Yi, Y.: Attention guided contextual feature fusion network for salient object detection. Image Vis. Comput. **117**, 104337 (2022)
36. Zhang, J., Xie, J., Barnes, N., Li, P.: Learning generative vision transformer with energy-based latent space for saliency prediction. Adv. Neural. Inf. Process. Syst. **34**, 15448–15463 (2021)
37. Zhang, P., Wang, D., Lu, H., Wang, H., Ruan, X.: Amulet: aggregating multi-level convolutional features for salient object detection. In: ICCV, pp. 202–211 (2017)
38. Zhang, P., Wang, D., Lu, H., Wang, H., Yin, B.: Learning uncertain convolutional features for accurate saliency detection. In: ICCV, pp. 212–221 (2017)
39. Zhang, X., Wang, T., Qi, J., Lu, H., Wang, G.: Progressive attention guided recurrent network for salient object detection. In: CVPR, pp. 714–722 (2018)
40. Zhao, R., Ouyang, W., Li, H., Wang, X.: Saliency detection by multi-context deep learning. In: Proceedings of the IEEE Conference on Computer Vision and Pattern Recognition, pp. 1265–1274 (2015)
41. Zhou, H., Xie, X., Lai, J.H., Chen, Z., Yang, L.: Interactive two-stream decoder for accurate and fast saliency detection. In: Proceedings of the IEEE/CVF Conference on Computer Vision and Pattern Recognition, pp. 9141–9150 (2020)
42. Zhou, Q., Zhou, C., Yang, Z., Xu, Y., Guan, Q.: Non-binary IOU and progressive coupling and refining network for salient object detection. Expert Syst. Appl. **230**, 120370 (2023)

Graphormer-Based Contextual Reasoning Network for Small Object Detection

Jia Chen[1,2], Xiyang Li[1], Yangjun Ou[1,2(✉)], Xinrong Hu[1,2], and Tao Peng[1,2]

[1] Wuhan Textile University, Wuhan 430200, Hubei, China
yjou@wtu.edu.cn
[2] Engineering Research Center of Hubei Province for Clothing Information, Wuhan 430200, Hubei, China

Abstract. Different from common objects, it is usually difficult to obtain fine enough visual information for small objects. Existing context-based learning methods only focus on whether the local contextual information of the object can be better extracted, but these methods easily lead to the loss of other effective information in the scene. To address this issue, this study proposes a Graphomer-based Contextual Reasoning Network (GCRN), which improves the feature expression of objects by modeling and reasoning about contextual information. Specifically, the method first establishes contextual information encoding using a context relation module, which is used to capture the local contextual features of objects as well as the dependencies among objects. Then, all the contextual information is aggregated by Graphormer in the context reasoning module to enrich the visual features of small objects. We conduct extensive experiments on two public datasets (*i.e.* MSCOCO and TinyPerson). The results show that this method can enhance the feature expression of small objects to a certain extent, and has good performance in feature information transmission.

Keywords: Small Objects · Contextual Information · Feature Expression · Context Relation · Context Reasoning

1 Introduction

The main task of small object detection is to accurately detect low-resolution objects in the image (*i.e.* objects below 32 pixels × 32 pixels), and to obtain the category and location information of the object. Small object detection is a research hotspot in the field of computer vision, with a wide range of practical applications, such as satellite remote sensing [23], maritime rescue [7], and security patrols [1], to name a few. However, since small objects have few visual features and cannot provide rich feature information, small object detection is also a research difficulty in the field of computer vision.

Supplementary Information The online version contains supplementary material available at https://doi.org/10.1007/978-981-99-8546-3_24.

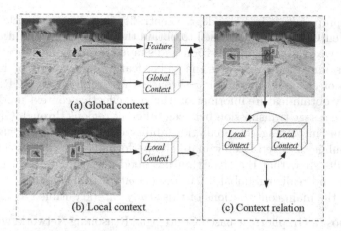

Fig. 1. (a) input the whole image into the model as contextual information. (b) extract the local contextual information around the object. (c) add the central relationship of the object based on extracting the local context.

The early context learning algorithm represented by Gist [2,3] uses the scene layout of the whole image to obtain the global context features, and uses it as the auxiliary knowledge of the detection model to identify the objects in the image, as shown in Fig. 1(a). Although this kind of method can enhance the feature representation of small objects, it is extremely dependent on the adaptation of the object to the scene [1]. When the object appears in an unexpected scene, the global context will lead to the opposite result. To make more accurate use of contextual information, subsequent context learning methods [2,11] usually generate a series of context windows around the target to extract the local features of the target itself and around the object, as shown in Fig. 1(b). Although these methods improve the accuracy of small object detection to some extent, a single local context feature will lead to the lack of important details of small objects to a great extent.

To prevent the model from paying too much attention to the local contextual information, we introduce the degree centrality relationship of the object, and model and reason the contextual relationship between the objects. We find that there is a similar constraint relationship between the object and the target in the image, which is similar to the thought of "social structure". As shown in Fig. 1(c), the more neighbors gathered around the object, the more important the object is [17]. We can use this social network central relationship and combine it with the local contextual information around the object to build a context relation model, to effectively enhance the feature expression of small objects.

In this work, we propose a Graphormer-based Contextual Reasoning Network (GCRN), which can improve the feature representation of objects by modeling and reasoning the contextual information. The method includes a context relation module and a context reasoning module. Specifically, the context relation module aims to extract effective contextual information to optimize the quality of regional proposals. It includes degree centrality encoding, semantic encoding, and spatial encoding, which are used to obtain the weight of each node, semantic relations between objects, and position and shape information, respectively. The context reasoning module updates the global feature information by integrating three kinds of encoding information through the Graphormer. GCRN method can not only obtain feature information through effective context modeling, but also realize message transmission between different regions through Graphormer network, thus improving the prediction ability of the model. Through experiments on public data sets, we evaluate the performance of GCRN. The experimental results show that the algorithm has good performance in fusing feature information and realizing global feature information transmission.

In sum, the main contributions of this study can be outlined as follows:

- We propose a Graphormer-based Contextual Reasoning Network, which could improve the efficiency of small object detection by purposefully propagating contextual information between regions.
- We design a degree centrality encoding to measure the importance of each object, which enhances the feature representation of small objects to some extent.
- We use Graphormer to aggregate all the contextual information, and this improvement can better realize the transmission and interaction of global feature information.

2   Related Work

The purpose of context learning is to enable the information between objects and scenes to interact, spread and change reasonably. By properly modeling the context, the performance of object detection can be improved, especially for those small objects whose appearance features are insignificant. Common context learning methods [2,11] usually only extract local context information, but this method can easily lead to the loss of other effective information in the scene. The study found that the graphic structure [17,21] shows an amazing ability to absorb external information. Xu et al. [21] proposed a Reasoning RCNN based on Faster RCNN [16], which encodes the knowledge graph by constructing the context relationship, and uses the a priori context relationship to affect object detection. However, due to the complex factors of the real world, artificial knowledge graph is usually not very popular. The attention mechanism of Transformer [19] makes it have a global receiving field, so that the model can pay attention to the global background information in the scene. However, for the data of graphic structure, the effect of using Transformer to aggregate node features is unsatisfactory.

Graphormer [25] is a model that applies Transformer to graph structures, and it proves that Transformer can message passing in non-Euclidean spaces. Inspired by this, we propose to use Transformer for feature aggregation and feature transformation operations, which solves the problem of the limited acceptance range of existing context learning methods, enabling the model to focus on object feature information in the entire scene.

3 Proposed Method

3.1 Overview

Figure 2 is an overall block diagram of GCRN. Taking FasterR-CNN [16] as the baseline, we use the region proposal generated in the first stage to build the graph structure model $G = \langle V, E \rangle$, where $V = \{v_1, v_2, ..., v_n\}$ is the number of proposal nodes. We obtain the edge $e_{ij} \in E$ from v_i to v_j by finding the K neighbor nodes of each node v_i. Then, degree centrality encoding, semantic encoding, and spatial encoding are designed to extract the implicit structure information of graph G. Finally, we use Transformer to aggregate the context coding information between nodes, and further integrate the initial features to update the node information. GCRN can capture the internal relationship between objects and enrich the characteristic information of small objects by imitating the mechanism of human vision.

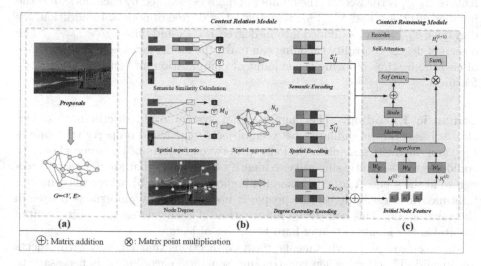

Fig. 2. This is our GCRN method. It includes two modules. The context relation module includes the degree centrality encoding constructed according to the number of adjacent nodes, the semantic encoding constructed according to the semantic similarity of the initial regional features, and the spatial encoding constructed according to the spatial position similarity between objects. The context reasoning module captures three kinds of encoding information through Encoder in Graphormer.

3.2 Context Relation Module

Degree Centrality Encoding. Celebrities who usually have a large number of fans have more influence in daily life, which is an important factor in predicting social networking trends. Similarly, we can associate that in the graph structure, the more neighbor nodes of a node, the greater its influence factor, as shown in Fig. 2(b). The influence of this degree centrality feature on the prediction results is easy to be ignored, but it is an indispensable factor in object prediction according to our research. In GCRN, we construct degree centrality encoding as auxiliary information when Transformer updates node features.

Specifically, we propose a degree centrality encoding that assigns weights to each node according to the number of degrees of nodes. Then, we use the weighted node feature together with the initial feature as the initial input of the Transformer. This process formula is expressed as follows:

$$H_i^{(0)} = x_i + z_{d(v_i)}, \tag{1}$$

$$z_{d(v_i)} = \frac{\sum_{j=1}^n y_{ij}}{n-1}(i \neq j), \tag{2}$$

where x_i is the node feature vector, $H_i^{(0)} = x_i$ represents the initial layer node feature, $z_{d(v_i)}$ is the weight (the number of degrees) learned by each node, n is the number of nodes, and $\sum_{j=1}^n y_{ij}$ represents the degree of node i. In addition, for digraphs, $z_{d(v_i)}$ can be divided into in-degree $z_{d+(v_i)}$ and out-degree $z_{d-(v_i)}$. The design of degree centrality encoding can make the model focus on the importance of each node, thus reducing the amount of computation and enriching the node features.

Semantic Encoding. The human visual system could discriminate object classes based on the similarity between objects, which is a property that proves the same intrinsic semantic relationship between objects of the same class. We could use this semantic similarity to identify the objects that are not easy to recognize in the scene. A concrete example is shown in Fig. 2(b), where the semantic information of easily detectable people is used to identify ambiguous people in the same scene. This contextual information of easily detectable objects often facilitates the identification of hard-to-detect objects.

In most complex environments, most connections that interact with objects are invalid. Therefore, when constructing semantic encoding, it is necessary to calculate the semantic correlation degree between nodes in graph G, retain the relationship with a high correlation degree, and cut off the relationship with a low correlation degree. The semantic correlation encoding after pruning introduces the semantic similarity feature into the multi-headed attention module through the matrix addition operation. Semantic relevance S'_{ij} is calculated as follows:

$$S'_{ij} = \varphi(i,j) \cdot g(r_i, r_j) = \varphi(i,j) \cdot \sigma(r_i) \cdot \sigma(r_j)^T, \tag{3}$$

where φ is an indicator function, which is used to determine whether the two regions overlap (overlap is 0, otherwise 1). $(r_i, r_j) \in R^{N_r \times D}$ is a given library of region proposals and D is the dimension of the initial region. $g(\cdot, \cdot)$ is a semantic correlation function, which is used to calculate the semantic similarity between the initial region proposals r. We use $\sigma(\cdot)$ function to project the region proposals into potential feature representations.

Spatial Encoding. In images, small objects tend to form clusters more easily, and objects belonging to the same class tend to have similar spatial aspect ratios. As shown in Fig. 2(b), in the same scene, humans and birds are more likely to gather together with the same kind in space, and the spatial proportions of humans and humans are also similar. According to these two characteristics of small objects, we design a spatial encoding to enrich the characteristic information of objects. The spatial encoding S_{ij}'' formula is expressed as:

$$S_{ij}'' = \varphi(i,j) \cdot f(C_i^o, C_j^o) = \varphi(i,j) \cdot M_{ij} \cdot N_{ij}, \tag{4}$$

where φ is an indicator function for judging whether regions overlap, $f(\cdot,\cdot)$ is a function used to calculate the spatial similarity between two proposals, $C_i = (x_i, y_i, w_i, h_i)$ is the coordinates of the ith proposal, and $C_j = (x_j, y_j, w_j, h_j)$ is the coordinate of the jth proposal. M_{ij} is used to calculate the spatial similarity between two proposals, and N_{ij} is used to measure the distance weight between proposals. The M_{ij}, N_{ij} formula is expressed as follows:

$$M_{ij} = \frac{min(w_i, w_j) \cdot min(h_i, h_j)}{w_i h_i + w_j h_j - min(w_i, w_j) \cdot min(h_i, h_j)}, \tag{5}$$

$$N_{ij} = exp(-\mu \cdot M_{ij}^d), \tag{6}$$

where μ is the scale parameter set, which is set to $5e^{-4}$ in this paper. M_{ij}^d is the spatial distance from two proposals to the center.

3.3   Context Reasoning Module

Graphormer. In the context reasoning module, we use the Graphormer network to aggregate degree centrality encoding, semantic encoding, and spatial encoding information. As shown in Fig. 2(c), degree centrality encoding is input into Encoder with the original feature, while semantic encoding and spatial encoding are input into Encoder as biased items of self-attention. Using Graphormer to aggregate three kinds of encoded information can promote messages to be better transmitted between nodes, thus enhancing the feature representation of small objects.

Specifically, we express the initial node feature V_i as x_i, and add it to the degree centrality encoding $z_{w(v_i)}$ matrix to obtain the hidden feature $H_i^{(l)} \in R^{1 \times d}$, where d is the hidden dimension. We take hidden feature $H_i^{(l)}$ of the i node and other node features $H_j^{(l)}$ in the graph G as the input of Encoder. Then, the values entered are multiplied by the three weight matrices $W_Q \in R^{d \times d_K}$, $W_K \in R^{d \times d_K}$ and $W_V \in R^{d \times d_V}$, respectively, which are projected to the corresponding representations Q, K, V as input to the self-attention module. We use vector Q and vector K to calculate the dot product to get the correlation weight A between each node, and then normalize the correlation weight A between each node. The main purpose of layer normalization is to stabilize the gradient during training. At the same time, we also add the semantic encoding S_{ij}' and the spatial encoding S_{ij}'' as the bias items of weight A.

Through the softmax function, the weight A between each node is converted into the probability distribution between $[0, 1]$, and then the probability distribution between each node and the corresponding Q value are calculated by dot product. Finally, through the accumulation of the calculation results of the dot product, the feature $H_i^{(l+1)}$ of the updated node i is obtained. Each node in graph G goes through the same process in parallel to update its features. Therefore, the whole calculation process of node update features could be expressed as follows:

$$Q = H_i^{(l)} W_Q, K = H_j^{(l)} W_K, V = H_j^{(l)} W_V, \tag{7}$$

$$A_{ij} = \frac{QK^T}{\sqrt{d_k}} + S'_{ij} + S''_{ij}, \tag{8}$$

$$H_i^{(l+1)} = \text{Attention} \left(H^{(l)} \right) = \sum_{j \in V} \text{softmax} \left(A_{ij} \right) V, \tag{9}$$

where $j \in V$, $V = \{v_1, v_2, \quad \ldots, \quad v_n\}$ represents the set of nodes, A_{ij} represents the correlation weight between node i and node j.

Graphormer Layer. Graphormer is built based on Transformer [19]. Each Transformer layer consists of two parts: Multi-Head Attention (MHA) and Feed Forward Network (FFN). To enable Graphormer to better capture the characteristics of nodes, we apply LayerNorm (LN) before MHA and FFN. The Graphomer layer can be represented as:

$$H'^{(l)} = MHA(LN(H^{(l-1)})) + H^{l-1}, \tag{10}$$

$$H^{(l)} = FFN(LN(H'^{(l)})) + H'^{(l)}, \tag{11}$$

where $H^{(l-1)}$ represents the node feature of layer $l - 1$, $H'^{(l)}$ represents the node feature of layer l after attention calculation, and $H^{(l)}$ represents the node feature of layer l after normalization.

4   Experiments

4.1   Datasets and Evaluation Metric

The MS COCO [14] dataset is a general dataset in object detection, including 140,000 images, 80 object categories, and 91 stuff categories. The images in the COCO dataset contain common scenes in daily life, with complex backgrounds and small object sizes, so it can provide more contextual information for small object detection. We use COCO standard evaluation indicators to evaluate our small target detection task. The COCO evaluation index is obtained by calculating the average value of the IoU threshold ($i.e$ $0.50 : 0.05 : 0.95$), including AP, AP_{50} ($IoU > 0.5$), AP_{75} ($IoU > 0.75$), and AP_S, AP_M, AP_L according to the target area. In this paper, we focus on the $mathitAP_S$ metric to determine the performance of the model.

The TinyPerson [26] dataset, which focuses on seaside pedestrian detection, includes 1010 images, each containing more than 200 people, with a total of 7,2561 instances. The dense goals in TinyPerson help us to build a context model. For the TinyPerson dataset, we use the evaluation criteria it provides to evaluate the detection accuracy of GCRN. TinyPerson calculates and divides AP (i.e. AP_{50}^{tiny}, AP_{25}^{tiny}), and AP_{75}^{tiny}, according to different object sizes and setting different IoU thresholds in the image. In this paper, we use $mathit{MR}$ (miss rate) and $mathit{AP}$ to evaluate the performance of the GCRN model.

Table 1. Detection results of different detectors on MS COCO test-dev. The best results of each AP are highlighted in bold.

	Method	Backbone	$AP(\%)$	AP_{50}	AP_{75}	AP_S	AP_M	AP_L
Two-stage	Faster RCNN [12]	ResNet-101-FPN	39.4	60.2	43.0	22.3	43.3	49.9
	Mask RCNN [6]	ResNet-101-FPN	40.0	60.5	44.0	22.6	44.0	52.6
	IR RCNN [4]	ResNet-101	39.7	62.0	43.2	22.9	42.4	50.2
	Cascade RCNN [27]	ResNet-101-FPN	42.8	62.1	46.3	23.7	45.5	55.2
	TridentNet [13]	ResNet-101	42.7	63.6	46.5	23.9	46.6	56.6
One-stage	CornerNet511 [10]	Hourglass-104	42.1	57.8	45.3	20.8	44.8	**56.7**
	FoveaBox [9]	ResNet-50	37.1	56.6	39.6	20.6	39.8	46.9
	Libra EBox [8]	ResNet-101-FPN	40.8	62.2	43.3	24.8	44.7	52.1
	QueryDet [22]	ResNet-50	37.6	58.8	39.7	23.3	41.9	49.5
	YOLOv4 [20]	CSPDarknet-53	43.0	64.9	46.5	24.3	46.1	55.2
Our	GCRN	ResNet-50-FPN	43.6	65.5	47.8	23.5	45.7	52.4
	GCRN	ResNet-101-FPN	**44.9**	**66.9**	**48.3**	**25.4**	**46.9**	54.3

4.2 Parameter Setting

Our training and verification experiments are carried out on an NVIDIA V100 GPU, the experimental environment is Pytorch1.7.0, CUDA10.2, and the training time is about 7 days. We use ResNet50 and ResNet-101 as the backbone of GCRN and Faster R-CNN as the baseline of GCRN. Our backbone networks are pre-trained on PASCAL VOC. The Graphormer size in the GCRN model is set to lump $L = 6$ and dumped $d = 512$.

On the MS COCO dataset, we use random gradient descent (SGD) for training. The initial learning rate is set to 0.005, the momentum is 0.9, the initial weight attenuates to 0.0001, and the total number of iterations is $218K$.

On the TinyPerson dataset, the partition ratio of the model training set and test set is 8:2, of which 8,792 images are used for training and 2,280 images are used for testing. In TinyPerson, the parameter setting of model training is the same as that of the COCO dataset, but the total number of iterations is $164K$.

4.3   Comparison with State-of-the-Art Models

Results on MS COCO. To verify the detection effect of the GCRN model, we carried out targeted experiments on the MS COCO dataset. Table 1 shows the detection performance of all detectors on the test dataset (5k images). From this table, we find that GCRN has obvious advantages in small object detection, and the AP_S index is improved by 1.1% compared with the best YOLOv4 [20]. This shows that we can effectively improve the visual features of small objects by using degree centrality encoding, semantic encoding, and spatial encoding to transmit feature information through Graphormer. It is worth noting that GCRN is designed for complex scenes with multiple small objects, which makes it flexible and portable for different detection systems, thus improving the performance of small object detection.

Table 2. Detection results for different detectors on TinyPerson. The best results for each *mathitAP* are marked in bold.

Method	Backbone	AP_{50}^{tiny}	AP_{25}^{tiny}	AP_{75}^{tiny}	MR_{50}^{tiny}	MR_{25}^{tiny}	MR_{75}^{tiny}
RetinaNet [24]	ResNet-101-FPN	33.53	61.51	2.28	88.31	76.33	98.76
Faster RCNN [12]	ResNet-101-FPN	47.35	63.18	5.83	87.57	76.59	98.39
FCOS [18]	ResNet-101	17.90	41.95	1.50	96.28	90.34	99.56
Libra RCNN [15]	ResNet-101-FPN	44.68	64.77	6.26	89.22	82.44	98.39
RetinaNet with S-α [5]	ResNet-101-FPN	48.34	71.18	5.34	87.73	74.85	98.57
FreeAnchor [28]	ResNet-101	44.26	67.06	4.35	89.66	79.61	98.78
Cascade RCNN [27]	ResNet-101-FPN	51.35	72.54	6.72	86.48	**73.40**	**98.18**
GCRN	ResNet-50-FPN	50.57	72.09	6.02	86.63	74.85	98.61
GCRN	ResNet-101-FPN	**53.72**	**74.68**	**7.90**	**85.31**	73.82	98.19

Results on TinyPerson. To further verify the effectiveness of GCRN, we carried out additional experiments on the TinyPerson dataset with many tiny images and compared them with other detectors. Table 2 shows the performance of different methods on *MP* and *AP* metrics. We find that there is a cross-domain gap between the aerial tiny character image and the natural scene image, so most general detectors can not achieve satisfactory performance in this case. We use Faster RCNN as the baseline and then use data enhancement and multi-scale training methods to enhance GCRN. In the detection task, our method achieves better results than Cascade RCNN [27].

4.4   Ablation Experiment and Qualitative Results

Effect of Each Module. We verify the impact of each encode in GCRN on detection performance on MS COCO val, and the ablation study data are shown in Table 3. The experimental results show that the proposed degree centrality encoding, semantic encoding, and spatial encoding improve the accuracy of small object detection by 0.5, 0.7, and 1.3, respectively. This shows that semantic encoding could capture the semantic information between targets more

Table 3. Different model in GCRN. DEG is degree centrality encoding, SEM is semantic encoding, and SPA is spatial encoding.

Module			$AP(\%)$	AP_S	AP_M	AP_L
DEG	SEM	SPA				
			41.8	22.3	43.3	50.3
		✔	42.3	22.5	43.9	51.3
	✔		42.5	22.6	44.3	51.9
✔			43.1	23.1	44.3	51.5
✔	✔	✔	**43.6**	**23.3**	**45.7**	**52.4**

Table 4. The effect of different numbers of neighbor nodes K on the detection accuracy.

K	$AP(\%)$	AP_S	AP_M	AP_L
6	41.8	21.6	43.7	50.5
12	42.7	22.4	44.5	51.3
16	43.1	23.0	**45.8**	52.1
18	43.4	**23.3**	45.2	**52.2**
21	**43.5**	23.0	45.0	52.2

effectively, spatial encoding could obtain spatial layout information more accurately, and degree centrality encoding is indispensable to enhance the feature expression of small objects.

Effect of Different Number of Neighbor Nodes. In the process of constructing the graph, we verify the influence of the number of neighbor nodes K on the model detection results. Table 4 shows the results in the range of 6 to 21. We can see that when the number of neighbor nodes is too small, the detection results are not ideal, but when there are too many neighbor nodes, the detection accuracy will decrease. When the number of neighbor nodes $K = 16$, the model detection accuracy will be maintained in a good state.

Qualitative Analysis on MSCOCO. The test results of GCRN and MASK RCNN [6] on the MS COCO dataset are shown in Fig. 3. From the first line of the figure, we can see that the detection performance of Mask RCNN will degrade when the background is complex and blurred. It mistakenly identifies the blurry background as a "bottle", but GCRN does a good job of avoiding this problem. This proves that GCRN can use context information to accurately identify specific targets. As can be seen from the test results in the second lines of Fig. 3, GCRN can effectively avoid false detection (the person on the billboard and the model by the door are identified as "people") and missed detection (blocked motorcycles are not identified). This shows that GCRN can improve the performance of object detection by using the interaction of context information between objects.

Origin   Mask RCNN   GCRN

Fig. 3. Comparison between Mask RCNN and GCRN on COCO dataset. The detection results with scores higher than 0.5 are shown. The first column is the input image, the detection results of Mask RCNN and GCRN are displayed in the second and third columns, respectively.

5   Conclusion

This study developed a Graphomer-based Contextual Reasoning Network with a context relation module and a context Reasoning module for small object detection. The contextual relationship module is used to capture the importance of object nodes, semantic relationship and the position shape information between objects, respectively. The contextual inference module aims to integrate three encoding information through Transformer to update the global feature information. The experimental results show that our model can enhance the feature representation of small objects and can use global information to improve the performance of the system. In the future, we use the method of graph optimization to aggregate features to achieve global representation.

References

1. Chen, G., et al.: A survey of the four pillars for small object detection: multiscale representation, contextual information, super-resolution, and region proposal. Int. J. Comput. Vis. **52**(2), 936–953 (2022)
2. Chen, Z., Huang, S., Tao, D.: Context refinement for object detection. In: ECCV, pp. 71–86 (2018)
3. Divvala, S.K., Hoiem, D., Hays, J.H., Efros, A.A., Hebert, M.: An empirical study of context in object detection. In: CVPR, pp. 1271–1278 (2009)
4. Fu, K., Li, J., Ma, L., Mu, K., Tian, Y.: Intrinsic relationship reasoning for small object detection. arXiv preprint arXiv:2009.00833 (2020)
5. Gong, Y., Yu, X., Ding, Y., Peng, X., Zhao, J., Han, Z.: Effective fusion factor in FPN for tiny object detection. In: WACV, pp. 1160–1168 (2021)
6. He, K., Gkioxari, G., Dollár, P., Girshick, R.: Mask R-CNN. In: ICCV, pp. 2961–2969 (2017)

7. Hong, M., Li, S., Yang, Y., Zhu, F., Zhao, Q., Lu, L.: SSPNet: scale selection pyramid network for tiny person detection from UAV images. IEEE Geosci. Remote Sens. Lett. **19**, 1–5 (2021)

8. Huang, S., Liu, Q.: Addressing scale imbalance for small object detection with dense detector. Neurocomputing **473**, 68–78 (2022)

9. Kong, T., Sun, F., Liu, H., Jiang, Y., Li, L., Shi, J.: Foveabox: beyound anchor-based object detection. IEEE Trans. Image Process. **29**, 7389–7398 (2020)

10. Law, H., Deng, J.: Cornernet: detecting objects as paired keypoints. In: ECCV, pp. 734–750 (2018)

11. Lei, J., Luo, X., Fang, L., Wang, M., Gu, Y.: Region-enhanced convolutional neural network for object detection in remote sensing images. IEEE Trans. Geosci. Remote Sens. **58**(8), 5693–5702 (2020)

12. Li, B., Yan, J., Wu, W., Zhu, Z., Hu, X.: High performance visual tracking with siamese region proposal network. In: CVPR, pp. 8971–8980 (2018)

13. Li, Y., Chen, Y., Wang, N., Zhang, Z.: Scale-aware trident networks for object detection. In: ICCV, pp. 6054–6063 (2019)

14. Lin, T.-Y., et al.: Microsoft COCO: common objects in context. In: Fleet, D., Pajdla, T., Schiele, B., Tuytelaars, T. (eds.) ECCV 2014. LNCS, vol. 8693, pp. 740–755. Springer, Cham (2014). https://doi.org/10.1007/978-3-319-10602-1_48

15. Pang, J., Chen, K., Shi, J., Feng, H., Ouyang, W., Lin, D.: Libra R-CNN: towards balanced learning for object detection. In: CVPR, pp. 821–830 (2019)

16. Ren, S., He, K., Girshick, R., Sun, J.: Faster R-CNN: towards real-time object detection with region proposal networks. IEEE Trans. Pattern Anal. Mach. Intell. **6**, 1137–1149 (2017)

17. Shetty, R.D., Bhattacharjee, S.: A weighted hybrid centrality for identifying influential individuals in contact networks. In: CONECCT, pp. 1–6. IEEE (2022)

18. Tian, Z., Shen, C., Chen, H., He, T.: FCOS: fully convolutional one-stage object detection. In: ICCV, pp. 9627–9636 (2019)

19. Vaswani, A., et al.: Attention is all you need. In: NeurIPS, vol. 30 (2017)

20. Wang, C.Y., Bochkovskiy, A., Liao, H.Y.M.: Scaled-YOLOv4: scaling cross stage partial network. In: CVPR, pp. 13029–13038 (2021)

21. Xu, H., Jiang, C., Liang, X., Lin, L., Li, Z.: Reasoning-RCNN: unifying adaptive global reasoning into large-scale object detection. In: CVPR, pp. 6419–6428 (2019)

22. Yang, C., Huang, Z., Wang, N.: Querydet: cascaded sparse query for accelerating high-resolution small object detection. In: CVPR, pp. 13668–13677 (2022)

23. Yang, F., Fan, H., Chu, P., Blasch, E., Ling, H.: Clustered object detection in aerial images. In: ICCV, pp. 8311–8320 (2019)

24. Yang, S., Luo, P., Loy, C.C., Tang, X.: Wider face: a face detection benchmark. In: CVPR, pp. 5525–5533 (2016)

25. Ying, C., et al.: Do transformers really perform badly for graph representation? In: NeurIPS, vol. 34, pp. 28877–28888 (2021)

26. Yu, X., Gong, Y., Jiang, N., Ye, Q., Han, Z.: Scale match for tiny person detection. In: WACV, pp. 1257–1265 (2020)

27. Zhang, W., Fu, C., Chang, X., Zhao, T., Li, X., Sham, C.W.: A more compact object detector head network with feature enhancement and relational reasoning. Neurocomputing **499**, 23–34 (2022)

28. Zhang, X., Wan, F., Liu, C., Ji, X., Ye, Q.: Learning to match anchors for visual object detection. IEEE Trans. Pattern Anal. Mach. Intell. **44**(6), 3096–3109 (2021)

PVT-Crowd: Bridging Multi-scale Features from Pyramid Vision Transformer for Weakly-Supervised Crowd Counting

Zhanqiang Huo, Kunwei Zhang, Fen Luo$^{(\boxtimes)}$, and Yingxu Qiao

School of Software, Henan Polytechnic University, Jiaozuo 454000, China
luofenjsj@hpu.edu.cn

Abstract. Weakly-supervised crowd counting does not require location-level annotations, but only relies on count-level annotations to achieve the task of crowd counting for images, which is becoming a new research hotspot in the field of crowd counting. Currently, weakly-supervised crowd counting networks based on deep learning mostly use Transformers to extract features and establish global contexts, ignoring feature information at different scales, resulting in insufficient feature utilization. In this paper, we propose a well-designed end-to-end crowd counting network named PVT-Crowd bridging multi-scale features from the Pyramid Visual Transformer Encoder for weakly-supervised crowd counting. Specifically, Adjacent-Scale Bridging Modules (ASBM) enable the interaction of high-scale semantic and low-scale detailed information from both channel and spatial dimensions. The Global-Scale Bridging Module (GSBM) performs a secondary fusion of multi-scale feature information. Extensive experiments show that our PVT-Crowd outperforms most weakly-supervised crowd counting networks and obtains competitive performance compared to fully-supervised ones. In particular, cross-dataset experiments confirm that our PVT-Crowd had a remarkable generality.

Keywords: crowd counting · multi-scale · transformer

1 Introduction

Crowd counting techniques are essential in crowd tracking, public transportation, and transfer counting. The existing crowd counting methods mainly rely on density estimation, point supervision, and weak supervision, where both density estimation and point supervision require location-level annotation. However, in situations with big and dense crowds, gathering location-level crowd annotations can become expensive and time-consuming. Therefore, weakly-supervised learning with count-level annotations is receiving increasing attention from researchers.

This work is supported by the National Science Foundation of China (No. 62273292).

Q. Liu et al. (Eds.): PRCV 2023, LNCS 14433, pp. 306–318, 2024.
https://doi.org/10.1007/978-981-99-8546-3_25

In crowd counting, significant variations in the scale of the human head due to factors such as distance, angle, and imaging equipment are key factors that constrain the application of crowd counting. The variations in head scale include two situations: the variations in head scale in the same image and the variations in head scale in different images. The first situation is shown in Fig. 1 left, where the three marked red boxes correspond to the three scales of heads, and the red boxes gradually decrease as the distance from the camera increases. The second situation is shown in Fig. 1 right, the density distribution of the crowd and the size of the heads are different between the two images. These problems make it quite challenging to estimate crowds from images quickly and accurately.

Fig. 1. Head scale variations in different situations. The left side gives the head scale variations in an image; The right side gives the head scale variations in different images.

Current partial weakly-supervised methods [6,22] use CNN to regress the total count of people from an image. However, the inherent limitation of CNN-based weakly-supervised methods is the limited receptive fields of contextual modeling, which frequently disregard the extraction of global receptive fields and multi-level information. The transformer is containing the global receptive fields, which is more advantageous than the CNN architecture. Many researchers are trying to use transformer architecture to solve weakly-supervised counting task.

Dosovitskiy et al. [2] introduced a visual transformer for image tasks called ViT. ViT divides the input image into patches before projecting these patches into a 1-D sequence with linear embedding, but the output feature map of ViT has only one scale. Recently, some stratified transformer architectures have been introduced [1,18,19], where the output of the Pyramidal Visual Transformer v2 [19] includes multi-scale feature maps. However, these networks do not adequately utilize multi-scale feature information, and there is a lack of integration between features at different scales. Therefore, we proposed PVT-Crowd network enhances the integration between different scale features by bridging adjacent

scale features and global scale features, and achieves effective information utilization of regression counting.

The contributions of this paper are the following:

- PVT-Crowd, a new end-to-end weakly-supervised crowd counting network,
  which uses a pyramid-structured visual transformer to extract multi-scale
  features to a global context.
- An efficient Adjacent-Scale Bridging Module (ASBM) is designed to bridge
  features from adjacent stages of the transformer encoder by the short aggregation to obtain rich, multi-level contextual crowd information. We also suggest
  a Global-Scale Bridging Module (GSBM) bridging all the outputs of ASBMs
  to achieve adequate utilization of multi-scale feature information.
- Extensive experiments on three benchmark crowd counting datasets including ShanghaiTech PartA, PartB and UCF-QNRF show that the proposed
  approach achieves excellent counting performance under weakly-supervised
  settings. Importantly, our model has greater generality and obtains state-
  of-the-art performance in cross-dataset evaluation, outperforming even most
  fully-supervised methods.

2   Related Work

2.1   Fully-Supervised Crowd Counting

The mainstream of crowd counting is fully-supervised crowd counting methods
based on density maps [4,7,23]. The multi-scale architecture is developed for
head scale inconsistency caused by the changes of view in crowd images. Specifically, MCNN [23] extracts multi-scale feature information using filters with different sizes of receptive fields. CP-CNN [14] is a contextual pyramid CNN that
fuses the features from different levels. TEDNet [5] combines multiple encoding-
decoding paths in a hierarchical assembly to obtain the supervised information
for accurate crowd counting. PACNN [12] integrates viewpoint information into
a multi-scale density map.

Attention mechanism is another helpful technique employed by many methods. SAAN [3] first introduces the attention mechanism for scale selection. Jiang
et al. [4] propose a network for generating attention masks associated with
regions of different density levels. MAN [9] proposes a multifaceted attention
network to improve the transformer model's performance on the local space.

2.2   Weakly-Supervised Crowd Counting

HA-CNN [15] adapts to new scenes by utilizing the image-level labels of crowd
images as supervision way. Yang et al. [22] propose a network that classifies a
given image based on the number of people. MATT [6] training models use only
a small amount of point-level annotations, mostly using count-level annotations.
TransCrowd [8] extracts the information directly from the Transformer architecture to the image, and the decoding structure directly maps the token to several
people, which works well.

In weakly-supervised crowd counting field, the counting capabilities of the above CNN-based weakly-supervised counting methods still cannot match results the results of fully-supervised counting methods. Transformer-based weakly-supervised counting method has difficulties in transferring transformer to dense pixel-level prediction and cannot learn to multi-scale feature maps. It is not conducive to the accurate estimation of the crowd number.

3 Our Method

The constructed crowd counting model is given in Fig. 2. It uses a Pyramid Vision Transformer v2 [19] to extract multi-scale features from the input images. The features are enhanced with information by the Adjacent-Scale Bridging Module, all features are bridged by the Global-Scale Bridging Module, and fed into the regression head to estimates the crowd number.

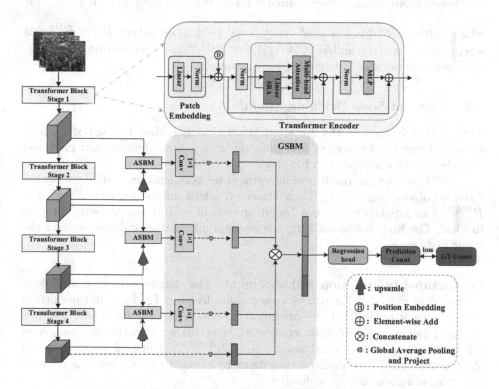

Fig. 2. The pipeline of PVT-Crowd.

3.1 Transformer Encoder

The Transformer encoder block comprises an Multi-head Attention (MHA) containing a linear spatial reduction attention ($LinearSRA$) and a feedforward layer. A standard Transformer layer is given in the top right corner of the Fig. 2, and for each layer l, contains layer normalization (LN) and residual connections. The output can be written as follows:

$$Z'_l = LinearSRA(LN(Z_{l-1})) + Z_{l-1} \qquad (1)$$

$$Z_l = MLP(LN(Z'_l)) + Z'_l \qquad (2)$$

where Z_l is the output of layer l. Here, the Multi-Layer Perceptron (MLP) is improved, PVTv2's MLP layer adds a layer of 3×3 convolution between the two linear layers. The processing of $LinearSRA$ can be written as follows:

$$LinearSRA(Q, K, V) = \left[head_0, \cdots, head_{N_i} \right] W^0 \qquad (3)$$

$$head_j = Attention \left(QW_j^Q, LinearSR(K) W_j^K, LinearSR(V) W_j^V \right) \qquad (4)$$

where $LinearSR(\cdot)$ is a spatial dimensional pooling operation, $W^0 \in R^{D \times D}$ is a reprojection matrix, and $W^Q/W^K/W^V \in R^{D \times \frac{D}{N}}$ are three learning matrices. Q does not perform a $LinearSR(\cdot)$ operation.

3.2   Adjacent-Scale Bridging Module

This section introduces the Adjacent-Scale Bridging Module (ASBM), which takes as input the features of four different scales. Now, describe our proposed module structure as shown in Fig. 3(a).

ASBM achieves information enhancement between features at different scales. Network of two neighboring Transformer encoder's output features F_i, $F_{i+1} \in R^{H \times W \times C}$ as input. The feature fusion process of ASBM is a two-step process in total. The first step is realizing the channel information enhancement of the input features F_i.

(a) Channel Information Enhancement. The channel attention module is displayed in Fig. 3(b). Firstly, we concatenate features F_i, F_{i+1} and perform a global average pooling (GAP) operation to obtain feature F_p. Then, features F_p are input in a set of fully connected layers to achieve the integration of information to obtain channel weights $w_c \in R^{H \times W \times C}$. Channel weights w_c and shallow features F_i are multiplied element by element to complete the channel information fusion process, defined as:

$$F' = f_c\big(G([F_i, F_{i+1}])\big) \otimes F_i \qquad (5)$$

where $G(\cdot)$ means global average pooling, $[\cdot]$ means the channel connection of two features, $f_c(\cdot)$ is a fully connected layer. The channel-enhanced feature F' is further processed by ASBM to achieve the second step of the information fusion process.

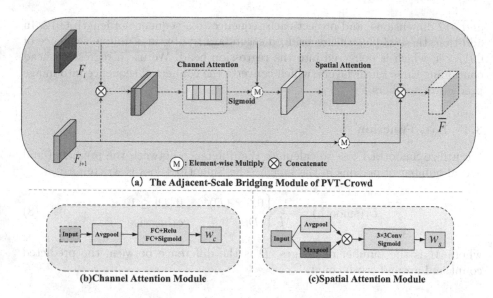

(a) The Adjacent-Scale Bridging Module of PVT-Crowd

(b)Channel Attention Module   (c)Spatial Attention Module

Fig. 3. (a) The Adjacent-Scale Bridging Module of PVT-Crowd. The information enhancement of channel direction and spatial direction is obtained after channel joining of the features of two adjacent stages. (b) Use channel attention to achieve channel information enhancement. (c) Use spatial attention to achieve spatial information enhancement.

(b) Spatial Information Enhancement. The space attention module is displayed in Fig. 3(c). First average pooling and max pooling of feature F' to get the feature F'_{avg} and F'_{max}. Then the two obtained features F'_{avg} and F'_{max} are channel connected, The connected features are passed through 3×3 convolution and Sigmoid activation function to generate spatial attention weights $w_s \in R^{H \times W \times 1}$, which are multiplied element by element with F_{i+1} to obtain F'', defined as:

$$F'' = \sigma \left(f^{3 \times 3} \left(\left[F'_{avg}, F'_{max} \right] \right) \right) \otimes F_{i+1} \qquad (6)$$

where $[\cdot]$ means the feature connection in the channel direction, $\sigma(\cdot)$ means Sigmoid activation function, $f^{3 \times 3}(\cdot)$ means 3×3 convolution. Finally, features F' and F'' are channel-connected to complete information fusion and obtain fused feature $\overline{F}$, defined as:

$$\overline{F} = [F', F''] \qquad (7)$$

3.3 Global-Scale Bridging Module and Regression Head

We use a global-scale bridging module to efficiently utilize the output after fused feature enhancement inspired by Crowdformer [11]. Then take the feature maps from each stage and use a single-point convolution kernel to reduce the number of channels while decreasing the computational effort. The features implement global averaging pooling to obtain one-dimensional sequences of 64, 128, 320 and 512 dimensions, and project each sequence to a sequence of length 6912. In addition, these sequences are bridged secondary to obtain a unique sequence of 6912×4, which is delivered into the regression head. We use a regression head containing three linear layers, which converts the connected sequence into image prediction numbers.

3.4 Loss Function

We utilize SmoothL1 loss to calculate the difference between the predicted and actual number of persons in the image. The SmoothL1 loss is written as:

$$L_1(\text{smooth}) = \frac{1}{M} \begin{cases} 0.5 \times (D)^2, & \text{if } |D| \leq 1 \\ |D| - 0.5, & \text{otherwise} \end{cases} \tag{8}$$

where M is the number of images, D is the difference between the predicted count and ground truth count.

4   Experiments

4.1   Implementation Details

We use the model weights of 'pvtv2-b5' in PVTv2 [19] to initialize the transformer encoder and follow the data implementation strategy employed in [8]. During training, we increase the amount of data using operations such as random cropping of images. Because image sizes vary within and between datasets, we resize every image to 1152×768. Each image is segmented into six equal-sized sub-images. We use Adam to optimize the model, both the learning rate and weight decay are set to 1e-5. The training batch size has been set to 8, and the V100 GPU is used for experiments.

4.2   Evaluation Metrics

We choose Mean Absolute Error (MAE) and Mean Square Error (MSE) as evaluation metrics:

$$\text{MAE} = \frac{1}{N} \sum_{i=1}^{N} |P_i - G_i| \tag{9}$$

$$\text{MSE} = \sqrt{\frac{1}{N} \sum_{i=1}^{N} |P_i - G_i|^2} \tag{10}$$

where N is the number of test images, P_i and G_i are the predicted and ground truth count of the ith image, respectively.

4.3   Performance on Three Datasets

Performance on ShanghaiTech. ShanghaiTech dataset has 1198 images. The dataset images are classified into two portions: PartA and PartB. To evaluate the performance of our method, We analyze the model with the dataset's eight fully-supervised and four weakly-supervised approaches. The performance is given in Table 1, and our model has the most advanced performance on this dataset. Specifically, compared to TransCrowd [8], our method boosts the MAE and MSE performance by 6.8% and 3.8% on PartA, and 7.5% and 11.8% on PartB, respectively. Although it is not fair to evaluate fully-supervised and weakly-supervised methods, our model still obtains competitive outperformance compared to fully-supervised methods.

Table 1. Performance comparison of our approach to other popular methods on Shanghai PartA, Shanghai PartB and UCF-QNRF datasets. * indicates the weakly-supervised method. Loc and Cnt are Location and Count, respectively.

Method	Publication	label		Shanghai PartA		Shanghai PartB		UCF-QNRF	
		Loc	Cnt	MAE	MSE	MAE	MSE	MAE	MSE
MCNN [23]	CVPR16	✓	✓	110.2	173.2	26.4	41.3	277.0	426.0
CSRNet [7]	CVPR18	✓	✓	68.2	115.0	10.6	16.0	-	-
PACNN [12]	CVPR19	✓	✓	66.3	106.4	8.9	13.5	-	-
BL [10]	ICCV19	✓	✓	62.8	101.8	7.7	12.7	88.7	154.8
DM-Count [17]	NeurIPS20	✓	✓	59.7	95.7	7.4	11.8	85.6	148.3
BCCT [16]	arxiv21	✓	✓	**53.1**	**82.2**	7.3	11.3	83.8	143.4
MAN [9]	CVPR22	✓	✓	56.8	90.3	-	-	**77.3**	**131.5**
CrowdFormer [21]	IJCAI22	✓	✓	56.9	97.4	**5.7**	**9.6**	78.8	136.1
Yang et al. [22]*	ECCV20	✗	✓	104.6	145.2	12.3	21.2	-	-
MATT [6]*	PR21	✗	✓	80.1	129.4	11.7	17.5	-	-
TransCrowd [8]*	SCIS22	✗	✓	66.1	105.1	9.3	16.1	97.2	**168.5**
Xiong et al. [20]*	arxiv22	✗	✓	76.9	113.1	-	-	133.0	218.8
ours*	-	✗	✓	**61.6**	**101.1**	8.6	14.2	**95.1**	175.0

Performance on UCF-QNRF. UCF-QNRF dataset contains 1535 crowd images, including 1201 images in the training set and 334 images in the test set. As given in Table 1, our method obtains one performance of MAE with 95.1 and MSE with 175.0 on the UCF-QNRF dataset. However, the MSE metric is unsatisfactory, and our method still needs improvement. Compared with most weak supervision methods, our proposed method still has significant improvements.

4.4   Cross-Dataset Performance

In the cross-dataset evaluation, the quantitative results are given in Table 2, we can witness that the proposed method leads to state-of-the-art (SOTA) performance in weakly-supervised crowd counting, except for ShA→ShB (Shanghai

PartA→Shanghai PartB). Although our approach in the weakly supervised setting, it performs even better than the fully-supervised method, which shows the transferability of the model. The results in Table 2 show that the model trained on a large dataset with complex scenes (QNRF) performs better when trained on a dataset with simple scenes (ShB). Just the opposite, the model train on ShB does not show very good performance on ShA.

Table 2. Performance of different methods for generalizability in cross-dataset experiments. * indicates the weakly-supervised method.

Method	Publication	ShA→ShB		ShB→ShA		QNRF→ShA		QNRF→ShB	
		MAE	MSE	MAE	MSE	MAE	MSE	MAE	MSE
MCNN [23]	CVPR16	85.2	142.3	221.4	357.8	-	-	-	-
D-ConvNet [13]	ECCV18	49.1	99.2	140.4	226.1	-	-	-	-
BL [10]	ICCV19	-	-	-	-	69.8	123.8	15.3	26.5
TransCrowd [8]*	SCIS22	18.9	31.1	141.3	258.9	78.7	122.5	13.5	21.9
Xiong et al. [20]*	arxiv22	-	-	-	-	88.7	127.8	-	-
ours*	-	19.9	31.6	130.2	221.0	66.4	109.2	11.1	17.5

For our model to show significant generalizability in cross-dataset evaluation, we give a probability analysis: (1) Due to its self-attentive-like architecture, long-term spatial dependence is established, and multi-scale features are available. It is favorable for information extraction of images at different scales. (2) Then, the extracted features are enhanced with information, and the bridging module enables complete learning of contrasting features, which facilitates the differentiation of various objects. (3) Finally, the model counting method is sequence-to-count mapping, which maps class tokens with rich classification information to obtain counting results.

4.5   Visualization

We show the visualized feature maps on Shanghai PartA dataset using the heat maps, as shown in Fig. 4. Our model focuses more on dense crowd areas and can correctly distinguish background areas from crowd areas, proving the validity of the proposed method.

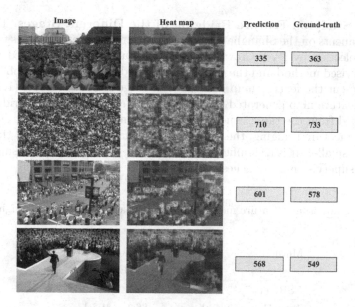

Fig. 4. Visualization of heat maps.

4.6   Ablation Study

Ablation Study on Model Modules. The ablation experiments of the constituents of PVT-Crowd on the Shanghai PartA and PartB datasets are given in Table 3. We can witness that ASBM and GSBM provide incremental improvements, respectively. Specifically, only using the transformer to extract the last layer of features of the image for counting has a MAE of 64.5 and MSE of 110.8 in the PartA dataset and a MAE of 9.9 and MSE of 18.8 in PartB. We observe that after adding the GSBM in PartA and PartB, the MAE and MSE improved only by a small amount; however, after adding the ASBM, the performance of MAE and MSE improved by 2.9 and 9.7 in PartA and 1.3 and 4.6 in PartB, respectively. This demonstrates the effectiveness of ASBM and GSBM for feature information enhancement.

Table 3. Effects of GSBM and ASBM in the proposed method.

Method	Label		Shanghai PartA		Shanghai PartB	
	Loc	Cnt	MAE	MSE	MAE	MSE
baseline	✗	✓	64.5	110.8	9.9	18.8
baseline+GSBM	✗	✓	64.3	105.5	9.2	15.5
baseline+ASBM+GSBM	✗	✓	**61.6**	**101.1**	**8.6**	**14.2**

Ablation Study on Feature Bridging of the Different Stages. The ablation experiments on the Shanghai PartA dataset under different stages of feature bridging selection by PVT-Crowd are given in Table 4 and compared with the fully-supervised method and the weakly-supervised method. The numbers in the table represent the feature map generated at stage i. For example, (1,2) represents the feature map generated at stage 1 and stage 2 for feature bridging. AS can be noted from the table, when the feature generated in adjacent stages are selected for feature bridging, the extracted information is richer, and the counting error is smaller. It is a significant improvement over both the full-supervision and weak-supervision approaches.

Table 4. Comparison of feature map bridging with each strategy on Shanghai PartA dataset.

Method	Label		Shanghai PartA	
	Loc	Cnt	MAE	MSE
CSRNet [7]	✓	✓	68.2	115.0
TransCrowd [8]*	✗	✓	66.1	105.1
(1,3)(2,3)(3,4)	✗	✓	67.0	112.1
(1,3)(2,4)(3,4)	✗	✓	63.9	108.4
(1,2)(2,3)(3,4)	✗	✓	**61.6**	**101.1**

Table 5. Performance of features secondary bridging for each stage on Shanghai PartA dataset.

Method	Label		Shanghai PartA	
	Loc	Cnt	MAE	MSE
(4)	✗	✓	64.5	110.8
(1,2,3)	✗	✓	69.1	107.8
(1,2,3,4)	✗	✓	**61.6**	**101.1**

Ablation Study on Feature Secondary Bridging of the Different Stages. The ablation experiments on the Shanghai PartA dataset under different stages of feature secondary bridging selected by PVT-Crowd are given in Table 5. The numbers in the methods in the table represent the feature map generated under stage i. For example, (1, 2, 3) represents the features generated by the ASBM in stages 1, 2 and 3 for pooling as tokens for aggregation. We can observe that the counting works best when bridging the tokens generated from four stages. This is because bridging the features of the other stages, which provide the rich information that in the previous stage, resulting in a better performance. Significantly, when we use only the last layer of features for crowd counting, the counting performance already exceeds that of most SOTA methods.

Ablation Study on Loss Function. We quantify the L1 loss and Smooth L1 loss for the Shanghai PartA and PartB dataset experiments in Table 6. Using Smooth L1 loss, MAE and MSE can be decreased from 63.4 and 102.6 to 61.6 and 101.1 in PartA. MAE and MSE can be decreased from 8.8 and 14.3 to 8.6 and 14.2 in PartB.

Table 6. Choice of loss function

Method	Label		Shanghai PartA	
	Loc	Cnt	MAE	MSE
L1	✗	✓	63.4	102.0
SmoothL1	✗	✓	**61.6**	**101.1**

5 Conclusion

In this work, we propose a Pyramidal Visual Transformer Network for weakly-supervised crowd counting, named PVT-Crowd. The Adjacent-Scale Bridging Module and Global-Scale Bridging Module are designed for the scale variation problem, and they can better extract image global features and enhance image crowd information. An extensive evaluation of three benchmark datasets shows that PVT-Crowd achieves superior counting performance compared to other mainstream weakly-supervised methods and competitive performance is achieved compared to some fully-supervised methods. In the future, we plan to research an excellent multi-scale regression head to challenge prediction scenes with huge scale changes. Furthermore, further achieve network lightweight.

References

1. Chu, X., et al.: Twins: revisiting the design of spatial attention in vision transformers. In: Advances in Neural Information Processing Systems, vol. 34, pp. 9355–9366 (2021)
2. Dosovitskiy, A., Beyer, L., Kolesnikov, A., Weissenborn, D., et al.: An Image is Worth 16x16 Words: Transformers for Image Recognition at Scale. arXiv e-prints arXiv:2010.11929 (2020)
3. Hossain, M., Hosseinzadeh, M., Chanda, O., Wang, Y.: Crowd counting using scale-aware attention networks. In: 2019 IEEE Winter Conference on Applications of Computer Vision (WACV), pp. 1280–1288 (2019)
4. Jiang, X., et al.: Attention scaling for crowd counting. In: 2020 IEEE/CVF Conference on Computer Vision and Pattern Recognition (CVPR), pp. 4705–4714 (2020)
5. Jiang, X., et al.: Crowd counting and density estimation by trellis encoder-decoder networks. In: 2019 IEEE/CVF Conference on Computer Vision and Pattern Recognition (CVPR), pp. 6126–6135 (2019)

6. Lei, Y., Liu, Y., Zhang, P., Liu, L.: Towards using count-level weak supervision for crowd counting. Pattern Recogn. **109**, 107616 (2021)
7. Li, Y., Zhang, X., Chen, D.: CSRNet: dilated convolutional neural networks for understanding the highly congested scenes. In: 2018 IEEE/CVF Conference on Computer Vision and Pattern Recognition, pp. 1091–1100 (2018)
8. Liang, D., Chen, X., Xu, W., Zhou, Y., Bai, X.: TransCrowd: weakly-supervised crowd counting with transformers. Sci. China Inf. Sci. **65**(6), 160104 (2022)
9. Lin, H., Ma, Z., Ji, R., Wang, Y., Hong, X.: Boosting crowd counting via multi-faceted attention. In: 2022 IEEE/CVF Conference on Computer Vision and Pattern Recognition (CVPR), pp. 19596–19605 (2022)
10. Ma, Z., Wei, X., Hong, X., Gong, Y.: Bayesian loss for crowd count estimation with point supervision. In: 2019 IEEE/CVF International Conference on Computer Vision (ICCV), pp. 6141–6150 (2019)
11. Savner, S.S., Kanhangad, V.: Crowdformer: weakly-supervised crowd counting with improved generalizability (2022). arXiv:2203.03768
12. Shi, M., Yang, Z., Xu, C., Chen, Q.: Revisiting perspective information for efficient crowd counting. In: 2019 IEEE/CVF Conference on Computer Vision and Pattern Recognition (CVPR), pp. 7271–7280 (2019)
13. Shi, Z., et al.: Crowd counting with deep negative correlation learning. In: 2018 IEEE/CVF Conference on Computer Vision and Pattern Recognition, pp. 5382–5390 (2018)
14. Sindagi, V.A., Patel, V.M.: Generating high-quality crowd density maps using contextual pyramid CNNs. In: 2017 IEEE International Conference on Computer Vision (ICCV), pp. 1879–1888 (2017)
15. Sindagi, V.A., Patel, V.M.: HA-CCN: hierarchical attention-based crowd counting network. IEEE Trans. Image Process. **29**, 323–335 (2020)
16. Sun, G., Liu, Y., Probst, T., Paudel, D.P., Popovic, N., Gool, L.V.: Boosting crowd counting with transformers (2021). arXiv:2105.10926
17. Wan, J., Chan, A.: Modeling noisy annotations for crowd counting. In: Advances in Neural Information Processing Systems, vol. 33, pp. 3386–3396 (2020)
18. Wang, W., et al.: Pyramid vision transformer: a versatile backbone for dense prediction without convolutions. In: 2021 IEEE/CVF International Conference on Computer Vision (ICCV), pp. 548–558 (2021)
19. Wang, W., et al.: PVT V2: improved baselines with pyramid vision transformer. Comput. Vis. Media **8**(3), 415–424 (2022)
20. Xiong, Z., Chai, L., Liu, W., Liu, Y., Ren, S., He, S.: Glance to count: Learning to rank with anchors for weakly-supervised crowd counting (2022). arXiv:2205.14659
21. Yang, S., Guo, W., Ren, Y.: Crowdformer: an overlap patching vision transformer for top-down crowd counting. In: Raedt, L.D. (ed.) Proceedings of the Thirty-First International Joint Conference on Artificial Intelligence, IJCAI 2022, pp. 1545–1551 (2022)
22. Yang, Y., Li, G., Wu, Z., Su, L., Huang, Q., Sebe, N.: Weakly-supervised crowd counting learns from sorting rather than locations. In: Vedaldi, A., Bischof, H., Brox, T., Frahm, J.-M. (eds.) ECCV 2020. LNCS, vol. 12353, pp. 1–17. Springer, Cham (2020). https://doi.org/10.1007/978-3-030-58598-3_1
23. Zhang, Y., Zhou, D., Chen, S., Gao, S., Ma, Y.: Single-image crowd counting via multi-column convolutional neural network. In: 2016 IEEE Conference on Computer Vision and Pattern Recognition (CVPR), pp. 589–597 (2016)

Multi-view Contrastive Learning Network for Recommendation

Xiya Bu[1,2] and Ruixin Ma[1,2(✉)]

[1] School of Software Technology, Dalian University of Technology, Dalian, China
maruixin@dlut.edu.cn
[2] The Key Laboratory for Ubiquitous Network and Service of Liaoning Province, Dalian, China

Abstract. Knowledge graphs (KGs) are being introduced into recommender systems in more and more scenarios. However, the supervised signals of the existing KG-aware recommendation models only come from the historical interactions between users and items, which will lead to the sparse supervised signal problem. Inspired by self-supervised learning, which can mine supervised signals from the data itself, we apply its contrastive learning framework to KG-aware recommendation, and propose a novel model named Multi-view Contrastive Learning Network (MCLN). Unlike previous contrastive learning methods that usually generate different views by ruining graph nodes, MCLN comprehensively considers four different views, including collaborative knowledge graph (CKG), user-item interaction graph (UIIG), and user-user graph (UUG) and item-item graph (IIG). We treat the CKG as a global-level structural view, and the other three views as local-level collaborative views. Therefore, MCLN performs contrastive learning between the four views at the local and global levels, aiming to mine the collaborative signals between users and items, between users, and between items, and the global structural information. Besides, in the construction of UUG and IIG, a receptive field is designed to capture important user-user and item-item collaborative signals. Extensive experiments on three datasets show that MCLN significantly outperforms state-of-the-art baselines.

Keywords: Knowledge graph · Recommender systems · Supervised signals · Self-supervised learning · Contrastive learning

1 Introduction

There are roughly three categories of existing KG-aware recommendation: embedding-based recommendation [1,15,25], path-based recommendation [6,24], and propagation-based recommendation [14,17,19,20].

This work is supported by the National Natural Science Foundation of China (61906030), the Science and Technology Project of Liaoning Province(2021JH2/10300064), the Youth Science and Technology Star Support Program of Dalian City (2021RQ057) and the Fundamental Research Funds for the Central Universities(DUT22YG241).

Among the embedding-based methods, the knowledge graph embedding (KGE) method [7,18,26] is applied to pre-train the KG to obtain items representations under a specific semantic space. For instance, CKE [25] employs TransR [8] on the KG to obtain entities representations containing structural information, and then the aligned items representations are obtained. KTUP [1] learns a translation-based model through TransH [21]. Although the KGE method can be applied to obtain items representations containing structural information, compared with the recommendation task, the KGE method is more suitable for in-graph applications(e.g. KG reasoning, KG completion), and the obtained items representations do not represent correlations between items.

In path-based methods, multiple connectivity patterns between users and items are mined based on KG to provide accurate guidance for recommendation. The path-based methods view the KG as a directed HIN, and extract its latent features based on meta-paths or meta-graphs to represent the relevancy between users and items. The path-based methods can effectively improve the accuracy and interpretability of recommendations, but to achieve the desired results, numerous high-quality meta-paths or meta-graphs need to be extracted and constructed from the KG. In addition, the path-based methods is not end-to-end frameworks, which leads to the fact that when the recommendation scenario changes, we have to re-extract and construct the meta-paths or meta-graphs based on the new scenario. Constructing high-quality meta-paths or meta-graphs is an extremely complex task in itself, which requires designers to have particularly rich domain knowledge.

Propagation-based methods iteratively propagate on the KG to integrate neighborhood information from different hops to enrich nodes representations. Their optimization goals are set to be consistent with the recommender system, thereby improving the recommendation effect. RippleNet [14] propagates the user's potential preferences on the KG, and utilizes entities representations on the KG to update the user representation. However, RippleNet does not consider the importance of relations in the propagation process, and the size of its ripple-set constructed based on KG may become out of control as the size of KG increases, resulting in substantial storage and computing losses. In order to avoid introducing more noise, KGCN [17] only samples part of the neighborhood information of the entities, and then applies the framework of the graph convolutional network (GCN) [23] for information propagation to obtain items representations containing important neighborhood information. However, KGCN does not take into account explicit collaborative signals, which will lead to missing information in the resulting items representations. KGAT [19] is a model for information propagation on CKG to enrich users and items representations. However, KGAT does not distinguish users nodes from items nodes in the information propagation process, which will result in the obtained users representations and items representations containing more noise. KGIN [20] models the user intent behind user-item interactions from a more fine-grained perspective, and executes GNN on CKG.

In general, the models we proposed above are all trained in a supervised learning manner, and their supervised signals only come from sparse historical interactions.

In order to solve the problem of sparse supervised signals in existing KG-aware recommendation models, it is crucial to explore a model that can efficiently mine supervised signals from the data itself to improve users and items representations learning. Besides, KGCL [22] is a contrastive learning-based model that generates different views by ruining graph nodes, which does not guarantee the quality of the generated views. Unlike KGCL, our proposed MCLN comprehensively considers and constructs four different views, which can guarantee the quality of the generated views.

All in all, our main contributions can be summarized as shown below:

- We emphasize the importance of exploiting the self-supervised learning mechanism to mine supervised signals from the data itself, which provides abundant auxiliary signals for KG-aware recommendation.
- We propose a novel model MCLN, which constructs a multi-perspective cross-view contrastive learning framework for KG-aware recommendation under the self-supervised learning mechanism. MCLN comprehensively considers four kinds of views, including global-level CKG that considers structural information, local-level UIIG that considers collaborative signals between users and items, local-level UUG that considers collaborative signals between users, and local-level IIG that considers collaborative signals between items.
- We apply the intersection-union ratio to determine the correlation between users and between items, to reasonably construct UUG and IIG. Besides, a receptive field is designed to avoid introducing more noise.
- We conduct extensive experiments on three datasets accompanying specific scenarios. Experimental results demonstrate that MCLN significantly outperforms state-of-the-art baselines.

2 Problem Formulation

The user-item interaction graph $I = \{(u, i_{uv}, v)|u \in U, v \in V\}$ represents sparse historical interactions between users and items, where $U = \{u_1, u_2, \ldots\}$ and $V = \{v_1, v_2, \ldots\}$ represent the user set and the item set respectively. If there is a historical interaction between the user u and the item v then $i_{uv} = 1$, otherwise $i_{uv} = 0$. In addition, knowledge graph $G = \{(h, r, t)|h, t \in E, r \in R\}$ indicates the correlation between items, where $E = \{e_1, e_2, \ldots\}$ represents the entity set, which has three different types of entities under the KG-aware recommendation: item entities, attribute entities and other entities, and $R = \{r_1, r_2, \ldots\}$ represents the set of relations between entities. The KG is composed of numerous entity-relation-entity triplets, and each triplet indicates that there is a certain relation between the head entity and the tail entity. Besides, the item-entity alignment set $A = \{(v, e)|v \in V, e \in E\}$ is used to represent the entities in the KG corresponding to the items in the UIIG.

Based on UIIG I and KG G, the task of KG-aware recommendation is to predict the possibility of interaction between user u and item v that user u have not historically interacted with. Specifically, the KG-aware recommendation model aims to learn a prediction function $\hat{i}_{uv} = F(u, v|\Theta, I, G)$, where $\hat{i}_{uv}$ represents the probability that the model predicts that user u will interact with item v, and Θ represents the set of model parameters.

3   Methodology

We now describe in detail our proposed MCLN framework. As shown in Fig. 1, MCLN consists of three main components: 1) Multi-view generation layer. 2) Local-level contrastive learning layer. 3) Global-level contrastive learning layer.

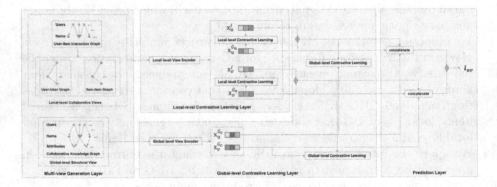

Fig. 1. Illustration of our proposed MCLN framework.

3.1   Multi-view Generation Layer

In this section, we consider and construct four different types of views in a very comprehensive and fine-grained way. First considering the global structural information, we construct the CKG $G_c = \{(h, r, t)|h, t \in U \cup E, r \in R \cup Interact\}$ through the item-entity alignment set A joint KG G and UIIG I, where $Interact$ indicates the set of interaction relations between users and items. UIIG is treated as a local-level collaborative view to explore collaborative signals between users and items. In addition, in order to explore collaborative signals between users and between items in a more fine-grained way, we construct two local-level collaborative views: UUG and IIG, which have not been considered in previous research.

As shown in Fig. 2, we can get the item set that interact with the user and the user set that interact with the item based on the UIIG, and then employ the intersection-union ratio to get the correlations between users, formulated as:

$$IoU_{uu^*} = \frac{|\{v \mid (u, v) \in I\} \cap \{v \mid (u^*, v) \in I\}|}{|\{v \mid (u, v) \in I\} \cup \{v \mid (u^*, v) \in I\}|} \qquad (1)$$

where the numerator represents the intersection size of two sets and the denominator represents the union size of two sets. Similarly, we can obtain the correlations IoU_{vv^*} between items. Intersection-union ratio can take into account the size of each set, so using the intersection-union ratio can reasonably and adequately obtain the correlation between the two sets, so as to obtain the correlation between the users or items behind the two sets.

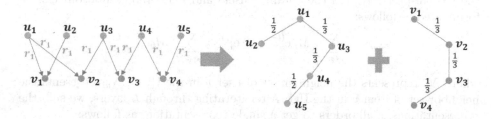

Fig. 2. User-user graph and item-item graph based on user-item interaction graph.

To avoid introducing too much noise and filter out a large number of irrelevant edges, we elaborately designed a respective field to preserve the most relevant connections, as follows:

$$\widehat{IoU}_{uu^*} = \begin{cases} IoU_{uu^*} & IoU_{uu^*} \in \text{top\_}K(IoU_u) \\ 0 & \text{otherwise} \end{cases} \tag{2}$$

where $\widehat{IoU}_{uu^*}$ is a sparse graph adjacency matrix. Similarly, another sparse graph adjacency matrix $\widehat{IoU}_{vv^*}$ can be obtained. Next, $\widehat{IoU}_{uu^*}$ and $\widehat{IoU}_{vv^*}$ are normalized as follows:

$$M_u = (D)^{-\frac{1}{2}}\widehat{IoU}_{uu^*}(D)^{-\frac{1}{2}}, \quad M_v = (D)^{-\frac{1}{2}}\widehat{IoU}_{vv^*}(D)^{-\frac{1}{2}} \tag{3}$$

where D is the diagonal degree matrix of $\widehat{IoU}_{uu^*}$ or $\widehat{IoU}_{vv^*}$.

3.2 Local-Level Contrastive Learning Layer

Local-Level Views Encoder. UIIG, UUG and IIG emphasize collaborative signals between users and items, between users, and between items, respectively. To capture these three types of collaborative signals, we perform information propagation on these three views based on the framework of GCN. To make the graph convolution operation more concise and efficient, we omit feature transformation and non-linear activation [5] in the graph convolution operation. First, we recursively perform information propagation in the UIIG, and the obtained users and items representations can be formulated as follows:

$$e_u^{(b)} = \sum_{v \in N_u} \frac{1}{\sqrt{|N_u||N_v|}} e_v^{(b-1)}, \quad e_v^{(b)} = \sum_{u \in N_v} \frac{1}{\sqrt{|N_u||N_v|}} e_u^{(b-1)} \tag{4}$$

where N_u and N_v represent the neighbor sets of user u and item v, respectively. After iterating through B layers, considering the connectivity information of different orders, we sum the representations of all orders to get a single representation, as follows:

$$x_u^I = e_u^{(0)} + e_u^{(1)} + \cdots + e_u^{(B)}, \quad x_v^I = e_v^{(0)} + e_v^{(1)} + \cdots + e_v^{(B)} \qquad (5)$$

Then we recursively perform the information propagation on the UUG and the IIG, respectively, and the obtained users and items representations can be formulated as follows:

$$e_u^{(l)} = \sum_{u^* \in N_u} M_u e_{u^*}^{(l-1)}, \quad e_v^{(l)} = \sum_{v^* \in N_v} M_v e_{v^*}^{(l-1)} \qquad (6)$$

where N_u represents the neighbor set of user u in the UUG, N_v represents the neighbor set of item v in the IIG. After iterating through L layers, we sum the representations of all orders to get a single representation, as follows:

$$x_u^{G_u} = e_u^{(0)} + e_u^{(1)} + \cdots + e_u^{(L)}, \quad x_v^{G_v} = e_v^{(0)} + e_v^{(1)} + \cdots + e_v^{(L)} \qquad (7)$$

Local-Level Contrastive Learning Optimization. For supervising the three local-level collaborative views to learn comprehensive users and items representations, we perform contrastive learning [4,13] between the specific representation x_u^I of user u in the UIIG and the specific representation $x_u^{G_u}$ of user u in the UUG, and between the specific representation x_v^I of item v in the UIIG and the specific representation $x_v^{G_v}$ of item v in the IIG. To map them to the same contrastive space for better contrastive learning between them, we apply the hidden layer in the neural network to perform spatial transformation on them, as shown below:

$$\tilde{x}_u^I = W_2 \sigma(W_1 x_u^I + b_1) + b_2, \quad \tilde{x}_v^I = W_2 \sigma(W_1 x_v^I + b_1) + b_2 \qquad (8)$$

where W_1, W_2 are trainable weight matrices; b_1, b_2 are biases; σ is LeakyReLU function [9]. Similarly, we can obtain $\tilde{x}_u^{G_u}$ and $\tilde{x}_v^{G_v}$. Inspired by previous contrastive learning work [27–29], we take the paired representations of the same node learned from the paired views as the positive sample, and other forms of paired representations are taken as negative samples, and then we define the contrastive loss as follow:

$$L_u^{local} = -\log \frac{e^{\cos\text{-}sim(P_u)}}{e^{\cos\text{-}sim(P_u)} + \sum e^{\cos\text{-}sim(S_u)}} \qquad (9)$$

where P_u is $\left(\tilde{x}_u^{G_u}, \tilde{x}_u^I\right)$, which represents the positive representation pair; S_u is the negative representation pair, where can be constructed based on the same view or different views; $\cos\text{-}sim$ represents the cosine similarity between two vectors. Similarly, we can obtain L_v^{local}. Then, we obtain the local-level contrastive loss by combining L_u^{local} and L_v^{local}, as follows:

$$L^{local} = L_u^{local} + L_v^{local} \qquad (10)$$

3.3 Global-Level Contrastive Learning Layer

Global-Level View Encoder To capture the global structural information on the CKG, we design a relation-aware GNN to recursively perform information propagation on the CKG, and the obtained users and items representations can be formulated as follows:

$$e_u^{(c)} = \frac{1}{|N_u|} \sum_{(u,r,v)\in N_u} e_v^{(c-1)}, \quad e_v^{(c)} = \sum_{(v,r,t)\in N_v} \pi(v,r,t)e_t^{(c-1)} \tag{11}$$

where $\pi(v,r,t)$ represents the importance of the neighbor entity t to the central entity v in the information propagation. We implement the $\pi(v,r,t)$ through the relation-aware attention mechanism [2,12], as follows:

$$\pi(v,r,t) = \cos\_\text{sim}(v+r,t) \tag{12}$$

We model attention with the cosine similarity between $v+r$ and t, which can fully preserve the structural information of subgraphs centered on v, so that the neighbors closer to v can get more attention. Then we adopt the softmax function [10] to normalize the coefficients of all neighbors adjacent to v:

$$\pi(v,r,t) = \frac{e^{\pi(v,r,t)}}{\sum_{(v,r',t')\in N_v} e^{\pi(v,r',t')}} \tag{13}$$

After iterating through C layers on the CKG, we sum the representations of all orders to get a single representation, as follows:

$$x_u^{G_c} = e_u^{(0)} + e_u^{(1)} + \cdots + e_u^{(C)}, \quad x_v^{G_c} = e_v^{(0)} + e_v^{(1)} + \cdots + e_v^{(C)} \tag{14}$$

Global-Level Contrastive Learning Optimization. As with the local-level contrastive learning optimization, we first map the global-level and local-level node representations into the same contrastive space, and then we can obtain $\widetilde{x}_u^{G_c}$, $\widetilde{x}_v^{G_c}$, $\widetilde{x}_u^l$, and $\widetilde{x}_v^l$.

Then we adopt the same positive and negative sample strategy as the local-level contrastive learning optimization to define its contrastive loss, as follows:

$$L_u^g = -\log\frac{e^{\cos\_\text{sim}(P_u)}}{e^{\cos\_\text{sim}(P_u)} + \sum e^{\cos\_\text{sim}(S_u^g)}} \tag{15}$$

$$L_u^l = -\log\frac{e^{\cos\_\text{sim}(P_u)}}{e^{\cos\_\text{sim}(P_u)} + \sum e^{\cos\_\text{sim}(S_u^l)}} \tag{16}$$

where L_u^g and L_u^l represent the contrastive loss calculated from the global view and from the local view, respectively. Similarly, we can obtain L_v^g and L_v^l. Then we obtain the global-level contrastive loss by combining $L_{u/v}^g$ and $L_{u/v}^l$:

$$L^{global} = \frac{1}{2M}\sum_{i=1}^{M}(L_u^g + L_u^l) + \frac{1}{2N}\sum_{i=1}^{N}(L_v^g + L_v^l) \tag{17}$$

3.4 Prediction Layer

After encoding the four views, we can obtain multiple representations $\{x_u^I, x_u^{G_u}, x_u^{G_c}\}$ of user u and multiple representations $\{x_v^I, x_v^{G_v}, x_v^{G_c}\}$ of item v. By combining the above representations, we have the final user representation and the final item representation, as follows:

$$x_u^* = x_u^{G_c} || x_u^{G_u} || \left(x_u^I + x_u^{G_u} \right), \quad x_v^* = x_v^{G_c} || x_v^{G_v} || \left(x_v^I + x_v^{G_v} \right) \tag{18}$$

where $||$ is the concatenation operator used to perform vector splicing. Next, we perform the inner product between user representation x_u^* and item representation x_v^* to predict the matching score between user u and item v:

$$\hat{i}_{uv} = {x_u^*}^\top x_v^* \tag{19}$$

For the KG-aware recommendation task, we optimize it with pairwise BPR loss [11], calculated by:

$$L_{\text{BPR}} = \sum_{(u,v_s,v_p) \in O} -\ln \sigma \left(\hat{i}_{uv_s} - \hat{i}_{uv_p} \right) \tag{20}$$

where $O = \{(u, v_s, v_p) \,|\, (u, v_s) \in O^+, (u, v_p) \in O^-\}$ indicates the training set, O^+ denotes the historical (positive) interaction set and O^- denotes the unobserved (negative) interaction set; σ is the sigmoid function [3]. Finally, we obtain the objective loss function by combining L^{local}, L^{global} and L_{BPR}:

$$L_{MCLN} = L_{\text{BPR}} + \alpha \left(\beta L^{local} + (1 - \beta) L^{global} \right) + \lambda ||\Theta||_2^2 \tag{21}$$

where Θ is the model parameter set; α and β are the hyper-parameters that control the strength of different losses; $\lambda ||\Theta||_2^2$, conducted to prevent overfitting, is a L_2 regularization parameterized by λ.

4 Experiments

We perform extensive experiments on three datasets accompanying specific scenarios: Book-Crossing, MovieLens-1M, and Last.FM which are published by MCCLK [29].

4.1 Experiments Setup

We implement our proposed MCLN model in Pytorch. For the sake of fairness, we set the size of ID embeddings to 64 for all models, and the parameters of all models are initialized with the Xavier initializer. Besides, each model is optimized by applying the Adam optimizer, and the batch size for each model is set to 2048. We confirm the optimal settings for each model by applying a grid search: the learning rate η is searched within $\{0.0003, 0.0001, 0.003, 0.001\}$, the coefficient of L_2 normalization λ is tuned in $\{10^{-7}, 10^{-6}, 10^{-5}, 10^{-4}, 10^{-3}\}$. The optimal settings of hyper-parameters of other baseline models are provided by their original papers.

4.2   Performance Comparison

We adopt two widely used metrics [14,17] AUC and $F1$ to evaluate CTR prediction and report the experimental performance of all models in Table 1.

As shown in Table 1, our proposed MCLN model consistently outperforms all baselines on three datasets accompanying different scenarios, which demonstrates the effectiveness of MCLN.

Table 1. Overall Performance Comparison.

	Book-Crossing		MovieLens-1M		Last.FM	
	AUC	$F1$	AUC	$F1$	AUC	$F1$
CKE [25]	0.6759	0.6235	0.9065	0.8024	0.7471	0.6740
PER [24]	0.6048	0.5726	0.7124	0.6670	0.6414	0.6033
RippleNet [14]	0.7211	0.6472	0.9190	0.8422	0.7762	0.7025
KGCN [17]	0.6841	0.6313	0.9090	0.8366	0.8027	0.7086
KGNN-LS [16]	0.6762	0.6314	0.9140	0.8410	0.8052	0.7224
KGAT [19]	0.7314	0.6544	0.9140	0.8440	0.8293	0.7424
KGIN [20]	0.7273	0.6614	0.9190	0.8441	0.8486	0.7602
MCCLK [29]	0.7572	0.6737	0.9343	0.8621	0.8723	0.7845
MCLN	**O.7684**	**0.6790**	**0.9384**	**0.8660**	**0.8832**	**0.8092**

4.3   Study of MCLN

Effect of Aggregation Depth. To study the effect of UIIG aggregation depth, the layer depth B is adjusted in the range of $\{1, 2, 3\}$. By analogy, the layer depth L in UUG and IIG and the layer depth C in CKG are adjusted within the range of $\{1, 2, 3\}$. Note that to avoid mutual influence between multiple layer depths, when changing a certain layer depth, the other layer depths are set to 1. In Table 2, we report the performance comparisons of MCLN on the Book-Crossing, MovieLens-1M, and Last.FM datasets as layer depth changes.

From Table 2, we can see that MCLN achieves the best performance when $L = 1$ or $L = 2$. This indicates that MCLN can capture enough neighborhood information in UUG and IIG with only one or two information propagations, which proves the effectiveness of UUG and IIG construction. Besides, in the combination of $B = 1$, $L = 1$, and $C = 2$, MCLN can achieve the best performance on the Book-Crossing dataset. In the combination of $B = 2$, $L = 1$, and $C = 2$, MCLN can achieve the best performance on the MovieLens-1M dataset. In the combination of $B = 3$, $L = 2$, and $C = 2$, MCLN can achieve the best performance on the Last.FM dataset. We report the corresponding results in Table 1.

Table 2. Effect of aggregation depth.

	Book-Crossing		MovieLens-1M		Last.FM	
	AUC	*F1*	*AUC*	*F1*	*AUC*	*F1*
$B = 1$	**0.7659**	**0.6797**	0.9366	0.8649	0.8723	0.7900
$B = 2$	0.7609	0.6769	**0.9378**	**0.8663**	0.8761	**0.7918**
$B = 3$	0.7559	0.6686	0.9353	0.8652	**0.8765**	0.7917
$L = 1$	**0.7659**	**0.6797**	**0.9366**	**0.8649**	0.8723	0.7900
$L = 2$	0.7517	0.6722	0.9352	0.8641	**0.8819**	**0.8008**
$L = 3$	0.7383	0.6584	0.9352	0.8634	0.8792	0.7983
$C = 1$	0.7659	**0.6797**	0.9366	0.8649	0.8723	0.7900
$C = 2$	**0.7684**	0.6790	**0.9381**	**0.8662**	**0.8755**	**0.8011**
$C=3$	0.7646	0.6696	0.9361	0.8607	0.8733	0.7915

Effect of User-User Graph and Item-Item Graph. To verify the impact of UUG and IIG, we apply $MCLN_{w/oUUG}$ to represent the MCLN model with UUG removed. We apply $MCLN_{w/oIIG}$ to represent the MCLN model with IIG removed. The experimental results are summarized in Table 3. From Table 3, we can see that the performance of MCLN on the three datasets decreases to varying degrees no matter whether the UUG is removed or the IIG is removed. This demonstrates the importance and effectiveness of modeling UUG and IIG, which can fully capture collaborative signals between users and between items.

Table 3. Effect of user-user graph and item-item graph.

	Book-Crossing		MovieLens-1M		Last.FM	
	AUC	*F1*	*AUC*	*F1*	*AUC*	*F1*
MCLN	**0.7684**	**0.6790**	**0.9384**	**0.8660**	**0.8832**	**0.8092**
$MCLN_{w/oUUG}$	0.7552	0.6737	0.9361	0.8643	0.8809	0.8006
$MCLN_{w/oIIG}$	0.7207	0.6568	0.9344	0.8610	0.8673	0.7790

5   Conclusion

In this work, we aim to apply the self-supervised learning mechanism to KG-aware recommendation to alleviate the sparse supervised signal problem. We propose a new model, MCLN, which applies the contrastive learning framework to learn better users and items representations. Its core lies in: (1) MCLN comprehensively considers four different views at the local and global levels. (2) Contrastive learning of MCLN at the local and global levels can fully capture the collaborative signals between users and items, between users and between items, and the global structural information.

References

1. Cao, Y., Wang, X., He, X., Hu, Z., Chua, T.S.: Unifying knowledge graph learning and recommendation: Towards a better understanding of user preferences. In: The World Wide Web Conference, pp. 151–161 (2019)
2. Chen, J., Zhang, H., He, X., Nie, L., Liu, W., Chua, T.S.: Attentive collaborative filtering: Multimedia recommendation with item-and component-level attention. In: Proceedings of the 40th International ACM SIGIR conference on Research and Development in Information Retrieval, pp. 335–344 (2017)
3. Han, J., Moraga, C.: The influence of the sigmoid function parameters on the speed of backpropagation learning. In: Mira, J., Sandoval, F. (eds.) IWANN 1995. LNCS, vol. 930, pp. 195–201. Springer, Heidelberg (1995). https://doi.org/10.1007/3-540-59497-3_175
4. Hassani, K., Khasahmadi, A.H.: Contrastive multi-view representation learning on graphs. In: International Conference on Machine Learning, pp. 4116–4126. PMLR (2020)
5. He, X., Deng, K., Wang, X., Li, Y., Zhang, Y., Wang, M.: Lightgcn: simplifying and powering graph convolution network for recommendation. In: Proceedings of the 43rd International ACM SIGIR conference on research and development in Information Retrieval, pp. 639–648 (2020)
6. Hu, B., Shi, C., Zhao, W.X., Yu, P.S.: Leveraging meta-path based context for top-n recommendation with a neural co-attention model. In: Proceedings of the 24th ACM SIGKDD International Conference on Knowledge Discovery & Data Mining, pp. 1531–1540 (2018)
7. Ji, G., He, S., Xu, L., Liu, K., Zhao, J.: Knowledge graph embedding via dynamic mapping matrix. In: Proceedings of the 53rd Annual Meeting of the Association for Computational Linguistics and the 7th International Joint Conference on Natural Language Processing (volume 1: Long papers). pp. 687–696 (2015)
8. Lin, Y., Liu, Z., Sun, M., Liu, Y., Zhu, X.: Learning entity and relation embeddings for knowledge graph completion. In: Proceedings of the AAAI Conference on Artificial Intelligence, vol. 29 (2015)
9. Maas, A.L., Hannun, A.Y., Ng, A.Y., et al.: Rectifier nonlinearities improve neural network acoustic models. In: Proceedings of ICML, Atlanta, Georgia, USA, vol. 30, p. 3 (2013)
10. Memisevic, R., Zach, C., Pollefeys, M., Hinton, G.E.: Gated softmax classification. In: Advances in Neural Information Processing Systems 23 (2010)
11. Rendle, S., Freudenthaler, C., Gantner, Z., Schmidt-Thieme, L.: Bayesian personalized ranking from implicit feedback. In: Proceedings of of Uncertainty in Artificial Intelligence, pp. 452–461 (2014)
12. Vaswani, A., et al.: Attention is all you need. In: Advances in Neural Information Processing Systems 30 (2017)
13. Velickovic, P., Fedus, W., Hamilton, W.L., Liò, P., Bengio, Y., Hjelm, R.D.: Deep graph infomax. ICLR (Poster) **2**(3), 4 (2019)
14. Wang, H., et al.: Ripplenet: propagating user preferences on the knowledge graph for recommender systems. In: Proceedings of the 27th ACM International Conference on Information and Knowledge Management, pp. 417–426 (2018)
15. Wang, H., Zhang, F., Xie, X., Guo, M.: Dkn: deep knowledge-aware network for news recommendation. In: Proceedings of the 2018 World Wide Web Conference, pp. 1835–1844 (2018)

16. Wang, H., et al.: Knowledge-aware graph neural networks with label smoothness regularization for recommender systems. In: Proceedings of the 25th ACM SIGKDD International Conference on Knowledge Discovery & Data Mining, pp. 968–977 (2019)
17. Wang, H., Zhao, M., Xie, X., Li, W., Guo, M.: Knowledge graph convolutional networks for recommender systems. In: The World Wide Web Conference, pp. 3307–3313 (2019)
18. Wang, Q., Mao, Z., Wang, B., Guo, L.: Knowledge graph embedding: a survey of approaches and applications. IEEE Trans. Knowl. Data Eng. **29**(12), 2724–2743 (2017)
19. Wang, X., He, X., Cao, Y., Liu, M., Chua, T.S.: Kgat: knowledge graph attention network for recommendation. In: Proceedings of the 25th ACM SIGKDD International Conference on Knowledge Discovery & Data Mining, pp. 950–958 (2019)
20. Wang, X., et al.: Learning intents behind interactions with knowledge graph for recommendation. In: Proceedings of the Web Conference 2021, pp. 878–887 (2021)
21. Wang, Z., Zhang, J., Feng, J., Chen, Z.: Knowledge graph embedding by translating on hyperplanes. In: Proceedings of the AAAI Conference on Artificial Intelligence, vol. 28 (2014)
22. Yang, Y., Huang, C., Xia, L., Li, C.: Knowledge graph contrastive learning for recommendation. In: Proceedings of the 45th International ACM SIGIR Conference on Research and Development in Information Retrieval, pp. 1434–1443 (2022)
23. Ying, R., He, R., Chen, K., Eksombatchai, P., Hamilton, W.L., Leskovec, J.: Graph convolutional neural networks for web-scale recommender systems. In: Proceedings of the 24th ACM SIGKDD International Conference on Knowledge Discovery & Data Mining, pp. 974–983 (2018)
24. Yu, X., et al.: Personalized entity recommendation: a heterogeneous information network approach. In: Proceedings of the 7th ACM International Conference on Web Search and Data Mining, pp. 283–292 (2014)
25. Zhang, F., Yuan, N.J., Lian, D., Xie, X., Ma, W.Y.: Collaborative knowledge base embedding for recommender systems. In: Proceedings of the 22nd ACM SIGKDD International Conference on Knowledge Discovery and data mining, pp. 353–362 (2016)
26. Zhu, M., Zhen, D., Tao, R., Shi, Y., Feng, X., Wang, Q.: Top-N collaborative filtering recommendation algorithm based on knowledge graph embedding. In: Uden, L., Ting, I.-H., Corchado, J.M. (eds.) KMO 2019. CCIS, vol. 1027, pp. 122–134. Springer, Cham (2019). https://doi.org/10.1007/978-3-030-21451-7_11
27. Zhu, Y., Xu, Y., Yu, F., Liu, Q., Wu, S., Wang, L.: Deep graph contrastive representation learning. arXiv preprint arXiv:2006.04131 (2020)
28. Zhu, Y., Xu, Y., Yu, F., Liu, Q., Wu, S., Wang, L.: Graph contrastive learning with adaptive augmentation. In: Proceedings of the Web Conference 2021, pp. 2069–2080 (2021)
29. Zou, D., et al.: Multi-level cross-view contrastive learning for knowledge-aware recommender system. In: Proceedings of the 45th International ACM SIGIR Conference on Research and Development in Information Retrieval, pp. 1358–1368 (2022)

Uncertainty-Confidence Fused Pseudo-labeling for Graph Neural Networks

Pingjiang Long, Zihao Jian, and Xiangrong Liu[✉]

Department of Computer Science and Technology, School of Informatics,
Xiamen University, Xiamen, China
xrliu@xmu.edu.cn

Abstract. Graph Neural Networks (GNNs) have achieved promising performance for semi-supervised graph learning. However, the training of GNNs usually heavily relies on a large number of labeled nodes in a graph. When the labeled data are scarce, GNNs easily over-fit rto the few labeled samples, resulting in the degenerating performance. To address this issue, we propose a novel graph pseudo-labeling framework. The proposed framework combines both the predication confidence and approximate Bayesian uncertainty of GNNs, resulting in a metric for generating more reliable and balanced pseudo-labeled nodes in graph. Furthermore, an iterative re-training strategy is employed on the extended training label set (including original labeled and pseudo-labeled nodes) to train a more generalized GNNs. Extensive experiments on benchmark graph datasets demonstrate that the proposed pseudo-labeling framework can enhance node classification performance of two alternative GNNs models by a considerable margin, specifically when labeled data are scarce.

Keywords: Graph neural networks · Pseudo-labeling · Model uncertainty · Node classification

1 Introduction

Recently, Graph neural networks(GNNs) [1–3] have attracted promising attention for its remarkable performance on various real world graph based applications [4–6]. However, most graph tasks are under semi-supervised scenario, where the labeled nodes are usually scarce, this make existing GNNs suffer from severe over-fitting problem, resulting in tremendous performance degeneration. For the extreme case, the performance of GNNs may even be inferior to the unsupervised methods. Specifically, for many real-world scenarios such as chemical molecule and drug, it is intractable to label the nodes in graph data. Thus, it is crucial to design a novel framework to enhance the generalization of existing GNNs especially when labeled data are scarce.

To alleviate this issue, the automatic generation of pseudo-labeling for the unlabeled nodes in a graph has become a common approach [7–9]. However, existing graph pseudo-labeling methods may suffer from the following limitations.

Q. Liu et al. (Eds.): PRCV 2023, LNCS 14433, pp. 331–342, 2024.
https://doi.org/10.1007/978-981-99-8546-3_27

On one hand, most methods generate pseudo labels according to the predication confidence output by softmax. However, a high confidence score cannot ensure the model prediction reliable due to the calibration error [10] of a unfavorably calibrated neural network especially when the training samples are scarce [11,12]. Hence, existing pseudo-labeling methods usually introduce too much noisy pseudo-labels misguide by the high confidence score in a under calibrated GNN model trained by few labeled samples, which provides incorrect supervised information thus hinder or degenerate their performance improvement.

On the other hand, existing pseudo-labeling methods face the challenge on balancing the proportion of the selected pseudo-labels of different classes and determining an appropriate number of pseudo-labels for different scale dataset and label rate. There are two commonly used strategies:(1) The *hard-threshold confidence* strategy that sets a fixed threshold τ to filter the pseudo-labels whose prediction confidence exceeds τ and add them into the expanded label set [9,13]. However this mechanism can easily result in the problem of severe class imbalance. (2) The *top-K confidence* methods that choose equally K pseudo-labels with the highest confidence score for each class respectively [7,8], while it lacks of flexibility on the number of pseudo-labels for it is challenging to manually decide a favorable K for different graph and label rate. And it remain us a question whether it is reasonable to equally assign K pseudo-labels for each class of nodes especially for the skewed graph(the number of nodes belong to one class may be large while another can be particularly small).

In this work, we realized the advance of uncertainty measure of neural network and the power of uncertainty to mitigate calibration error of prediction [10,14]. The common approach to estimate the uncertainty of neural network is to model the uncertainty of the model via probabilistic graphical model [14]. MC-dropout [10] develops a theoretical framework casting dropout training in deep neural networks as approximate Bayesian inference in deep Gaussian processes which provides us with a tool to model uncertainty of neural networks with dropout. Inspired by this, we introduce uncertainty into our graph pseudo-labeling framework to decrease calibration error and noisy pseudo-labels leading to a fused uncertainty-confidence metric to guide pseudo-labeling. To tackle the problem of imbalance and numbers of pseudo-labels, we develop a 'soft confidence threshold' technique to determine suitable amount of pseudo-labels for each class adaptively and automatically. Based on this, the number of pseudo-labels of each class is assigned according to their approximate label proportion. After pseudo-labeling, an iterative re-training scheme is proposed to further enhance the generalization of GNNs model.

Our main contributions of this paper are summarized as follows:(1) The proposed technique combines the uncertainty of neural network and predication confidence to guide pseudo-labeling, which effectively reduce the incorrect pseudo-labels and noisy training as well as ensure the balance and adaptiveness of pseudo-labels. (2) The proposed framework can be flexibly equipped with the diverse GNNs models, which significantly outperform the baseline GNNs methods by a considerable large margin on semi-supervised graph learning.

2   Methodology

2.1   Preliminaries

Let $\mathcal{G} = (\mathcal{V}, \mathcal{E}, X)$ be a graph, where $\mathcal{V}$ is the node set and $|\mathcal{V}| = n$ means that the graph $\mathcal{G}$ has n nodes, $\mathcal{E}$ is the edge set, where each edge connects two nodes and $X \in R^{n \times d}$ is the features matrix. Each node $v_i \in \mathcal{V}$ has a d-dimensional feature vector $X_i \in R^d$ that is the i-th row of X. We use $L \subset \mathcal{V}$ to denote the labeled nodes in graph $\mathcal{G}$ and $\mathcal{U} = \mathcal{V} - L$ denotes the unlabeled node set. Each node is categorized into one of C classes. Each of the labeled node $v_i \in \mathcal{L}$ has a one-hot label $Y_i = [y_i^0 ... y_i^{C-1}] \in \{0, 1\}^C$, where $y_i^c = 1$ indicates that v_i belongs to the c-th class.

2.2   Framework Overview

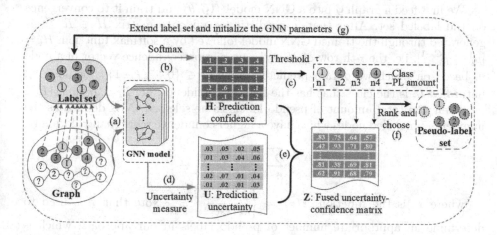

Fig. 1. Overview of UCP-GNN framework. First, a GNNs model is trained to convergence (a); then based on the confidence aware pseudo-labeling scheme (b,c) the number of pseudo-labels of each class is determined; after that, uncertainty aware pseudo-labeling (d) method is introduced to obtain a uncertainty measure for each prediction; subsequently, a fused matrix is construct to guide pseudo-labeling; finally, an iterative retraining framework (f,g) is developed to expand training samples and retrain a more generalized GNN.

The proposed UCP-GNN consists of three components: (1) Confidence aware pseudo-labeling. (2) Uncertainty aware pseudo labeling. (3) Uncertainty-confidence fused pseudo-labeling. (4) Retraining of the GNNs model. The overview of the proposed framework is illustrated in Fig. 1.

2.3   Confidence Aware Pseudo-labeling

For common GNNs, the confidence of the prediction of model varies greatly on various datasets, diverse label rate and model performance. For instance, when the training nodes are scarce, the model performance is inferior or the graph dataset is relatively small, most prediction nodes are in a weak confidence state which indicates that it is difficult to keep accuracy of pseudo labels. To this end, it reminds us to select relatively small amount of pseudo labels to ensure the label purity. In contrast, for a more generalized GNN model trained on larger graph or label set, there exist more high confidence nodes which requires to choose more pseudo labels to further enhance the performance of network.

To achieve this, we first use a confidence threshold $\tau \in (0, 1)$ to filter the high confidence prediction to determine the total number of pseudo labels which will to be selected across all classes(denoted as N). Then we can further draw the amount of pseudo labels for each class of node (denoted as n_c) according to the total amount of pseudo labels N and predicted label proportion.

We first feed a graph $\mathcal{G}$ into a GNN model $f(\mathcal{G}, \theta)$ and train it to convergence on the labeled set. After that, a prediction confidence matrix $H \in R^{n \times C}$ is generated through the trained GNN model followed by a softmax function. $H_i = [p_i^0 \cdots p_i^{C-1}]$, i.e. the i-th row of H is the prediction confidence vector of a node v_i based on which we obtain predicted label $\widetilde{L}_i \in \{0 \cdots C - 1\}^n$ for node v_i, naturally, $\widetilde{L}_i$ is the class that has the highest confidence, i.e. $\widetilde{L}_i = argmax(H_i)$.

Then, the total amount of pseudo labels to be selected, N is defined as the number of unlabeled nodes which have a higher confidence score than τ, formally:

$$N = \sum_{i \in \mathcal{U}} \mathbb{1}[H_i^{\widetilde{L}_i} \geq \tau] \tag{1}$$

where $\mathbb{1}()$is the function: $\mathbb{1}(\bullet) = \begin{cases} 1 \ if \ \bullet = true \\ 0 \ else \end{cases}$ Note that τ is used to determine an appropriate number of pseudo labels in our approach which is essentially different from the 'hard confidence threshold' methods where the threshold is designed to directly choose specific pseudo labels. Intuitively, τ can be regarded as a confidence valve which controls the amount of pseudo labels to be selected. Specifically, for a well trained GNN model it allows to generate more pseudo labels, otherwise less pseudo labels are selected to mitigate noise. Then it remains us a question that how to assign amount of pseudo labels for each class, i.e., n_c. We employ the strategy of allocating them based on the proportion of each class of (predicted)labels.

We first derive the proportion of each class of predicted labels in the entire predicted label set $\widetilde{L}$: $r_c = \frac{\sum_{i=0}^{n-1} \mathbb{1}[\widetilde{L}_i = c]}{n}$. Then, n_c is calculated as the proportion r_c in N: $\hat{n}_c = [N \times r_c]$ where '$[\ ]$' is the integer symbol. The number of pseudo labels produced by this method prevent from the proportion imbalance of training labels and is beneficial for providing reasonable supervised information for each class respectively.

2.4 Uncertainty Aware Pseudo-labeling

In the last subsection, we obtain the amount of pseudo labels n_c that will be selected for each class c respectively, then we need a metric to guide pseudo-labeling process. Naturally, the prediction confidence H is a prevalent measure to achieve pseudo-labeling by sorting on each class of predicted labels based on its confidence score. This setting makes our approach approximate the variation of the combination of 'hard confidence threshold' and 'top-K confidence'. However, as we analysed in Sect. 1, the existing confidence based methods face the challenge of introducing much noisy pseudo labels due to the network calibration error, especially when training labels are scarce. To alleviate this, the uncertainty of neural network is introduced to achieve a more satisfactory accuracy and decrease noise of selected pseudo labels.

The key issue to leverage uncertainty to pseudo labeling for GNNs is to compute the uncertainty of neural networks efficiently. In our approach, for convenience and considering computation complexity, we employ a general and computationally cheap approach (MC-dropout [10]) to calculate the uncertainty measure of each prediction of GNNs. MC-dropout uncertainty is the standard deviation of 10 predictions with the random dropout regularization of a trained neural network model. Specifically, assuming $H(k) \in [0,1]^{n \times C}$ is the k-th prediction probability of the trained model with dropout available, $i.e.$,

$$H_{(K)} = f(G, \theta)_{dropout} \tag{2}$$

Then, the uncertainty matrix U is formulated as:

$$U = \sigma(H_{(0)} \cdots H_{(10)} \in R^{n \times C}) \tag{3}$$

where σ is the standard deviation. Thus, the uncertainty of a node v_i predicted to class c, $i.e.$ $u(p_i^c)$ is the i-th row and c-th column of U. As a result, a larger uncertainty indicates that this prediction is unstable thus has higher risk of calibration error which remind us to prudently select this pseudo label although it has a high confidence. In the contrary, a smaller value of uncertainty means the prediction is more reliable. Therefore, we are more inclined to select the pseudo labels which have small uncertainty.

2.5 Uncertainty-Confidence Fused Pseudo-labeling

In the prior two subsections, we get the confidence matrix H and uncertainty matrix U as well as the amount of pseudo labels that will be selected for each class. At this point, to decrease the noisy pseudo labels, a method of uncertainty-confidence fused pseudo-labeling is developed to guide pseudo labeling by constructing a fused matrix $Z \in R^{n \times C}$ which combines confidence matrix H and uncertainty matrix U. The fused matrix Z can be used to capture a pseudo label with relatively high confidence and lower uncertainty to improve its reliability.

Here we provide the details of the construction of Z. We generate the value of Z for each class respectively, $i.e.$, Z is constructed in columns (shown Fig. 2).

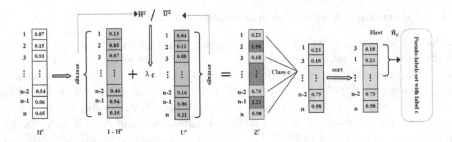

Fig. 2. The construction of the fused uncertain-confidence measure matrix Z. H^c and U^c denote the predication confidence and network uncertainty on class c for all the n nodes in graph, respectively. Z^c is the fused metric.

For each class $c \in \{0 \cdots C - 1\}$, Z^c, the c-th column of Z is the weighted sum of H^c and U^c:

$$Z^c = H^c + \lambda_c \cdot U^c \qquad (4)$$

where $\lambda_c > 0$ is a weighted coefficient. However, this setting exists a problem that the value of Z is not able to simultaneously reflect the trend of the increasing of H and the decreasing of U, more specifically, this can not achieve the objective that the variation of Z is the same when H increases and U decrease since a lager value of H and a smaller value of U indicate that the prediction is reliable. To make the Z meaningful, we rewrite Eq. 4 as

$$Z^c = 1 - H^c + \lambda_c \cdot U^c \qquad (5)$$

Thus, a smaller value of Z_i^c (the i-th row and c-th column of Z) means it is more reliable that node v_i predicted to the c-th class, for it has a higher H and lower U simultaneously. And a larger λ_c means the uncertainty contributes more in the pseudo labeling, otherwise the confidence is more significant. Note that when λ_c is set to 0, the proposed framework degenerates to the only confidence based pseudo labeling strategy.

Then it remains us a question that how to set the value of λ_c. Especially, we empirically observed that the distribution of the value of H^c and U^c varies significantly in different graph datasets, label rates and class of predictions and there is a gap between the value of H^c and U^c. For convenience and effectiveness, the uncertainty and confidence are defined to have the equal contribution resulting in λ_c to be the quotient of the average of $1 - H_c$ and U_c, formally:

$$\lambda_c = \frac{\sum_{i=1}^{n}(1 - H_i^c)/n}{\sum_{i=1}^{n} U_i^c/n}. \qquad (6)$$

This design has the advantage that λ_c is generated automatically to fit diverse prediction output qualities thus avoids manual adjustment. After that, we are able to select n_c pseudo labels for each class c based on the constructed Z. For all nodes predicted to class c, we order them based on their values of Z^c and select the n_c nodes with the lowest value and add them into the pseudo-label set $\hat{L}$ with pseudo-label c

2.6   Model Retraining

Algorithm 1 : UCP-GNN

Input: a graph $\mathcal{G}$, a labeled node set L, a set of unlabeled node set $\mathcal{U}$, a parameterized GNNs $f(\mathcal{G}, \theta)$.

Output: a trained GNNs model $f(\mathcal{G}, \theta)$.

1: Train a GNNs model $f(\mathcal{G}, \theta, 0)$ using the labeled data L.
2: **for** i from 1 to $MaxIterations$ **do**
3:    $H := f(\mathcal{G}, L) \in [0,1]^{n \times C}$ is the prediction confidence output by GNN model $f(\mathcal{G}, \theta, i-1)$.
4:    Pseudo labels set $\hat{L} = \emptyset$.
5:    Compute the total number of pseudo-labels to be selected N according to Eq. (1)
6:    Compute the uncertainty matrix U according to Eq. (3).
7:    **for** each class c **do**
8:       Get n_c, the number of pseudo labels to be selected in class c.
9:       Construct Z^c according to Eqs. (5) and (6).
10:       Sort the unlabeled nodes predicted to class c according to their value in Z^c.
11:       Select n_c unlabeled nodes with pseudo label c and add them into $\hat{L}$.
12:    **end for**
13:    Initialize a GNN $f(\mathcal{G}, \theta, i)$
14:    Train $f(\mathcal{G}, \theta, i)$ by minimize the cross-entropy loss on $\mathcal{L} \cup \hat{\mathcal{L}}$.
15:    $f(\mathcal{G}, \theta) = f(\mathcal{G}, \theta, i)$
16: **end for**
17: return $f(\mathcal{G}, \theta)$

After pseudo-labeling, we expand the given label set with the selected pseudo labels to retrain a more generalized GNN model. To mitigate the iterative propagation and accumulation of noisy information in retraining stage caused by the introduced incorrect pseudo labels, we initialize the parameters of GNN model before each retraining. Specifically, we optimize the GNN model by minimizing the cross entropy loss on the extended label set.

As a result, the GNN model is more generalized retrained on expanded label set, thus it can generate superior prediction output based on which our pseudo labeling mechanism is able to select more and reliable pseudo labels. After that, the performance of the GNN model can be further improved by using the newly generated and preferable expanded label set. This is actually a mutually reinforcing process, the number of selected pseudo labels increase gradually as well as the test accuracy in the retraining stage until reach the peak and retain stable. The main procedure of the proposed algorithm is listed in Algorithm 1.

3   Experiments

3.1   Experiment Setup

Dataset. Our experiment uses 2 citation networks: Cora, Citeseer and 2 coauthor network: Coauthor-CS, Coauthor-Phy.

Baselines. To evaluate the performance of the proposed UCP-GNN, we compare it to the following recently proposed baselines: (1) Graph neural network methods: GCN [1], GAT [15], and DAGNN [2]. (2) Unsupervised-methods: LINE [16] and GMI [17]. (3) Pseudo-labeling GNN methods: Self-training [7], Co-training [7], Union [7], Intersection [7], M3S [8], and DSGNN [9].

Implement Details. For each graph dataset, we randomly select {1, 3, 5, 10, 15, 20} nodes from per class for training. Among the remaining nodes, 500 nodes are randomly selected as validation, and all the rest of nodes are used for test. We conduct 10 times random samples for each data split and report the average classification accuracy on the test set of 10 runs for every data split. To validate the efficacy of the proposed UCP-GNN framework, we take GCN [1] and DAGNN [2] which is the shallow and deep graph learning model respectively as backbone GNN termed as UCP-GCN and UCP-DAG. The proposed method only involves two hyper-parameters including τ and $MaxIterations$. We set τ to be 0.92 in all experiments and set $MaxIterations$ to be 10 for Cora and CiteSeer and 5 for Coauthor-CS and Coauthor-Phy. For other hyper-parameters, we adopt the parameter setting same as the original GCN and DAGNN methods.

3.2   Comparison with the Baselines

Tables 1 list the average classification accuracy (%) of the compared methods on different datasets and training label rates. We highlight the best performer by **bold** and the second best performer is marked by underline.

As shown in Tables 1 and 2, the proposed framework almost achieves the best performance in different label rates on all the four graph datasets. For citation network, UCP-GCN achieves 15.8%, 12.1%, 8%, 6.7%, 4%, 3.8% performance gain and UCP-DAG achieves 13.2%, 10.9%, 7.4%, 4.8%, 3.6%, 3% performance gain compared against their underlying GNN model GCN and DAGNN with given 1,3,5,10,15,20 labels per class for training.

When labeled nodes are scarce, the performance enhancement obtained by UCP-GNN are much more evident, which demonstrates that the proposed framework can achieve more promising performance with few labeled nodes. For instance, the performance of UCP-DAG with given 3 labeled nodes per class for training(79.2%) nearly matches the accuracy of GCN(79.7%) when provided 20 labels per class on Cora dataset. On Citeseer dataset, the result of UCP-DAG with 3 labels per class (69.5%) even exceeds performance of GCN(66.3%) and DAGNN(67.2%) with provided 20 labeled nodes per class. We observed that the unsupervised method, GMI, achieve competitive performance on the citation networks, especially when few labels are available, that inspires us to further explore the novel unsupervised approach as our future works.

Table 1. Average accuracy(%), we exclude some competitors which significantly under-performs others and out of memory on 24G GPU on Coauthor dataset.

Dataset	Cora						Citeseer					
Labels per class	1	3	5	10	15	20	1	3	5	10	15	20
LINE [16]	49.4	62.3	64.0	70.2	72.6	74.5	27.9	35.8	38.6	42.3	44.7	49.4
GMI [17]	51.3	68.4	72.7	75.9	77.5	79.1	47.3	58.6	63.5	64.7	67.0	68.2
GCN [1]	42.8	63.9	71.7	74.5	78.7	79.7	40.3	53.6	57.3	63.5	65.8	66.3
GAT [15]	42.3	60.1	71.5	75.1	79.2	81.1	35.6	49.1	55.6	60.8	65.3	66.4
DAGNN [2]	58.5	70.2	75.8	78.2	80.9	82.5	44.5	56.7	60.0	65.4	66.4	67.2
Self-training [7]	50.7	66.3	72.1	76.4	78.5	79.9	38.4	51.3	57.3	65.1	68.5	68.6
Co-training [7]	46.6	65.2	71.9	74.2	77.3	79.5	42.7	51.6	59.4	62.4	64.3	64.6
Union [7]	51.5	67.5	73.2	76.6	79.2	80.6	41.8	52.9	58.9	64.7	66.2	67.2
Intersection [7]	40.2	64.9	72.1	75.8	77.1	78.3	37.9	52.7	60.4	66.5	69.1	68.9
M3S [8]	53.1	68.6	72.0	75.5	78.7	79.6	39.4	54.3	64.2	66.8	67.4	68.3
DSGCN [9]	60.2	71.3	74.5	77.7	80.1	81.9	49.3	63.3	65.2	67.4	68.1	68.4
UCP-GCN (**ours**)	<u>61.5</u>	<u>74.8</u>	<u>79.3</u>	<u>81.4</u>	<u>82.1</u>	<u>83.0</u>	<u>52.7</u>	<u>66.8</u>	<u>67.7</u>	<u>70.0</u>	<u>70.4</u>	<u>70.6</u>
UCP-DAG (**ours**)	**69.5**	**79.2**	**80.9**	**82.6**	**83.2**	**84.1**	**59.8**	**69.5**	**69.6**	**70.5**	**71.2**	**71.6**
Dataset	**Coauthor-CS**						**Coauthor-Phy**					
Labels per class	1	3	5	10	15	20	1	3	5	10	15	20
GCN [1]	74.1	86.6	87.9	90.1	90.5	90.6	83.1	89.1	91.0	91.7	92.2	92.5
DAGNN [2]	76.5	87.2	88.6	90.3	91.2	91.4	87.0	90.6	91.6	92.7	93.1	93.4
Self-training [7]	72.7	84.2	88.0	89.5	89.4	89.8	84.3	90.2	90.4	92.0	92.3	92.6
Co-training [7]	73.2	83.1	84.6	85.4	86.2	87.7	83.4	87.5	91.2	91.0	91.2	92.0
Union [7]	74.5	84.5	85.4	86.3	86.6	87.8	84.5	88.3	90.7	91.3	91.1	91.3
Intersection [7]	75.6	87.0	87.2	88.5	89.5	89.7	79.5	89.3	91.3	91.2	91.4	92.0
M3S [8]	75.2	87.2	88.3	90.9	91.3	91.5	87.6	90.4	<u>92.4</u>	92.8	92.8	93.4
DSGCN [9]	**79.8**	87.4	89.1	89.6	89.7	90.0	85.4	88.9	90.4	90.4	91.9	92.7
UCP-GCN (**ours**)	<u>79.0</u>	<u>88.7</u>	<u>89.8</u>	**91.4**	<u>91.6</u>	<u>91.7</u>	<u>89.9</u>	<u>91.7</u>	92.3	<u>93.0</u>	**93.6**	<u>93.8</u>
UCP-DAG (**ours**)	78.7	**89.1**	**90.1**	<u>91.1</u>	**91.5**	**91.9**	**91.8**	**92.6**	**92.8**	**93.5**	<u>93.5</u>	**94.2**

3.3 Pseudo-labels Analysis

We compare the pseudo-labels generated by our framework and conventional methods on Cora dataset with 3 labels per class: (1) the top-K confidence method [7,8], denoted as 'Top-K','K' is set as 1057 according to [7]; (2) hard confidence method [9,13], denoted as 'H-Threshold'. Figure 3 shows the number, class proportion distribution and accuracy of pseudo labels generated by different methods with different retraining iterations, as well as test accuracy of model retrained on corresponding pseudo labels.

As shown in Fig. 3, 'Top-K' only chooses the fixed number of pseudo-labels in each re-train stage, resulting in a low accuracy of the pseudo-labels and lacking of adaptability. While 'H-Threshold' and our UCP-GNN are able to select more pseudo labels with the retraining iteration increases, which has more flexibility. It can be observed from Fig. 3(b) that the distribution of pseudo labels

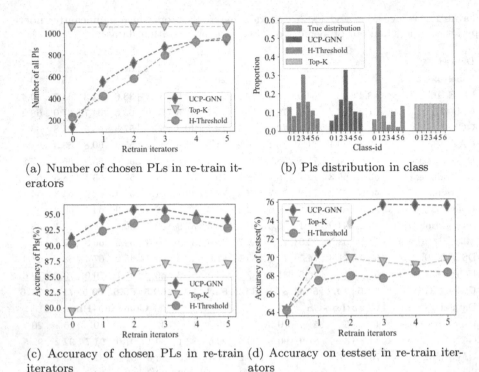

(a) Number of chosen PLs in re-train iterators

(b) Pls distribution in class

(c) Accuracy of chosen PLs in re-train (d) Accuracy on testset in re-train iter-
iterators   ators

Fig. 3. Pseudo labels analysis on Cora-3

across classes generated by UCP-GNN approximate to the true label distribution in graph, however the 'h-Threshold' suffer from severe imbalance problem.
Benefit from the fused uncertainty-confidence strategy, the pseudo labels generated by UCP-GNN exhibit higher accuracy thus introduce less noise. With these
advantages, UCP-GNN achieves superior test performance with the increase of
retraining iterations.

3.4 Ablation Studies

To explore the contribution of different components of our UCP-GNN, we evaluate the performances of the variants of UCP-GNN with some components
removed on Cora: (1)**UCP-w/o-B**. Without pseudo label balance mechanism.
We directly select pseudo labels whose value of Z is smaller than $2 \times (1 - \tau)$.
(2) **UCP-w/o-A**. Without adaptability on the number of pseudo labels. We
directly select equal number of pseodo labels. (3)**UCP-w/o-U**. Without uncertainty. We perform pseudo labeling only based on confidence H without fusing
uncertainty.

Figure 2 reports the ablation results of UCP-GCN and UCP-DAG on Cora
dataset with different label rates. Observe that each component of UCP-GNN

has its own contribution for all variants of UCP-GNN achieve inferior results compared with the full UCP-GNN. An observation is that the balance of pseudo-labels greatly affect the classification performance especially only few labeled data is available. The contribution of adaptability of the number of pseudo labels is relatively stable on different label rates. It can be realized that uncertainty contribute more as the fewer labels are provided, the main reason is that the GNN model is poorly calibrated when trained with few labels thus most predictions are not stable and have higher uncertainty resulting in more noisy information without considering uncertainty.

Table 2. Ablation results of UCP-GCN on Cora dataset.

training nodes	1	3	5	10	15	20
UCP-GCN	**61.5**	**74.8**	**79.3**	**81.4**	**82.1**	**83.0**
UCP-w/o-U	53.5 ($\downarrow$ 8.0)	70.0($\downarrow$ 4.8)	78.0($\downarrow$ 1.3)	79.6($\downarrow$ 1.8)	81.0($\downarrow$ 1.1)	81.6($\downarrow$ 1.4)
UCP-w/o-A	50.3($\downarrow$ 4.8)	67.0($\downarrow$ 11.2)	75.4($\downarrow$ 7.8)	76.7($\downarrow$ 3.9)	78.3($\downarrow$ 3.8)	79.1($\downarrow$ 3.9)
UCP-w/o-B	39.7($\downarrow$ 21.8)	50.3($\downarrow$ 24.5)	66.8($\downarrow$ 12.5)	77.9($\downarrow$ 3.5)	79.0($\downarrow$ 3.1)	80.6($\downarrow$ 2.4)

4   Conclusion

In this paper, we propose a novel uncertainty-confidence fused pseudo-labeling framework for graph semi-supervised learning. We first adaptively and proportionally determine the number of pseudo labels to be selected for each class via a 'soft confidence threshold'. Then the uncertainty measure is generated through MC-dropout. After that, a fused matrix is constructed which combined prediction confidence and uncertainty to guide pseudo-labeling. Finally, an iterative retraining framework is proposed to retrain a more generalized GNNs. Extensive experiments demonstrate that our approach successfully enhance the performance in a significant margin when applied on GNNs, specifically when labeled data are scarce.

Acknowledgement. This work was supported by the National Natural Science Foundation of China (Nos.62072384, 61872309, 62072385, 61772441), the Zhejiang Lab (No. 2022RD0AB02) and XMU Undergraduate Innovation and Entrepreneurship Training Programs (No.202310384191).

References

1. Kipf, T.N., Welling, M.: Semi-supervised classification with graph convolutional networks, arXiv preprint arXiv:1609.02907 (2016)
2. Zhou, K., Huang, X., Li, Y., Zha, D., Chen, R., Hu, X.: Towards deeper graph neural networks with differentiable group normalization. Adv. Neural. Inf. Process. Syst. **33**, 4917–4928 (2020)

3. Jin, T., et al.: Deepwalk-aware graph convolutional networks. Sci. China Inf. Sci. **65**(5), 152104 (2022)
4. Chang, L., Dan, G.: Encoding social information with graph convolutional networks forpolitical perspective detection in news media. In: Proceedings of the 57th Annual Meeting of the Association for Computational Linguistics (2019)
5. Bai, T., Zhang, Y., Wu, B., Nie, J.-Y.: Temporal graph neural networks for social recommendation. In: 2020 IEEE International Conference on Big Data (Big Data), pp. 898–903 (2020)
6. Fout, A.M.: Protein interface prediction using graph convolutional networks (2018)
7. Li, Q., Han, Z., Wu, X.-M.: Deeper insights into graph convolutional networks for semi-supervised learning. In: Thirty-Second AAAI Conference on Artificial Intelligence (2018)
8. Sun, K., Lin, Z., Zhu, Z.: Multi-stage self-supervised learning for graph convolutional networks on graphs with few labeled nodes. In: Proceedings of the AAAI Conference on Artificial Intelligence, vol. 34, pp. 5892–5899 (2020)
9. Zhou, Z., Shi, J., Zhang, S., Huang, Z., Li, Q.: Effective semi-supervised node classification on few-labeled graph data, arXiv preprint arXiv:1910.02684 (2019)
10. Gal, Y., Ghahramani, Z.: Dropout as a bayesian approximation: representing model uncertainty in deep learning. In: International Conference on Machine Learning, pp. 1050–1059 (2016)
11. Rizve, M.N., Duarte, K., Rawat, Y.S., Shah, M.: In defense of pseudo-labeling: an uncertainty-aware pseudo-label selection framework for semi-supervised learning, arXiv preprint arXiv:2101.06329 (2021)
12. Lakshminarayanan, B., Pritzel, A., Blundell, C.: Simple and scalable predictive uncertainty estimation using deep ensembles. In: Advances in Neural Information Processing Systems 30 (2017)
13. Yang, H., Yan, X., Dai, X., Chen, Y., Cheng, J.: Self-enhanced gnn: improving graph neural networks using model outputs. In: 2021 International Joint Conference on Neural Networks (IJCNN), pp. 1–8. IEEE (2021)
14. Gawlikowski, J., et al.: A survey of uncertainty in deep neural networks, arXiv preprint arXiv:2107.03342 (2021)
15. Veličković, P., Cucurull, G., Casanova, A., Romero, A., Lio, P., Bengio, Y.: Graph attention networks, arXiv preprint arXiv:1710.10903 (2017)
16. Tang, J., Qu, M., Wang, M., Zhang, M., Yan, J., Mei, Q.: Line: large-scale information network embedding. In: Proceedings of the 24th International Conference on World Wide Web, pp. 1067–1077 (2015)
17. Peng, Z., et al.: Graph representation learning via graphical mutual information maximization. In: Proceedings of the Web Conference 2020, pp. 259–270 (2020)

FSCD-Net: A Few-Shot Stego Cross-Domain Net for Image Steganalysis

Xiangwei Lai and Wei Huang[✉]

School of Informatics, Xiamen University, Xiamen, China
whuang@xmu.edu.cn

Abstract. Image steganalysis plays a critical role in various applications, aiming to detect hidden data within images. However, its effectiveness is significantly impeded by the cover source mismatch, where training and testing images come from different sources. This challenge is particularly pronounced in few-shot scenarios with limited training samples. In this paper, we propose a novel steganalysis model explicitly tailored to address the cover source mismatch in few-shot settings. By leveraging the pair constraint and incorporating features from accessible frequency domain images, our model enhances transferability and generalization, effectively improving steganalysis performance. To capture intricate relationships among embedded image regions across diverse domains, we introduce the FixedEMD module, which focuses on high-dimensional channels. Extensive experiments validate the efficacy of our model, demonstrating its superiority in image steganalysis under cover source mismatch conditions. Our findings underscore the potential of our model to overcome cover source mismatch challenges, making significant contributions to the advancement of image steganalysis.

Keywords: Image steganalysis · Few-shot · Cover source mismatch

1 Introduction

Image steganography, a captivating fusion of science and art, revolves around the concealment of data within images. This is typically accomplished by subtly modifying pixel values or discrete cosine transform (DCT) coefficients in the JPEG domain. With the advent of content-adaptive schemes, the security of this process has been significantly enhanced. These methods ingeniously embed covert data into regions of intricate content, thereby intensifying the challenge of detection. Notable examples of such content-adaptive approaches include techniques like HUGO [17], WOW [9], and S-UNIWARD [10].

Steganalysis, the complementary process of uncovering hidden data within images, has witnessed significant advancements. Initially, the field relied on traditional machine learning algorithms such as Support Vector Machines (SVM). However, there has been a gradual shift towards leveraging the power of Convolutional Neural Networks (CNNs), which have demonstrated superior capabilities

Q. Liu et al. (Eds.): PRCV 2023, LNCS 14433, pp. 343–355, 2024.
https://doi.org/10.1007/978-981-99-8546-3_28

in extracting relevant features. As a result, CNN-based steganalysis networks such as Xu-Net [23], Ye-Net [25], Yedroudj-Net [26], Zhu-Net [29], and Luo-Net [14] have achieved impressive performances.

Despite significant progress, deploying steganalysis tools in real-world scenarios presents substantial challenges due to Cover Source Mismatch (CSM) [3]. CSM occurs when a detector, trained on a specific cover source, encounters images from an alternate source, leading to increased detection errors. To address this problem, researchers have developed various algorithms. For example, Hou et al. [11] utilize a density-based unsupervised outlier detection algorithm, while Zhang et al. [30] propose J-Net, a deep adaptive network specifically designed for mitigating cover-source mismatch. Zhang et al. [28] introduce a feature-guided subdomain adaptation steganalysis framework to enhance deep learning-based models. These approaches offer effective solutions to the CSM problem and achieve impressive detection results. However, their effectiveness may vary considerably in real-world scenarios with limited samples.

In this paper, we propose FSCD-Net, a novel few-shot steganalysis network specifically designed to address the CSM issue in few-shot scenarios. To enhance the network's generalization and transferability capabilities from the frequency to the spatial domain, we adopt a pre-training process using a large dataset of JPEG images. Subsequently, we employ FixedEMD with channel attention to measure high-dimensional features and align the local regions where steganographic information is typically embedded across different data domains. Our network consists of two extractors that sequentially capture steganographic features from the support set, which are then input to a linear classifier for prediction. During the pre-training phase, we carefully avoid embedding information into the chrominance channels to eliminate potential biases in color distortion detection. Moreover, we adhere to the pair constraint principle [2], ensuring that each mini-batch contains both the cover image and its corresponding stego variant. This strategy effectively mitigates traditional convergence issues that often arise during network training from scratch.

Through exhaustive experimental validation, this paper substantiates the effectiveness of our proposed algorithm. Our findings not only confirm FSCD-Net's exceptional adaptability but also highlight its marked improvement in detection performance relative to established steganalysis models and alternative CSM solutions. The study constitutes a significant breakthrough in the realm of image steganalysis, offering a robust solution to the challenges associated with CSM in few-shot learning scenario.

2 Related Work

2.1 Image Steganalysis

Steganalysis techniques can be broadly categorized into handcrafted feature-based methodologies and deep learning-driven approaches. Handcrafted feature-based methods rely on manually designed features, with channel selection-based features showing promise in enhancing detection performance. However, these methods have limitations, as the performance is constrained by the manual selection of features.

In recent years, there has been a shift towards employing deep learning methods to address steganalysis challenges. For example, Ye-Net [25] utilizes a set of high-pass filters, similar to the computation of Spatial Rich Models (SRM) [7], to generate residual maps. It incorporates channel selection knowledge or pixel change probabilities to facilitate feature learning. Luo-Net [14] introduces a convolutional visual transformation-based image steganography algorithm [6], capturing local and global correlations among noise features. Similarly, the approach proposed by Butora et al. [2] employs pretraining to enhance model convergence and generalization.

Despite achieving remarkable performance through extensive training on large-scale datasets and evaluations on same-source training and testing sets, effectively addressing the CSM problem remains a challenge, particularly in real-world applications with limited training samples or few-shot scenarios.

2.2 Few-Shot Learning

Traditional machine learning techniques often struggle with suboptimal performance when dealing with small-scale datasets due to their reliance on large amounts of data for accurate model training. Few-shot learning (FSL) methodologies offer a promising approach to address this constraint. FSL strategies can generally be categorized into three main types [22]: data-based [5], model-based [13,24], and algorithm-based approaches [20].

In the context of model-based strategies, metric learning plays a crucial role as it leverages various metrics to improve the formation of intricate structured representations. The Earth Mover's Distance (EMD) metric [18] has demonstrated effectiveness in this regard, and methods like DeepEMD [27] have successfully mitigated noise from image background regions.

Furthermore, recent studies [19] have proposed end-to-end pre-training frameworks for acquiring feature extractors or classifiers on base classes. However, it is important to note that the covert nature of steganographic embedding may limit the applicability of these methods in steganalysis.

3   Method

In Sect. 3.1, the framework of the proposed method is presented, addressing these challenges effectively. Section 3.2 introduces an intelligent image cropping strategy designed to enhance feature extraction. Detailed insights into the FixedEMD method are provided in Sect. 3.3.

3.1   Framework

As illustrated in Fig. 1, the framework is comprised of multiple stratified layers. Each of these layers will be elaborated upon subsequently.

Preprocessing Layer. Inspired by Ye-Net and Yedroudj-Net, the approach employs 30 filter kernels for modeling residuals in steganalysis. This optimizes the network's focus on unique steganographic artifacts, leading to a preprocessed image dimension of $30 \times 128 \times 128$.

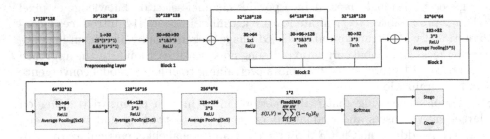

Fig. 1. The detailed architecture of the proposed network

Blocks. The framework includes three blocks, each serving a specific purpose in feature extraction.

The framework comprises three specialized blocks for feature extraction. Block1 employs a 1×1 point convolution coupled with a 3×3 depth convolution to capture spatial and channel-specific correlations, yielding an output of $30 \times 128 \times 128$. Block2 focuses on inter-channel relationships and localized attributes, enhancing the representation to dimensions of $192 \times 128 \times 128$. Block3 fine-tunes noise residual extraction, accommodating various channel and embedding conditions, and results in a compact feature set with dimensions of $256 \times 8 \times 8$.

FixedEMD: Drawing from the principles of EMD [18] and DeepEMD [27], a new module, FixedEMD, is introduced. This component employs a cross-reference mechanism to aggregate nuanced texture details and compute node weights, thereby mitigating the influence of irrelevant areas. As Fig. 2 exemplifies, FixedEMD excels at isolating intricate texture regions in steganographic images and directly contrasting steganographic regions between support and query sets. Key regions like the dog's eyes, ears, and neck textures in Fig. 2, as well as the complex textures of the jellyfish's tentacles and surrounding marine vegetation, underscore its ability to detect substantial steganographic content. For a thorough discussion of FixedEMD, readers are directed to Sect. 3.3.

3.2   Smart Crop

Random cropping, without considering the distribution of steganographic information, can lead to misleading model predictions. To tackle this issue, an intelligent cropping technique is introduced, drawing inspiration from the methodology outlined in [2]. The technique takes into account image texture complexity and standardizes the image size to 128×128 pixels.

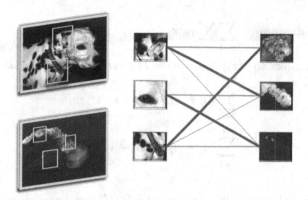

Fig. 2. The FixedEMD method focuses on regions of images with higher texture complexity to learn universal steganographic features. The images used in our demonstration are sourced from [21].

For a given luminance channel represented by an $M \times N$ DCT matrix, we partition it into 8×8 blocks denoted by (i, j), where $i, j \in \{1, \ldots, \frac{M}{8}\} \times \{1, \ldots, \frac{N}{8}\}$. The block energy B_{ij} is defined as the count of DCT coefficients C_{kl}^{ij}, where $k, l \in \{1, \ldots, 8\}$, that satisfy $|C_{kl}^{ij}| > T$. Mathematically, $B_{ij} = \sum_{k,l=0}^{7} [|C_{kl}^{ij}| > T]$. To select a 128×128 crop, the algorithm aims to maximize the sum of block energies over a 16×16 region. The starting coordinates of the crop are denoted by m and n, and are determined by optimizing the following sum: $\sum_{i=0}^{15} \sum_{j=0}^{15} B_{\frac{n}{8}+i, \frac{m}{8}+j}$.

3.3 FixedEMD

In the FixedEMD module, the Channel Attention Model is employed to emphasize critical channels relevant for steganalysis. This is complemented by the use of CrossEMD, which is designed to capture region-specific features indicative of steganographic embedding across diverse images.

EMD. Earth Mover's Distance (EMD) [18] is an effective tool used to evaluate the disparity between two collections of weighted entities or distributions. The principle behind it lies in the concept of the transportation problem (TP) from linear programming.

Consider two sets, a source set $S = \{s_i \mid i = 1, 2, ..., m\}$ providing goods and a target set $D = \{d_j \mid j = 1, 2, ..., k\}$ receiving these goods. Here, s_i and d_j correspond to the amount of goods available at the ith source and required at the jth destination, respectively. The cost per unit for transporting goods from source i to destination j is denoted as c_{ij}, while x_{ij} signifies the units transferred.

The TP aims to find an optimal distribution $\widetilde{X} = \tilde{x}_{ij} \mid \{i = 1, ..., m, j = 1, ..., k\}$ that minimizes the total cost, given specific constraints:

$$\text{minimize} \quad \sum_{i=1}^{m}\sum_{j=1}^{k} c_{ij}x_{ij},$$

$$\text{subject to} \quad x_{ij} \geq 0, \quad \forall i = 1, \ldots, m, \quad \forall j = 1, \ldots, k,$$

$$\sum_{j=1}^{k} x_{ij} = s_i, \quad \forall i = 1, \ldots, m, \tag{1}$$

$$\sum_{i=1}^{m} x_{ij} = d_j, \quad \forall j = 1, \ldots, k.$$

Channel Attention Model. Inspired by [8,15], we leverage the higher-level features, where each channel mapping serves as a distinctive representation for accurately localizing the cover and stego classes.

Distinct from the position attention module, the channel attention map $M \in \mathbb{R}^{C \times C}$ is directly derived from the original features $B \in \mathbb{R}^{C \times H \times W}$, where H and W denote the spatial dimensions of the feature map. This process involves reshaping B into $\mathbb{R}^{C \times N}$, where $N = H \times W$, followed by a matrix multiplication between B and its transpose. Subsequently, a softmax layer is applied to obtain the channel attention map $M \in \mathbb{R}^{C \times C}$:

$$m_{ji} = \frac{\exp(B_i \cdot B_j)}{\sum_{i=1}^{C} \exp(B_i \cdot B_j)}. \tag{2}$$

In this context, m_{ji} quantifies the influence of the i-th channel on the j-th channel. The channel attention map M is then used in a matrix multiplication with the transpose of M and B. The resultant matrix is reshaped back to $\mathbb{R}^{C \times H \times W}$, aligning with the dimensions of the original features. This matrix undergoes a scaling operation by a factor β and is summed element-wise with B to yield the final output

$$F_j = \beta \sum_{i=1}^{C} (m_{ji}B_i) + B_j. \tag{3}$$

By incorporating these inter-channel dependencies, feature discriminability is effectively enhanced, which contributes to improved performance in steganalysis tasks.

CrossEMD. In steganalysis, when the same steganographic algorithm is used to hide information in a dataset, a significant amount of information is embedded in regions with higher texture complexity.

The approach under investigation emphasizes the exploitation of discriminative local attributes. We decompose an image into a set of local representations and measure the similarity between two images using the optimal matching cost. The output of the Channel Attention model, denoted as F, is transposed to yield P with dimensions $H \times W \times C$. Each image representation in P consists of a collection of local feature vectors $[p_1, p_2, \ldots, p_{HW}]$. Each vector p_i can be viewed as a node in a set. Consequently, the similarity between two images is quantified as the optimal matching cost between the two sets of vectors.

Following the original formulation of Earth Mover's Distance (EMD), pairwise distances between the embedded nodes p_i and q_j from the two image features are computed. This allows for the derivation of the cost per unit. The equation for calculating c_{ij} is given as:

$$c_{ij} = 1 - \frac{|\text{cov}(p_i, q_j)|}{\sigma_{p_i} \sigma_{q_j}}. \tag{4}$$

Here, nodes with similar representations tend to generate fewer matching costs between each other. If the texture complexity is higher, the value of c_{ij} becomes smaller, indicating the region where steganographic embedding occurs.

To this end, we propose a cross-reference mechanism for obtaining the weights s_i and d_j. This mechanism leverages the dot product between a node feature and the average node feature in the other structure to compute a relevance score, which serves as the weight value

$$s_i = \max\{p_i^T \cdot \frac{1}{HW} \sum_{j=1}^{HW} q_j, 0\}. \tag{5}$$

Here, p_i and q_j denote the vectors from the two feature maps. The weights are constrained to be non-negative using the $\max(\cdot)$ function. It should be noted that the calculation for s_i is presented as an example, and obtaining d_i follows a similar procedure.

Once the optimal matching flows $\tilde{X}$ are obtained, the similarity score S between the image representations is computed as:

$$S(U, V) = \sum_{i=1}^{HW} \sum_{j=1}^{HW} (1 - c_{ij}) \tilde{x}_{ij}. \tag{6}$$

4 Experiment

4.1 Datasets

To assess the efficacy of FSCD-Net, experimental evaluations were carried out on four distinct datasets.

1. **mini-ImageNet** [21]: The dataset contains 100 classes, with each class consisting of 600 randomly selected images.
2. **BOSSBase** [1]: The BOSSBase database consists of 10,000 cover images captured with seven different cameras.
3. **BOWS2** [16]: The BOWS2 database comprises 10,000 cover images taken with seven different cameras.
4. **ALASKA2** [4]: The ALASKA2 database contains 80,000 cover images with varying sizes.

4.2   Evaluation Metrics

In steganalysis, model performance is evaluated quantitatively as a binary classi-fication problem. Steganographic images are considered positive cases (S), while original images are considered negative cases (C). The correct classification of cover instances is represented by N, while the correct classification of stego instances is represented by P. The evaluation metrics used are:

1. False Alarm Rate (FA Rate): The fraction of misclassified cover images, given by $\frac{C-N}{C}$.
2. Missed Detection Rate (MD Rate): The fraction of misclassified stego images, calculated as $\frac{S-P}{S}$.
3. Accuracy: The ratio of correctly classified samples to the total number of samples, represented by $\frac{P+N}{S+C}$.

4.3   Experiment Setting

A pre-training dataset consisting of 10,000 random images from mini-ImageNet was employed as the cover set. The J-UNIWARD [10] steganography tool with embedding rates between 0.2 and 0.4 bpnzac was used to generate the stego set, resulting in a pre-training model of 20,000 images.

During the training phase, images were sourced from spatial-domain datasets BOSSBase and BOWS2. From each of these datasets, a random subset of 3,000 images was selected. These images were processed using the newly proposed 'Smart Crop' method, which is designed to focus on regions with complex tex-tures. They were resized to a resolution of 128×128 pixels to form the cover set. Within this subset of 3,000 images, a specific set of 200 images was isolated for the purpose of steganographic embedding. To test the robustness and general-izability of the method, auxiliary experiments were conducted using alternative sample sizes of 1,500, 4,500, and 6,000 images. Empirical findings across these varying scales remained consistent, validating the decision to present results exclusively from the 3,000-image subset for clarity and conciseness.

In the experiments, three steganography tools-HUGO, WOW, and S-UNI-WARD were employed with designated embedding rates of 0.1, 0.2, 0.3, 0.4, and 0.5 bpp. This generated a stego dataset comprising 3,000 images. Both cover and stego images were categorized into 30 classes, distinguished by the steganography method and the embedding rate, for the purpose of spatial domain training. For the evaluation phase, a traditional binary classification approach commonly used in steganalysis was adopted.

For evaluation, grayscale images from the ALASKA2 dataset were utilized, selected for their diverse range of camera sources to better simulate real-world conditions. A subset of 3,000 images was randomly chosen to form the cover set, and an additional 200 images were earmarked for steganographic embedding. Tools such as S-UNIWARD, WOW, and HUGO were employed with specified embedding rates of 0.1, 0.2, 0.3, 0.4, and 0.5 bpp. This led to the creation of a comprehensive stego dataset comprising 3,000 images.

4.4 Hyperparameters

FSCD-Net was trained using mini-batch stochastic gradient descent (SGD) with a momentum decay of 0.9 and a weight decay of 0.001. A batch size of 16 accommodated 8 cover/stego pairs for the pre-training phase. The initial learning rate was set at 0.005 and underwent dynamic adjustments based on network performance metrics.

Table 1. Performance comparison with existing steganographic models in the Few-shot scenario

k-Shot	Steganalysis Network	Metrics		
		Accuracy	FA Rate	MD Rate
1	Xu-Net	55.53%	0.4377	0.4527
	Ye-Net	57.57%	0.4177	0.4310
	Yedroudj-Net	58.18%	0.4027	0.4337
	Luo-Net	59.78%	0.3850	0.4193
	FSCD-Net	**65.23%**	**0.3383**	**0.3577**
5	Xu-Net	59.17%	0.3890	0.4277
	Ye-Net	61.82%	0.3753	0.3883
	Yedroudj-Net	64.53%	0.3617	0.3480
	Luo-Net	66.73%	0.3520	0.3133
	FSCD-Net	**71.19%**	**0.3020**	**0.2743**

4.5 Compared with Other Steganalysis Methods

We compared our proposed FSCD-Net with five popular steganalysis networks: Xu-Net [23], Ye-Net [25], Yedroudj-Net [26] and Luo-Net [14].

As shown in Table 1, existing models such as Xu-Net and Ye-Net rely on simple high-pass filtering techniques and lack deep steganalysis feature extraction. This limits their ability to generalize across different domains. In contrast, models like Yedroudj-Net and Luo-Net are more proficient in learning residual information, especially Luo-Net, which effectively captures noise residuals from both local and global perspectives.

In comparison, our proposed FSCD-Net achieves promising results by rapidly localizing local features associated with steganographic embeddings and effectively adapting to different data domains.

4.6   Compared with Other Models Addressing the CSM Issue

Table 2. Performance comparison with other models addressing the CSM in $N = 2$, $K = 5$ few-shot scenarios

Steganalysis Network	Accuracy	FA Rate	MD Rate
H-LOF [11]	58.37%	0.3877	0.4350
J-Net [30]	66.43%	0.3243	0.3470
FG-SAS [28]	69.82%	**0.2983**	0.3053
FSCD-Net	**71.18%**	0.3020	**0.2743**

Experiments were conducted for a range of K values, including $K = 1$, $K = 2$, $K = 5$, $K = 8$, and $K = 10$. Suboptimal performance was observed for both FSCD-Net and competing algorithms at $K = 1$ and $K = 2$. However, FSCD-Net consistently outperformed other methods for $K = 5$, $K = 8$, and $K = 10$. The $K = 5$ setting, which most closely resembles real-world conditions and imposes stringent constraints, is highlighted in Table 2 to showcase FSCD-Net's performance.

As shown in Table 2, FSCD-Net outperformed three contemporary steganalysis models: H-LOF, J-Net, and FG-SAS, which were designed to address the Cover Source Mismatch (CSM) problem. H-LOF struggled to distinguish outliers when test samples and auxiliary carrier samples closely resembled each other, leading to compromised performance. J-Net and FG-SAS showed effectiveness in assimilating target domain features but heavily relied on a large quantity of training samples.

In contrast, FSCD-Net demonstrated its proficiency in extracting salient features even with limited training samples. The superior performance of FSCD-Net showcases its effectiveness in steganalysis tasks.

4.7   Ablation Study

Table 3. Results of ablation experiments in $N = 2$, $K = 5$ few-shot scenarios

Steganalysis Network	Accuracy	FA Rate	MD Rate
Base-Net	63.35%	0.3613	0.3717
Base-Net with CA	63.93%	0.3567	0.3647
Base-Net without PC	64.27%	0.3487	0.3660
Base-Net with CrossEMD	67.33%	0.3410	0.3457
FSCD-Net without PC	68.12%	0.3173	0.3203
FSCD-Net without CA	69.53%	0.3127	0.2967
FSCD-Net without CrossEMD	66.87%	0.3187	0.3440
FSCD-Net	**71.18%**	**0.3020**	**0.2743**

Extensive experiments were conducted to assess the contributions of individual components within the proposed network, as outlined in Table 3. The baseline model, referred to as Base-Net, incorporates Pair Constraints (PC) but excludes both Channel Attention (CA) and CrossEMD.

The results underscore the pivotal role each component plays in the network's efficacy. For instance, implementing the pair constraint during training negatively affected generalization performance by approximately 3%. A performance decline of 1.6% was observed when the channel attention mechanism was not utilized. Most notably, excluding the CrossEMD module led to a substantial performance dip of around 4.5%. Interestingly, the impact of the pair constraint appears to be intricately linked with the network's ability to efficiently extract features. These results accentuate the importance of these components for achieving optimal FSCD-Net performance.

5 Conclusion

In this paper, we proposed FSCD-Net, a novel approach to tackle the CSM problem in image steganalysis. By employing the pair constraint during pre-training and utilizing the FixedEMD module, we successfully enhanced the model's generalization ability and focused on regions with hidden steganographic embeddings across different data domains. Experimental results demonstrated the effectiveness of our approach, as FSCD-Net exhibited rapid adaptation and achieved significant improvements in detection performance compared to state-of-the-art steganalysis models and other methods addressing the CSM problem. Our findings highlight the importance of addressing CSM in steganalysis and offer a valuable solution for improving the reliability of steganalysis models in real-world scenarios.

Acknowledgments. The work on this paper was supported by Natural Science Foundation of China under Grant 61402390, and Fujian Natural Science Foundation Program Youth Innovation Project under Grant 2018J05112.

References

1. Bas, P., Filler, T., Pevný, T.: "Break our steganographic system": the ins and outs of organizing boss. In: Filler, T., Pevný, T., Craver, S., Ker, A. (eds.) IH 2011. LNCS, vol. 6958, pp. 59–70. Springer, Heidelberg (2011). https://doi.org/10.1007/978-3-642-24178-9_5
2. Butora, J., Yousfi, Y., Fridrich, J.: How to pretrain for steganalysis. In: Proceedings of the 2021 ACM Workshop on IH & MMSEC, pp. 143–148 (2021)
3. Cancelli, G., Doërr, G., Barni, M., Cox, I.J.: A comparative study of±steganalyzers. In: 2008 IEEE 10th Workshop on Multimedia Signal Processing, pp. 791–796. IEEE (2008)
4. Cogranne, R., Giboulot, Q., Bas, P.: Alaska# 2: challenging academic research on steganalysis with realistic images. In: 2020 WIFS, pp. 1–5. IEEE (2020)

5. Chu, W.H., Li, Y.J., Chang, J.C., Wang, Y.C.F.: Spot and learn: a maximum-entropy patch sampler for few-shot image classification. In: CVPR, pp. 6251–6260 (2019)
6. Dosovitskiy, A., et al.: An image is worth 16x16 words: transformers for image recognition at scale. arXiv preprint arXiv:2010.11929 (2020)
7. Fridrich, J., Kodovsky, J.: Rich models for steganalysis of digital images. TIFS **7**(3), 868–882 (2012)
8. Fu, J., et al.: Dual attention network for scene segmentation. In: CVPR, pp. 3146–3154 (2019)
9. Holub, V., Fridrich, J.: Designing steganographic distortion using directional filters. In: WIFS, pp. 234–239. IEEE (2012)
10. Holub, V., Fridrich, J., Denemark, T.: Universal distortion function for steganography in an arbitrary domain. EURASIP J. Inform. Sec., 1–13 (2014)
11. Hou, X., Zhang, T.: Simple and effective approach for construction of universal blind steganalyzer. Multimedia Tools Appli. **77**, 27829–27850 (2018)
12. Huiskes, M.J., Lew, M.S.: The mir flickr retrieval evaluation. In: Proceedings of the 1st ACM International Conference on Multimedia Information Retrieval, pp. 39–43 (2008)
13. Li, B., Yang, B., Liu, C., Liu, F., Ji, R., Ye, Q.: Beyond max-margin: Class margin equilibrium for few-shot object detection. In: CVPR, pp. 7363–7372 (2021)
14. Luo, G., Wei, P., Zhu, S., Zhang, X., Qian, Z., Li, S.: Image steganalysis with convolutional vision transformer. In: ICASSP, pp. 3089–3093. IEEE (2022)
15. Ma, Y., Ji, J., Sun, X., Zhou, Y., Ji, R.: Towards local visual modeling for image captioning. Pattern Recogn. **138**, 109420 (2023)
16. Bas, P., Furon, T.: BOWS-2 (Jul 2007). http://bows2.gipsa-lab.inpg.fr
17. Pevný, T., Filler, T., Bas, P.: Using high-dimensional image models to perform highly undetectable steganography. In: Böhme, R., Fong, P.W.L., Safavi-Naini, R. (eds.) IH 2010. LNCS, vol. 6387, pp. 161–177. Springer, Heidelberg (2010). https://doi.org/10.1007/978-3-642-16435-4_13
18. Rubner, Y., Tomasi, C., Guibas, L.J.: The earth mover's distance as a metric for image retrieval. Int. J. Comput. Vision **40**(2), 99 (2000)
19. Shen, Z., Liu, Z., Qin, J., Savvides, M., Cheng, K.T.: Partial is better than all: revisiting fine-tuning strategy for few-shot learning. In: AAAI, vol. 35, pp. 9594–9602 (2021)
20. Sun, Q., Liu, Y., Chua, T.S., Schiele, B.: Meta-transfer learning for few-shot learning. In: CVPR, pp. 403–412 (2019)
21. Vinyals, O., Blundell, C., Lillicrap, T., Wierstra, D., et al.: Matching networks for one shot learning. In: Advances in NIPS 29 (2016)
22. Wang, Y., Yao, Q., Kwok, J.T., Ni, L.M.: Generalizing from a few examples: a survey on few-shot learning. ACM Comput. Surv. (csur) **53**(3), 1–34 (2020)
23. Xu, G., Wu, H.Z., Shi, Y.Q.: Structural design of convolutional neural networks for steganalysis. IEEE Signal Process. Lett. **23**(5), 708–712 (2016)
24. Yang, B., et al.: Dynamic support network for few-shot class incremental learning. TPAMI (2022)
25. Ye, J., Ni, J., Yi, Y.: Deep learning hierarchical representations for image steganalysis. TIFS **12**(11), 2545–2557 (2017)
26. Yedroudj, M., Comby, F., Chaumont, M.: Yedroudj-net: an efficient cnn for spatial steganalysis. In: ICASSP, pp. 2092–2096. IEEE (2018)
27. Zhang, C., Cai, Y., Lin, G., Shen, C.: Deepemd: few-shot image classification with differentiable earth mover's distance and structured classifiers. In: CVPR, pp. 12203–12213 (2020)

28. Zhang, L., Abdullahi, S.M., He, P., Wang, H.: Dataset mismatched steganalysis using subdomain adaptation with guiding feature. Telecommun. Syst. **80**(2), 263–276 (2022)

29. Zhang, R., Zhu, F., Liu, J., Liu, G.: Depth-wise separable convolutions and multi-level pooling for an efficient spatial cnn-based steganalysis. TIFS **15**, 1138–1150 (2019)

30. Zhang, X., Kong, X., Wang, P., Wang, B.: Cover-source mismatch in deep spatial steganalysis. In: Wang, H., Zhao, X., Shi, Y., Kim, H.J., Piva, A. (eds.) IWDW 2019. LNCS, vol. 12022, pp. 71–83. Springer, Cham (2020). https://doi.org/10.1007/978-3-030-43575-2_6

Preference Contrastive Learning
for Personalized Recommendation

Yulong Bai, Meng Jian[✉], Shuyi Li, and Lifang Wu

Faculty of Information Technology, Beijing University of Technology, Beijing, China
`jianmeng648@163.com`

Abstract. Recommender systems play a crucial role in providing personalized services but face significant challenges from data sparsity and long-tail bias. Researchers have sought to address these issues using self-supervised contrastive learning. Current contrastive learning primarily relies on self-supervised signals to enhance embedding quality. Despite performance improvement, task-independent contrastive learning contributes limited to the recommendation task. In an effort to adapt contrastive learning to the task, we propose a preference contrastive learning (PCL) model by contrasting preferences of user-items pairs to model users' interests, instead of the self-supervised user-user/item-item discrimination. The supervised contrastive manner works in a single view of the interaction graph and does not require additional data augmentation and multi-view contrasting anymore. Performance on public datasets shows that the proposed PCL outperforms the state-of-the-art models, demonstrating that preference contrast betters self-supervised contrast for personalized recommendation.

Keywords: Recommender System · Contrastive Learning · Interest Propagation · Graph Convolution

1 Introduction

In recent years, recommender systems are prevalent in social media platforms to address the great challenge of information overload [22,24]. As key evidence, the collaborative filtering (CF) signals hidden in user-item interactions are mined to predict users' interests [8,16]. Graph convolutional networks (GCN) models [7,17] aggregate high-order CF signals on the user-item interaction graph and have made progress for recommendation [6]. However, these models suffer much from the inherent data sparsity and bias of interactions [1,2]. To alleviate the issues, contrastive learning [18,23] is applied to interest modeling.

Existing contrastive learning-based recommendation models [23] mainly employ self-supervised manner, which contrasts nodes between different views to perform self-discrimination for embedding learning. The embeddings of a node in different views aggregate CF signals from different interaction graphs, therefore, contrast between them enhances the representation stability by self-discrimination [18]. The contrast between nodes distorts the original interest

Q. Liu et al. (Eds.): PRCV 2023, LNCS 14433, pp. 356–367, 2024.
https://doi.org/10.1007/978-981-99-8546-3_29

distribution and makes them distinct from each other, which tends to violate the CF mechanism for the recommendation. Although self-discrimination improves the recommendation as an auxiliary task, it inherently conflicts with the pairwise user-item matching, and even the self-supervised learning itself is irrelevant to the recommendation task. Besides, the aggregation of CF signals by graph convolution takes much computational cost [14] and the multi-view aggregation for self-supervised contrast doubles the burden [23]. We argue that task-oriented contrastive learning [10] is promising to meet pairwise user-item matching for personalized recommendation and avoid multi-view embedding learning saving computational cost.

This work strives to study task-oriented contrastive learning to model users' interests in a supervised manner for recommendation task. In this paper, we propose a preference contrastive learning (PCL) model to promote the classical pairwise ranking with an additional task-oriented contrastive objective to jointly optimize recommendation. The preference contrast enhances interest consistency between users and items in interactions compared to those of the unobserved pairs. Beyond the conventional self-supervised node discrimination, pairwise matching, as the contrastive unit, brings inherently supervision signals of interactions for interest modeling. PCL contrasts user-item pairs with the aggregated user/item embeddings in a single channel of graph convolution, which does not need data augmentation and edge dropout to support multi-view embedding learning and contrasting between the views. The contributions of this work are summarized as follows:

- We adapt contrastive learning to the personalized recommendation by intuitively introducing pairwise user-item contrast, resulting in task-oriented contrastive learning.
- We propose a preference contrastive learning (PCL) model for personalized recommendation by contrasting preferences of user-items pairs to model users' interests, instead of the self-supervised user-user/item-item discrimination.
- The proposed PCL works in a single view of the interaction graph which is free of the general data augmentation and multi-view contrasting.

The experimental results on benchmark datasets demonstrate that the proposed PCL has apparent advantages compared to the baselines and reduces the computational burden of contrastive learning for the recommendation.

2   Related Work

2.1   GCN-Based Recommendation

In recent years, the GCN-based recommendation models have been developed rapidly [3,7]. These models build a graph on the user-item interaction records and learn embeddings through graph convolution to make recommendations. NGCF [16] cascades multiple graph convolutional layers to explore high-order interactions from the graph structure and encode deep CF signals to embed users' interests. LightGCN [7] removes redundant feature transformation and nonlinear activation in graph convolution, improving training efficiency and recommendation performance. UltraGCN [14] employs an ultra-simplified GCN by approximating the limit of multi-layer graph convolutions for efficient interest propagation. DHCF [9] employs hypergraph convolution to encode hybrid connectivity between users and items for interest learning. KADM [13] takes external knowledge clues to augment the interaction graph for graph convolution and leverages both knowledge and interactions to model users' interests. These works effectively improve the recommendation performance by graph convolution but still suffer greatly from the bias of the interactions.

2.2 Contrastive Learning-Based Models

Contrastive learning mechanism [18,23,25] improves the robustness of models in a self-supervised manner, which enhances the embedding consistency between different views of a node to alleviate the inherent data bias. SGL [18] introduces contrastive learning as an auxiliary task into the recommender learning and generates multi-view embeddings with randomly cropped graphs to improve robustness in modeling users' interests. SimGCL [23] substitutes the graph augmentation by adding a perturbation in graph convolution simulating the discrepancy between different views of graph structure, which successfully improves the recommendation performance. MCCLK [25] employs the knowledge graph to build embeddings from a semantic view in contrast to the collaborative view of interactions, which effectively merges semantic information into interest modeling. HCCF [20] establishes hypergraph contrasting between global embeddings on a hypergraph and local embeddings on the interaction graph to emphasize embedding consistency. GCCE [12] derives an auxiliary view on users' negative interests by complementary embeddings to contrast with the view of positive interests for debiased recommendation. Although these models have enhanced the representations, the auxiliary contrastive task pursuing embedding consistency between different views seems irrelevant to the recommendation task. In addition, multi-view learning for contrasting bears a heavy computational burden to optimize recommendations.

3 Methodology

In this section, we introduce the details of the proposed PCL, as shown in Fig. 1, including interaction graph and embedding initialization, multi-layer interest propagation, pairwise contrast *vs* self-supervised contrast, and recommendation and optimization.

3.1 Interaction Graph and Embedding Initialization

Let $U = \{u_1, u_2, \dots, u_M\}$ and $I = \{i_1, i_2, \dots, i_N\}$ represent the set of M users and N items, respectively. Users' interactions on items are recorded by

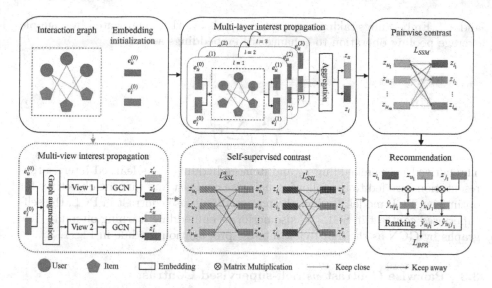

Fig. 1. Framework of the proposed preference contrastive learning (PCL) for personalized recommendation.

$Y = [y_{ui}]_{M \times N}$, $y_{ui} = 1$ if user u has interacted with item i, otherwise $y_{ui} = 0$. A user-item interaction graph $\mathcal{G} = \{\mathcal{V}, \mathcal{E}\}$ is formed with the given interactions, where vertex set $\mathcal{V} = \{U \cup I\}$ and edges in set $\mathcal{E}$ are formed by interactions between U and I, $|\mathcal{V}| = M + N$. On the graph, a lookup table [5,15] is initialized to encode user ID and item ID by $\{\mathbf{e}_u^{(0)}|u \in U\}$ and $\{\mathbf{e}_i^{(0)}|i \in I\}$, as initialized embeddings of user u and item i, respectively.

3.2   Multi-layer Interest Propagation

As illustrated in Fig. 1, the proposed PCL updates embeddings of user u and item i with a multi-layer interest propagation. It employs a standard graph convolutional network (GCN) [7] to aggregate multi-order CF signals from the graph structure to address data sparsity with graph neighbors. The graph convolution performs interest propagation between neighbors by

$$e_u^{(l)} = \frac{1}{\sqrt{|\mathcal{N}_u|}} \sum_{i \in \mathcal{N}_u} \frac{1}{\sqrt{|\mathcal{N}_i|}} e_i^{(l-1)},$$
$$e_i^{(l)} = \frac{1}{\sqrt{|\mathcal{N}_i|}} \sum_{u \in \mathcal{N}_i} \frac{1}{\sqrt{|\mathcal{N}_u|}} e_u^{(l-1)}, \tag{1}$$

where $e_u^{(l)}$ and $e_i^{(l)}$ represent embeddings of user u and item i on the lth convolutional layer, $l = 1, 2, ..., L$ $\mathcal{N}_u$ and $\mathcal{N}_i$ represent the set of interaction neighbors for user u and item i, respectively. With multi-layer propagation, high-order CF signals hidden in the graph structure would be aggregated into embeddings $e_u^{(l)}$

and $e_i^{(l)}$. Embeddings taking varying CF signals are then aggregated by a simple average pooling operation to enhance the embeddings as

$$
z_u = \frac{1}{L+1} \sum_{l=0}^{L} e_u^{(l)},
$$

$$
z_i = \frac{1}{L+1} \sum_{l=0}^{L} e_i^{(l)},
$$

(2)

The derived user embedding z_u and item embedding z_i as learned interest representations are fed to the following pairwise contrast and ranking objectives for optimizing PCL model. Besides the proposed pairwise contrast in PCL, the self-supervised contrast in SGL [18] also performs embedding learning on multi-view graphs by GCN as the multi-view interest propagation module in Fig. 1.

3.3   Pairwise Contrast *vs* Self-supervised Contrast

With the learned embeddings z_u and z_i, we strive to build a contrastive objective to enhance the representation ability for recommendation task. Speaking of applying contrastive learning to recommendation task, the famous SGL [18] elaborates a self-supervised contrast to improve robustness of embeddings to structural noises. It benefits from a graph augmentation operation generating multi-view subgraphs, which supports the multi-view interest propagation as Fig. 1 to learn multi-view embeddings z_u', z_u'' and z_i', z_i'' for the same user u and item i. The corresponding embeddings of a node between different views are regarded as a positive pair, while those of different nodes are as a negative pair. Self-supervised contrast leverages InfoNCE loss [4] on the positive and negative pairs to enhance embedding consistency as

$$
\mathcal{L}_{ssl}^u = -\sum_{u \in U} \log \frac{\exp(s(z_u', z_u'')/\tau)}{\sum_{v \in U} \exp(s(z_u', z_v'')/\tau)},
$$

$$
\mathcal{L}_{ssl}^i = -\sum_{i \in I} \log \frac{\exp(s(z_i', z_i'')/\tau)}{\sum_{j \in I} \exp(s(z_i', z_j'')/\tau)},
$$

(3)

where $s()$ evaluates the cosine similarity between embeddings, and τ is the temperature hyper-parameter to control the penalty discrimination on negative samples. A smaller τ makes the model focus more on hard negatives [18], therefore, τ affects the performance varying to different datasets.

Although the self-discrimination in Eq. (3) enhances representation ability by introducing self-supervised signals, it emphasizes embedding independence which inherently conflicts with the core CF theory for interest prediction. CF leverages users' similar interests to recommend items. The independence requirement in Eq. (3) can not meet the interest filtering for recommendation. It seems significant to adapt contrastive learning to the recommendation task by taking pairwise user-item matching as the core. Therefore, following [19], we modify the self-supervised contrastive objective in SGL into a task-oriented preference contrastive objective as

$$\mathcal{L}_{cl} = -\sum_{u \in \mathcal{U}} \log \frac{\exp(s(z_u, z_i)/\tau)}{\sum_{j \in \mathcal{I}} \exp(s(z_u, z_j)/\tau)}, \tag{4}$$

where cosine similarity $s(\cdot)$ measures the matching degree between users and items, τ controls the penalty on preference differentiation, which not only focuses on hard negatives of user interests' but also improves the adaptability to the varying density of datasets. We follow a similar InfoNCE form as Eq. (3) taking pairwise user-item interactions as positive samples and unobserved pairs as negatives. The optimization of pairwise contrast in Eq. (4) maximizes the pairwise user-item matching in interaction records while minimizing the matching in those unobserved pairs. By contrasting the matching preferences of user-item pairs and reusing supervision signals, the contrast task is adapted to the recommendation task, and the computational burden from multi-view contrasting of self-supervised learning is avoided. The preference contrast turns the self-supervised auxiliary task into the main task and implements joint learning as SGL for recommendation.

3.4 Recommendation and Optimization

For the main task of recommendation, we feed the learned embeddings z_u and z_i into an interaction function to predict the pairwise interaction score of user u and item i. PCL employs a simple inner product as the interaction function.

$$\hat{y}_{ui} = z_u^\top z_i, \tag{5}$$

where $\hat{y}_{ui}$ is the predicted interaction score between user u and item i. By ranking the predicted scores, the Top-N item list is recommended for users. Following SGL, we preserve the main ranking loss BPR [15] for recommendation as

$$\mathcal{L}_{bpr} = \sum_{(u,i,j) \in \mathcal{O}} -\log \sigma(\hat{y}_{ui} - \hat{y}_{uj}), \tag{6}$$

where $\mathcal{O} = \{(u, i, j) \mid y_{u,i} = 1, y_{u,j} = 0\}$, $\sigma(\cdot)$ is the sigmoid activation function. The recommendation task therefore is jointly optimized combining both ranking loss $\mathcal{L}_{bpr}$ and contrastive loss $\mathcal{L}_{cl}$ as

$$\mathcal{L} = \mathcal{L}_{bpr} + \lambda_1 \mathcal{L}_{cl} + \lambda_2 ||\Theta||_2^2. \tag{7}$$

where λ_1 and λ_2 are hyperparameters to control the contrastive loss and the parameter regularization, respectively, and Θ is the set of trainable parameters. In this way, we adapt SSL to the pairwise user-item matching scenario, which strengthens the learning of users' interests and eliminates the distortion of self-discrimination on the recommendation task.

3.5   Model Analysis

We further discuss the design of the proposed PCL by jointly optimizing the ranking loss $\mathcal{L}_{bpr}$ and contrastive loss $\mathcal{L}_{cl}$ to illustrate its rationality and advantage for the recommendation. There are two core differences observed between the ranking loss in Eq. (6) and contrastive loss in Eq. (4).

- On the pairwise user-item matching, $\mathcal{L}_{bpr}$ employs the inner product of user and item, while $\mathcal{L}_{cl}$ conducts cosine similarity between user and item. $\mathcal{L}_{bpr}$ dedicates to ranking positive and negative samples directly on embeddings in a high-dimensional representation space to optimize interest representation. In complementary, $\mathcal{L}_{cl}$ tends to focus on optimizing the direction consistency between user and item on interest modeling.
- For negative sampling, $\mathcal{L}_{bpr}$ adopts the pairwise ranking with positive and negative samples, i.e., $1:1$ sampling, while $\mathcal{L}_{cl}$ utilizes the all-in form with batch-size negative samples resulting in $1:N$ sampling. $\mathcal{L}_{cl}$ selectively optimizes the embeddings by $1:N$ positive and negative samples which performs learning on a continuous large scope in the embedding space. It also implements a penalty mechanism with a temperature coefficient to adaptively learn from positive and negative samples and enhance learning on hard negative samples.

In summary, the ranking loss $\mathcal{L}_{bpr}$ ensures the quality of the recommendations with local pairwise optimization, and the contrastive loss $\mathcal{L}_{cl}$ individually learn users' interests in large-scale global optimization. Therefore, the ranking loss and contrastive loss work collaboratively to improve the recommendation performance.

4   Experiments

4.1   Datasets

Table 1. Statistics of the AmazonBook and LastFM datasets.

Dataset	#Users	#Items	#Interactions	Density
AmazonBook	70,679	24,915	846,434	0.048%
LastFM	1,884	17,632	92,826	0.279%

To verify the effectiveness of the proposed PCL, we conduct experiments on two public datasets, including AmazonBook [6], and LastFM [25]. The statistics of these datasets are summarized in Table 1. They have different interaction sizes and density levels, allowing us to verify the role of the proposed PCL in different scenarios. On each dataset, we use 80% interactions for training and the remaining for testing.

4.2 Experimental Settings

Following the conventional setup, we utilize the Xavier normal distribution for embedding initialization and Adam optimizer for model training. The embedding size is fixed to 64, with a batch size of 2048 and a learning rate of 10^{-2}. Interest propagation performs 3-layer graph convolution and graph dropout is enabled by default. For hyper-parameters tuning in CL, we search λ_1 and temperature τ in $\{0.1, 0.2, \ldots, 1\}$. The baseline models similarly perform parameter searches to achieve the best performance. In order to ensure the fairness of comparison, LightGCN [7] is used as the backbone network for graph convolution, and the same parameter settings are maintained for interest propagation. Evaluation metrics adopt Recall@N, NDCG@N, and Precision@N, commonly used in recommender systems [7,21]. In addition, the average training time for each epoch is calculated to evaluate the computational cost of each model. The top-N recommendations for each user are taken to evaluate the recommendation performance. The experiments for PCL and baselines are implemented using $Python$3.8 and $Pytorch$2.0, and the code will be open-sourced on https://github.com/BJUTRec/PCL after acceptance.

4.3 Performance Comparison

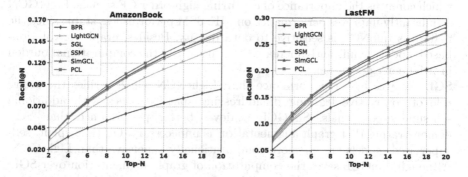

Fig. 2. Performance comparison of Top-N recommendations by Recall@N on Amazon-Book and LastFM datasets, $N = 2, 4, 6, \ldots, 20$.

We compare the performance of the proposed PCL with several baseline models, including the classic CF model MF [11], the graph convolution model Light-GCN [7], the self-supervised contrastive learning model SGL [18], contrastive recommendation model SSM [19], and SimGCL [23]. The Top-N performance curves of all comparisons are shown in Fig. 2, and the specific performance at Top-20 and the time(ms) spent per epoch is given in Table 2, where the best performance and the strongest baseline are bolded and underlined, respectively. The experimental results show that

Table 2. Performance comparison of Top-20 recommendations by Recall, NDCG, Precision, and Time(ms) spent per epoch on AmazonBook and LastFM datasets.

	AmazonBook				LastFM			
	Recall	NDCG	Precision	Time	Recall	NDCG	Precision	Time
MF	0.0905	0.0905	0.0096	1141	0.2197	0.2155	0.1085	344
LightGCN	0.1395	0.1082	0.0147	8855	0.2729	0.2741	0.1345	566
SGL	0.1530	0.1191	0.0162	21895	0.2752	0.2744	0.1353	1149
SSM	0.1575	0.1229	0.0169	8820	0.2527	0.2563	0.1243	605
SimGCL	0.1555	0.1211	0.0167	16619	0.2835	0.2836	0.1396	658
PCL	**0.1622**	**0.1243**	**0.0172**	9048	**0.2894**	**0.2919**	**0.1426**	615
Improv	2.98%	1.14%	1.78%	–	2.08%	2.93%	2.15%	–

- The proposed PCL consistently performs better than other baselines in all cases. More specifically, it represents an improvement over the strongest baseline w.r.t. Recall@20 by 2.98% and 2.08% on AmazonBook and LastFM, respectively. PCL also has an apparent advantage in time spent. The experimental results demonstrate the effectiveness of PCL in modeling users' interests by preference contrastive learning.
- Compared with MF, the graph convolution-based models perform better, which confirms the importance of capturing high-order CF signals. LightGCN significantly improves performance on MF. SSM performs less stable since the main loss InfoNCE is sensitive to data density. Besides, graph convolution-based models significantly increase the time cost of the computational burden in neighbor aggregation.
- SGL achieves better performance than LightGCN which verifies the active role of contrastive learning in addressing sparsity issues with additional self-supervised signals. SimGCL achieves better performance than SGL demonstrating that graph augmentation is unnecessary. On time spent, self-supervised models are far more computationally expensive than LightGCN. Although SimGCL saves the computation of graph augmentation over SGL, it still preserves multi-view embedding learning and cross-view contrasting with huge computation.
- The proposed PCL consistently outperforms the strongest baselines, demonstrating the importance of contrasting pairwise user-item preference for interest modeling. The computational cost has been highly reduced compared with the SGL since the multi-view contrasting is adapted to preference contrasting.

4.4   Long-Tail Recommendation Analysis

We conduct experiments to evaluate the role of the proposed PCL in addressing the long-tail bias problem [1,2]. We group items according to their popularity by ranking the size of interactions they received and dividing them into 5 groups

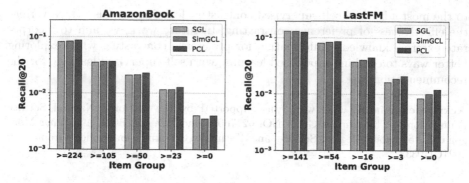

Fig. 3. Performance comparison of Top-20 recommendations in item groups of varying popularity by Recall@20 on AmazonBook, and LastFM datasets.

with relatively balanced interactions. For instance, the AmazonBook dataset is divided into 5 item groups with minimum interaction sizes 224, 105, 50, 23, and 0 in order. Figure 3 provides a performance comparison of SGL, SimGCL, and the proposed PCL, on the item groups. In Fig. 3, the horizontal axis displays the minimum interaction sizes of item groups in descending order from left to right, and the vertical axis shows the Recall@20 on each group. It can be observed from the results that

- The proposed PCL outperforms SGL and SimGCL in most of the groups. It yields apparent improvement on long-tail items, even unpopular items, compared with SGL and SimGCL. PCL preserves the performance level with the other methods on popular items. Due to the adaption from the self-supervised task to the recommendation task, PCL leverages the given interactions to supervise the contrastive learning leading to superior performance on long-tail items.
- SGL performs basically the same as SimGCL on AmazonBook, however, on LastFM, SGL performs weaker than SimGCL on the unpopular item group. It implies the sensitive role of graph structure for learning embeddings. Furthermore, the performance indicates that perturbed graph convolution works better than graph augmentation to alleviate the long-tail issue.

5   Conclusion

This study explores the adaptation of contrastive learning to the recommendation task and investigates the task confliction from self-supervised learning. We have proposed a preference contrastive learning (PCL) model which utilizes supervision signals to adapt contrastive learning to interest modeling by contrasting pairwise user-item preferences as an additional objective on the recommendation task. Experiments conducted on public datasets demonstrate that PCL exhibits excellent performance. It addresses the long-tail problem compared to the most advanced self-supervised contrastive learning models, which verifies the effectiveness of preference contrasting. In future work, we aim to incorporate external knowledge information for preference contrasting while exploring better ways to combine supervised learning with self-supervised learning for the recommendation task.

Acknowledgement. This work was supported by the National Natural Science Foundation of China under Grant NO. 62176011 and NO. 62306021, and Inner Mongolia Autonomous Region Science and Technology Foundation under Grant NO. 2021GG0333.

References

1. Chen, J., Dong, H., Wang, X.: Bias and debias in recommender system: a survey and future directions. ACM Trans. Inform. Syst. **41**(3), 1–39 (2023)
2. Clauset, A., Shalizi, C.R., Newman, M.E.: Power-law distributions in empirical data. SIAM Rev. **51**(4), 661–703 (2009)
3. Gasteiger, J., Bojchevski, A., Günnemann, S.: Combining neural networks with personalized pagerank for classification on graphs. In: International Conference on Learning Representations (2019)
4. Gutmann, M., Hyvärinen, A.: Noise-contrastive estimation: a new estimation principle for unnormalized statistical models. In: Proceedings of the Thirteenth International Conference on Artificial Intelligence and Statistics, pp. 297–304. JMLR Workshop and Conference Proceedings (2010)
5. He, K., Zhang, X., Ren, S., Sun, J.: Deep residual learning for image recognition. In: Proceedings of the IEEE Conference on Computer Vision and Pattern Recognition, pp. 770–778 (2016)
6. He, R., McAuley, J.: Ups and downs: modeling the visual evolution of fashion trends with one-class collaborative filtering. In: Proceedings of the 25th International Conference on World Wide Web, pp. 507–517 (2016)
7. He, X., Deng, K., Wang, X., Li, Y., Zhang, Y., Wang, M.: LightGCN: simplifying and powering graph convolution network for recommendation. In: Proceedings of the 43rd International ACM SIGIR conference on research and development in Information Retrieval, pp. 639–648 (2020)
8. He, X., Liao, L., Zhang, H., Nie, L., Hu, X., Chua, T.S.: Neural collaborative filtering. In: Proceedings of the 26th International Conference on World Wide Web, pp. 173–182 (2017)
9. Ji, S., Feng, Y., Ji, R., Zhao, X., Tang, W., Gao, Y.: Dual channel hypergraph collaborative filtering. In: Proceedings of the 26th ACM SIGKDD International Conference on Knowledge Discovery & Data Mining, pp. 2020–2029 (2020)
10. Khosla, P., et al.: Supervised contrastive learning. Adv. Neural. Inf. Process. Syst. **33**, 18661–18673 (2020)
11. Koren, Y., Bell, R., Volinsky, C.: Matrix factorization techniques for recommender systems. Computer **42**, 30–37 (2009)
12. Liu, M., Jian, M., Shi, G., Xiang, Y., Wu, L.: Graph contrastive learning on complementary embedding for recommendation. In: Proceedings of the 2023 International Conference on Multimedia Retrieval (2023)

13. Liu, X., Wu, S., Zhang, Z.: Unify local and global information for top-n recommendation. In: Proceedings of the 45th International ACM SIGIR Conference on Research and Development in Information Retrieval, pp. 1262–1272 (2022)
14. Mao, K., Zhu, J., Xiao, X., Lu, B., Wang, Z., He, X.: Ultragcn: ultra simplification of graph convolutional networks for recommendation. In: Proceedings of the 30th ACM International Conference on Information & Knowledge Management, pp. 1253–1262 (2021)
15. Rendle, S., Freudenthaler, C., Gantner, Z., Schmidt-Thieme, L.: Bpr: bayesian personalized ranking from implicit feedback. In: Proceedings of the Twenty-Fifth Conference on Uncertainty in Artificial Intelligence, pp. 452–461 (2009)
16. Wang, X., He, X., Wang, M., Feng, F., Chua, T.S.: Neural graph collaborative filtering. In: Proceedings of the 42nd international ACM SIGIR conference on Research and development in Information Retrieval, pp. 165–174 (2019)
17. Wu, F., Souza, A., Zhang, T., Fifty, C., Yu, T., Weinberger, K.: Simplifying graph convolutional networks. In: Proceedings of the 36th International Conference on Machine Learning, pp. 6861–6871 (2019)
18. Wu, J., et al.: Self-supervised graph learning for recommendation. In: Proceedings of the 44th International ACM SIGIR Conference on Research and Development in Information Retrieval, pp. 726–735 (2021)
19. Wu, J., et al.: On the effectiveness of sampled softmax loss for item recommendation. arXiv preprint arXiv:2201.02327 (2022)
20. Xia, L., Huang, C., Xu, Y., Zhao, J., Yin, D., Huang, J.: Hypergraph contrastive collaborative filtering. In: Proceedings of the 45th International ACM SIGIR Conference on Research and Development in Information Retrieval, pp. 70–79 (2022)
21. Yang, Y., Huang, C., Xia, L., Li, C.: Knowledge graph contrastive learning for recommendation. In: Proceedings of the 45th International ACM SIGIR Conference on Research and Development in Information Retrieval, pp. 1434–1443 (2022)
22. Ying, R., He, R., Chen, K., Eksombatchai, P., Hamilton, W.L., Leskovec, J.: Graph convolutional neural networks for web-scale recommender systems. In: Proceedings of the 24th ACM SIGKDD International Conference on Knowledge Discovery & Data Mining, pp. 974–983 (2018)
23. Yu, J., Yin, H., Xia, X.: Are graph augmentations necessary? simple graph contrastive learning for recommendation. In: Proceedings of the 45th International ACM SIGIR Conference on Research and Development in Information Retrieval, pp. 1294–1303 (2022)
24. Yuan, F., He, X., Karatzoglou, A., Zhang, L.: Parameter-efficient transfer from sequential behaviors for user modeling and recommendation. In: Proceedings of the 43rd International ACM SIGIR Conference on Research and Development in Information Retrieval, pp. 1469–1478 (2020)
25. Zou, D., et al.: Multi-level cross-view contrastive learning for knowledge-aware recommender system. In: Proceedings of the 45th International ACM SIGIR Conference on Research and Development in Information Retrieval, pp. 1358–1368 (2022)

GLViG: Global and Local Vision GNN May Be What You Need for Vision

Tanzhe Li[1], Wei Lin[1], Xiawu Zheng[2], and Taisong Jin[1(✉)]

[1] Key Laboratory of Multimedia Trusted Perception and Efffcient Computing, Ministry of Education of China, School of Informatics, Xiamen University, Xiamen, China
jintaisong@xmu.edu.cn
[2] Peng Cheng Laboratory, Shenzhen, China

Abstract. In this article, we propose a novel vision architecture termed GLViG, which leverages graph neural networks (GNNs) to capture local and important global information in images. To achieve this, GLViG represents image patches as graph nodes and constructs two types of graphs to encode the information, which are subsequently processed by GNNs to enable efficient information exchange between image patches, resulting in superior performance. In order to address the quadratic computational complexity challenges posed by high-resolution images, GLViG adaptively samples the image patches and optimizes computational complexity to linear. Finally, to enhance the adaptation of GNNs to the 2D image structure, we use Depth-wise Convolution dynamically generated positional encoding as a solution to the fixed-size and static limitations of absolute position encoding in ViG. The extensive experiments on image classification, object detection, and image segmentation demonstrate the superiority of the proposed GLViG architecture. Specifically, the GLViG-B1 architecture achieves a significant improvement on ImageNet-1K when compared to the state-of-the-art GNN-based backbone ViG-Tiny (80.7% vs. 78.2%). Additionally, our proposed GLViG model surpasses popular computer vision models such as Convolutional Neural Networks (CNNs), Vision Transformers (ViTs), and Vision MLPs. We believe that our method has great potential to advance the capabilities of computer vision and bring a new perspective to the design of new vision architectures.

Keywords: Graph Neural Networks · Image Classification · Object Detection · Image Segmentation

1 Introduction

Over the past decade, deep learning has emerged as the dominant learning model for a variety of vision tasks. As a result, researchers have investigated numerous vision architectures to propel advancements in computer vision research. These vision architectures address key aspects such as locality, translation equivariance, and global receptive field. These properties have played a crucial role in

Q. Liu et al. (Eds.): PRCV 2023, LNCS 14433, pp. 368–382, 2024.
https://doi.org/10.1007/978-981-99-8546-3_30

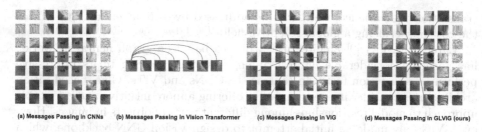

(a) Messages Passing in CNNs   (b) Messages Passing in Vision Transformer   (c) Messages Passing in ViG   (d) Messages Passing in GLVIG (ours)

Fig. 1. Illustration of messages passing in different backbones. Blue patches are message receivers, red and green patches are messages to be propagated. CNNs have locality and translation equivariance, Vision Transformer [7] and ViG [9] have the global receptive field, and the proposed GLViG can have a good receptive field while having locality and translation equivariance, which combines the advantages of CNNs, Vision Transformer, and ViG. (Color figure online)

the remarkable achievements of various computer vision models. For example, Convolutional Neural Networks (CNNs) have long been the preferred choice of the computer vision (CV) community, owing to their exceptional ability to model locality and translation equivariance. However, CNNs architectures leverage small convolution kernels with a limited receptive field, which limits the model's performance. To this end, Transformer [40], known for capturing long-range dependencies in natural language processing (NLP), was introduced to computer vision to capture the global information of images. Vision Transformer [7] has demonstrated its expressive capability across a broad range of vision tasks. Various variants and extensions of Vision Transformer such as DeiT [38], PVT [41,42] and Swin Transformer [26], have demonstrated exceptional performance and surpassed CNNs in a variety of vision tasks. Meanwhile, the Transformer architecture without attention has been shown to make a significant impact, recent studies [36,37] have shown that Multi-Layer Perception (MLP), utilizing the Transformer architecture, is capable of competing with both Convolutional Neural Networks (CNNs) and Vision Transformers. Despite the increasing popularity of Vision Transformers, CNNs remain competitive, as demonstrated by study like ConvNeXt [27]. The study highlights CNNs' expressive capabilities through larger local receptive fields achieved via 7×7 convolution kernels. Very recently, considering the potential of graph structure for capturing irregular objects, ViG [9] made the first attempt to leverage graph neural networks (GNNs) to employ large-scale vision tasks, which shows the potential of graph representations for large-scale vision tasks.

On one hand, the success of CNNs in image recognition relies on their translation equivariance and locality, which are crucial elements. However, CNNs' performance enhancement is limited by their local receptive field property. While larger Depth-wise Convolution kernels [15] can expand the receptive field, their performance decreases beyond a certain size. On the other hand, the success of the Transformer's [40] architecture in vision is closely linked to its global receptive field. However, the Vision Transformer [7] suffers from a lack of locality and translation equivariance, which can be addressed by local attention [26], but at the cost of sacrificing a global receptive field.

Lately, the ViG [9] architecture, which is based on graph neural networks, has garnered considerable interest among researchers owing to its impressive performance in vision tasks. In contrast to CNNs and ViTs, GNN-based vision architectures consider images as graphs, offering a more intuitive and adaptable approach to model relationships among different parts of objects in images. However, ViG only made an initial attempt to design Vision GNN backbone, which involves the following three issues: (1) ViG utilizes a global graph to represent relationships among image patches, disregarding the contiguity of features in adjacent regions of the image. (2) ViG's graph construction approach, which has quadratic complexity, is computationally expensive and unsuitable for processing high-resolution images. (3) The fixed size of ViG's positional encoding restricts its flexibility to handle high-resolution images, while its static nature adversely affects the overall model performance. The aforementioned drawbacks constrain further performance improvements.

To address the limitations of CNNs, ViTs, and ViG, we propose a novel vision architecture for vision tasks, termed GLViG. Our proposed method effectively unifies the locality, translation equivariance of CNNs, and the global receptive field of ViTs and ViG, providing a promising alternative to CNNs, ViTs, and ViG in numerous image-level and pixel-level dense prediction tasks. (For a clear visualization of the distinctions between our proposed method and CNNs, ViTs, and ViG, please see Fig. 1.)

The contributions of this article are summarized as follows:

(1) To enable locality and translation equivariance, spatially adjacent image patches are selected for each patch as local neighbors, allowing it to receive information from the corresponding neighborhood. Furthermore, to achieve a global receptive field, the K-Nearest-Neighboring strategy (KNN) is used to dynamically identify related image patches globally.
(2) To improve the computational efficiency of graph-based image processing with high-resolution inputs, an adaptive graph node downsampling strategy is employed, resulting in linear computational complexity during global graph construction.
(3) By incorporating dynamic position encoding information generated through Depth-wise Convolution [15] into graph nodes instead of the fixed-size, static absolute position encoding used in ViG, we achieve greater flexibility in processing high-resolution images and enhance the ability of GNNs to adapt to the 2D image structure.

2   Related Work

2.1   CNNs

CNN [20] was proposed very early to employ handwritten digit recognition. Due to limitations of the computational cost and data, CNNs architectures have received little attention from researchers. Until AlexNet [19] won the ImageNet image classification challenge in 2012, the potential of CNNs was the first widely acknowledged by researchers. Since then, different extensions and variants of CNNs have been proposed to employ various vision tasks including but not limited to VGG [32], GoogLeNet [33], ResNet [12], MobileNet [15], and EfficientNet [35]. Recently, ConvNeXt [27,45] has obtained a promising performance competitive with Vision Transformer [40]. CNNs are still an essential vision architecture applied to solve many vision tasks in real applications.

2.2 Vision Transformers

When Transformer [40] has achieved remarkable results in the field of NLP, computer vision also benefits from the powerful expressive capability of the model. ViT [7] has demonstrated that Transformer architecture can effectively handle the vision tasks. However, ViT still has drawbacks, such as lacking inductive bias and requiring vast training data. Due to the computational complexity of self-attention, ViT is challenging to apply to images with large sizes. To solve the drawbacks of ViT, DeiT [38] enables ViT to be trained effectively on ImageNet-1K [31] by knowledge distillation [13]. Swin Transformer [26] and PVT [41,42] introduce a pyramid architecture and optimize the computational complexity of self-attention.

2.3 Vision MLPs

With the encouraging progress made by Transformer in computer vision, MLP-Mixer [36], ResMLP [37], and gMLP [25] verified that Attention and Convolution are not necessary for image recognition and that using only MLPs can achieve fantastic performance. Vision MLPs are once again recognized in computer vision, which has attracted many researchers.

2.4 Graph Neural Networks

Graph is a ubiquitous structure in the real world, whether images, text, complex social networks, molecular structure, etc., they can all be represented as graphs. GNNs have become very popular in real applications. GNNs are also used extensively in the field of computer vision, such as multi-label classification [3], object detection [46], and image segmentation [51]. Recently, Vision GNN (ViG) [9] applies graph neural network to large-scale image recognition for the first time, connecting parts of objects for image recognition through graph structure, showing the potential of graph neural network as the vision backbone.

3   Methodology

3.1   Graph Level Processing of an Image

Graph Structured Image Representation. For an image with input size $H \times W \times 3$, we first divide the image into patches and each patch is treated as a node in a graph, resulting in a set of nodes $\mathcal{V} = \{\mathbf{v}_1, \mathbf{v}_2, ..., \mathbf{v}_n\}$. Next, we apply learnable parameters W to transform the initial features of the graph nodes into vectors, resulting in a set of graph node features $\mathbf{X} = \{\mathbf{x}_1, \mathbf{x}_2, ..., \mathbf{x}_n\}$, where $\mathbf{x}_i \in \mathbb{R}^D$, D is the feature dimension and $i = 1, 2, \cdots, n$.

For the proposed GLViG framework, the edges are carefully constructed to capture the local and global receptive field for each image patch. Thus, the primary objective is to design an appropriate methodology for linking each image patch with its corresponding patch. For each image patch $\mathbf{v}_i \subseteq \mathcal{V}$ in an image, we find the related K_G image patches globally using KNN while finding K_L neighboring image patches from the local region. The global and local neighbors of $\mathbf{v}_i$ are denoted by $\mathcal{N}(\mathbf{v}_i)$, then we add an edge e_{ji} directed from $\mathbf{v}_j$ to $\mathbf{v}_i$ for all $\mathbf{v}_j \in \mathcal{N}(\mathbf{v}_i)$ and obtain a set of directed edges $\mathcal{E} = \{\mathbf{e}_1, \mathbf{e}_2, ..., \mathbf{e}_n\}$. $\mathcal{V}$ and $\mathcal{E}$ make up the graph structure representation $\mathcal{G} = (\mathcal{V}, \mathcal{E})$ of the image.

Our approach involves a Local-Graph-Construction strategy where the linked image patches are treated as neighbors, providing support for both locality and translation. Meanwhile, We also apply a Global-Graph-Construction technique to enable the global receptive field of each image patch, leveraging the K-Nearest-Neighboring strategy to identify the K most similar patches covering the entire image as neighbors.

The Efficient Graph Construction. While KNN is a widely used method for graph construction, its reliance on calculating Euclidean distances between all graph nodes results in a computational complexity proportional to the square of the number of nodes in the graph, denoted as:

$$\Omega(\text{KGC}) = n^2 C + 2nC \tag{1}$$

where n is the number of graph nodes, C is the feature dimension of graph nodes, and KGC is the KNN graph construction.

When ViG [9] significantly first introduced GNN to large-scale image recognition, considering the complexity required for large-scale image recognition model training, ViG performed 4× sampling and 2× sampling in the first two stages (stage1 and stage2), respectively. Take the 4× sampling as an example, and its computational complexity is as follows:

$$\Omega(\text{KGC}) = \frac{n^2}{16}C + nC + \frac{n}{16}C \tag{2}$$

where n is the number of graph nodes, C is the feature dimension of the graph nodes, and KGC is the KNN graph construction.

The above graph node sampling strategy is employed at 224×224 image resolutions, its computational complexity is intractable when handling the issues of object detection, instance segmentation, and other downstream tasks with high-resolution images. To this end, our GLViG adaptively samples the number of graph nodes to a fixed size, termed adaptive graph node sampling. The computational complexity is:

$$\Omega(KGC) = knC + nC + kC \tag{3}$$

where n is the number of graph nodes, C is the feature dimension of the graph nodes, k is the number of sampled graph nodes, which is set to 196, and KGC is the KNN graph construction.

Our adaptive graph node sampling improves the quadratic complexity into linear complexity, significantly reducing the computational complexity of the model in high-resolution images, which makes GLViG compatible with a wide range of vision tasks.

Adding Positional Information to Graph Nodes. Graph-based image structures do not incorporate 2D spatial information, which limits their performance for visual tasks. To encode the 2D spatial position information of graph nodes, ViG [9] incorporates a positional encoding vector for each graph node feature:

$$\mathbf{x}_i \leftarrow \mathbf{x}_i + \mathbf{e}_i, \tag{4}$$

where $\mathbf{e}_i \in \mathbb{R}^D$, $\mathbf{x}_i$ is the feature of input graph node i.

However, ViG [9] uses fixed-size position encoding, necessitating interpolation algorithms when handling images of different sizes. This leads to imprecise position information and model inflexibility, negatively impacting model performance. At the same time, the position encoding of ViG is static and remains unchanged after training, potentially limiting its generalization ability and robustness.

Previous works [4,17,22,42] have shown that convolution with padding can incorporate positional encoding, so we employed Depth-wise Convolution [15] with padding to dynamically generate position encoding in the graph-based image representation:

$$\mathbf{e}_i = DepthwiseConv(\mathbf{x}_i) \tag{5}$$

$$\mathbf{x}_i \leftarrow \mathbf{x}_i + \mathbf{e}_i, \tag{6}$$

where $\mathbf{e}_i \in \mathbb{R}^D$, $\mathbf{x}_i$ is the feature of input graph node i.

In addition, we also integrated Depth-wise Convolution [15] into the FFN [40] layer to enhance positional information.

The positional encoding generated by convolution varies dynamically with graph node features and image resolution, resulting in flexible processing of images with different sizes and enriched positional encoding information, making it more robust. Meanwhile, the depth-separable convolution [15] can weight to sum the neighboring nodes' information, which benefits the diversity of nodes' information aggregation methods and promotes the information exchange among graph nodes.

Messages Passing. For the graph-based image representation $\mathcal{G} = (\mathcal{V}, \mathcal{E})$ of the image, we conduct the messages passing among the image patches as

$$\begin{aligned}\mathcal{G}' &= Update(GCN(\mathcal{G}, W_1), W_2)\\ &= Update(GCN(\mathcal{G}_l, \mathcal{G}_g, W_1), W_2),\end{aligned} \tag{7}$$

where $\mathcal{G}'$ is the updated graph structure representation, $\mathcal{G}_l$ is the local graph structure representation, $\mathcal{G}_g$ is the global graph structure representation, GCN is the graph convolution, $Update$ is the operation to update the graph structure representation, and W_1, W_2 is the learnable parameters.

Furthermore, we use max-relative graph convolution [21]. The operation is defined as

$$\begin{aligned}\mathbf{x}'_i &= Mrconv(\mathbf{x}_i, \mathcal{N}(\mathbf{x}_i), W_1)\\ &= [\mathbf{x}_i, \max(\mathbf{x}_g - \mathbf{x}_i), \max(\mathbf{x}_l - \mathbf{x}_i)]W_1,\end{aligned} \tag{8}$$

where $Mrconv$ is the max-relative graph convolution, $\mathcal{N}(\mathbf{x}_i)$ is the neighbor node feature of the graph node i, $\mathbf{x}_i$ is the feature of the graph node i, $\mathbf{x}_g$ is the global neighbor feature of node i, $\mathbf{x}_l$ is the local neighbor feature of node i.

3.2   GNN-Based Vision Architecture

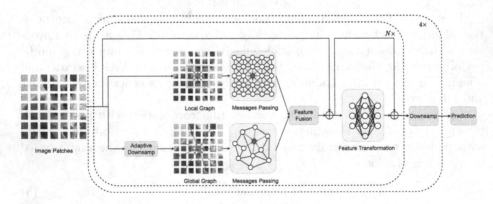

Fig. 2. The overall framework of the proposed method. The proposed method splits an image into fixed-sized image patches. Based on the image patches, a locality and globality-aware graph is constructed. And then, the constructed image graphs are fed into the Messages Passing module, and neighbor nodes pass information through the red edges. The circles in Feature Transformation are neurons of the multi-layer perceptron, and the lines in them represent the neural network's weights. (Color figure online)

The details of the network are shown in Fig. 2. Taking the GLViG-B1 version as an example, it first slices the input RGB images into overlapping patches through the stem layer. Each image patch is considered a node in a graph, and each node has a feature dimension of 48.

Our GLViG consists of four stages, where each stage aggregates local and global information of the linked image patches in an image using the GLViG building block. To learn a hierarchical representation, the image patches in adjacent image regions are merged as the stage increases into a new graph node. This strategy reduces the number of graph nodes to a quarter of their original size while increasing the feature dimension of the nodes. According to the design principle of the previous pyramid structure, we stack the building blocks mainly in the third stage.

4 Experiments

4.1 Image Classification on ImageNet-1K

Experimental Settings: We utilized ImageNet-1K [31] for our image classification experiments. ImageNet-1K, also known as the ISLVRC 2012 dataset, consists of 1,281,167 training images and 50,000 validation images, encompassing 1000 classes. Our proposed method was implemented using PyTorch [29] and Timm [44]. To ensure a fair comparison, we adhered to the experimental parameter settings widely employed in DeiT [38]. The proposed model was trained on 224^2-resolution images for 300 epochs. We employed the AdamW [28] optimizer with a cosine decay learning rate and 20 epochs of linear warm-up. Data augmentation and regularization techniques incorporated in our approach included RandAugmentation [6], Repeated augmentation [14], Mixup [50], CutMix [49], Random Erasing [52], Weight Decay, Label Smoothing [34], and Stochastic Depth [16]. We chose not to use Exponential Moving Average (EMA) [30] as it did not yield any improvement in the final model's performance.

Results: Table 1 lists the comparison of the proposed GLViG with other CNNs, Vision Transformers, Vision MLPs, and Vision GNN on Imagenet-1K. GLViG outperforms common CNNs (RSB-ResNet [12], ConvMixer [39], ConvNeXt [27, 45]), Vision Transformers (PVTv2 [42], ViT [7] and DeiT [38]), MLPs (MLP-Mixer [36], gMLP [25]) and Vision GNN (ViG [9]) with similar parameters and FLOPs, which reveals that our GLViG can handle vision tasks well and have much potential in vision.

4.2 Object Detection and Instance Segmentation

Settings: We conducted experiments on object detection and instance segmentation using the COCO 2017 dataset [24], which comprises 118,000 training images, 5,000 validation images, and 20,000 test images. To implement these tasks, we utilized MMDetection [1]. To ensure a fair comparison, we employed the model pre-trained on ImageNet-1K [31] as the backbone. Specifically, we used the RetinaNet [23] framework for object detection and the Mask R-CNN [11] framework for instance segmentation.

Table 1. Results of GLViG and other backbones on ImageNet-1K [31].

Model	Resolution	Type	Params (M)	FLOPs (G)	Top-1 (%)
PVTv2-B0 [42]	224×224	Attention	3.4	0.6	70.5
T2T-ViT-7 [48]	224×224	Attention	4.3	1.1	71.7
TNT-Ti [10]	224×224	Attention	6.1	1.4	73.9
DeiT-Tiny/16 [38]	224×224	Attention	5.7	1.3	72.2
gMLP-Ti [25]	224×224	MLP	6	1.4	72.3
ConvMixer-512/16 [39]	224×224	Conv	5.4	–	73.8
GLViG-B0 (ours)	224×224	GNN	5.4	0.9	**77.1**
PVTv2-B1 [42]	224×224	Attention	13.1	2.1	78.7
ViT-B/16 [7]	224×224	Attention	86	55.5	77.9
PoolFormer-S12 [47]	224×224	Pooling	12	1.8	77.2
RSB-ResNet-18 [12,43]	224×224	Conv	12	1.8	70.6
ConvMixer-1024/12 [39]	224×224	Conv	14.6	-	77.8
ConvNeXt V2-P [27,45]	224×224	Conv	9.1	1.4	79.7
RSB-ResNet-50 [12,43]	224×224	Conv	25.6	4.1	79.8
ViG-Ti [9]	224×224	GNN	10.7	1.7	78.2
GLViG-B1 (ours)	224×224	GNN	10.5	1.8	**80.7**

Results: According to Table 2, we find that GLViG outperforms different vision architectures including ResNet [12] (CNN) and PVT [41] (Vision Transformer) under RetinaNet [23] 1× and Mask R-CNN [11] 1× settings. These experimental results demonstrate GLViG's more expressive capability to employ the vision downstream tasks.

4.3 Ablation Study

The ablation results of GLViG are listed in Table 3. We observe that all three designs can enhance the performance of the model.

Global and Local Graph Construction: Global and Local Graph Construction (GLGC) is a crucial step of the proposed method. Compared to ViG [9] that considers only Global Graph Construction, GLViG can achieve better Top-1 accuracy (77.4% vs 76.1%) on ImageNet-1K. It is attributed that GLViG enables image patches to aggregate the local and global features of different patches in an image by combining the global receptive field, locality, and translation equivariance.

Depth-Wise Convolution: Introducing a Depth-wise Convolution with padding in the FFN layer can implicitly add the position information of image patches to the network, which is beneficial for the corresponding image patches to perceive their positions and enrich the way of information aggregation. The

Table 2. Object detection and instance segmentation results on COCO val2017 [24]. The proposed GLViG is compared with the other backbones using RetinaNet [23] and Mask R-CNN [11] frameworks.

Backbone	RetinaNet 1×						
	Param (M)	AP	AP_{50}	AP_{75}	AP_S	AP_M	AP_L
ResNet18 [12]	21.3	31.8	49.6	33.6	16.3	34.3	43.2
PVT-Tiny [41]	23.0	36.7	56.9	38.9	22.6	38.8	50.0
PoolFormer-S12 [47]	21.7	36.2	56.2	38.2	20.8	39.1	48.0
GLViG-B1 (ours)	**19.0**	**40.5**	**61.2**	**43.3**	**24.1**	**44.3**	**53.5**
Backbone	Mask R-CNN 1×						
	Param (M)	AP^b	AP_{50}^b	AP_{75}^b	AP^m	AP_{50}^m	AP_{75}^m
ResNet18 [12]	31.2	34.0	54.0	36.7	31.2	51.0	32.7
PVT-Tiny [41]	32.9	36.7	59.2	39.3	35.1	56.7	37.3
PoolFormer-S12 [47]	31.6	37.3	59.0	40.1	34.6	55.8	36.9
GLViG-B1 (ours)	**29.2**	**41.9**	**63.7**	**45.7**	**38.5**	**60.8**	**41.3**

Table 3. Ablation experiments of GLViG, where "Local Graph", "Global Graph", "DConv", and "AGNS" represent Local Graph Construction, Global Graph Construction, Depth-wise Convolution, and Adaptive Graph Node Sampling respectively. The basic architecture is GLViG-B1 and the number of training epochs is set to 100.

Method	Params	FLOPs (224 × 224)	FLOPs (1280 × 800)	ImageNet top-1 acc
Global Graph	10.37 M	1.72 G	93.5 G	76.1%
Local Graph	10.37 M	1.57 G	32.1 G	75.9%
Global Graph+Local Graph	10.53 M	1.75 G	94.0 G	77.4%
++DConv	10.51 M	1.78 G	94.8 G	78.5%
+++AGNS	10.51 M	1.78 G	37.1 G	78.5%

model with Depth-wise Convolution obtains the superior Top-1 accuracy (78.5% vs 77.4%) on ImageNet-1K.

Adaptive Graph Node Sampling: As shown in Table 2 and Table 3, compared to ViG's graph node sampling strategy, our Adaptive Graph Node Sampling (AGNS) significantly decreases the computational cost by 60%, and GLViG can also achieve superior performance in high-resolution images tasks like object detection and instance segmentation.

5 Conclusion

In this paper, we have proposed a GNN-based vision architecture. Specifically, the proposed method leverages global and local graph construction to model the locality, translation equivariance, and the global receptive field. For the quadratic complexity of global graph construction, we achieve linear computational complexity by adaptive graph node sampling. Finally, Depth-wise Convolution is adopted to add positional information to the graph nodes implicitly. We demonstrate the potential of GNNs in image classification, object detection, and image segmentation tasks. We hope our work will inspire future research.

Acknowledgments. This work was supported by National Key R&D Program of China (No.2022ZD0118202) and the National Natural Science Foundation of China (No. 62072386, No. 62376101).

A   Appendix

A.1   Semantic Segmentation

Settings: We chose the ADE20K dataset [53] to evaluate the semantic segmentation performance of GLViG. The ADE20K dataset consists of 20,000 training images, 2,000 validation images, and 3,000 test images, covering 150 semantic categories. For our framework, we utilized MMSEG [5] as the implementation framework and Semantic FPN [18] as the segmentation head. All experiments were conducted using 4 NVIDIA 3090 GPUs. During the training phase, the backbone was initialized with weights pre-trained on ImageNet-1K, while the newly added layers were initialized with Xavier [8]. We optimized our model using AdamW [28] with an initial learning rate of 1e-4. Following common practices [2,18], we trained our models for 40,000 iterations with a batch size of 32. The learning rate was decayed using a polynomial decay schedule with a power of 0.9. In the training phase, we randomly resized and cropped the images to a size of 512×512. During the testing phase, we rescaled the images to ensure the shorter side was 512 pixels.

Table 4. Results of semantic segmentation on ADE20K [53] validation set. We calculate FLOPs with input size 512×512 for Semantic FPN.

Backbone	#P(M)	GFLOPs	mIoU (%)
ResNet18 [12]	16.0	32.0	32.9
PVT-Tiny [41]	17.0	33.0	35.7
PoolFormer-S12 [47]	16.0	31.0	37.2
GLViG-B1	**13.3**	**30.1**	**40.3**

Results: As shown in Table 4, GLViG outperforms the representative backbones in Semantic Segmentation, including ResNet [12] (CNN), Attention-based method PVT [41] (Vision Transformer). Considering no relevant data is available, we didn't compare ViG [9] to the proposed method. For example, GLViG-B1 outperforms ResNet-18 by 7.4% mIoU (40.3% vs. 32.9%) and PVT-Tiny by

4.6% mIoU (40.3% vs.35.7%) under comparable parameters and FLOPs. In general, GLViG performs competently in semantic segmentation and consistently outperforms various backbone models (Tables 5 and 6).

A.2   Details of the GLViG Architecture

Table 5. Detailed settings of GLViG series.

Models	Channels	Blocks	Params (M)	FLOPs (G)
GLViG-B0	[32, 64, 160, 256]	[2, 2, 6, 2]	5.4	0.84
GLViG-B1	[48, 96, 240, 384]	[2, 2, 6, 2]	10.5	1.8

Table 6. Training hyper-parameters of GLViG for ImageNet-1K [31].

GLViG	B0 B1
Epochs	300
Optimizer	AdamW [28]
Batch size	1024
Start learning rate (LR)	2e-3
Learning rate schedule	Cosine
Warmup epochs	20
Weight decay	0.05
Label smoothing [34]	0.1
Stochastic path [16]	0.1 0.1
Repeated augment [14]	✓
RandAugment [6]	✓
Mixup prob. [50]	0.8
Cutmix prob. [49]	1.0
Random erasing prob. [52]	0.25

References

1. Chen, K., et al.: Mmdetection: Open mmlab detection toolbox and benchmark. arXiv preprint arXiv:1906.07155 (2019)
2. Chen, L.C., Papandreou, G., Kokkinos, I., Murphy, K., Yuille, A.L.: Deeplab: semantic image segmentation with deep convolutional nets, atrous convolution, and fully connected crfs. IEEE Trans. Pattern Anal. Mach. Intell. **40**(4), 834–848 (2017)
3. Chen, Z.M., Wei, X.S., Wang, P., Guo, Y.: Multi-label image recognition with graph convolutional networks. In: Proceedings of the IEEE/CVF Conference on Computer Vision and Pattern Recognition, pp. 5177–5186 (2019)

4. Chu, X., et al.: Conditional positional encodings for vision transformers. arXiv: Computer Vision and Pattern Recognition (2021)
5. Contributors, M.: Mmsegmentation: Openmmlab semantic segmentation toolbox and benchmark (2020)
6. Cubuk, E.D., Zoph, B., Shlens, J., Le, Q.V.: Randaugment: Practical automated data augmentation with a reduced search space. In: Proceedings of the IEEE/CVF Conference on Computer Vision and Pattern Recognition Workshops, pp. 702–703 (2020)
7. Dosovitskiy, A., et al.: An image is worth 16x16 words: transformers for image recognition at scale. arXiv preprint arXiv:2010.11929 (2020)
8. Glorot, X., Bengio, Y.: Understanding the difficulty of training deep feedforward neural networks. In: Proceedings of the Thirteenth International Conference on Artificial Intelligence and Statistics, pp. 249–256. JMLR Workshop and Conference Proceedings (2010)
9. Han, K., Wang, Y., Guo, J., Tang, Y., Wu, E.: Vision gnn: an image is worth graph of nodes. arXiv preprint arXiv:2206.00272 (2022)
10. Han, K., Xiao, A., Wu, E., Guo, J., Xu, C., Wang, Y.: Transformer in transformer. Adv. Neural. Inf. Process. Syst. **34**, 15908–15919 (2021)
11. He, K., Gkioxari, G., Dollár, P., Girshick, R.: Mask r-cnn. In: Proceedings of the IEEE International Conference on Computer Vision, pp. 2961–2969 (2017)
12. He, K., Zhang, X., Ren, S., Sun, J.: Deep residual learning for image recognition. In: Proceedings of the IEEE Conference on Computer Vision and Pattern Recognition, pp. 770–778 (2016)
13. Hinton, G., Vinyals, O., Dean, J.: Distilling the knowledge in a neural network. arXiv preprint arXiv:1503.02531 (2015)
14. Hoffer, E., Ben-Nun, T., Hubara, I., Giladi, N., Hoefler, T., Soudry, D.: Augment your batch: Improving generalization through instance repetition. In: Proceedings of the IEEE/CVF Conference on Computer Vision and Pattern Recognition, pp. 8129–8138 (2020)
15. Howard, A.G., et al.: Mobilenets: efficient convolutional neural networks for mobile vision applications. arXiv preprint arXiv:1704.04861 (2017)
16. Huang, G., Sun, Yu., Liu, Z., Sedra, D., Weinberger, K.Q.: Deep networks with stochastic depth. In: Leibe, B., Matas, J., Sebe, N., Welling, M. (eds.) ECCV 2016. LNCS, vol. 9908, pp. 646–661. Springer, Cham (2016). https://doi.org/10.1007/978-3-319-46493-0_39
17. Islam, A., Jia, S., Bruce, N.D.B.: How much position information do convolutional neural networks encode. arXiv: Computer Vision and Pattern Recognition (2020)
18. Kirillov, A., Girshick, R., He, K., Dollár, P.: Panoptic feature pyramid networks. In: Proceedings of the IEEE/CVF Conference on Computer Vision and Pattern Recognition, pp. 6399–6408 (2019)
19. Krizhevsky, A., Sutskever, I., Hinton, G.E.: Imagenet classification with deep convolutional neural networks. Commun. ACM **60**(6), 84–90 (2017)
20. LeCun, Y., Bottou, L., Bengio, Y., Haffner, P.: Gradient-based learning applied to document recognition. Proc. IEEE **86**(11), 2278–2324 (1998)
21. Li, G., Muller, M., Thabet, A., Ghanem, B.: Deepgcns: can gcns go as deep as cnns? In: Proceedings of the IEEE/CVF International Conference on Computer Vision, pp. 9267–9276 (2019)
22. Li, Y., Zhang, K., Cao, J., Timofte, R., Gool, L.V.: Localvit: bringing locality to vision transformers. Comput. Vis. Pattern Recog. (2021)

23. Lin, T.Y., Goyal, P., Girshick, R., He, K., Dollár, P.: Focal loss for dense object detection. In: Proceedings of the IEEE International Conference on Computer Vision, pp. 2980–2988 (2017)
24. Lin, T.-Y., et al.: Microsoft COCO: common objects in context. In: Fleet, D., Pajdla, T., Schiele, B., Tuytelaars, T. (eds.) ECCV 2014. LNCS, vol. 8693, pp. 740–755. Springer, Cham (2014). https://doi.org/10.1007/978-3-319-10602-1_48
25. Liu, H., Dai, Z., So, D., Le, Q.V.: Pay attention to mlps. Adv. Neural. Inf. Process. Syst. **34**, 9204–9215 (2021)
26. Liu, Z., et al.: Swin transformer: hierarchical vision transformer using shifted windows. In: Proceedings of the IEEE/CVF International Conference on Computer Vision, pp. 10012–10022 (2021)
27. Liu, Z., Mao, H., Wu, C.Y., Feichtenhofer, C., Darrell, T., Xie, S.: A convnet for the 2020s. In: Proceedings of the IEEE/CVF Conference on Computer Vision and Pattern Recognition, pp. 11976–11986 (2022)
28. Loshchilov, I., Hutter, F.: Decoupled weight decay regularization. arXiv preprint arXiv:1711.05101 (2017)
29. Paszke, A., et al.: Pytorch: an imperative style, high-performance deep learning library. In: Advances in Neural Information Processing Systems 32 (2019)
30. Polyak, B.T., Juditsky, A.B.: Acceleration of stochastic approximation by averaging. SIAM J. Control. Optim. **30**(4), 838–855 (1992)
31. Russakovsky, O., et al.: Imagenet large scale visual recognition challenge. Int. J. Comput. Vision **115**, 211–252 (2015)
32. Simonyan, K., Zisserman, A.: Very deep convolutional networks for large-scale image recognition. arXiv preprint arXiv:1409.1556 (2014)
33. Szegedy, C., et al.: Going deeper with convolutions. In: Proceedings of the IEEE Conference on Computer Vision and Pattern Recognition, pp. 1–9 (2015)
34. Szegedy, C., Vanhoucke, V., Ioffe, S., Shlens, J., Wojna, Z.: Rethinking the inception architecture for computer vision. In: Proceedings of the IEEE Conference on Computer Vision and Pattern Recognition, pp. 2818–2826 (2016)
35. Tan, M., Le, Q.: Efficientnet: rethinking model scaling for convolutional neural networks. In: International Conference on Machine Learning, pp. 6105–6114. PMLR (2019)
36. Tolstikhin, I.O., et al.: Mlp-mixer: an all-mlp architecture for vision. Adv. Neural. Inf. Process. Syst. **34**, 24261–24272 (2021)
37. Touvron, H., et al.: Resmlp: feedforward networks for image classification with data-efficient training. IEEE Trans. Pattern Anal. Mach. Intell. (2022)
38. Touvron, H., Cord, M., Douze, M., Massa, F., Sablayrolles, A., Jégou, H.: Training data-efficient image transformers & distillation through attention. In: International Conference on Machine Learning, pp. 10347–10357. PMLR (2021)
39. Trockman, A., Kolter, J.Z.: Patches are all you need? arXiv preprint arXiv:2201.09792 (2022)
40. Vaswani, A., et al.: Attention is all you need. In: Advances in neural information processing systems 30 (2017)
41. Wang, W., et al.: Pyramid vision transformer: a versatile backbone for dense prediction without convolutions. In: Proceedings of the IEEE/CVF International Conference on Computer Vision, pp. 568–578 (2021)
42. Wang, W., et al.: Pvt v2: improved baselines with pyramid vision transformer. Comput. Vis. Media **8**(3), 415–424 (2022)
43. Wightman, R., Touvron, H., Jégou, H.: Resnet strikes back: an improved training procedure in timm. arXiv preprint arXiv:2110.00476 (2021)

44. Wightman, R., et al.: Pytorch image models (2019)
45. Woo, S., et al.: Convnext v2: co-designing and scaling convnets with masked autoencoders. arXiv preprint arXiv:2301.00808 (2023)
46. Xu, H., Jiang, C., Liang, X., Li, Z.: Spatial-aware graph relation network for large-scale object detection. In: Proceedings of the IEEE/CVF Conference on Computer Vision and Pattern Recognition, pp. 9298–9307 (2019)
47. Yu, W., et al.: Metaformer is actually what you need for vision. In: Proceedings of the IEEE/CVF Conference on Computer Vision and Pattern Recognition, pp. 10819–10829 (2022)
48. Yuan, L., et al.: Tokens-to-token vit: training vision transformers from scratch on imagenet. In: Proceedings of the IEEE/CVF International Conference on Computer Vision, pp. 558–567 (2021)
49. Yun, S., Han, D., Oh, S.J., Chun, S., Choe, J., Yoo, Y.: Cutmix: regularization strategy to train strong classifiers with localizable features. In: Proceedings of the IEEE/CVF International Conference on Computer Vision, pp. 6023–6032 (2019)
50. Zhang, H., Cisse, M., Dauphin, Y.N., Lopez-Paz, D.: Mixup: beyond empirical risk minimization. arXiv preprint arXiv:1710.09412 (2017)
51. Zhang, L., Li, X., Arnab, A., Yang, K., Tong, Y., Torr, P.H.: Dual graph convolutional network for semantic segmentation. arXiv preprint arXiv:1909.06121 (2019)
52. Zhong, Z., Zheng, L., Kang, G., Li, S., Yang, Y.: Random erasing data augmentation. In: Proceedings of the AAAI Conference on Artificial Intelligence, vol. 34, pp. 13001–13008 (2020)
53. Zhou, B., Zhao, H., Puig, X., Fidler, S., Barriuso, A., Torralba, A.: Scene parsing through ade20k dataset. In: Proceedings of the IEEE Conference on Computer Vision and Pattern Recognition, pp. 633–641 (2017)

SVDML: Semantic and Visual Space Deep Mutual Learning for Zero-Shot Learning

Nannan Lu[✉], Yi Luo, and Mingkai Qiu

School of Information and Control Engineering, China University of Mining and
Technology, Xuzhou 221100, China
lnn_921@126.com

Abstract. The key challenge of zero-shot learning (ZSL) is how to identify novel objects for which no samples are available during the training
process. Current approaches either align the global features of images to
the corresponding class semantic vectors or use unidirectional attentions
to locate the local visual features of images via semantic attributes to
avoid interference from other noise in the image. However, they still have
not found a way to establish a robust correlation between the semantic
and visual representation. To solve the issue, we propose a Semantic and
Visual space Deep Mutual Learning (SVDML), which consists of three
modules: class representation learning, attribute embedding, and mutual
learning, to establish the intrinsic semantic relations between visual features and attribute features. SVDML uses two kinds of prototype genera-
tors to separately guide the learning of global and local features of images
and achieves interaction between two learning pipelines by mutual learning, so that promotes the recognition of the fine-grained features and
strengthens the knowledge generalization ability in zero-shot learning.
The proposed SVDML yields significant improvements over the strong
baselines, leading to the new state-of the-art performances on three popular challenging benchmarks.

Keywords: Zero-shot Learning · Semantic Representation · Visual
Representation · Mutual Learning

1 Introduction

In recent years, the field of computer vision has seen significant advances in the
development of zero-shot learning (ZSL) [12] techniques. ZSL is a challenging
problem that aims to recognize objects from unseen classes without any training
examples. This is achieved by leveraging auxiliary information, such as attributes
[6], text descriptions [5] and word embeddings [15] to establish a relationship
between seen and unseen classes.

Supported by National Natural Science Foundation of China (No. 62006233), National
Key RD Program of China (2019YFE0118500) and the Open Competition Mechanism
to Select the Best Candidates Fundation of Shanxi Province.

Q. Liu et al. (Eds.): PRCV 2023, LNCS 14433, pp. 383–395, 2024.
https://doi.org/10.1007/978-981-99-8546-3_31

One of the key challenges in ZSL is the domain shift problem [7], where the data distribution between the seen and unseen classes can be significantly different. Since the model is only trained on seen classes in the ZSL setting, the domain shift phenomenon can easily lead to a decline in model performance during testing on unseen classes. This is even more evident in Generalized Zero-Shot Learning (GZSL) [17], where the testing space of the model includes both seen and unseen classes, making it more likely to bias towards seen classes, resulting in a decrease in the classification accuracy for unseen classes. Although the difference in data distribution in visual space leads to domain shift problems, the consistency of semantic information such as human annotated attributes between different categories provides possibilities for eliminating the impact of domain shift. Existing methods have attempted to address this issue through various approaches, such as using generative models [16] to synthesize unseen class samples, calibrating the model's prediction scores [11], and learning a shared latent space between visual and semantic features [13]. However, these methods still have limitations and there is room for improvement.

Recently, there has been growing interest in techniques of prototype learning [21] and deep mutual learning [23]. Prototype learning refers to the process of learning a discriminative representation for each class in the form of a prototype, which can be used to classify new instances based on their similarity to the prototypes. Deep mutual learning, on the other hand, involves training multiple networks simultaneously, allowing them to learn from each other and improve their performance, which has been shown to improve the performance of individual models and can be applied to ZSL by training multiple models to learn from each other's predictions [1].

Recent research has explored the method of combining these techniques with ZSL models. For example, APN [20] uses prototypes to improve the attribute localization ability of image representations, thereby enhancing the performance of ZSL classification. MSDN [1] combines two attention sub-nets with a semantic distillation loss that aligns each other's class posterior probabilities. This encourages the mutual learning between the two mutual attention sub-nets, which allows them to learn collaboratively and teach each other throughout the training process. However, even though these models have used techniques such as prototypes or deep mutual learning, they still have not found a way to establish a robust correlation between the semantic and visual representation, and therefore cannot overcome the influence of domain shift problem.

To address this issue, we propose a novel ZSL model that combines attention mechanisms, prototype learning and deep mutual learning. Our model, called Semantic and Visual Space Deep Mutual Learning (SVDML), leverages the strengths of these technologies while harnessing their interactions to inspire greater capabilities. By utilizing attention and prototypes, SVDML learns class semantic prototypes and attribute semantic prototypes separately to align with global and local features of images. The engagement of deep mutual learning keeps the two formally different but semantically related pipelines intrinsically consistent under mutual supervision, thereby enabling the model to master the power of semantic consistency against domain shift. Through the elegant collaboration and pairing of three common but rarely coexisting mechanisms of attention, prototypes and deep mutual learning, the SVDML model can more efficiently achieve robust alignment of attributes and visual representations to alleviate the impact of domain shift and improve its performance on ZSL tasks.

SVDML consists of three main components: class representation learning module, local attribute embedding and mutual learning module. The class representation learning and local attribute embedding are guided by the prototype generator to learn the global and local features of images with respect to different semantics. Through mutual learning, the two modules are able to learn from and teach each other effectively. The mutual learning loss aligns the posterior probabilities between class representation learning module and local attribute embedding module, facilitating effective knowledge transfer between visual features and semantic information.

In summary, the SVDML model presents an innovative approach to ZSL by effectively aligning visual and semantic information through its unique framework. The combination of prototypes, deep mutual learning, and attention mechanisms allows for effective knowledge transfer between visual features and semantic information. Extensive experiments on three benchmark datasets show that our SVDML achieves consistent improvement over the state-of-the-art in both ZSL and GZSL settings.

2   Methodology

2.1   Problem Setting

Assume that we have training set $D^s = (x_i^s, y_i^s)$ with C^s seen classes, where $x_i^s \in \mathcal{X}^s$ denotes the i-th image and $y_i^s \in \mathcal{Y}^s$ is the corresponding label. Testing set $D^u = (x_i^u, y_i^u)$ with C^u unseen classes, where $x_i^u \in \mathcal{X}^u$ is the i-th unseen sample and $y_i^u \in \mathcal{Y}^u$ is the corresponding label. There are K human annotated attributes shared between all classes $C = C^s \cup C^u$. We define class semantic as $\mathcal{CS} = \{cs_i\}_{i=1}^{C}$, attribute semantic as $\mathcal{AS} = \{as_j\}_{j=1}^{K}$. In the conventional Zero-Shot Learning setting, there is no overlap between training and testing label, i.e., $C^s \cap C^u = \varnothing$. While in the Generalized Zero-Shot Learning (GZSL) setting, testing samples can be drawn from either seen or unseen classes.

2.2   Overview

The overview framework of the proposed SVDML is shown in Fig. 1, including local attribute embedding, class representation learning and mutual learning, respectively. The two primary data streams in the proposed framework consist of images with pixel-level features and attributes with semantic information, which are inputs of feature embedding and prototype generation. ResNet serves as the backbone to learn feature embeddings. Within the framework, the class representation learning and local attribute embedding modules guided by the

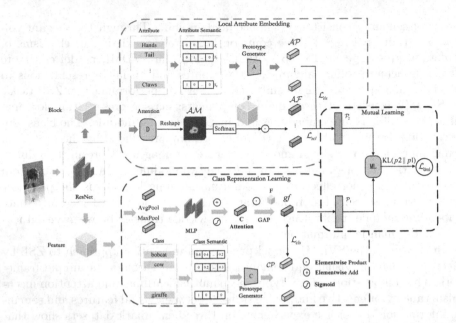

Fig. 1. Framework overview. Our SVDML method consists of class representation learning module, local attribute embedding module, and mutual learning module. The prototypes for both classes and attributes are generated from different generator with shared parameters. By comparing the prototypes with global features and local features, we obtain their respective predicted probabilities. Through the mutual learning mechanism, the two networks learn from each other and share information.

prototype generator, respectively, can learn the local and global features of the images with respect to different semantics. Through mutual learning, the two modules are able to learn from and teach each other effectively. Next, we will provide an introduction to each module within our framework.

2.3 Class Representation Learning

This module not only learns the global feature representation of image, but also design a class semantic prototype generator to guide more accurate mapping from class semantic space to visual feature space.

In the module, we use a bi-attention mechanism to extract the global visual feature representation from the feature tensor $F \in \mathbb{R}^{H \times W \times K}$ that are learned from image $\mathcal{X}$ by the backbone network, so that the feature with size ($1 \times 1 \times$ K) is obtained. Firstly, the feature tensor is calculated by the channel attention mechanism, formulated as Eq. 1.

$$att(\mathcal{F}) = (\text{MLP}(\text{GAP}(\mathcal{F}) + \text{GMP}(\mathcal{F}))) \tag{1}$$

where, GAP denotes the operation of global average pooling, GMP is the operation of global maximum pooling and MLP is the Multilayer Perceptron. The bi-attention mechanism promotes higher-level abstracted features, and generate the reconstructed visual feature F'. On the other hand, the class semantics are taken as input and mapped to the visual feature space by the prototype generator. By aligning each feature $F' = \{f_i\}_{i=1}^C$ with the class semantic prototypes CP, we can obtain the semantic attention weight for the i-th channel of the m-th class, formulated as:

$$att_m^i = \frac{\exp\left(cp_c^\top W f_i\right)}{\sum_{c=1}^{C^s} \exp\left(cp_c^\top W f_i\right)} \tag{2}$$

where, att_m^i is the attention to quantify the importance of the m-th class label for the i-th channel ranging from 1 to K for an image. W is an embedding matrix to measure the similarity between each semantic and visual vector. Then, a product is performed between att_m^i and the reconstructed visual features F', and the result is delivered to a GAP operation to obtain the global feature gf.

We compare the global feature gf and its corresponding class prototype CP and employ cosine similarity to represent their divergence, so that do the classification. And the prediction probability can be defined as Eq. 3.

$$p_i = \frac{\exp\left(\alpha \cdot \cos\left(gf, cp_i\right)\right)}{\sum_{j=1}^C \exp\left(\alpha \cdot \cos\left(gf, cp_j\right)\right)} \tag{3}$$

where, α is scaling factor, p_i denotes the prediction probability that image $\mathcal{X}$ belongs to the i-th class which is a scaling factor. Based on this, the cross entropy loss $\mathcal{L}_{cls}$ is used to optimize the class representation learning, which is defined as:

$$\mathcal{L}_{cls} = -\log p_i \tag{4}$$

2.4 Local Attribute Embedding

Local attribute embedding focuses on learning local features which refers to discriminative regions of image matched with attribute semantics. As the backbone network outputs the visual feature tensor F that are global features learned from image, we need to learn attribute-related features that are friendly to unseen classes recognition, so that a region attention mechanism is used to guide the region learning of image, and a prototype generator designed for attribute semantic promotes the region attention approaching to the attribute semantic prototypes.

The local attribute embedding module does not use region detection methods to find the discriminative regions, while utilizes the implicit attribute localization ability within feature extractor (such as CNN) to locate discriminative regions [24]. In the backbone network, we set the convolutions after residual blocks of ResNet to highlight parts of visual features, which can be taken as an attention mask. The specific formula is defined as:

$$AM = \text{conv}(F) \tag{5}$$

where, $F \in \mathbb{R}^{H \times W \times K}$ and $\mathrm{conv}(\cdot)$ is the 1×1 convolution with $1 \times 1 \times K$ $\times$ C parameters. $AM \in \mathbb{R}^{H \times W \times C}$ denotes the attention tensor which serves as a soft mask indicating the implicit localization of the attributes. Thus, to slice the attribute mask AM by channels, the soft mask $\{am_j\}_{j=1}^{C}$ is multiplied with the image feature tensor F. Then, the attribute masks are obtained through the sigmoid function that normalizes values into a range from 0 to 1, where the value approaching to 1 stands for high confidence of having attribute-related visual features in the location while that approaching to 0 is the opposite. By iterating over am_j for each j, we eventually obtain the attribute features $AF = \{af_j\}_{j=1}^{C}$.

As shown in Fig. 1, the upper stream in local attribute embedding module represents a prototype generator for attribute semantics, similar with that in class representation learning, which promotes the local features learned via region localization to align the attribute semantic prototypes. Afterwards, we define a triplet loss function $\mathcal{L}_{\mathrm{tls}}$, to achieve the alignment between feature and its prototype, which is specifically formulated as Eq. 6.

$$\mathcal{L}_{tls} = \sum_{j=1}^{K} \left[d\left(af_j^n, ap_j\right) - \beta \min_{i \neq j} d\left(af_j^n, ap_i\right) \right]_{+} \tag{6}$$

where, af_j^n denotes the j-th attribute features of the n-th images in a batch. $d(\cdot, \cdot)$ is the cosine distance $(1 - \cos(\cdot))$, and $[\cdot]_{+}$ represents the ReLU function.

In previous work [14], it has been observed that only highly discriminative features (anchor points) among different features of the same attribute have a guiding effect on prototype generation. Conversely, features that deviate from the prototype (reciprocal points) [9] result in domain shift phenomena. Therefore, we introduce the concept of contrastive learning, for a batch of images, to pull close enhanced attribute features af_j^n across images corresponding to the same attribute features, and push away those corresponding to different attribute features. The contrastive loss function $\mathcal{L}_{scl}$ is represented as:

$$\mathcal{L}_{scl} = -\frac{1}{K} \sum_{j=1}^{K} \log p_j \tag{7}$$

where, p_i is formulated as Eq. 8.

$$p_j = \frac{\sum_{j=1}^{K} \exp\left(\cos\left(af_j^n, af_j^{n+}\right)/\tau\right)}{\sum_{j=1}^{K} \exp\left(\cos\left(af_j^n, af_j^{n+}\right)/\tau\right) + \sum_{j=1}^{K} \exp\left(\cos\left(af_j^n, af_j^{n-}\right)/\tau\right)} \tag{8}$$

where, af_j^+ denotes the anchor points which are greater than τ and af_j^+ denotes the reciprocal points which are smaller than τ, representing features corresponding to different attributes. By computing the similarity with af_j^+, the similar image features are pulled closer together while being pushed away from features of different attributes. The $\mathcal{L}_{scl}$ encourages the related features to gather together.

2.5   Mutual Learning

The critical challenge of Zero-Shot Learning is to infer the latent semantic knowledge between the visual features of seen classes and the attribute features, allowing effective knowledge transfer to unseen classes. Therefore, it is crucial to extract the underlying semantic relations between visual features and attribute features from seen classes and transfer to unseen classes. Deep mutual learning [23] can collaboratively train an ensemble of networks and teach with each other, so that we introduce mutual learning to achieve the interaction between class representation learning and local attribute embedding during the training process and promote the learning of visual features and attribute features to uncover the latent relations within the visual and attribute features.

Here, we use probability distance and Jensen-Shannon divergence (JSD) to measure difference between visual features and attribute features. Therefore, the mutual learning loss, denoted as $\mathcal{L}_{dml}$, is defined based on the probability distance ℓ_d and JSD ℓ_{JSD} [1] between the predictions of the class representation learning module and the local attribute embedding module, p_1 and p_2. The loss is specified as:

$$\mathcal{L}_{dml} = \frac{1}{n_b} \sum_{i=1}^{n_b} [\underbrace{\frac{1}{2} \left(D_{KL} \left(p_1 \left(x_i \right) \| p_2 \left(x_i \right) \right) + D_{KL} \left(p_2 \left(x_i \right) \| p_1 \left(x_i \right) \right) \right)}_{\ell_{JSD}}$$

$$+ \underbrace{\| p_1 \left(x_i \right) - p_2 \left(x_i \right) \|_2^2]}_{\ell_d}, \tag{9}$$

where, n_b is the number of samples in a batch, $\| \cdot \|$ is the Euclidean distance, and x_i is the training image. The KL divergence from q to p can be computed as:

$$D_{KL}(p\|q) = \sum_{x \in \mathcal{X}^s} p(x) \log \left(\frac{p(x)}{q(x)} \right)$$

where, $x \in \mathcal{X}^s$ denotes the image x with seen classes.

Eventually, the total loss of the proposed SVDML is formulated as:

$$\mathcal{L}_{total} = \mathcal{L}_{cls} + \lambda_{tls} \mathcal{L}_{tls} + \lambda_{scl} \mathcal{L}_{scl} + \lambda_{dml} \mathcal{L}_{dml} \tag{10}$$

where λ_{tls}, λ_{scl} and λ_{dml} are the balance factors for this three losses.

2.6   Zero-Shot Classification

As to zero-shot classification, an unseen image x_i is inputted to the class representation learning module and the local attribute embedding module, respectively, and the two modules output the soft prediction p_1 and p_2. Afterwards, the prediction p_1 and p_2 are explicitly calibrated and fused using coefficients α_1 and α_2, so that we can get the corresponding label for the sample x_i. The label can be predicted by the following equation:

$$C^* = \arg \max_{C \in \mathcal{C}^u / \mathcal{C}} \left(\alpha_1 p_1 \left(x_i \right) + \alpha_2 p_2 \left(x_i \right) \right) \tag{11}$$

where, $\mathcal{C}^u / \mathcal{C}$ corresponds to the ZSL/GZSL setting.

3 Experiments

Datasets: The proposed method is evaluated on the zero-shot image datasets AWA2, CUB, and SUN. To facilitate comparison with the state-of-the-art, we adopt the proposed split suggested by Xian et al. [19] as the class division standard.

Evaluation Metrics: We consider both conventional ZSL and GZSL settings in all three datasets. Mean Class Accuracy (MCA) is adopted as the evaluation indicator for ZSL setting, which takes average value of Top-1 accuracy on unseen classes. And harmonic mean (H) is employed to evaluate the performance of GZSL setting, which is the most comprehensive metric to reflect model performance by taking accuracies of both seen and unseen classes into consideration, which is defined as follows:

$$H = \frac{2 \times MCA_S \times MCA_U}{MCA_S + MCA_U} \tag{12}$$

where, MCA_s and MCA_u are $MCAs$ for seen classes and unseen classes, respectively.

Implementation Details: The deep networks in this paper are built using the PyTorch deep learning framework. The experiments are conducted on Ubuntu 18.04 with hardware consisting of 16 vCPU Intel Xeon Processor (Skylake, IBRS) and 2 T T4 GPUs. By default, the models use ResNet101 pretrained on the ImageNet1k as the backbone network, without fine-tuning. The input image resolution is 448 × 448 and global feature gf is of 2048 dimensions. The prototype generator consists of 2 FC layers with shared paraments, 2 ReLU layers, and 1 FC layer, where the hidden size of prototype generator is set to 1024. The training is performed using the SGD optimizer with a momentum of 0.9, weight decay 0.0001. The learning rate is decayed every 10 epochs, with the decay factor of 0.5. The maximum number of epoch is set to 32. The loss weights λ_{dlm}, λ_{tls}, and λ_{scl} are set to 0.01, 0.001, and 0.001, respectively.

3.1 Comparison with State-of-the-Arts

We compare our experiments with the recent state-of-the-arts in Table 1. On the conventional ZSL side, SVDML shows very gratifying results in all three datasets. As it follows the state-of-the-art models on CUB, AWA2 and SUN. In these dataset, SVDML obtains competitive performance with 74.7%, 73.0% and 61.2% for MCA, respectively. The reason for the lower accuracy of our model compared to MSDN, which is also an mutual learning model, is that MSDN uses word vectors pre-trained by the GloVe so that it can utilize more priori knowledge

Table 1. Results (%) of the state-of-the-art ZSL and GZSL modes on CUB, SUN and AWA2, including generative methods, common space based methods, and embedding-based methods. The symbol "-" indicates no results.

Methods	CUB				SUN				AWA2			
	Acc	GZSL			Acc	GZSL			Acc	GZSL		
		U	S	H		U	S	H		U	S	H
Generative Methods												
GCM-CF (CVPR 21) [22]	-	61.0	59.7	60.3	-	47.9	37.8	42.2	-	60.4	75.1	67.0
FREE (ICCV 21) [3]	-	55.7	59.9	57.7	-	47.4	37.2	41.7	-	60.4	75.4	67.1
Embedding-based Methods												
DAZLE (CVPR 20) [11]	66.0	56.7	59.6	58.1	59.4	52.3	24.3	33.2	67.9	60.3	75.7	67.1
HSVA (NeurIPS 21) [4]	62.8	52.7	58.3	55.3	63.8	48.6	39.0	43.3	-	59.3	76.6	66.8
MSDN (CVPR 22) [1]	76.1	68.7	67.5	68.1	65.8	52.2	34.2	41.3	70.1	60.7	74.5	67.7
SLGR^A (AAAI 23) [8]	78.7	52.3	71.1	60.3	-	-	-	-	-	-	-	-
SVDML (Ours)	74.7	69.3	68.6	69.0	61.2	43.6	38.9	41.1	73.0	58.5	87.7	70.2

in seen classes. We use one-hot semantic vectors in attributes because it is more suitable basis for the class semantics to get higher performance in GZSL.

On the GZSL side, our SVDML generalizes well to unseen classes, achieving the best accuracies of 69.0% 70.2% on CUB and AWA2, respectively. This benefits come from the semantic and visual feature mutual learning of SVDML, enabling to search of the intrinsic semantic representations for effective knowledge transfer from seen to unseen classes. At the same time, our model achieved a competitive accuracy of 41.1% on SUN. A possible reason for this result could be that certain abstract attributes in the SUN dataset have uncertain sizes in images (e.g., sky), which increases the difficulty of accurately localizing these attributes using attention maps.

3.2 Abaltion Studies

To provide further insight into SVDML, we design ablation studies to evaluate the effectiveness. According to the experiment of each setting as shown in Table 2, each variant tends to be superior than the previous model. Compared to the baseline ($\mathcal{L}_{cls}$), the method employs the local attribute features (e.g., $\mathcal{L}_{cls} + \mathcal{L}_{scl}$ and $\mathcal{L}_{cls} + \mathcal{L}_{tls}$), and achieves the gains of acc/H by 3.7%/1.4% and 6.5%/9.8% on CUB, respectively. These results show that the local attribute feature module discover and utilize richer fine-grained information. Compared to the method without mutual learning module ($\mathcal{L}_{cls} + \mathcal{L}_{scl} + \mathcal{L}_{tls}$), the SVDML ensembles the complementary embeddings learned, achieving acc/H improvements of 2.6/0.9% on CUB.

3.3 Hyperparameter Analysis

Effects of Combination Coefficients. We conduct experiments to determine the effectiveness of the combination coefficients ($\alpha1$, $\alpha2$) between classes learning module and local feature module in Eq. 11. As shown in Fig. 2, SVDML performs

Table 2. Ablation study of loss terms in total losses on cub

$\mathcal{L}_{cls}$	$\mathcal{L}_{tls}$	$\mathcal{L}_{scl}$	$\mathcal{L}_{dml}$	$Acc_T(\%)$	$H(\%)$
✓				64.0	55.4
✓	✓			67.7	56.8
✓		✓		70.5	65.2
✓	✓	✓		72.1	68.1
✓	✓	✓	✓	74.7	69.0

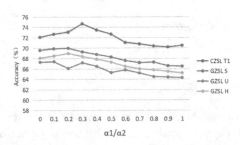

Fig. 2. The effectiveness of the combination coefficients ($\alpha 1$, $\alpha 2$)

poorly when $\alpha 1/\alpha 2$ is set too small or large, because the class representation learning module and local attribute embedding module are complementary for discriminative semantic embedding representations. Thus, we set $\alpha 1/\alpha 2 = 0.3$.

Effects of Scaling Factor. Then we show the impact of scaling factor α in Eq. 3 by varying it from 20 to 30 in Fig. 3 (a). SVDML achieves the best performance on CUB at $\alpha = 25$ for both T1 and H. In the impact of scaling factor β in Eq. 6 by varying it from 0 to 1 in Fig. 3 (b), achieving the best performance on CUB at $\beta = 0.4$ for both T1 and H. We also show the impact of scaling factor τ in Eq. 7 by varying it from 0 to 1 in Fig. 3 (c), achieving the best performance on CUB at $\tau = 0.6$ for both T1 and H.

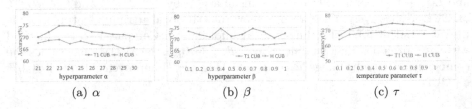

| (a) α | (b) β | (c) τ |

Fig. 3. Ablation study for hyperparameter α, β, τ.

Effects of Loss Weight. We study how to set the related loss weights of SVDML: λ_{tls}, λ_{scl} and λ_{dml}. Based on the results in Fig. 4 (a), (b) and (c). We choose λ_{tls}, λ_{scl} and λ_{dml} as 0.001, 0.001, 0.01 for CUB, respectively.

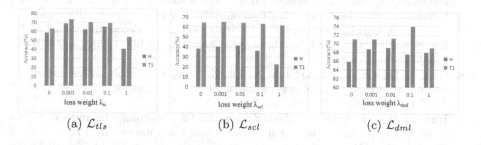

(a) $\mathcal{L}_{tls}$ (b) $\mathcal{L}_{scl}$ (c) $\mathcal{L}_{dml}$

Fig. 4. Ablation study for loss weight $\mathcal{L}_{tls}$, $\mathcal{L}_{scl}$, $\mathcal{L}_{dml}$.

4   Conclusion

In this paper, we propose a novel semantic and visual space deep mutual learning (SVDML) network for ZSL. SVDML consists of class representation learning and local attribute embedding module, which learns attribute based global features and enhanced attribute features for semantic embedding representations, respectively. To encourage interaction between the two modules, we introduce a mutual learning module that aligns each other's posterior probabilities. Thus, SVDML discovers the intrinsic semantic representations between visual and attribute features for effective knowledge transfer of ZSL. Extensive experiments on three popular benchmarks show the superiority of SVDML.

References

1. Chen, S., et al.: MSDN: mutually semantic distillation network for zero-shot learning. In: Proceedings of the IEEE/CVF Conference on Computer Vision and Pattern Recognition, pp. 7612–7621 (2022)
2. Chen, S., et al.: DSP: dynamic semantic prototype for generative zero-shot learning (2022)
3. Chen, S., et al.: Free: feature refinement for generalized zero-shot learning. In: Proceedings of the IEEE/CVF International Conference on Computer Vision, pp. 122–131 (2021)
4. Chen, S., et al.: HSVA: hierarchical semantic-visual adaptation for zero-shot learning. Adv. Neural. Inf. Process. Syst. **34**, 16622–16634 (2021)
5. Elhoseiny, M., Zhu, Y., Zhang, H., Elgammal, A.: Link the head to the "beak": zero shot learning from noisy text description at part precision. In: Proceedings of the IEEE Conference on Computer Vision and Pattern Recognition, pp. 5640–5649 (2017)

6. Farhadi, A., Endres, I., Hoiem, D., Forsyth, D.: Describing objects by their attributes. In: 2009 IEEE Conference on Computer Vision and Pattern Recognition, pp. 1778–1785. IEEE (2009)
7. Fu, Y., Hospedales, T.M., Xiang, T., Fu, Z., Gong, S.: Transductive multi-view embedding for zero-shot recognition and annotation. In: Fleet, D., Pajdla, T., Schiele, B., Tuytelaars, T. (eds.) Computer Vision–ECCV 2014: 13th European Conference, Zurich, Switzerland, 6–12 September 2014, Proceedings, Part II 13. LNCS, vol. 8690, pp. 584–599. Springer, Cham (2014). https://doi.org/10.1007/978-3-319-10605-2_38
8. Guo, J., Guo, S., Zhou, Q., Liu, Z., Lu, X., Huo, F.: Graph knows unknowns: reformulate zero-shot learning as sample-level graph recognition. Proc. AAAI Conf. Artif. Intell. **37**(6), 7775–7783 (2023)
9. Hur, S., Shin, I., Park, K., Woo, S., Kweon, I.S.: Learning classifiers of prototypes and reciprocal points for universal domain adaptation. In: Proceedings of the IEEE/CVF Winter Conference on Applications of Computer Vision (WACV), pp. 531–540, January 2023
10. Huynh, D., Elhamifar, E.: Compositional zero-shot learning via fine-grained dense feature composition. Adv. Neural. Inf. Process. Syst. **33**, 19849–19860 (2020)
11. Huynh, D., Elhamifar, E.: Fine-grained generalized zero-shot learning via dense attribute-based attention. In: Proceedings of the IEEE/CVF Conference on Computer Vision and Pattern Recognition, pp. 4483–4493 (2020)
12. Larochelle, H., Erhan, D., Bengio, Y.: Zero-data learning of new tasks. In: AAAI, vol. 1, p. 3 (2008)
13. Li, X., Xu, Z., Wei, K., Deng, C.: Generalized zero-shot learning via disentangled representation. In: Proceedings of the AAAI Conference on Artificial Intelligence, vol. 35, pp. 1966–1974 (2021)
14. Li, X., Yang, X., Wei, K., Deng, C., Yang, M.: Siamese contrastive embedding network for compositional zero-shot learning. In: Proceedings of the IEEE/CVF Conference on Computer Vision and Pattern Recognition, pp. 9326–9335 (2022)
15. Mikolov, T., Chen, K., Corrado, G., Dean, J.: Efficient estimation of word representations in vector space. arXiv preprint arXiv:1301.3781 (2013)
16. Pourpanah, F., et al.: A review of generalized zero-shot learning methods. IEEE Trans. Pattern Anal. Mach. Intell. **45**(4), 4051–4070 (2022)
17. Scheirer, W.J., de Rezende Rocha, A., Sapkota, A., Boult, T.E.: Toward open set recognition. IEEE Trans. Pattern Anal. Mach. Intell. **35**(7), 1757–1772 (2012)
18. Wan, Z., et al.: Transductive zero-shot learning with visual structure constraint. In: Advances in Neural Information Processing Systems, vol. 32 (2019)
19. Xian, Y., Schiele, B., Akata, Z.: Zero-shot learning-the good, the bad and the ugly. In: Proceedings of the IEEE Conference on Computer Vision and Pattern Recognition, pp. 4582–4591 (2017)
20. Xu, W., Xian, Y., Wang, J., Schiele, B., Akata, Z.: Attribute prototype network for zero-shot learning. Adv. Neural. Inf. Process. Syst. **33**, 21969–21980 (2020)
21. Yang, H.M., Zhang, X.Y., Yin, F., Liu, C.L.: Robust classification with convolutional prototype learning. In: Proceedings of the IEEE Conference on Computer Vision and Pattern Recognition, pp. 3474–3482 (2018)
22. Yue, Z., Wang, T., Sun, Q., Hua, X.S., Zhang, H.: Counterfactual zero-shot and open-set visual recognition. In: Proceedings of the IEEE/CVF Conference on Computer Vision and Pattern Recognition, pp. 15404–15414 (2021)

23. Zhang, Y., Xiang, T., Hospedales, T.M., Lu, H.: Deep mutual learning. In: Proceedings of the IEEE Conference on Computer Vision and Pattern Recognition (CVPR), June 2018
24. Zhou, B., Khosla, A., Lapedriza, A., Oliva, A., Torralba, A.: Learning deep features for discriminative localization. In: Proceedings of the IEEE Conference on Computer Vision and Pattern Recognition, pp. 2921–2929 (2016)

Heterogeneous Graph Attribute Completion via Efficient Meta-path Context-Aware Learning

Lijun Zhang[1,2], Geng Chen[1(✉)], Qingyue Wang[1], and Peng Wang[1,2]

[1] School of Computer Science, Northwestern Polytechnical University, Xi'an, China
lijunzhang@mail.nwpu.edu.cn, geng.chen.cs@gmail.com, peng.wang@nwpu.edu.cn
[2] Ningbo Institute, Northwestern Polytechnical University, Xi'an, China

Abstract. Heterogeneous graph attribute completion (HGAC) is an emerging research direction and has drawn increasing research attention in recent years. Although making significant progress, existing HGAC methods suffer from three limitations, including (i) the ignorance of valuable meta-path information during the graph node encoding, (ii) insufficient graph context learning as only neighboring nodes are considered during the completion process, and (iii) low efficiency due to the use of heterogeneous graph neural network for downstream tasks. To address these limitations, we propose an efficient meta-path context-aware learning framework to improve the completion of missing attributes of heterogeneous graphs. In our framework, we introduce a collaborative meta-path-driven embedding scheme, which incorporates valuable meta-path prior knowledge during node sampling. Furthermore, we devise a context-aware attention mechanism to complete the missing node attributes, which captures the information of non-neighbor nodes. Finally, we adopt graph attention networks for downstream tasks, effectively improving computational efficiency. Extensive experiments on three real-world heterogeneous graph datasets demonstrate that our framework outperforms state-of-the-art HGAC methods remarkably.

Keywords: Graph neural networks · Attribute completion ·
Heterogeneous graph learning · Graph node classification

1 Introduction

Graph is a common data structure that has been widely used in various fields, such as social network analysis, recommendation systems, etc. According to the number of node types and edge types, graphs can be divided into homogeneous graphs and heterogeneous graphs. As shown in Fig. 1, the graph structure data is incomplete in IMDB dataset collected from the real-world, with only a small fraction of samples containing complete attributes. This information is often not fully available in practical situations due to the unavailability of private information of many individuals in applications, such as social networks. There are

Q. Liu et al. (Eds.): PRCV 2023, LNCS 14433, pp. 396–408, 2024.
https://doi.org/10.1007/978-981-99-8546-3_32

generally two kinds of methods to handle missing data. Methods in the first category utilize conventional machine learning techniques to impute the absent data, subsequently leveraging the completed data for downstream applications. Typical representative methods include mean imputation [6], matrix factorization [12], multiple imputation [16], etc. Methods in the second category combine attribute completion and downstream tasks of heterogeneous graphs using end-to-end graph neural networks. These methods have shown promising results and have gained increasing popularity. Both supervised [1,10,18] and unsupervised methods [7] based on graph neural networks have been proposed.

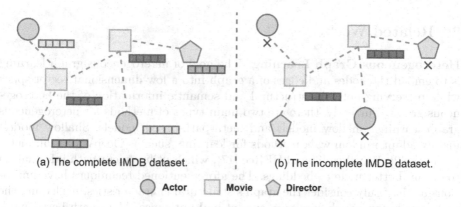

(a) The complete IMDB dataset.   (b) The incomplete IMDB dataset.

◯ Actor   ▢ Movie   ⬠ Director

Fig. 1. Comparison of heterogeneous graphs with complete and incomplete attributes. (a) The IMDB dataset with complete attributes. (b) The IMDB dataset with incomplete attributes.

Despite their progress, existing methods suffer from three major limitations. Firstly, they have primarily utilized random walks for embedding learning and have not taken into account the significance of meta-path. This approach leads to the loss of valuable prior knowledge due to the presence of diverse node types in heterogeneous graphs. Secondly, attribute completion in existing methods has been heavily reliant on information from neighboring nodes, leading to the exclusion of crucial information about non-neighbor nodes. Lastly, existing methods have employed a heterogeneous graph neural network for downstream tasks after attribute completion, which results in low computational efficiency.

To address these limitations, we propose an efficient meta-path context-aware learning framework (i.e., EMC-Net) for improved HGAC. Our EMC-Net employs a meta-path-driven random walk embedding approach that incorporates prior knowledge of meta-path while sampling nodes. To complete the attributes, we utilize a context-aware attention mechanism to prevent non-neighbor information loss and fill in the missing node data. Furthermore, we directly employ improved graph attention networks for downstream tasks such as classification to improve computational efficiency. In summary, our contributions are asfollows:

- We utilize a collaborative random walk embedding scheme based on meta-path, which provides richer prior information for node embedding.
- We adopt a context-aware attention mechanism for attribute completion, which avoids the information loss between non-adjacent nodes.
- We improve the model efficiency by directly using improved graph attention networks for downstream tasks.
- Experiments conducted on three real-world datasets demonstrate that our EMC-Net has better performance and can be used as a plug-and-play plugin for any baseline model. Our code is publicly available at https://github.com/LijunZhang01/EMC_Net.

2 Related Work

Heterogeneous Graph Learning. The goal of heterogeneous graph learning is to embed the nodes and edges of a graph into a low-dimensional vector space while preserving both the structural and semantic information of the heterogeneous graph. Currently, there are two main types of methods for heterogeneous graph learning: shallow models and meta-path-based models. Shallow models mainly adopt random walk methods for learning, such as DeepWalk [15], meta-path2vec [2], HIN2Vec [3] and HERec [17], which utilize meta-path information to obtain better node embeddings. The aforementioned techniques have limitations as they only consider the graph's structural characteristics, neglecting the node attributes, which results in suboptimal outcomes. Meta-path-based models [4,5,8,21], address this limitation by dividing the heterogeneous graph into multiple different homogeneous subgraphs and learning them separately using meta-path information.

Graph Attribute Completion. For dealing with common missing attributes in the real world, there are two types of methods: traditional machine learning methods and neural network-based methods. Traditional methods include mean interpolation [6] and matrix factorization [12], while most neural network-based methods adopt an end-to-end network structure that learns both the completion of incomplete attributes and downstream tasks together, some of which use graph neural networks as the underlying infrastructure. These methods [1,7,9,10] have shown good performance and potential, and thus, there are more and more works based on graph neural networks. The HGNNAC [10] method used a combination of attribute completion and heterogeneous graph neural networks to handle heterogeneous graph data with missing attributes.

3   Methodology

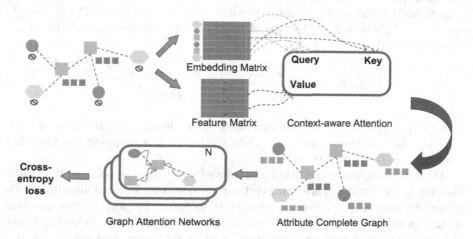

Fig. 2. The framework of our EMC-Net. It mainly includes three modules: Node Embedding Module, Attribute Completion Module, and Heterogeneous Graph Neural Network.

3.1   Problem Formulation

Given a graph $\mathcal{G} = (\mathcal{V}, \mathcal{E}, A, X)$ where $\mathcal{V} = \{v_0, v_1, ..., v_N\}$ represents the set of N nodes, $\mathcal{E}$ represents the set of edges, $A \in \mathcal{R}^{N \times N}$ represents the adjacency matrix of the graph, and $X \in \mathcal{R}^{N \times d}$ represents the node feature matrix, where d is the dimension of node features.

Since the attributes of the graph are incomplete, the nodes can be divided into $\mathcal{V}_{cp}$ and $\mathcal{V}_{ms}$, where $\mathcal{V}_{cp}$ represents the nodes with complete attributes and $\mathcal{V}_{ms}$ represents the nodes with missing attributes. The feature matrix corresponding to $\mathcal{V}_{cp}$ is denoted as X_{cp}. Our goal is to complete the missing features for nodes $v \in \mathcal{V}_{ms}$ by combining the known node attributes and graph structural properties. We define X_{ac} as the completed node attributes, then $X_{new} = [X_{cp}|X_{ac}]$ is the new node feature matrix and then input it into a graph neural network for downstream tasks.

3.2   Overall Framework

Figure 2 illustrates the proposed EMC-Net. Our model consists of three main components: node embedding module, attribute completion module, and heterogeneous graph neural network. Initially, we utilize the graph structure information to encode the nodes, followed by employing context-aware attention to fill in missing attributes. The completed attributes are subsequently fed into the heterogeneous graph neural network to facilitate downstream tasks. In the following sections, we will provide a detailed explanation of each module.

3.3 Meta-path-Driven Node Embedding

The meta-path-driven random walk mainly involves computing the embedding of each node based on the graph structure. Given a meta-path $P : T_1 \longrightarrow T_2 \longrightarrow \dots \longrightarrow T_l$, where T_i is the type of node v_i. The adoption probability of the i step is defined as:

$$
p(v_i|v_{i-1}, P) = \begin{cases} \frac{1}{|\mathcal{N}_{T_i(v_{i-1})}|} & A_{i-1,i} = 1 \text{ and } t(v_i) = T_i \\ 0 & \text{others} \end{cases},
\tag{1}
$$

where $|\mathcal{N}_{T_i(v_{i-1})}|$ represents the number of nodes in the neighborhood of node v_{i-1} that have a type equal to T_i, while $t(v_i) = T_i$ is a mapping function that maps node v_i to its own type.

Then, the skip-gram algorithm is employed to encode the nodes of the graph. The skip-gram algorithm is a word vector representation learning algorithm used in natural language processing. Its goal is to learn the vector representation of each word, so that words with similar semantics are closer in vector space. The sampled node sequences can be viewed as sentences, and each node is a word in the sentence. Ultimately, for each node v_i, the embedding we obtain is represented as E_i, E_{cp} represents the embedding with complete attributes nodes and E_{ms} represents the embedding with missing attributes nodes.

3.4 Attribute Completion Module

In the attribute completion module, we mainly adopt attention mechanism to complete missing attributes. Attention mechanism is a commonly used technique in neural network that combines different parts of information with different weights. In detail, the attention mechanism computes attention scores for each key-value pair in response to a query. It then combines the values using a weighted sum, where the weights are determined by the attention scores. The calculation formula for this process is as shown below:

$$
\text{Attention}(Q, K, \overline{V}) = \text{softmax}(QK^T)\overline{V},
\tag{2}
$$

where Q represents the query vector, K represents the key vector, $\overline{V}$ represents the value vector.

Additionally, inspired by the work of Transformer [19], we have incorporated the use of multi-head attention in our model. The mathematical representation for multi-head attention is as follows:

$$
\text{MultiHead}(Q, K, \overline{V}) = \text{mean}(\text{head}_1, \dots, \text{head}_h)W^O,
\tag{3}
$$

$$
\text{head}_i = \text{Attention}(QW_i^Q, KW_i^K, \overline{V}W_i^V),
\tag{4}
$$

where W_i^Q, W_i^K, W_i^V, and W^O are learnable projection matrices. During attribute completion, our query is the embedding of the node with missing attributes E_{ms}, the key is the embedding of the node with complete attributes

E_{cp}, and the value is the attributes of the node with complete attributes X_{cp}. So, the calculation formula for this process is as shown below:

$$X_{\text{ac}} = \text{MultiHead}(E_{\text{ms}}, E_{\text{cp}}, X_{\text{cp}}), \qquad (5)$$

$$X_{\text{new}} = [X_{\text{cp}}|X_{\text{ac}}]. \qquad (6)$$

Before completing the attributes imputation, we initialize using one-hot encoding and learn the differences relative to the initialization. This helps accelerate the convergence speed.

3.5 Heterogeneous Graph Neural Network

Upon acquisition of the attributes X_{new}, since the meta-path information of the heterogeneous graph has already been incorporated into the node embedding process, inspired by SimpleHGN [13], we have employed improved GAT for the purpose of graph node classification, thereby enhancing the model's computational efficiency.

The core idea of GAT [20] is to calculate the attention weights between node pairs in the neighborhood of each node, and then use these weights as the weighted sum of the feature vectors of neighboring nodes to obtain a new representation for each node. Specifically, for node i and its neighbor node j, GAT calculates the attention weight e_{ij} between them and normalizes it to obtain w_{ij}. Then, for node i, GAT weights the feature vectors x_j of all its neighbor nodes with w_{ij} and sums them up to obtain the new feature vector x_i. More complex graph neural networks can be constructed by stacking multiple GAT layers. The formula for the traditional GAT is as follows:

$$x_i^{l+1} = \sigma\left(\sum_{j \in \mathcal{N}_i} w_{ij}^l W^l x_j^l\right), \qquad (7)$$

where x_i^{l+1} represents the feature vector of node i in the $l+1$ layer, x_j^l represents the feature vector of node j in the l layer, $\mathcal{N}_i$ is the set of neighboring nodes of node i, W^l is a learnable weight matrix, and σ is an activation function.

Inspired by SimpleHGN [13], we make some improvements to GAT to handle heterogeneous graphs. Next, I will provide a detailed explanation. Firstly, we incorporate edge type information into the attention calculation. Specifically, the formula for our updated attention weight w_{ij} calculation is as follows:

$$w_{ij}^l = \frac{\exp\left(\text{LeakyReLU}(a^\top[Wx_i^l|Wx_j^l|W_r r_{(i,j)}^l])\right)}{\sum_{k \in \mathcal{N}_i} \exp\left(\text{LeakyReLU}(a^\top[Wx_i^l|Wx_k^l|W_r r_{(i,k)}^l])\right)}, \qquad (8)$$

where a is a learnable weight vector, LeakyReLU is a leaky rectified linear unit function, and $[\cdot|\cdot]$ denotes the concatenation operation, W_r represents a learnable transformation matrix, where $r_{(i,j)}^l$ denotes the feature of the edge between node i and node j in the l layer.

402 L. Zhang et al.

Furthermore, to mitigate the problem of excessive smoothing in graph neural network, a potential solution is to incorporate residual connections. These connections are added between the node representations of different layers. Consequently, the node aggregation formula for the $l+1$ layer is adjusted as follows:

$$x_i^{l+1} = \sigma \left(\sum_{j \in \mathcal{N}_i} w_{ij}^l W^l x_j^l + W_{res}^{l+1} x_i^l \right), \tag{9}$$

where W_{res}^{l+1} represents a transformation matrix used in the residual connection, which is applied when the dimensions of the node representations in the l and $l+1$ layers are different.

Table 1. Statistics of datasets.

Dateset	# Nodes	# Node Types	# Edges	# Edge Types	Target
DBLP	26,128	4	239,566	6	Author
ACM	10,942	4	547,872	8	Paper
IMDB	21,420	4	86,642	6	Movie

In addition, we employ multi-head attention, as described in Sect. 3.4, and implemented L2 regularization on the output embedding, i.e.,

$$z_i = \frac{x_i^l}{||x_i^l||}, \tag{10}$$

where z_i is the output embedding of node i and x_i^l is the final representation from Eq. (9).

4 Experiments

4.1 Datasets

In order to assess the effectiveness of our proposed framework, we opted to utilize three well-known heterogeneous graph datasets: ACM, IMDB, and DBLP. Additionally, we employed the same data partitioning approach as that employed in the SimpleHGN [13]. Below, we provide comprehensive descriptions for each of the aforementioned datasets.

- DBLP: The DBLP dataset is a bibliography website for computer science. We use a commonly used subset for our analysis. It includes four types of nodes. The target node for our classification is the author, and there are four classes.
- ACM: The ACM dataset is a citation website for papers, and we extract a subset of it using the same principle as HAN [21].

- IMDB: The IMDB dataset is collected from a movie website and includes three types of nodes. The target node for our classification is the movie.

We have designed two types of experiments to validate our method. The first type involves nodes with missing attributes that require classification. The representative dataset for this type is DBLP. The second type involves nodes with existing attributes that require classification, while attributes for other node types are missing. The representative datasets for this type are ACM and IMDB. For the node classification task, which is a supervised learning task, we used 24% of the nodes for training, 6% for validation, and 70% for testing for each dataset. The detailed data for the three datasets are summarized in Table 1.

4.2 Comparison Methods

We compared our proposed framework with existing state-of-the-art methods to validate its effectiveness and selected representative methods from three categories: neural network methods for homogeneous graph processing, neural network methods for heterogeneous graph processing, and methods combining attribute completion with graph representation learning. Specifically, we selected GCN [11] and GAT [20] as the representative methods for homogeneous graph processing, HetGNN [22], HAN [21], MAGNN [5], and SimpleHGN [13] as the representative methods for heterogeneous graph processing, and HGNNAC [10] as the representative method for combining attribute completion with graph representation learning. Some of the baseline experimental results come from [23]. These comparisons help us better understand the current neural network technologies for processing graph-structured data and evaluating the effectiveness of our proposed framework.

4.3 Implementation Details

During training, we used the Adam optimizer with a learning rate of $1e-3$ and weight decay of $1e-4$ for ACM and DBLP datasets. The training process was run for 500 epochs. For the IMDB dataset, we used the Adam optimizer with a learning rate of $1e-3$ and weight decay of $2e-4$, and trained for 300 epochs. The parameter negative_slope of LeakyReLU is 0.05.

4.4 Experimental Results

Table 2 presents a performance evaluation of EMC-Net in comparison to other graph neural network techniques. The study initially compared EMC-Net with the homogeneous neural network, including GCN and GAT, and observed a significant improvement of 0.79%–14.31% across three datasets. Furthermore, when compared to heterogeneous neural network, such as HetGNN, HAN, MAGNN, and SimpleHGN, EMC-Net demonstrated superior performance with an improvement of 0.52%–3.37% compared to the best-performing method, SimpleHGN. This indicates the effectiveness of the completion module we added to our method. When comparing our method with HGNNAC, which only considers neighbor completion, we found that our context-aware attention completion significantly improves stability and effectiveness, achieving at least 1.82%–14.96% improvement on all three datasets. All of them effectively demonstrate the effectiveness of our EMC-Net.

Additionally, by observing the performance of the DBLP dataset, we found that our EMC-Net can significantly improve node classification when the target nodes lack original attributes. This indicates that our model can effectively learn the features of known attributes nodes and graph structure features to make correct classifications. Furthermore, for datasets such as ACM and IMDB that already have target node attributes, our method can still improve performance by completing non-target nodes.

Table 2. Quantitative comparison of EMC-Net and sate-of-the-art methods.

Dataset	DBLP		ACM		IMDB	
Model	Metrics					
	Macro-F1	Micro-F1	Macro-F1	Micro-F1	Macro-F1	Micro-F1
GCN	90.54 ± 0.27	91.18 ± 0.25	92.63 ± 0.23	92.60 ± 0.22	59.95 ± 0.72	65.35 ± 0.35
GAT	92.96 ± 0.35	93.46 ± 0.35	92.41 ± 0.84	92.39 ± 0.84	56.95 ± 1.55	64.24 ± 0.55
HetGNN	92.77 ± 0.24	93.23 ± 0.23	84.93 ± 0.78	84.83 ± 0.76	47.87 ± 0.33	50.83 ± 0.26
HAN	93.17 ± 0.19	93.64 ± 0.17	87.68 ± 1.94	87.73 ± 1.81	59.70 ± 0.90	65.61 ± 0.54
MAGNN	93.16 ± 0.38	93.65 ± 0.34	91.06 ± 1.44	90.95 ± 1.43	56.92 ± 1.76	65.11 ± 0.59
SimpleHGN	93.83 ± 0.18	94.25 ± 0.19	92.92 ± 0.67	92.85 ± 0.68	62.98 ± 1.66	67.42 ± 0.42
HGNNAC	92.97 ± 0.72	93.43 ± 0.69	90.89 ± 0.87	90.83 ± 0.87	56.63 ± 0.81	63.85 ± 0.85
Ours	**94.74 ± 0.13**	**95.13 ± 0.12**	**93.42 ± 0.36**	**93.33 ± 0.38**	**65.10 ± 0.30**	**68.21 ± 0.31**

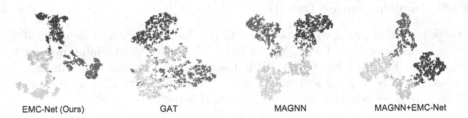

EMC-Net (Ours)   GAT   MAGNN   MAGNN+EMC-Net

Fig. 3. Visualization of the embeddings on DBLP. Each point represents an author, with colors indicating research fields.

4.5   Visualization of Embeddings

To provide a more intuitive comparison between different models, we conducted a visualization task on the DBLP dataset, learning node embeddings for MAGNN,

GAT, MAGNN+EMC-Net, and EMC-Net. MAGNN+EMC-Net refers to applying our proposed completion method to the MAGNN baseline network. We used t-SNE [14] to project the embeddings into two-dimensional space and colored the nodes based on their class. The results are shown in Fig. 3.

From the visualizations, it can be observed that GAT performed the worst, and some authors belonging to different classes mixed with blurred boundaries. MAGNN performed relatively better than GAT, but authors of the same class were scattered and lacked sufficient clustering. When MAGNN was combined with our proposed framework, MAGNN+EMC-Net exhibited denser cluster structures, enabling better differentiation of different classes in the visualizations. Our EMC-Net produced clearly defined boundaries and dense cluster structures after clustering, indicating that our model significantly improved the performance of the clustering task and could effectively differentiate authors from different research fields.

4.6 Ablation Studies

In order to verify the effectiveness of each module, we conducted the following ablation experiments.

Effectiveness of Network Design. We divided the model into three modules: initialization module (INIT), completion module (AC), and heterogeneous graph network module (HIN). Then we removed these three modules separately to observe their impact on performance. "-AC" represents to use of zero imputation. "-HIN" represents the use of MAGNN instead of improved GAT for heterogeneous graph network module. "-INIT" represents the non-use of one-hot initialization. The experimental results are presented in Table 3. From the experimental results, it can be seen that among these three modules, INIT has the greatest impact.

Table 3. Effectiveness of network design. The effectiveness of our components is demonstrated by removing different components from our model.

Dataset	DBLP		ACM		IMDB	
Model	Metrics					
	Macro-F1	Micro-F1	Macro-F1	Micro-F1	Macro-F1	Micro-F1
-AC	93.83 ± 0.18	94.25 ± 0.19	92.92 ± 0.67	92.85 ± 0.68	62.98 ± 1.66	67.42 ± 0.42
-HIN	93.53 ± 0.31	94.00 ± 0.31	92.14 ± 0.92	92.044 ± 0.95	58.00 ± 1.10	65.53 ± 1.10
-INIT	81.00 ± 0.23	81.52 ± 0.24	92.17 ± 0.84	92.10 ± 0.83	53.25 ± 1.26	61.30 ± 0.51
Ours	**94.74 ± 0.13**	**95.13 ± 0.12**	**93.42 ± 0.36**	**93.33 ± 0.38**	**65.10 ± 0.30**	**68.21 ± 0.31**

Effectiveness of Attribute Completion Module. We conducted ablation studies on different completion methods, "ZERO-AC" refers to completion using zero filling, "ONE-HOT" refers to completion using one-hot encoding, "HGNNAC" refers to completion using the method described in the

HGNNAC [10]. The experimental results are presented in Table 4. It is evident
that our completion method surpasses the performance of the three preceding
completion methods across all three datasets, thereby showcasing the efficacy
and reliability of our completion approach.

Table 4. Attribute completion module effectiveness analysis. Compare our completion
method with other completion methods.

Dataset	DBLP		ACM		IMDB	
Model	Metrics					
	Macro-F1	Micro-F1	Macro-F1	Micro-F1	Macro-F1	Micro-F1
ZERO-AC	93.83 ± 0.18	94.25 ± 0.19	92.92 ± 0.67	92.85 ± 0.68	62.98 ± 1.66	67.42 ± 0.42
ONE-HOT	93.50 ± 0.56	94.00 ± 0.53	93.15 ± 0.73	93.06 ± 0.38	63.72 ± 0.72	67.52 ± 0.32
HGNNAC	93.24 ± 0.49	93.73 ± 0.45	93.16 ± 0.24	93.09 ± 0.23	64.44 ± 1.13	67.67 ± 0.39
Ours	**94.74 ± 0.13**	**95.13 ± 0.12**	**93.42 ± 0.36**	**93.33 ± 0.38**	**65.10 ± 0.30**	**68.21 ± 0.31**

Validation of Model Plug-and-Play Capability. We applied our proposed
completion method and the completion method from the HGNNAC to MAGNN
and SimpleGNN. Table 5 results demonstrate that our completion method has
a significant improvement on both MAGNN and SimpleHGN. Moreover, the
experimental results also indicate that our completion module can be used as
a plug-and-play method for any graph neural network that deals with graph
representation learning, which is one of the important advantages of our method.

Table 5. Validation of model plug-and-play capability. Applying our model to different
heterogeneous graph neural network benchmark models.

Dataset	DBLP		ACM		IMDB	
Model	Metrics					
	Macro-F1	Micro-F1	Macro-F1	Micro-F1	Macro-F1	Micro-F1
MAGNN	93.16 ± 0.38	93.65 ± 0.34	91.06 ± 1.44	90.95 ± 1.43	56.92 ± 1.76	65.11 ± 0.59
MAGNN-HGNNAC	92.97 ± 0.72	93.43 ± 0.69	90.89 ± 0.87	90.83 ± 0.87	56.63 ± 0.81	63.85 ± 0.85
MAGNN-Ours	**93.53 ± 0.31**	**94.00 ± 0.31**	**92.14 ± 0.92**	**92.44 ± 0.95**	**58.00 ± 1.10**	**65.53 ± 1.10**
SimpleHGN	93.83 ± 0.18	94.25 ± 0.19	92.92 ± 0.67	92.85 ± 0.68	62.98 ± 1.66	67.42 ± 0.42
SimpleHGN-HGNNAC	93.24 ± 0.49	93.73 ± 0.45	93.16 ± 0.24	93.09 ± 0.23	64.44 ± 1.13	67.67 ± 0.39
SimpleHGN-Ours	**94.74 ± 0.13**	**95.13 ± 0.12**	**93.42 ± 0.36**	**93.33 ± 0.38**	**65.10 ± 0.30**	**68.21 ± 0.31**

Validation of Computational Efficiency. To valid the efficiency of our
model, we created an ablated version that employs a heterogeneous graph neural
network for downstream tasks after attribute completion, as in [13]. The results
show that our EMC-Net is about six times faster than the ablated version,
demonstrating its high efficiency.

5   Conclusion

In this work, we have proposed a effective model, EMC-Net, to address the issue of attributes missing in heterogeneous graphs. Our EMC-Net adopts a collaborative random walk embedding scheme based on meta-path, which incorporates prior knowledge of meta-path during node sampling. During the attribute completion process, it utilizes a context-aware attention mechanism to complete the information of nodes with missing attributes and avoids information loss. Finally, it directly employs improved graph attention networks for downstream tasks such as classification, effectively improving the computational efficiency. Extensive experiments on three real-world heterogeneous graph datasets demonstrate that EMC-Net is effective in HGAC and outperforms state-of-the-art methods.

Acknowledgements. This work was supported by the National Natural Science Foundation of China (No. U19B2037), National Key R&D Program of China (No.2020AAA0106900), Shaanxi Provincial Key R&D Program (No. 2021KWZ-03), and Natural Science Basic Research Program of Shaanxi (No. 2021JCW-03).

References

1. Chen, X., Chen, S., Yao, J., Zheng, H., Zhang, Y., Tsang, I.W.: Learning on attribute-missing graphs. IEEE Trans. Pattern Anal. Mach. Intell. **44**(2), 740–757 (2020)
2. Dong, Y., Chawla, N.V., Swami, A.: metapath2vec: scalable representation learning for heterogeneous networks. In: Proceedings of the 23rd ACM SIGKDD International Conference on Knowledge Discovery and Data Mining, pp. 135–144 (2017)
3. Fu, T.Y., Lee, W.C., Lei, Z.: HIN2Vec: explore meta-paths in heterogeneous information networks for representation learning. In: Proceedings of the 2017 ACM on Conference on Information and Knowledge Management, pp. 1797–1806 (2017)
4. Fu, X., King, I.: MECCH: metapath context convolution-based heterogeneous graph neural networks. arXiv preprint arXiv:2211.12792 (2022)
5. Fu, X., Zhang, J., Meng, Z., King, I.: MAGNN: metapath aggregated graph neural network for heterogeneous graph embedding. In: Proceedings of The Web Conference 2020, pp. 2331–2341 (2020)
6. García-Laencina, P.J., Sancho-Gómez, J.L., Figueiras-Vidal, A.R.: Pattern classification with missing data: a review. Neural Comput. Appl. **19**, 263–282 (2010)
7. He, D., et al.: Analyzing heterogeneous networks with missing attributes by unsupervised contrastive learning. IEEE Trans. Neural Networks Learn. Syst. (2022)
8. Hu, Z., Dong, Y., Wang, K., Sun, Y.: Heterogeneous graph transformer. In: Proceedings of the Web Conference 2020, pp. 2704–2710 (2020)
9. Jiang, B., Zhang, Z.: Incomplete graph representation and learning via partial graph neural networks. arXiv preprint arXiv:2003.10130 (2020)
10. Jin, D., Huo, C., Liang, C., Yang, L.: Heterogeneous graph neural network via attribute completion. In: Proceedings of the Web Conference 2021, pp. 391–400 (2021)
11. Kipf, T.N., Welling, M.: Semi-supervised classification with graph convolutional networks. In: International Conference on Learning Representations (2017)

12. Koren, Y., Bell, R., Volinsky, C.: Matrix factorization techniques for recommender systems. Computer **42**(8), 30–37 (2009)
13. Lv, Q., et al.: Are we really making much progress? Revisiting, benchmarking and refining heterogeneous graph neural networks. In: Proceedings of the 27th ACM SIGKDD Conference on Knowledge Discovery & Data Mining, pp. 1150–1160 (2021)
14. Van der Maaten, L., Hinton, G.: Visualizing data using t-SNE. J. Mach. Learn. Res. **9**(11), 2579–2605 (2008)
15. Perozzi, B., Al-Rfou, R., Skiena, S.: DeepWalk: online learning of social representations. In: Proceedings of the 20th ACM SIGKDD International Conference on Knowledge Discovery and Data Mining, pp. 701–710 (2014)
16. Rubin, D.B.: Multiple Imputation for Nonresponse in Surveys, vol. 81. Wiley, New York (2004)
17. Shi, C., Hu, B., Zhao, W.X., Philip, S.Y.: Heterogeneous information network embedding for recommendation. IEEE Trans. Knowl. Data Eng. **31**(2), 357–370 (2018)
18. Taguchi, H., Liu, X., Murata, T.: Graph convolutional networks for graphs containing missing features. Futur. Gener. Comput. Syst. **117**, 155–168 (2021)
19. Vaswani, A., et al.: Attention is all you need. In: Advances in Neural Information Processing Systems, vol. 30 (2017)
20. Veličković, P., Cucurull, G., Casanova, A., Romero, A., Liò, P., Bengio, Y.: Graph attention networks. In: International Conference on Learning Representations (2017)
21. Wang, X., et al.: Heterogeneous graph attention network. In: The World Wide Web Conference, pp. 2022–2032 (2019)
22. Zhang, C., Song, D., Huang, C., Swami, A., Chawla, N.V.: Heterogeneous graph neural network. In: Proceedings of the 25th ACM SIGKDD International Conference on Knowledge Discovery & Data Mining, pp. 793–803 (2019)
23. Zhu, G., Zhu, Z., Wang, W., Xu, Z., Yuan, C., Huang, Y.: AutoAC: towards automated attribute completion for heterogeneous graph neural network. arXiv preprint arXiv:2301.03049 (2023)

Fine-Grain Classification Method of Non-small Cell Lung Cancer Based on Progressive Jigsaw and Graph Convolutional Network

Zhengguang Cao and Wei Jia[(✉)]

School of Information Engineering, Ningxia University, Yinchuan 750021, China
jiawnx@163.com

Abstract. The classification of non-small cell lung cancer (NSCLC) is a challenging task that faces two main problems that limit its performance. The first is that the feature regions used to discriminate NSCLC types are usually scattered, requiring classification models with the ability to accurately locate these feature regions and capture contextual information. Secondly, there are a large number of redundant regions in NSCLC pathology images that interfere with classification, but existing classification methods make it difficult to deal with them effectively. To solve these problems, we propose a fine-grained classification network, Progressive Jigsaw and Graph Convolutional Network (PJGC-Net). The network consists of two modules. For the first problem, we designed the GCN-Based multi-scale puzzle generation (GMPG) module to achieve fine-grained learning by training separate networks for different scale images, which can help the network identify and localize feature regions used to discriminate NSCLC types as well as their contextual information. For the second problem, we propose the jigsaw supervised progressive training (JSPT) module, which removes a large number of redundant regions and helps the proposed model focus on the effective discriminative regions. Our experimental results and visualization results on the LC25000 dataset demonstrate the feasibility of this method, and our method consistently outperforms other existing classification methods.

Keywords: Non-small cell lung cancer · Fine-grained classification · Jigsaw · Graph Convolutional Network

1 Introduction

According to the World Health Organization, approximately 1.8 million people worldwide die from lung cancer each year. Non-small cell lung cancer (NSCLC) is the most prevalent type, and it is often diagnosed at advanced stages, resulting

This work was supported by the National Natural Science Foundation of China (No. 62062057).

in a less favorable prognosis [1]. Different subtypes of NSCLC differ in etiology, pathology, clinical manifestations, and treatment, so their accurate classification into benign lung tissue (Normal), lung adenocarcinoma (LUAD), and lung squamous cell carcinoma (LUSC) is important for clinical practice and lung cancer research [2].

Several methods have been proposed in the work on the NSCLC classification. Hou et al. [3] proposed a convolutional neural network (CNN) based method to divide the whole slice image into hundreds of small pieces, train a CNN model for each piece, and finally integrate the model results of these small pieces to obtain the detection results of the whole slice. Xu et al. [4] used a deep learning method for NSCLC classification and detection, which used CNN and attention mechanisms for feature extraction and data enhancement. The existing methods represented by these methods have two main problems: the lack of targeted identification of discriminative regions in the images and the difficulty of effectively dealing with redundant regions in the images that interfere with the classification. These problems can lead to poor classification of NSCLC pathology images.

To overcome the limitations of current methods, we propose a fine-grained classification method using multi-scale puzzles inspired by Progressive Multi-Granularity (PMG) [5]. The method applies the idea of puzzles to the classification task to complete fine-grained learning to improve the recognition of discriminative regions and add a Graph convolution neural network (GCN) [6,7] to capture their contextual information. Our progressive learning method is realized through a Two-stream Swin Transformer, enabling efficient handling of redundant image regions. The main contributions of this paper are as follows:

1. We propose a network of progressive learning using a Two-stream Swin Transformer, a graph convolutional network, and a jigsaw puzzle, referred to as PJGC-Net, for the fine-grained classification work of the NSCLC.
2. We designed the GCN-Based multi-scale puzzle generation (GMPG) module to implement the recognition capability against feature regions. The module consists of two parts: a multi-scale puzzle generator (MPG) and a graph convolutional network-based feature fusion network (GCNF-Net). GCNF-Net is used to build graph topology from the connection relationships between cells to better capture cellular structure and contextual information in cancer images. MPG generates random puzzles at multiple scales to improve the recognition of scattered feature regions.
3. We designed the jigsaw supervised progressive training (JSPT) module using Two-stream Swin Transformer with MPG-generated multi-scale puzzles to handle a large number of redundant regions that interfere with classification in images. The module can remove redundant regions that interfere with classification in pathological images by supervising the learning regions of small-scale puzzles with large-scale puzzles.

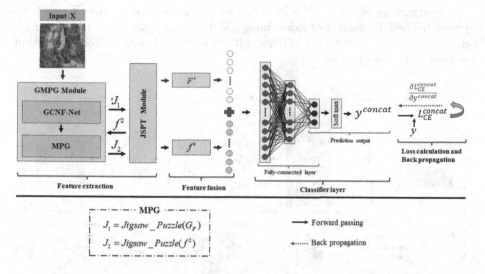

Fig. 1. Overall framework of the proposed PJGC-Net. *Jigsaw_Puzzle* is a random puzzle generator consisting of *jigsaw_generator*() and *shuffle*(), where *jigsaw_generator* () indicates that the input data is used to generate a puzzle, and *shuffle*() is a random number function for randomly disrupting the puzzle; the classification layer consists of a fully connected layer and a *Softmax* function; back propagation is computed by computing the partial derivative of loss L_{CE}^{concat} with respect to y^{concat}.

2 Method

2.1 Network Architecture

The overall structure of the proposed method is shown in Fig. 1, which is mainly composed of the GMPG module and the JSPT module. The input image X is first passed through GCNF-Net and MPG of the GMPG module, then passed into the JSPT module, and the feature map is passed into MPG again to generate the random puzzle J_2. After learning, we get the features learned twice, stitch them into new features $concat(F^4, f^4)$, and pass them into the classification layer to get the predicted value y^{concat}.

2.2 GMPG Module

GCNF-Net. Subtle cell structures and their surrounding contextual information are important for classification, and some existing methods lack the modeling of subtle structures and contextual information in images and thus may miss important information. To solve this problem, this paper introduces GCNF-Net. Piece of information on the spatial region and its channels as a node, and convolution operations in GCN can effectively propagate features and relationships between cells, thus better capturing cellular structures and contextual information in cancer images.

The structure of GCNF-Net is shown in Fig. 2, and the encoder's role is to aggregate the information of input image X. In the decoder part, the encoder's output feature representation is fused with the original image after integration for later feature extraction.

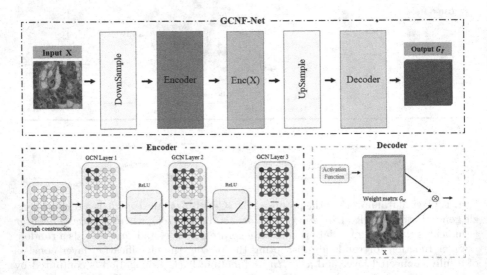

Fig. 2. GCN-Based Feature Fusion Network Architecture.

In the encoder, to reduce the computational burden of information aggregation in the GCN, we designed a network consisting of three convolutional layers and an average pooling layer, Avg Pool, to perform feature extraction while downsampling.

The process of downsampling and graph construction are shown in Fig. 3. The three convolutional layers correspond to convolutional kernel sizes of (5,5,3) and the number of channels output is (16,32,64). Finally, the pooling layer outputs a feature map of size $d \times d$ with 64 channels. The features generated by the above process are denoted as:

$$H^0 = DownSample(d, X) \tag{1}$$

$DownSample(\cdot)$ indicates to downsample the image X to the size of $d \times d$. Then, a pixel region is constructed as a node together with the channel location information, and the process is abbreviated as $\eta = d \times d$, indicating that the number of nodes is generated with $d \times d$. The image pixels in Fig. 3 are assumed to be $d = 4$, and use this to generate a 16×16 adjacency matrix $A \epsilon R^{N \times N}$, N represents node. If nodes are spatial neighbors, then $A_{ij} = 1$, otherwise $A_{ij} = 0$, so that the graph is constructed.

Each layer of GCN is normalized using the $ReLU()$, and each layer of GCN can be represented as $H^{l+1} = f(H^l, A)$, after employing the convolutional oper-

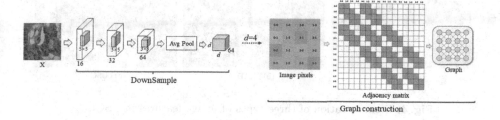

Fig. 3. Down Sampling and Graph construction Architecture.

ation [7], every GCN layer can be written as:

$$f(H^l, A) = \sigma(\widetilde{D}^{-\frac{1}{2}} * \widetilde{A} * \widetilde{D}^{-\frac{1}{2}} * H^l * W^l + b^l) \qquad (2)$$

Degree matrix $D_{ii} = \sum_j A_{ij}$, $\widetilde{A} = A + I_n$ is the adjacency matrix of the undirected graph with added self-connections, I_n is the identity matrix. $\widetilde{D}_{ii} = \sum_j \widetilde{A}_{ij}$, W^l is a layer-specific trainable weight matrix, b^l is a learnable bias vector. σ denotes an activation function such as $Relu()$, $H^l = R^{N \times D}$ is the matrix of activations in the l-th layer, $l = 0, 1, 2$.

The above operation is the encoder part, and the encoded output is represented as $Enc(X)$. Before entering the decoder, the feature map derived from the encoder is upsampled to the same size as the original image by an adaptive global pooling layer of 448 × 448. In the decoder part, the incoming information is passed through an activation function layer σ to generate the weights G_W, which are shown in Eq. (3):

$$G_W = \sigma(AdaptivePool((448, Enc(X))) \qquad (3)$$

After that, it is merged with the original image, and the above process is the decoding part $Dec()$ to get the graph convolutional feature map G_F. The process is shown in Eqs. (4) and (5):

$$Dec(Enc(x)) = G_F \qquad (4)$$

$$G_F = G_W \otimes X \qquad (5)$$

MPG. From left to right, Fig. 4 shows Normal, LUAD, and LUSC, and the red deposits and color of Normal make it easy to identify the features, so we mainly focus on LUAD and LUSC. The part marked by the red rectangular box in Fig. 4 is the vesicular structure, which is one of the common organelles in LUAD and is usually round or oval in shape. Vesicular structures in LUAD are usually larger, up to several microns to tens of microns in diameter, while vesicular structures in LUSC are usually smaller, usually 1–3 microns in diameter.

In this paper, the Jigsaw Puzzle method proposed by Du et al. [5] is extended to multiple scales for capturing vesicular features of different sizes in cell nests by

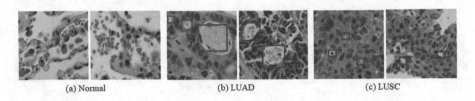

(a) Normal　　　　(b) LUAD　　　　(c) LUSC

Fig. 4. Demonstration of three types of image features in NSCLC.

targeting the vesicular structure features of LUAD and LUSC mentioned above. The MPG can be seen in Fig. 1, where the input data is divided into puzzle shapes that can accommodate vesicular structures and randomly disturbed in the first learning, denoted by J_1, so that the network focuses only on the region within the size of the puzzle when performing learning, thus capturing the features of vesicular structures. The process of generating a puzzle using the puzzle generator $Jigsaw\_Puzzle$ is represented as follows:

$$J_1 = shuffle(jigsaw\_generator(G_f, m)) \tag{6}$$

The above process is abbreviated as $\mu = m \times m$, which means generating a random jigsaw puzzle in the form of $m \times m$. After performing the first feature extraction, the second learning of the puzzle in the network with finer granularity can improve the performance of the model. f^l denotes the feature map obtained in the second learning stage, $l = 1,2,3,4$. Before proceeding to the third stage, the feature map is generated by $Jigsaw\_Puzzle$ in the form of a random $e \times e$ puzzle with smaller granularity J_2 to enhance feature extraction at finer details and to eliminate a large number of interfering regions. The process is represented in Eq. (7):

$$J_2 = shuffle(jigsaw\_generator(f^2, e)) \tag{7}$$

2.3　JSPT Module

In order to improve the learning efficiency, identify the vesicular structure of NSCLC more accurately, and deal with a large number of redundant regions efficiently, this paper uses the idea of progressive learning [8,9] to accomplish fine-grained learning.

　　The key in the fine-grained classification task is to find the most discriminative regions, and early work was usually done using manual annotation [10,11]. However, manual annotation is too costly and error-prone, and such methods are being replaced by fine-grained image detection methods [12–15], which are based on deep learning. Our method is to improve Swin Transformer [16,17] to Two-stream Swin Transformer, which implements the JSPT module by generating multi-scale puzzles through MPG to accomplish fine-grained classification tasks.

　　The structure of JSPT-Net is shown in Fig. 5. $F^1 - F^4$ and $f^1 - f^4$ of Two-stream Swin Transformer in the figure represent the feature maps of two branches of Swin Transformer after learning in stage1-stage4. $F^l \epsilon R^{H_l * W_l * C_l}$ denotes the upper part of the network i.e. the feature map after each stage of learning in the first learning, and $f^l \epsilon R^{h_l * w_l * c_l}$ denotes the lower part of the network i.e. the feature map after each stage of learning in the second learning, where H_l/h_l, W_l/w_l, C_l/c_l are the height, width and number of channels of the feature map at l-th stage, and $l = 1, 2, 3, 4$.

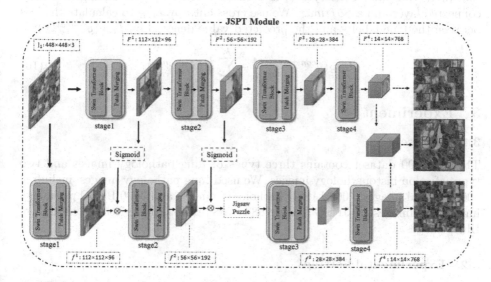

Fig. 5. Illustration of the JSPT module.

Specifically, the same image is fed into two separate branches of the network, representing two learning sessions. In the first learning, the feature map J_1 is divided into non-overlapping chunks of $\frac{H}{s} \times \frac{W}{s}$ by the Patch Partition layer, s is the size of the patch. The chunked image is mapped to the C dimension through the Linear Embedding layer. Then the number of Swin Transformer blocks contained in each stage is (1,1,3,1) through stage1-stage4 in turn, and the first three stages halve the size of the input and increase the depth through the Patch Merging layer, disguised to improve the perception of the patch and window. The first learning has a larger puzzle granularity and captures the larger size of the bubble structure. The difference in the second learning is that the first two stages are supervised by the first two stages of the first learning to exclude a large number of interference regions, which is achieved by incorporating F^1 and F^2 into weights into f^1 and f^2 through the activation function $Sigmoid$. The first two stage feature maps in the second learning are represented by Eq. (8):

$$f^l = Sigmoid(F^l) \otimes f^l \tag{8}$$

where $l = 1, 2$. The smaller granularity of the puzzle J_2 is generated by MPG before entering the third stage. As the network goes deeper, smaller bubble structure can be captured in the latter two stages. Finally, the features learned twice are combined and passed into the classification layer, and the resulting classification results are expressed in Eq. (9):

$$y^{concat} = Classifier(concat(F^4, f^4))$$ (9)

y^{concat} is the result of combining the feature maps learned twice for classification prediction, and $Classifier()$ is a classification layer consisting of a fully connected layer and a $Softmax$. We use cross-entropy L_{CE} to calculate the loss between the true label y and the predicted value denoted by $L_{CE}(y^{concat}, y)$:

$$L_{CE}(y^{concat}, y) = -\sum_{i=1}^{k} y_i log y_i^{concat}$$ (10)

3   Experiment

3.1   Datasets

The LC25000 dataset contains three types of lung pathology images and two types of colon histopathology images. We used lung pathology images, including 5000 Normal images, 5000 LUAD pathology images, and 5000 LUSC pathology images. All images are JPEG files of 768 × 768 pixels and are RGB (Red, Green, Blue) images.

3.2   Experimental Settings

We choose the relatively simple and efficient deep learning framework PyTorch to implement our proposed classification method, and the model is trained and tested in a GPU environment using an NVIDIA GeForce RTX 3080Ti graphics card.

The three types of images of NSCLC in the LC25000 dataset are 15000 in total, and we assign 12000 images to the training set and 3000 images to the validation set and test set. Randomly crop them to 448 × 448 sizes, and Gaussian blur and random level flipping are used to enhance the data. The classification loss is calculated using the cross-entropy loss function, and AdamW is used as the optimizer. The initial learning rate of the model is 0.0001, and the weight decay is 4e−2. The classification performance of the model is evaluated using several evaluation metrics commonly used in medical image classification, including accuracy, precision, recall, and F1-score.

3.3   Parameter Settings

In order to obtain the optimal performance of GMPG module, it is necessary to find the optimal pairing of MPG and GCNF-Net, and we combine their parameters two by two to form 16 configurations. We use the LC25000 dataset and Two-stream Swin Transformer as our backbone by training 100 epochs. From Table 1, we can see that $\eta = 4 \times 4$, $\mu = 6 \times 6$ has the best performance, so we choose the size of the puzzle as 6 × 6 and the number of nodes as 4 × 4.

Table 1. The combination of parameters η and μ of the GMPG module.

η	μ	Accuracy	Precision	Recall	F1-score
3×3	4×4	91.82	92.67	91.83	92.25
	6×6	89.68	91.27	89.67	90.46
	8×8	92.50	92.98	92.50	92.74
	10×10	91.67	91.87	91.67	91.77
4×4	4×4	93.19	93.20	93.17	92.25
	6×6	**95.35**	**95.41**	**95.35**	**95.37**
	8×8	90.66	91.47	90.67	91.07
	10×10	91.68	92.71	91.67	92.18
5×5	4×4	92.20	92.67	92.17	92.42
	6×6	92.32	92.34	92.33	92.33
	8×8	91.20	91.11	91.17	91.14
	10×10	92.70	92.74	92.67	92.71
6×6	4×4	92.34	92.71	92.33	92.52
	6×6	91.82	91.93	91.83	91.88
	8×8	91.80	91.85	91.83	91.84
	10×10	91.17	91.20	91.17	91.18

3.4 Ablation Experiment

We use LC25000 as the dataset and the Two-stream Swin Transformer as the backbone for the ablation experiment. The experimental results are shown in Table 2: The overall performance of backbone is around 91.18%, and when we add MPG to backbone, the overall performance is around 92.78%, which is 1.6% better than our backbone. The overall performance of backbone combined with GCNF-Net is around 93.33%. When we add the GMPG module to backbone, it reaches 95.37%. Afterwards, the puzzle generated by the GMPG module is extended to multiple scales on backbone to form the JSPT module, which in turn forms the final PJGC-Net, with a performance of about 99.0%. The performance was improved by 7.82% compared to backbone.

Table 2. Ablation experiment results.

Method	Accuracy	Precision	Recall	F1-score
Backbone	91.16	91.21	91.17	91.19
Backbone + MPG	92.70	92.93	92.67	92.80
Backbone + GCNF-Net	93.31	93.34	93.33	93.33
Backbone + GMPG module	95.35	95.41	95.35	95.37
PJGC-Net (ours)	**99.0**	**99.01**	**99.0**	**99.01**

To better visualize the effectiveness of our proposed method against NSCLC, we visualized the results using Gram-CAM [18]. The modules enable the network to focus on the areas shown in Fig. 6, and (a) and (b) in the figure are two images of the LUAD. From the figure, we can see that the regions focused on by backbone are confusing and unfocused. Although it does not focus on the feature regions we pointed out when we added GCNF-Net, it can strengthen the network's focus on the cellular regions, but the focus is still not comprehensive. The combination of backbone and MPG can make the network focus on the vesicular results in the figure, but the focus effect is not good. When we use the GMPG module containing GCNF-Net and MPG, the network has been able to focus well on the vesicular regions but also focuses on many redundant regions that interfere with the classification. After adding the JSPT module to form our final PJGC-Net, the network focuses on the feature regions we pointed out and removes the redundant regions.

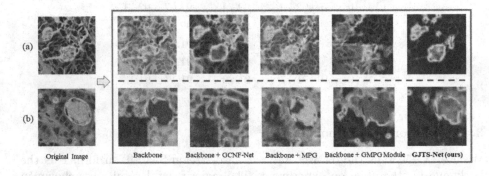

Fig. 6. Comparison of the effects produced by each part of the network.

3.5   Comparisons with Existing Classification Method

By using four different evaluation metrics, we compare the proposed method with existing fine-grained classification methods [12–15], puzzle-based methods [5,19] and classification methods for NSCLC [4,20] trained on the LC25000 dataset with 200 epochs. Table 3 shows their performance comparisons on the validation set, and we can see that our proposed PJGC-Net achieves a high accuracy of 99.0% and an F1-score of 99.01%, which is about 0.12% better than the second-place method, indicating that it outperforms the existing methods. In terms of precision and recall, our method shows a relatively excellent performance. Specifically, it has a precision of 99.01%, which is slightly lower than the PMG method, but a recall of 99.0%, which is an improvement of 0.06% over the second-place method. In addition, PJGC-Net shows good consistency in all evaluation metrics, which indicates that its classification performance is stable and reliable. The above shows that it can not only classify images accurately but also achieve a good balance between accuracy and recall.

Table 3. Comparative results of different methods on the LC25000 dataset (%).

Method	Accuracy	Precision	Recall	F1-score
Bilinear-CNN [13]	80.14	79.21	79.74	79.47
MA-CNN [12]	94.67	94.89	94.66	94.78
MC-Loss [14]	85.53	83.34	83.11	83.22
TDSA-Loss [15]	86.67	87.22	86.23	86.72
PMG [5]	98.83	**99.09**	98.83	98.95
Graph Jigsaw [19]	95.83	96.29	95.83	96.06
ISANET [4]	98.16	98.24	98.17	98.2
Deep MMSA [20]	98.67	98.70	98.67	98.68
PJGC-Net (ours)	**99.0**	99.01	**99.00**	**99.01**

4   Conclusion

In this study, we propose a new image classification method using the Two-stream Swin Transformer network as the main network framework, with the addition of the GMPG module and the JSPT module. The GMPG module can interact with image contextual information and generate multiple scales of images from the same original image for training. The JSPT module allows larger legends to act as supervisors for smaller legends, encouraging them to learn more specific functions. Finally, we merge these images back into the original image to create a new image with enhanced fine-grained features. Our experimental results show that our method significantly outperforms existing image classification methods on the LC25000 dataset.

Although our hybrid model performs well in classification tasks, it requires a lot of computational resources and labeled data. In the future, we will improve the model and explore new techniques to increase performance while reducing computational and data requirements to fit a wider range of application scenarios.

References

1. Subramanian, V., Do, M.N., Syeda-Mahmood, T.: Multimodal fusion of imaging and genomics for lung cancer recurrence prediction. In: 2020 IEEE 17th International Symposium on Biomedical Imaging (ISBI), pp. 804–808. IEEE, Iowa City (2020)
2. Qiu, Z., Bi, J., Gazdar, A.F., Song, K.: Genome-wide copy number variation pattern analysis and a classification signature for non-small cell lung cancer. Genes Chromosom. Cancer **56**, 559–569 (2017)
3. Hou, L., Samaras, D., Kurc, T.M., Gao, Y., Davis, J.E., Saltz, J.H.: Patch-based convolutional neural network for whole slide tissue image classification. In: 2016 IEEE Conference on Computer Vision and Pattern Recognition (CVPR), pp. 2424–2433. IEEE, Las Vegas (2016)

4. Xu, Z., Ren, H., Zhou, W., Liu, Z.: ISANET: non-small cell lung cancer classification and detection based on CNN and attention mechanism. Biomed. Signal Process. Control **77**, 103773 (2022)
5. Du, R., et al.: Fine-grained visual classification via progressive multi-granularity training of jigsaw patches. In: Vedaldi, A., Bischof, H., Brox, T., Frahm, J.-M. (eds.) ECCV 2020. LNCS, vol. 12365, pp. 153–168. Springer, Cham (2020). https://doi.org/10.1007/978-3-030-58565-5_10
6. Velikovi, P., Cucurull, G., Casanova, A., Romero, A., Liò, P., Bengio, Y.: Graph attention networks. arXiv preprint arXiv: 1710.10903 (2017)
7. Kipf, T., Welling, M.: Semi-supervised classification with graph convolutional networks. arXiv preprint arXiv: 1609.02907 (2016)
8. Zhang, T., Chang, D., Ma, Z., Guo, J.: Progressive co-attention network for fine-grained visual classification. In: 2021 International Conference on Visual Communications and Image Processing (VCIP), pp. 1–5. IEEE, Munich (2021)
9. Zheng, M., et al.: Progressive training of a two-stage framework for video restoration. In: 2022 IEEE/CVF Conference on Computer Vision and Pattern Recognition Workshops (CVPRW), pp. 1023–1030. IEEE, New Orleans (2022)
10. Lei, J., Duan, J., Wu, F., Ling, N., Hou, C.: Fast mode decision based on grayscale similarity and inter-view correlation for depth map coding in 3D-HEVC. IEEE Trans. Circuits Syst. Video Technol. **28**, 706–718 (2018)
11. Huang, S., Xu, Z., Tao, D., Zhang, Y.: Part-stacked CNN for fine-grained visual categorization. In: 2016 IEEE Conference on Computer Vision and Pattern Recognition (CVPR), pp. 1173–1182. IEEE, Las Vegas (2016)
12. Zheng, H., Fu, J., Mei, T., Luo, J.: Learning multi-attention convolutional neural network for fine-grained image recognition. In: 2017 IEEE International Conference on Computer Vision (ICCV), pp. 5219–5227. IEEE, Venice (2017)
13. Lin, T.Y., RoyChowdhury, A., Maji, S.: Bilinear CNN models for fine-grained visual recognition. In: 2015 IEEE International Conference on Computer Vision (ICCV), pp. 1449–1457. IEEE, Santiago (2015)
14. Chang, D., et al.: The devil is in the channels: mutual-channel loss for fine-grained image classification. IEEE Trans. Image Process. **29**, 4683–4695 (2020)
15. Chang, D., Zheng, Y., Ma, Z., Du, R., Liang, K.: Fine-grained visual classification via simultaneously learning of multi-regional multi-grained features. arXiv preprint arXiv: 2102.00367 (2021)
16. Liu, Z., et al.: Swin transformer: hierarchical vision transformer using shifted windows. In: 2021 IEEE/CVF International Conference on Computer Vision (ICCV), pp. 9992–10002. IEEE, Montreal (2021)
17. Liu, Z., et al.: Swin Transformer V2: scaling up capacity and resolution. In: 2022 IEEE/CVF Conference on Computer Vision and Pattern Recognition (CVPR), pp. 11999–12009. IEEE, New Orleans (2022)
18. Selvaraju, R.R., Cogswell, M., Das, A., Vedantam, R., Parikh, D., Batra, D.: Grad-CAM: visual explanations from deep networks via gradient-based localization. In: 2017 IEEE International Conference on Computer Vision (ICCV), pp. 618–626. IEEE, Venice (2017)
19. Li, Y., Lao, L., Cui, Z., Shan, S., Yang, J.: Graph jigsaw learning for cartoon face recognition. IEEE Trans. Image Process. **31**, 3961–3972 (2022)
20. Wu, Y., Ma, J., Huang, X., Ling, S.H., Su, S.W.: DeepMMSA: a novel multimodal deep learning method for non-small cell lung cancer survival analysis. In: 2021 IEEE International Conference on Systems. Man, and Cybernetics (SMC), pp. 1468–1472. IEEE, Melbourne (2021)

Improving Transferability of Adversarial Attacks with Gaussian Gradient Enhance Momentum

Jinwei Wang[1,2,3], Maoyuan Wang[1,3], Hao Wu[1,3], Bin Ma[4],
and Xiangyang Luo[2(✉)]

[1] Department of Computer and Software, Nanjing University of Information Science
and Technology, Nanjing 210044, China
[2] State Key Laboratory of Mathematical Engineering and Advanced Computing,
Zhengzhou 450001, China
luoxy_ieu@sina.com
[3] Engineering Research Center of Digital Forensics, Ministry of Education,
Nanjing University of Information Science and Technology, Nanjing 210044, China
[4] School of Cyberspace Security, Qilu University of Technology, Jinan 250353, China

Abstract. Deep neural networks (DNNs) can be susceptible to subtle perturbations that may mislead the model. While adversarial attacks are successful in the white-box setting, they are less effective in the black-box setting. To address this issue, we propose an attack method that simulates a smoothed loss function by sampling from a Gaussian distribution. We calculated the Gaussian gradient to enhance the momentum based on the smoothing loss function to improve the transferability of the attack. Moreover, We further improve transferability by changing the sampling range to make the Gaussian gradient prospective. Our method has been extensively tested through experiments, and the results show that it achieves higher transferability compared to state-of-the-art (SOTA) methods.

Keywords: Deep neural networks · adversarial examples · black-box attack · transferability · Gaussian gradient

1 Introduction

Deep Neural Networks (DNNs) are widely used in computer vision, playing a critical role in fields such as image classification and autonomous driving [7,13]. However, DNNs are vulnerable to imperceptible perturbations that can cause them to misclassify input, known as adversarial examples [4,18]. In recent years, adversarial examples have become an important research topic, as they can help improve the robustness of classification models [9,12].

Adversarial attacks can be categorized as either white-box or black-box attacks depending on whether the attacker has access to information about the targeted model [13,15]. In the white-box setting, the attacker has knowledge of

Original Image　　　MI-FGSM　　　VMI-FGSM　　　GMI-FGSM(ours)

Fig. 1. Comparison of the image with the original image after adding perturbations ($\epsilon = 16$). With the same image quality as MI-FGSM and VMI-FGSM, our method GMI-FGSM black box attack has a higher success rate.

the targeted model's parameters and framework and can craft adversarial examples accordingly [12]. There are several effective white-box attacks, including the gradient-based fast gradient sign method (FGSM) and its iterative version (I-FGSM) [4,9], as well as the C&W attack [1]. Black-box attacks are a realistic scenario where the attacker has no knowledge of the target model [15]. Improving the transferability of white-box attacks is one of the primary methods to perform black-box attacks [13,15,25,26]. There are two main categories of attacks used to improve the transferability of attacks on the model. The first category involved momentum-based gradient optimization methods, including basic momentum MI-FGSM [2], Nesterov accelerated gradient-based NI-FGSM [11], and variance-tuning VMI-FGSM [20]. The second category involved enhancing the transferability of adversarial examples by transforming the input, such as DI-FGSM [24], TI-FGSM [3], SI-FGSM [11].

Dong's MI-FGSM study concluded that poor local maxima during gradient ascent negatively affect the transferability of adversarial examples [2]. The iterative version of FGSM (I-FGSM) is more prone to falling into poor local maxima than the one-step FGSM, which reduces its transferability [4,9]. Although the black-box attack success rate of FGSM is higher than that of I-FGSM, its white-box attack success rate is lower. To tackle this issue, MI-FGSM proposes a momentum-based approach that stabilizes the update direction of adversarial examples, escaped poor local maxima, and improved transferability [2]. Nesterov acceleration is implemented in NI-FGSM [11] to efficiently look ahead, escape poor local maxima faster, and improve transferability. VMI-FGSM improved transferability by uniformly distributing the variance obtained through sampling, which enables a more optimal update direction and helps escape local maxima [20].

In this paper, we propose an attack method that simulates a smoothed loss function by sampling from a Gaussian distribution and calculates the gradient-enhanced momentum of the smoothed loss function. Previous methods of attack by Gaussian sampling have simply replaced the original point gradient with the gradient of the sampled points [21]. The Gaussian sampling-based smoothing operation allows non-convex loss functions to be approximated as convex functions after smoothing, which facilitates gradient ascent (descent) [14]. Therefore, the gradient direction calculated on the smoothed loss function can more effectively escape poor local maxima and find a more reliable update direction. For convenience, we refer to the gradient calculated on the smoothed loss function as the Gaussian gradient. Meanwhile, we improve the transferability by moving the sampling range by changing the sampling centroid to make the Gaussian gradient perspective. Compared to other methods, our method using Gaussian gradient smoothing can find better optimization directions and more effectively escape poor local maxima, leading to improved transferability of the generated adversarial examples. Figure 1 shows the adversarial examples for various attack methods at the same perturbation level ($\epsilon = 16$). The quality of the image (PSNR) is similar for the same perturbation size [24]. The experimental results demonstrate that we have higher transferability compared to the same type of methods without affecting the visual effect of the image.

2   Related Work

2.1   Adversarial Attacks

Deep neural networks are susceptible to small perturbations, and adversarial examples exploit this feature for adversarial attacks. Specifically, adversarial attacks are divided into two classes: white-box and black-box attacks. An attack in which the attacker knows in advance and exploits the specific structure and parameters of the attacked neural network is a white-box attack. On the contrary, the attacked model structure and parameters are not available in a black-box attack. Black-box attacks are more adaptive compared to white-box attacks, and therefore more suitable for real-world applications, and correspondingly more difficult to succeed.

Improving adversarial example transferability is one of the directions of black-box attacks [13]. Among the approaches to attack classification neural network models, they can be broadly classified into two classes, some methods based on gradient optimization [2,11,20] and methods based on input transformation [3, 11,24]. The input transformation approach is to preprocess the images entering the model before the attack to scramble the image features. Gradient-based methods basically stabilize the gradient update direction by optimization to prevent overfitting. In this paper, we are improving the method based on the optimized gradient.

2.2 Defend Against Adversarial Attacks

The adversarial defense topic has gradually become popular with the emergence of various methods for adversarial attacks [4,8]. At present, adversarial defense is mainly divided into two classes, one is adversarial training [19], in which the adversarial examples are also entered into the training process as a dataset during the training of the model, so that the adversarially trained model learns to resist perturbations in the gradient direction of the loss function, allowing the model to be more robust [22]. The other class is preprocessing before entering the model to remove most of the unfavorable perturbations so that the model has a lower probability of misclassifying the adversarial examples [5,10,23].

3   Methodology

Given a classifier $f(x;\theta)$ with output prediction, θ is the parameter, x is the input image and y is the output result. The goal of adversarial attacks is to seek an example x^{adv} in the vicinity of x but is misclassified by the classifier. x^{adv} satisfies $\left|\left|x - x^{adv}\right|\right|_p \leq \epsilon$ and $f(x;\theta) \neq f(x^{adv};\theta)$. Here p denotes the L_p norm and we focus on $p = \infty$.

3.1   Gaussian Gradient Enhanced Momentum

Our main objective is to obtain a more efficient update direction, which we achieve by simulating a smoother loss function through Gaussian sampling. This generates a Gaussian gradient that enhances momentum and enables us to get rid of local maxima, ultimately improving the transferability of the attack. The main idea is to blur the original loss function in order to construct smoother loss functions. Smoothing allows some non-convex functions to be approximated as convex, making it easier to perform gradient ascent (descent) [14]. This simulated process can be approximated through sampling, and it has been found that using Gaussian distribution sampling yields better results. In the Extended Method [14] the smoothing theory of Gaussian distribution sampling is formulated as Eq. 1:

$$J_f\left(x_t{}^{adv}, y, \theta\right) = E_{x \sim N((x_t{}^{adv}, \sigma^{(i)2})} J(x_t{}^{adv}, y, \theta). \tag{1}$$

Equation 1 can be approximated as:

$$J_f\left(x_t{}^{adv}, y, \theta\right) \approx \frac{1}{N} \sum_{i=1}^{n} J_f\left(x_i, y, \theta\right), \tag{2}$$

$$\nabla_{x_t{}^{adv}} J_f\left(x_t{}^{adv}, y, \theta\right) = \frac{\partial J_f\left(x_t{}^{adv}, y, \theta\right)}{\partial x_t{}^{adv}}$$

$$\approx \frac{1}{N} \sum_{i=1}^{n} \frac{\partial J_f\left(x_i, y, \theta\right)}{\partial x_t{}^{adv}} \tag{3}$$

$$\approx \frac{1}{N} \sum_{i=1}^{n} \nabla_{x_i} J_f\left(x_i, y, \theta\right).$$

Here $x_i = x_t^{adv} + r_i$, $r_i \sim N(0, \sigma^2)^d$, $N(0, \sigma^2)^d$ stands for the Gaussian distribution in d dimensions, σ is the variance of the Gaussian distribution. We define the Gaussian gradient as e_t, which can be calculated as Eq. 4.

$$e_t = E\left(x_t^{adv}\right) = \frac{1}{N} \sum_{i=1}^{n} \nabla_{x_i} J_f\left(x_i, y, \theta\right). \tag{4}$$

Next, we use a Gaussian gradient to enhance momentum, which helps us obtain a more efficient update direction and improve transferability, as in Eq. 5.

$$g_t = \mu \cdot g_{t-1} + \frac{\nabla_{x_t^{adv}} J_f\left(x_t^{adv}, y, \theta\right)}{\left\|\nabla_{x_t^{adv}} J_f\left(x_t^{adv}, y, \theta\right)\right\|_1} + \omega \cdot e_t. \tag{5}$$

Here x_t^{adv} is the input image of the t-th iteration, and ω is the weight factor of the Gaussian gradient.

3.2 Enhanced Prospective

The impact of prospective on adversarial examples is demonstrated in Lin's NI-FGSM [11], which uses anticipatory updates to provide the previously accumulated gradient with a correction that enables it to effectively look ahead. Building on this idea, we have improved our sampling operation. Specifically, we make the Gaussian gradient prospective by deviating the centroid of each sampling forward. This prospective can help the loss function value escape from poor local maxima more easily and quickly, resulting in improved transferability. The Gaussian gradient direction is more reliable, and we choose the direction of the last Gaussian gradient for the centroid deviation process, as shown in Eq. 6.

$$x_{center}^{adv} = x_t^{adv} + e_{t-1} \cdot \beta, \tag{6}$$

We are calculating the gradient of the current data point as well as the Gaussian gradient e_t, using the newly determined centroid x_{center}^{adv}. The algorithm of Gaussian smoothing loss function and enhanced prospective MI-FGSM denoted as GMI-FGSM is summarized in Algorithm 1. We can easily extend GMI-FGSM to Gaussian smoothing loss function and enhanced prospective NI-FGSM (GNI-FGSM).

Algorithm 1. GMI-FGSM

Require: A classifier f with loss function J; an original image x and ground-truth label y;

Require: The size of perturbation ϵ; iterations T and decay factor μ.

Require: The variance of Gaussian distribution σ,the distribution centroid distance parameter β and number of example N for Gaussian smoothing.

Ensure: An adversarial example x^{adv} with $\|x - x^{adv}\|_\infty \leq \epsilon$.

1: $\alpha = \epsilon/T$;
2: $g_0 = 0$; $x_0^{adv} = x$;
3: **for** $t = 0$ to $T - 1$ **do**
4:    Input x_t^{adv} calculate x_{center}^{adv} by Eq 6.
5:    Calculate the gradient $\nabla_{x_{center}^{adv}} J_f(x_{center}^{adv}, y, \theta)$.
6:    Update $e_t = E(x_{center}^{adv})$ by Eq 4.
7:    Update g_{t+1} by Gaussian smoothing

$$g_{t+1} = \mu \cdot g_t + \frac{\nabla_{x_{center}^{adv}} J_f(x_{center}^{adv}, y, \theta)}{\|\nabla_{x_{center}^{adv}} J_f(x_{center}^{adv}, y, \theta)\|_1} + \omega \cdot e_t. \qquad (7)$$

8:    Update x_{t+1}^{adv} by applying the sign gradient as

$$x_{t+1}^{adv} = \boldsymbol{x}_t^{adv} + \alpha \cdot \text{sign}(g_{t+1}). \qquad (8)$$

9: **end for**return $x^{adv} = x_T^{adv}$.

---

4   Experiments

4.1   Experimental Setup

Models and Dataset. To evaluate the performance of the proposed method, we chose six classification models, including Inception-v3 (Inc-v3) [17], Inception-v4 (Inc-v4), and Inception-Resnet-v2 (Incrs-v2) [16], Resnet-152 (Res-152) [6] and two adversarially trained models, namely Inc-v3$_{adv}$ and IncRes-v2$_{ens}$ [19]. For the dataset used in the experiments, we randomly selected 1000 clean images from the ILSVRC 2012 validation set. Those images were selected to be correctly classified by all tested models in order to accurately represent the effectiveness of the attack method.

Attack Methods and Hyper-parameters. We use MI-FGSM, NI-FGSM, VMI-FGSM, and VNI-FGSM as comparison algorithms. The variance-tuning VMI-FGSM is our baseline.

We follow the attack setting in [6] with the maximum perturbation of $\epsilon = 16$, number of iteration $T = 10$, and step size $\alpha = 1.6$. For MI-FGSM and NI-FGSM, we set the decay factor $\mu = 1.0$. For VMI-FGSM, we set the number of sampling points to $N = 20$ and the uniform distribution bound $\mu = 1.5$. For the proposed method, based on our experiments, we have set our parameters to $N = 200$, $\sigma = 2$, $\omega = 1.6$, and $\beta = 8.0$.

Table 1. The success rates (%) of non-targeted adversarial attacks against the six models we study. The adversarial examples are produced against Inc-v3, Inc-v4, IncRes-v2, and Res-152 using MI-FGSM, NI-FGSM, VMI-FGSM, VNI-FGSM, GMI-FGSM, and GNI-FGSM, respectively. * indicates the white-box attacks.

Model	Attack	Inc-v3	Inc-v4	IncRes-v2	Res-152	Inc-v3$_{adv}$	Incres-v2$_{ens}$
Inc-v3	MI-FGSM	98.8*	43.4	38.2	28.1	29.6	13.0
	NI-FGSM	99.1*	54.1	46.2	39.3	29.4	14.2
	VMI-FGSM	99.1*	61.1	56.1	45.5	40.0	22.8
	VNI-FGSM	99.5*	67.6	61.8	47.9	39.0	21.6
	GMI-FGSM (ours)	99.6*	83.2	80.2	66.1	**63.8**	46.1
	GNI-FGSM (ours)	99.9*	84.6	81.2	67.1	63.7	**48.1**
Inc-v4	MI-FGSM	40.6	98.1*	34.8	25.8	28.6	13.2
	NI-FGSM	43.9	99.1*	40.1	27.5	29.7	12.9
	VMI-FGSM	65.7	98.9*	57.0	43.0	38.1	24.3
	VNI-FGSM	68.7	99.6*	61.5	45.0	40.2	24.9
	GMI-FGSM (ours)	80.6	99.6*	74.6	54.7	**55.7**	43.1
	GNI-FGSM (ours)	81.7	99.8*	76.1	56.3	55.7	**44.8**
IncRes-v2	MI-FGSM	44.0	46.4	98.4	27.3	33.9	20.7
	NI-FGSM	44.5	48.9	99.3*	29.8	32.6	19.4
	VMI-FGSM	67.8	64.7	98.9*	44.8	45.0	36.6
	VNI-FGSM	70.0	68.0	99.5*	48.1	45.6	38.1
	GMI-FGSM (ours)	80.0	79.8	99.5*	58.9	65.4	58.2
	GNI-FGSM (ours)	81.4	80.1	99.9*	59.8	66.8	59.6
Res-152	MI-FGSM	48.0	41.8	34.3	99.9*	28.0	15.1
	NI-FGSM	52.6	44.3	37.2	99.9*	27.8	15.5
	VMI-FGSM	67.7	59.8	54.8	100.0*	39.4	25.6
	VNI-FGSM	71.0	65.4	57.1	100.0*	39.1	26.4
	GMI-FGSM (ours)	82.0	80.0	78.0	100.0*	64.1	56.9
	GNI-FGSM (ours)	83.6	81.5	78.6	100.0*	66.8	59.8

4.2 Attacking Normally Trained Model and Adversarially Trained Model

We first perform the six attack methods under a single normally trained neural network, and the generated adversarial examples are tested on the other three normally trained models, and two adversarial-trained models to test the transferability of attacks. The test results are shown in Table 1. The columns in the table are the attack model and the behavior test model.

Table 1 demonstrates that our attacks have a significantly higher success rate in black-box attacks than our competitive benchmark. For example, the adversarial examples generated by GNI-FGSM at Inc-v3 for the attack model all achieve 84.6%, 81.2%, and 67.1% attack success rates for the other three normally trained models, respectively. This is higher than the 67.6%, 61.8%, and 47.9% of our baseline method VNI-FGSM. The advantage of our method is more obvious in the adversarial training model, with 63.7% and 48.1% for GNI-FGSM compared to 39.0% and 21.6% for the baseline method VNI-FGSM. In short, our method leads to higher transferability.

Our experimental results demonstrate that the combination of our proposed Gaussian gradient enhanced momentum and enhanced prospective can more

Table 2. The attack success rates (%) of various gradient-based iterative attacks with or without DTS under the multi-model setting. The best results are marked in bold.

Model	Attack	Inc-v3	Inc-v4	IncRes-v2	Res-152	Inc-v3$_{adv}$	IncRes-v2$_{ens}$
Incv3	MI-DTS-FGSM	97.3	76.1	68.5	52.1	66.1	54.4
	VM-DTS-FGSM	99.7*	84.6	81.6	68.0	59.7	46.6
	GMI-DTS-FGSM(ours)	**99.8***	**90.3**	**90.9**	**73.4**	**76.6**	**66.0**
Incv4	MI-DTS-FGSM	75.3	96.1	71.8	46.4	50.6	38.0
	VM-DTS-FGSM	80.6	94.7*	77.1	61.6	52.9	43.1
	GMI-DTS-FGSM(ours)	**92.6**	**99.8***	**90.3**	**72.9**	**72.4**	**62.9**
IncRes-v2	MI-DTS-FGSM	54.8	55.9	87.6	35.5	47.8	40.4
	VM-DTS-FGSM	74.0	73.7	96.0*	51.0	53.2	46.1
	GMI-DTS-FGSM(ours)	**81.0**	**81.7**	**99.1***	**62.8**	**74.1**	**67.7**
Res-152	MI-DTS-FGSM	84.0	81.7	80.5	99.8	63.0	58.3
	VM-DTS-FGSM	90.3	88.5	84.7	99.9*	61.5	56.1
	GMI-DTS-FGSM(ours)	**96.0**	**94.5**	**93.8**	**100***	**82.6**	**80.6**

effectively escape poor local maxima and significantly improve the transferability of adversarial examples, while significantly outperforming SOTA attacks.

4.3 Attacking with Combined Input Transformations

Three popular input transformation methods, DIM, TIM, and SIM, have been selected to be combined with the gradient-based method. For the momentum-based method MI-FGSM, it is combined with the three input transformations DIM, TIM, and SIM, reflecting the improvement of the transferability of these three input transformation methods. There are experiments combining our baseline VMI-FGSM with this kind of input transformation method; and finally, the experiments combining our proposed method with the three input transformations.

As can be seen in Table 2, the attack method of optimized gradient and the attack method of input transformation can be naturally combined to improve the transferability. For the adversarial examples generated on the Inc-v3 model, the black-box attack success rate of our method on the normally trained model outperforms that of VMI-FGSM, and the attack success rate of our proposed method on the adversarially trained model is more than 7% higher than that of the other three methods. This convincing data performance fully demonstrates the effectiveness of our proposed method, which still outperforms the competitive benchmarks in combination with the attack methods for input transformation.

4.4 Ablation Study

In the ablation study of the number of sampling points, we calculated the Gaussian gradient in combination with the momentum-based MI-FGSM, and our method degenerates to MI-FGSM when $N = 0$. The number of sampling points N is positively correlated with the overall transferability, as shown in Fig. 2(a), and there is almost no significant change in transferability when the number of sampling points is between 200 and 500. However, an excessive number of sampling points can lead to information saturation and increase computational complexity. Therefore, we set $N = 200$ to achieve a trade-off between complexity and transferability.

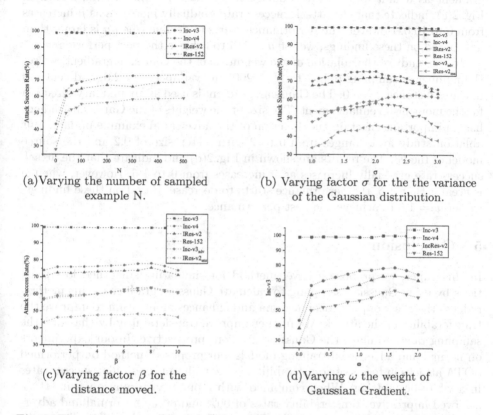

(a) Varying the number of sampled example N.

(b) Varying factor σ for the the variance of the Gaussian distribution.

(c) Varying factor β for the distance moved.

(d) Varying ω the weight of Gaussian Gradient.

Fig. 2. The success rates (%) with different hyper-parameters on the different models with adversarial examples generated by GMI-FGSM on the Inc-v3 model.

In the ablation study on the variance of the Gaussian distribution σ, we keep the number of sampling points fixed at $N = 200$. Variance is a crucial parameter for adjusting the spread of the Gaussian distribution and taking more or fewer data points in certain ranges. As expected from the nature of the Gaussian distribution, the sampling points are more widely dispersed when the variance is large and more tightly concentrated when the variance is small. Figure 2(b) shows the attack success rate of our proposed method for different values of σ. We observe that the attack success rate gradually increases as σ increases from 1.0 to 2.0. However, the performance begins to decrease when σ exceeds 2.0. Therefore, we set $\sigma = 2.0$ to achieve optimal performance.

In the ablation study of the distribution, centroid deviates distance μ, we set the number of sampling points to $N = 200$ and variance $\sigma = 2.0$. The location of the centroid of the distribution plays an important role in the prospective of the Gaussian gradient. To select the deviation direction, we use the direction of the previous Gaussian gradient e_t. We perform a simultaneous ablation experiment to study the effect of the deviation distance. We take the previous Gaussian gradient as a unit and vary the deviation distance μ. The results, shown in Fig. 2(c), indicate that the attack success rate gradually increases as μ increases from 0 to 10.0. However, the performance starts to decline when μ is larger than 8.0. Based on these findings, we set $\mu = 8.0$ to achieve the best performance.

In the study of the ablation of the weights ω of the Gaussian gradient, we set the number of sampling points to $N = 200$, the variance $\sigma = 2.0$, and centroid deviates distance $\beta = 8.0$. The Gaussian gradient is used as an auxiliary direction for the momentum enhancement. The size of the weights of the Gaussian gradient has an important effect on the direction of the adversarial example update. Our ablation study for ω ranges from 0 to 2 with a step size of 0.2 and the source model of Inc-v3. The results are shown in Fig. 2(d) and indicate that the attack success rate gradually increases as ω increases from 0 to 1.6. However, when ω is greater than 1.6, the performance starts to decrease. Based on these findings, we set $\omega = 1.6$ to achieve the best performance.

5   Conclusion

In this paper, we propose a novel method for simulating smoothed loss functions by using Gaussian sampling to calculate Gaussian gradients. This method reduces the effect of poor local maxima and enhances momentum to improve the transferability of the attack. We further improve transferability by changing the sampling range to make the Gaussian gradient prospective. In our experiments on normal and adversarial training models, our proposed method outperformed SOTA attacks in terms of transferability, while maintaining similar success rates in a white-box setting. When combined with input transformation, our attack achieved impressive transferability rates of 90% and 80% for normal and adversarially trained models, respectively. The use of momentum-based gradient optimization is gaining popularity in the black-box attack community and can be effectively combined with other methods, such as input transformation and feature attacks. We believe our method can make a significant contribution to the field of adversarial attacks and defenses.

References

1. Carlini, N., Wagner, D.: Towards evaluating the robustness of neural networks. In: 2017 IEEE Symposium on Security and Privacy (SP), pp. 39–57. IEEE (2017)
2. Dong, Y., et al.: Boosting adversarial attacks with momentum. In: Proceedings of the IEEE Conference on Computer Vision and Pattern Recognition, pp. 9185–9193 (2018)
3. Dong, Y., Pang, T., Su, H., Zhu, J.: Evading defenses to transferable adversarial examples by translation-invariant attacks. In: Proceedings of the IEEE/CVF Conference on Computer Vision and Pattern Recognition, pp. 4312–4321 (2019)

4. Goodfellow, I.J., Shlens, J., Szegedy, C.: Explaining and harnessing adversarial examples. arXiv preprint arXiv:1412.6572 (2014)
5. Guo, C., Rana, M., Cisse, M., Van Der Maaten, L.: Countering adversarial images using input transformations. arXiv preprint arXiv:1711.00117 (2017)
6. He, K., Zhang, X., Ren, S., Sun, J.: Identity mappings in deep residual networks. In: Leibe, B., Matas, J., Sebe, N., Welling, M. (eds.) ECCV 2016. LNCS, vol. 9908, pp. 630–645. Springer, Cham (2016). https://doi.org/10.1007/978-3-319-46493-0_38
7. Krizhevsky, A., Sutskever, I., Hinton, G.E.: ImageNet classification with deep convolutional neural networks. Commun. ACM 60(6), 84–90 (2017)
8. Kurakin, A., Goodfellow, I., Bengio, S.: Adversarial machine learning at scale. arXiv preprint arXiv:1611.01236 (2016)
9. Kurakin, A., Goodfellow, I.J., Bengio, S.: Adversarial examples in the physical world. In: Artificial Intelligence Safety and Security, pp. 99–112. Chapman and Hall/CRC (2018)
10. Liao, F., Liang, M., Dong, Y., Pang, T., Hu, X., Zhu, J.: Defense against adversarial attacks using high-level representation guided denoiser. In: Proceedings of the IEEE Conference on Computer Vision and Pattern Recognition, pp. 1778–1787 (2018)
11. Lin, J., Song, C., He, K., Wang, L., Hopcroft, J.E.: Nesterov accelerated gradient and scale invariance for adversarial attacks. arXiv preprint arXiv:1908.06281 (2019)
12. Liu, A., et al.: Perceptual-sensitive GAN for generating adversarial patches. In: Proceedings of the AAAI Conference on Artificial Intelligence, vol. 33, pp. 1028–1035 (2019)
13. Liu, Y., Chen, X., Liu, C., Song, D.: Delving into transferable adversarial examples and black-box attacks. arXiv preprint arXiv:1611.02770 (2016)
14. Mobahi, H., Fisher III, J.W.: A theoretical analysis of optimization by Gaussian continuation. In: Twenty-Ninth AAAI Conference on Artificial Intelligence (2015)
15. Papernot, N., McDaniel, P., Goodfellow, I., Jha, S., Celik, Z.B., Swami, A.: Practical black-box attacks against machine learning. In: Proceedings of the 2017 ACM on Asia Conference on Computer and Communications Security, pp. 506–519 (2017)
16. Szegedy, C., Ioffe, S., Vanhoucke, V., Alemi, A.A.: Inception-v4, Inception-ResNet and the impact of residual connections on learning. In: Thirty-First AAAI Conference on Artificial Intelligence (2017)
17. Szegedy, C., Vanhoucke, V., Ioffe, S., Shlens, J., Wojna, Z.: Rethinking the inception architecture for computer vision. In: Proceedings of the IEEE Conference on Computer Vision and Pattern Recognition, pp. 2818–2826 (2016)
18. Szegedy, C., et al.: Intriguing properties of neural networks. arXiv preprint arXiv:1312.6199 (2013)
19. Tramèr, F., Kurakin, A., Papernot, N., Goodfellow, I., Boneh, D., McDaniel, P.: Ensemble adversarial training: attacks and defenses. arXiv preprint arXiv:1705.07204 (2017)
20. Wang, X., He, K.: Enhancing the transferability of adversarial attacks through variance tuning. In: Proceedings of the IEEE/CVF Conference on Computer Vision and Pattern Recognition, pp. 1924–1933 (2021)
21. Wu, L., Zhu, Z.: Towards understanding and improving the transferability of adversarial examples in deep neural networks. In: Asian Conference on Machine Learning, pp. 837–850. PMLR (2020)
22. Wu, W., Su, Y., Lyu, M.R., King, I.: Improving the transferability of adversarial samples with adversarial transformations. In: Proceedings of the IEEE/CVF Conference on Computer Vision and Pattern Recognition, pp. 9024–9033 (2021)

23. Xie, C., Wang, J., Zhang, Z., Ren, Z., Yuille, A.: Mitigating adversarial effects through randomization. arXiv preprint arXiv:1711.01991 (2017)
24. Xie, C., et al.: Improving transferability of adversarial examples with input diversity. In: Proceedings of the IEEE/CVF Conference on Computer Vision and Pattern Recognition, pp. 2730–2739 (2019)
25. Zhang, J., et al.: Improving adversarial transferability via neuron attribution-based attacks. In: Proceedings of the IEEE/CVF Conference on Computer Vision and Pattern Recognition, pp. 14993–15002 (2022)
26. Zhao, Z., et al.: SAGE: steering the adversarial generation of examples with accelerations. IEEE Trans. Inf. Forensics Secur. **18**, 789–803 (2023)

Boundary Guided Feature Fusion Network for Camouflaged Object Detection

Tianchi Qiu, Xiuhong Li$^{(\boxtimes)}$, Kangwei Liu, Songlin Li, Fan Chen,
and Chenyu Zhou

School of Information Science and Engineering, Xinjiang University, Xinjiang, China
qiutc@stu.xju.edu.cn, xjulxh@xju.edu.cn

Abstract. Camouflaged object detection (COD) refers to the process of detecting and segmenting camouflaged objects in an environment using algorithmic techniques. The intrinsic similarity between foreground objects and the background environment limits the performance of existing COD methods, making it difficult to accurately distinguish between the two. To address this issue, we propose a novel Boundary Guided Feature fusion Network (BGF-Net) for camouflaged object detection in this paper. Specifically, we introduce a Contour Guided Module (CGM) aimed at modeling more explicit contour features to improve COD performance. Additionally, we incorporate a Feature Enhancement Module (FEM) with the goal of integrating more discriminative feature representations to enhance detection accuracy and reliability. Finally, we present a Boundary Guided Feature Fusion Module (BGFM) to boost object detection capabilities and perform camouflaged object predictions. BGFM utilizes multi-level feature fusion for contextual semantic mining, we incorporate the edges extracted by the CGM into the fused features to further investigate semantic information related to object boundaries. By adopting this approach, we are able to better integrate contextual information, thereby improving the performance and accuracy of our model. We conducted extensive experiments, evaluating our BGF-Net method using three challenging datasets.

Keywords: Camouflaged object detection · Contextual information · Feature fusion · Feature enhancement

1 Introduction

Camouflaged object detection aims to detect objects in images that are camouflaged. Camouflaged objects typically refer to objects with similar background colors, textures, or shapes, making them difficult to be detected by the naked eye or traditional object detection algorithms. Camouflaged objects can be roughly

---

Supported by the Xinjiang Natural Science Foundation (No. 2020D01C026).

Q. Liu et al. (Eds.): PRCV 2023, LNCS 14433, pp. 433–444, 2024.
https://doi.org/10.1007/978-981-99-8546-3_35

divided into two types: natural camouflaged objects and artificial camouflaged objects. Either naturally or artificially, this technology has applications in military, agriculture, and medical imaging. Currently, the scientific value and applications of camouflaged object detection have attracted the attention and exploration of numerous researchers.

Traditional COD algorithms [16] have used handcrafted features for detection and segmentation, but recent advances in deep learning have led to the development of more sophisticated methods [11]. For instance, the design of the SINet [7] framework is inspired by the first two stages of the hunting process. BGNet [2] guides COD representation learning by analyzing the edge semantics between valuable and relevant objects. Despite the significant progress made by these recently proposed COD methods, there are still some drawbacks. Existing methods are often affected by factors such as edge disruption or object contour blurring, making it difficult for them to accurately identify object structures and details, resulting in predictions with rough boundaries. Moreover, there is still considerable room for improvement in COD. First, some COD methods only extract high-level global information while ignoring low-level local edge information; enhancing the extraction of object edge information can help improve COD performance. Second, feature extraction for camouflaged objects is one of the main challenges in COD, and further innovation in feature enhancement is needed. Lastly, the fusion of multi-level features is a crucial step in detecting camouflaged objects; exploring how to aggregate multi-level features for top-down fusion to predict camouflaged objects is also our goal.

In this paper, we first introduce a novel Boundary Guided Feature fusion Network (BGF-Net) to better model and perceive camouflaged objects. We introduce a Contour-Guided Module (CGM) that integrates low-level local edge information and high-level global position information, resulting in a more explicit contour representation. Furthermore, we employ a Feature Enhancement Module (FEM) to integrate more discriminative feature representations, guiding the COD representation learning. Third, we propose a Boundary Guided Feature fusion Module (BGFM), BGFM utilizes multi-level feature fusion for contextual semantic mining, enabling better integration of contextual information and improving model performance and accuracy. Finally, we incorporate the edges extracted by the CGM into our BGFM to further investigate semantic information related to object boundaries, guiding and strengthening COD representation learning. Our main contributions are as follows:

- We propose a novel Boundary Guided Feature fusion Network (BGF-Net), to tackle the COD task. This network comprises the Contour-Guided Module (CGM), Feature Enhancement Module (FEM), and Boundary Guided Feature fusion Module (BGFM). Our network exploits and aggregates various boundary semantic information and further employs edge cues to reinforce feature learning for COD.
- We design a Contour-Guided Module (CGM) that retains multi-level low-level edge information and high-level global position information while employing a

series of convolutions to model a more explicit contour representation, thereby improving COD performance.

- We propose a Boundary Guided Feature fusion Module (BGFM), which uses edges extracted by the CGM to explore semantic information related to object boundaries, guiding and strengthening COD representation learning.

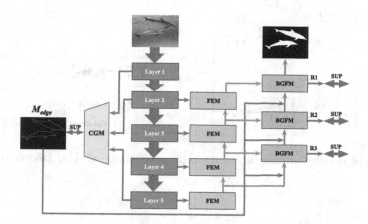

Fig. 1. This article introduces the overall structure of BGF-Net, which employs Res2Net as its backbone network and combines three key modules: CGM, FEM [18], and BGFM.

2 Our Method

2.1 Overall Architecture

The BGF-Net structure proposed in this paper is shown in Fig. 1. We adopt Res2Net-50 [10] as the backbone network and first input the RGB image into the backbone network to extract multi-level features, denoted as f_i ($i = 1,2,...,5$). According to research [26], low-level features from shallow convolutional layers retain the spatial information required to construct object edges, while high-level features from deep convolutional layers contain more semantic information about localization. Therefore, we design a Contour Guided Module (CGM). In this module, we combine low-level features (f_1, f_2) with high-level features (f_5) to model object-related edge information. By leveraging the rich semantic information in high-level features, we assist in mining edge information from low-level features, thus obtaining the boundary map M_{edge}. To acquire reliable multi-scale joint perceptual domain information, we employ a Feature Enhancement Module (FEM). By using convolutional kernels of different sizes and various dilation rates, we provide information at different scales. We input the backbone features f_i into the FEM module to obtain enhanced features X_i ($i = 2,3,4,5$). Finally, we use four Boundary Guided Feature fusion Modules (BGFM) to inject edge clues related to boundaries into X_i, enhancing the feature representation with object structure semantics. We continuously increase the accuracy of camouflaged object detection by iteratively fusing cross-level features. In testing, we choose the prediction from the last BGFM as the final result.

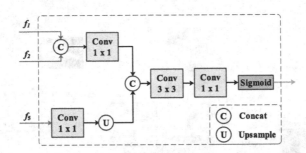

Fig. 2. The detailed architecture of the proposed Contour Guided Module (CGM).

2.2 Contour Guided Module

Edge priors play a positive role in object detection for segmentation and localization tasks [24]. Although low-level features contain rich edge information, they also introduce numerous non-object edges. Therefore, high-level semantic or spatial information is needed to drive the research of object-related edge features in camouflage. In this module, we combine low-level features (f_1, f_2) with high-level features (f_5) to model object-related edge information, as shown in Fig. 2. Specifically, we first concatenate f_1 and f_2 to obtain f_{12}. Then, we use two 1 $\times$ 1 convolutional layers to adjust the channel numbers of f_{12} and f_5, resulting in 64 (f_{12}) and 256 (f_5) channels, respectively. Next, we integrate the features f_{12} with the upsampled f_5 through a cascading operation. Finally, after passing through a 3 $\times$ 3 convolutional layer and a 1 $\times$ 1 convolutional layer, we use a Sigmoid function to obtain the boundary map M_{edge}. The process is defined as follows:

$$f_{12} = B_{\text{Conv}\,1\times1}\left(\text{Concat}\left(f_1, f_2\right)\right) \tag{1}$$

$$f_{125} = \text{Concat}\left(f_{12}, U\left(B_{\text{Conv}\,1\times1}\left(f_5\right)\right)\right) \tag{2}$$

$$M_{edge} = \text{Sig}\left(B_{\text{Conv}\,1\times1}\left(B_{\text{Conv}\,3\times3}\left(f_{125}\right)\right)\right) \tag{3}$$

where $B_{\text{Conv}k\times k}$ denotes a convolutional kernel of size k, stride of 1, padding of $\frac{k-1}{2}$, and finally batch normalized. $\text{Sig}(\cdot)$ denotes Sigmoid function. $Concat(\cdot)$ denotes concatenation operation, U denotes upsampling.

2.3   Boundary Guided Feature Fusion Module

According to the research paper [9], multi-level feature fusion can effectively improve the accuracy of object detection. Therefore, we introduce a boundary-guided feature fusion module based on the global context module of BBS-Net, as shown in Fig. 3. We consider the semantic correlation between different branches and scale features, and achieve the fusion of features at various levels through boundary guidance to enhance the expression ability of boundary features. Taking X_4 and X_5 as examples, we first upsample X_5, then concatenate it with X_4 along the first dimension, and connect them through a 1×1 convolution layer to obtain the initial aggregated feature f_m. Subsequently, we divide f_m along the channel dimension into four feature maps $(f_m^1, f_m^2, f_m^3, f_m^4)$ and perform cross-scale interactive learning, integrating neighboring branch features through a series of dilated convolutions to extract multi-scale contextual features. This process can be summarized as follows:

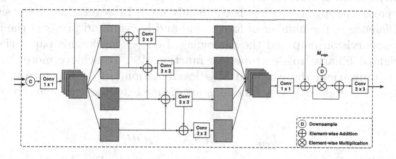

Fig. 3. The detailed architecture of the proposed Boundary Guided Feature Fusion Module(BGFM), "M" represents the extracted "M_{edge}" in CGM.

$$f_m^{j'} = \mathbf{F}_{conv}^{n_j} \left(f_m^{j-1'} \oplus f_m^j \oplus f_m^{j+1} \right), j \in \{1,2,3,4\} \tag{4}$$

In this context, $F_{conv}^{n_j}$ represents a 3×3 dilated convolution with a dilation rate of n_j. In the experiments, we set $n_j = (1,2,3,4)$. Additionally, when $i = 1$ there are only f_m^1 and f_m^2; when $i = 4$, there are only f_m^4 and $f_m^{3'}$. Next, we concatenate these four multi-scale features $f_m^{j'}$, followed by a 1×1 convolution, residual connection, and a 3×3 convolution operation, which can be represented as:

$$f_i^c = \mathbf{F}_{conv} \left(\mathbf{F}_{conv1} \left(\left[f_m^{j'} \right] \right) \oplus f_m \right) \tag{5}$$

Building upon this foundation, we integrate the boundary information from the boundary maps generated by the contour guidance module into the features, guiding the feature learning process and enhancing the expression of boundary characteristics. The representation method is as follows:

$$f_i^e = \mathbf{F}_{conv}\left((f_i^c \otimes D\left(f_e\right)) \oplus f_i^c\right) \tag{6}$$

where D represents downsampling, F_{conv} is a 3×3 convolution, and F_{conv1} is a 1×1 convolution. $\otimes$ denotes element-wise multiplication, while $\oplus$ signifies element-wise addition. [∗] stands for the concatenation operation, and f_i^c is the output of BGFM. It is important to note that for $i = (2,3)$, the output of the previous BGFM (f_{i+1}^c) is combined with f_i^a as the input for the next BGFM to obtain f_i^c. By adjusting the channel number of the feature f_i^c through another 1×1 convolution, we can obtain the prediction P_i for the camouflaged object (where $i \in 2, 3, 4$).

2.4 Overall Loss Function

In this study, the model generates four output results, including the boundary map (M_{edge}), and three pseudo-target masks (P_i, $i \in 2, 3, 4$) produced the boundary-guided feature fusion module. For pixel binary classification tasks, binary cross-entropy (BCE) loss is commonly used. However, due to the significant difference in the number of foreground and background pixels in the localization supervision map and the camouflaged object mask, this paper chooses the weighted binary cross-entropy loss function [19]. To achieve more effective global optimization, the weighted IOU loss function [19] is also incorporated. Therefore, the following loss functions are defined for the camouflaged object mask:

$$L_{em} = L_w BCE + L_w IOU \tag{7}$$

Regarding the boundary map, this paper adopts the Dice loss (Ldice) introduced in [20]. Consequently, the overall loss function is defined as follows:

$$L_{all} = \sum_{i=2}^{4} L_{em}\left(P_i, G\right) + \lambda L_{dice}\left(M_{edge}, G_e\right) \tag{8}$$

In this definition, G represents the ground truth of the camouflaged object, G_e represents the ground truth of the camouflaged object boundary, and λ is set to 3.

3   Experiments

3.1   Implementation Details

In BGF-Net, we adopt the Res2Net50 [10] pre-trained on ImageNet as the backbone network for our study. Our model is implemented in PyTorch, with input image sizes of 416×416. During the training process, our model optimization uses the Adam algorithm. The initial learning rate is set to 1e−4, the batch size is set to 26 during model training, and the number of epochs is set to 100. The training is performed on an NVIDIA 3090 GPU, and the entire training process takes approximately 3 h.

Table 1. Four evaluation metrics are employed in this study, namely S_α [4], F_β^ω [13], E_ϕ [6], and M [15]. The symbols "↑" and "↓" indicate that larger and smaller values are better, respectively. The best results are highlighted in bold.

Method	Year	COD10K-Test				CAMO-Test				NC4K-Test			
		$S_\alpha\uparrow$	$E_\phi\uparrow$	$F_\beta^\omega\uparrow$	$M\downarrow$	$S_\alpha\uparrow$	$E_\phi\uparrow$	$F_\beta^\omega\uparrow$	$M\downarrow$	$S_\alpha\uparrow$	$E_\phi\uparrow$	$F_\beta^\omega\uparrow$	$M\downarrow$
CPD	2019	0.750	0.853	0.640	0.053	0.716	0.796	0.658	0.113	0.717	0.793	0.638	0.092
EGNet	2019	0.737	0.810	0.608	0.056	0.662	0.766	0.612	0.124	0.767	0.850	0.719	0.077
UCNet	2020	0.776	0.857	0.681	0.042	0.739	0.787	0.700	0.095	0.813	0.872	0.777	0.055
SINet	2020	0.776	0.864	0.631	0.043	0.745	0.829	0.644	0.092	0.808	0.871	0.723	0.058
PraNet	2020	0.789	0.861	0.629	0.045	0.769	0.837	0.663	0.094	0.822	0.876	0.724	0.059
C2FNet21	2021	0.813	0.890	0.686	0.036	0.796	0.864	0.719	0.080	0.838	0.897	0.762	0.049
PFNet	2021	0.800	0.877	0.660	0.040	0.782	0.842	0.695	0.085	0.829	0.887	0.745	0.053
R-MGL	2021	0.814	0.852	0.666	0.035	0.775	0.847	0.673	0.088	0.833	0.867	0.731	0.052
LSR	2021	0.804	0.880	0.673	0.037	0.787	0.854	0.696	0.080	0.840	0.895	0.766	0.048
SINet-v2	2022	0.815	0.887	0.680	0.037	0.820	0.882	0.743	0.070	0.847	0.903	0.770	0.048
C2FNet22	2022	0.808	0.882	0.683	0.037	0.764	0.825	0.677	0.090	0.835	0.893	0.763	0.050
FAPNet	2022	0.820	0.887	0.693	0.036	0.805	0.857	0.724	0.080	0.849	0.900	0.773	0.047
BSANet	2022	0.817	0.887	0.696	0.035	0.804	0.860	0.728	0.079	0.842	0.897	0.771	0.048
PreyNet	2022	0.813	0.881	0.697	0.034	0.790	0.842	0.709	0.077	-	-	-	-
BGNet	2022	0.831	0.901	0.722	0.033	0.816	0.871	0.751	0.069	0.851	0.907	0.788	0.044
Ours	-	**0.838**	**0.903**	**0.732**	**0.031**	**0.833**	**0.887**	**0.771**	**0.066**	**0.858**	**0.909**	**0.795**	**0.043**

3.2 Datasets

The camouflaged object detection experiments in this article are validated on three public datasets. CAMO [11] involves 1,250 camouflaged images across eight categories. COD10K [7] is currently the largest camouflaged object detection dataset, containing label and bounding box information for each image. To support the localization and ranking of camouflaged objects, a ranking test dataset called NC4K is proposed in [12]. This dataset contains 4,121 images, each providing additional localization annotations and ranking annotations.

3.3 Evaluation Metrics

We employ four popular metrics to evaluate COD performance, allowing for a comprehensive comparison of our proposed model with other state-of-the-art methods. Below are the details of each metric.

Structure Measure(S_α) [4]: The evaluation metric for assessing structural similarity is defined as follows:

$$S_\alpha = \alpha S_o + (1 - \alpha)S_r. \tag{9}$$

where S_o represents object-aware structural similarity, while S_r represents region-aware structural similarity. The parameter $\alpha \in [0,1]$ is a weighting factor that balances the two measures, with a default value of 0.5 [3].

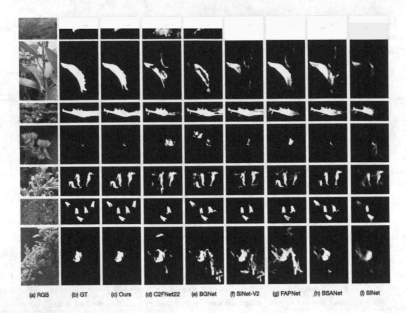

Fig. 4. A qualitative visual comparison between our model and the state-of-the-art models is presented. (a) input images, (b) ground truth, (c) our method, and (d)–(I) the state-of-the-art models.

E-measure (E_ϕ) [6]: It combines pixel-level matching with image-level statistical information to provide accurate evaluation of the overall and local performance of object detection. The definition is as follows:

$$E_\phi = \frac{1}{W \times H} \sum_{i=1}^{W} \sum_{j=1}^{H} \phi(S(x,y), G(x,y)). \tag{10}$$

where G denotes the ground truth with each pixel located at (x, y), ϕ represents the enhanced alignment matrix, and H and W are used to denote the length and width of G, respectively.

Mean Absolute Error ($\mathcal{M}$) [15]: This metric is used to evaluate the accuracy of the corresponding pixels between the predicted image and the ground truth image. The definition is as follows:

$$\mathcal{M} = \frac{1}{W \times H} \sum_{i=1}^{W} \sum_{j=1}^{H} |S(i,j) - G(i,j)|. \tag{11}$$

where S and G represent the predicted value and ground truth, respectively.

Weighted F-measure (F_β^ω) [13]: It is a weighted harmonic mean of precision and recall, which allows for a more accurate evaluation of the performance of object detectors. The definition of F_β^ω is given as follows:

Table 2. Four evaluation metrics are employed in this study, namely S_α [4], F_β^ω [13], E_ϕ [6], and $\mathcal{M}$ [15]. The symbols "↑" and "↓" indicate that larger and smaller values are better, respectively. The best results are highlighted in bold.

Method	Flying (714 images)				Terrestrial (699 images)				Aquatic (474 images)			
	S_α↑	E_ϕ↑	F_β^ω↑	$\mathcal{M}$↓	S_α↑	E_ϕ↑	F_β^ω↑	$\mathcal{M}$↓	S_α↑	E_ϕ↑	F_β^ω↑	$\mathcal{M}$↓
CPD	0.777	0.792	0.561	0.041	0.714	0.747	0.462	0.053	0.746	0.779	0.558	0.075
EGNet	0.771	0.795	0.568	0.040	0.711	0.737	0.467	0.049	0.693	0.729	0.497	0.088
UCNet	0.807	0.886	0.675	0.030	0.742	0.830	0.566	0.042	0.767	0.843	0.649	0.060
SINet	0.798	0.828	0.580	0.040	0.743	0.778	0.492	0.050	0.758	0.803	0.570	0.073
PraNet	0.819	0.888	0.669	0.033	0.756	0.835	0.565	0.046	0.781	0.848	0.644	0.065
C2FNet21	0.837	0.913	0.712	0.026	0.779	0.870	0.616	0.039	0.802	0.876	0.689	0.053
PFNet	0.824	0.903	0.691	0.030	0.773	0.854	0.606	0.041	0.793	0.868	0.675	0.055
R-MGL	0.835	0.886	0.872	0.028	0.781	0.872	0.571	0.037	0.801	0.874	0.621	0.054
LSR	0.830	0.906	0.707	0.027	0.772	0.855	0.611	0.038	0.803	0.875	0.694	0.053
SINet-v2	0.839	0.908	0.709	0.027	0.785	0.862	0.618	0.040	0.817	0.891	0.708	0.048
C2FNet22	0.835	0.905	0.721	0.026	0.780	0.859	0.628	0.038	0.800	0.874	0.691	0.052
FAPNet	0.846	0.907	0.730	0.025	0.789	0.865	0.634	0.038	0.818	0.889	0.713	0.049
BSANet	0.846	0.911	0.742	0.024	0.785	0.862	0.634	0.036	0.807	0.881	0.700	0.053
PreyNet	0.843	0.907	0.740	0.024	0.777	0.851	0.628	0.036	0.813	0.882	0.718	0.049
BGNet	0.859	**0.925**	0.764	0.022	0.801	0.877	0.664	0.034	0.820	0.891	0.723	0.045
Ours	**0.865**	**0.925**	**0.771**	**0.021**	**0.811**	**0.883**	**0.679**	**0.033**	**0.833**	**0.898**	**0.742**	**0.044**

$$\text{Precision}^w = \frac{TP^w}{TP^w + FP^w} \quad \text{Recall}^w = \frac{TP^w}{TP^w + FN^w}, \tag{12}$$

$$F_\beta^w = (1+\beta^2)\frac{\text{Precision}^w \cdot \text{Recall}^w}{\beta^2 \cdot \text{Precision}^w + \text{Recall}^w}. \tag{13}$$

where β represents the weight between precision and recall, with β^2 typically set to 0.3 to increase the weight of precision.

3.4 Comparison with State-of-the-Arts

Comparison Methods: We will compare our method with 18 state-of-the-art approaches to demonstrate the effectiveness of our method. These include, EGNet [25], UCNet [22], SINet [7], PraNet [8], C2FNet21 [17], PFNet [14], R-MGL [21], LSR [27], SINet-v2 [5], C2FNet22 [1], FAPNet [28], BSANet [29], PreyNet [23] and BGNet [18]. All tests mentioned in the text were conducted using the authors' opensource code to retrain the models or provided by the article authors for a fair comparison.

Quantitative Comparison: Table 1 presents the quantitative results of our model compared to the 15 competitors on the four widely-accepted evaluation metrics. As can be seen from the table, our method outperforms all other models in terms of the metrics, achieving the best performance on the datasets.

Table 3. Four evaluation metrics are employed in this study, namely S_α [4], F_β^ω [13], E_ϕ [6], and $\mathcal{M}$ [13]. The symbols "↑" and "↓" indicate that larger and smaller values are better, respectively. The best results are highlighted in bold.

Method	COD10K-Test				CAMO-Test				NC4K-Test			
	$S_\alpha\uparrow$	$E_\phi\uparrow$	$F_\beta^\omega\uparrow$	$\mathcal{M}\downarrow$	$S_\alpha\uparrow$	$E_\phi\uparrow$	$F_\beta^\omega\uparrow$	$\mathcal{M}\downarrow$	$S_\alpha\uparrow$	$E_\phi\uparrow$	$F_\beta^\omega\uparrow$	$\mathcal{M}\downarrow$
(a) B	0.831	0.896	0.712	0.032	0.818	0.868	0.745	0.074	0.853	0.903	0.782	0.046
(b) B+CGM+B1	0.834	0.900	0.719	**0.031**	0.823	0.880	0.759	0.069	0.856	0.907	0.788	0.044
(c) B+CGM+B1+C	0.835	0.898	0.722	0.032	0.828	0.879	0.763	0.067	0.857	0.907	0.791	**0.043**
(d) Ours	**0.838**	**0.903**	**0.732**	**0.031**	**0.833**	**0.887**	**0.771**	**0.066**	**0.858**	**0.909**	**0.795**	**0.043**

Qualitative Comparison: Fig. 4 showcases the superior detection results of our model compared to six leading COD methods. Our model excels in recognizing camouflaged objects, even when they share similar textures with their backgrounds, as seen in the first and second rows. It effectively handles both large and small camouflaged objects, providing clear contours and precise localization, outperforming C2FNet22, BGNet, SINet-V2, and FAPNet. The results demonstrate that our model can achieve better camouflaged object detection performance under various challenging factors.

Super-class Performance Comparison: To further validate the effectiveness of our proposed COD model, we report the quantitative superclass results, as shown in Table 2. We achieve the best performance in the three superclasses "Flying", "Terrestrial", and "Aquatic" across all four evaluation metrics. Specifically, compared to the well-performing BGNet model, our model improves the average performance of S_α by 1.18%, F_β^ω by 1.94%. Overall, our model demonstrates excellent performance in various superclass experiments.

3.5 Ablation Study

To further substantiate the effectiveness of our model's key components, we conducted an ablation study, incrementally adding key modules and presenting the results in Table 3. The baseline model (B) was stripped of the CGM and BGFM modules, where BGFM includes the edge-guided module B1 and the feature fusion module C, retaining only the 1×1 convolution in CAM [18] for feature aggregation. We then added the CGM and B1 modules to enhance edge features. Finally, we incorporated module C, resulting in further performance improvement.

Effectiveness of CGM and B1. According to Table 3, adding CGM and B1 to the baseline model (B) enhances performance. CGM fuses multi-scale features to generate more accurate localization maps. The performance improvement after adding CGM and B1 attests to our model's effectiveness.

Effectiveness of module C in BGFM. Adding module C to (b) aims to mine contextual semantics to improve object detection. Table 3 shows that the (c) model outperforms the baseline (b) model.

4 Conclusion

We present BGF-Net, an efficient camouflaged object detection framework incorporating a Feature Enhancement Module (FEM) and a Boundary Guided Feature fusion Module (BGFM). BGF-Net leverages object edges and semantic information through FEM, while BGFM achieves context-aware semantic mining using multi-level feature fusion. By integrating target edges from the Contour-Guided Module (CGM) into the fused features, the framework continuously refines object structure and boundaries, resulting in more comprehensive and precise object information. Our approach surpasses state-of-the-art methods in three benchmark tests, demonstrating its effectiveness.

References

1. Chen, G., Liu, S.J., Sun, Y.J., Ji, G.P., Wu, Y.F., Zhou, T.: Camouflaged object detection via context-aware cross-level fusion. IEEE Trans. Circuits Syst. Video Technol. **32**(10), 6981–6993 (2022)
2. Chen, T., Xiao, J., Hu, X., Zhang, G., Wang, S.: Boundary-guided network for camouflaged object detection. Knowl.-Based Syst. **248**, 108901 (2022)
3. Cheng, M.M., Fan, D.P.: Structure-measure: a new way to evaluate foreground maps. Int. J. Comput. Vision **129**, 2622–2638 (2021)
4. Fan, D.P., Cheng, M.M., Liu, Y., Li, T., Borji, A.: Structure-measure: a new way to evaluate foreground maps. In: Proceedings of the IEEE International Conference on Computer Vision, pp. 4548–4557 (2017)
5. Fan, D.P., Ji, G.P., Cheng, M.M., Shao, L.: Concealed object detection. IEEE Trans. Pattern Anal. Mach. Intell. **44**(10), 6024–6042 (2021)
6. Fan, D.P., Ji, G.P., Qin, X., Cheng, M.M.: Cognitive vision inspired object segmentation metric and loss function. Scientia Sinica Informationis **6**(6) (2021)
7. Fan, D.P., Ji, G.P., Sun, G., Cheng, M.M., Shen, J., Shao, L.: Camouflaged object detection. In: Proceedings of the IEEE/CVF Conference on Computer Vision and Pattern Recognition, pp. 2777–2787 (2020)
8. Fan, D.-P., et al.: PraNet: parallel reverse attention network for polyp segmentation. In: Martel, A.L., et al. (eds.) MICCAI 2020. LNCS, vol. 12266, pp. 263–273. Springer, Cham (2020). https://doi.org/10.1007/978-3-030-59725-2_26
9. Fan, D.-P., Zhai, Y., Borji, A., Yang, J., Shao, L.: BBS-Net: RGB-D salient object detection with a bifurcated backbone strategy network. In: Vedaldi, A., Bischof, H., Brox, T., Frahm, J.-M. (eds.) ECCV 2020. LNCS, vol. 12357, pp. 275–292. Springer, Cham (2020). https://doi.org/10.1007/978-3-030-58610-2_17
10. Gao, S.H., Cheng, M.M., Zhao, K., Zhang, X.Y., Yang, M.H., Torr, P.: Res2Net: a new multi-scale backbone architecture. IEEE Trans. Pattern Anal. Mach. Intell. **43**(2), 652–662 (2019)
11. Le, T.N., Nguyen, T.V., Nie, Z., Tran, M.T., Sugimoto, A.: Anabranch network for camouflaged object segmentation. Comput. Vis. Image Underst. **184**, 45–56 (2019)
12. Lv, Y., et al.: Simultaneously localize, segment and rank the camouflaged objects. In: Proceedings of the IEEE/CVF Conference on Computer Vision and Pattern Recognition, pp. 11591–11601 (2021)
13. Margolin, R., Zelnik-Manor, L., Tal, A.: How to evaluate foreground maps? In: IEEE Conference on Computer Vision and Pattern Recognition (CVPR) (2014)

14. Mei, H., Ji, G.P., Wei, Z., Yang, X., Wei, X., Fan, D.P.: Camouflaged object segmentation with distraction mining. In: Proceedings of the IEEE/CVF Conference on Computer Vision and Pattern Recognition, pp. 8772–8781 (2021)
15. Perazzi, F., Krähenbühl, P., Pritch, Y., Hornung, A.: Saliency filters: contrast based filtering for salient region detection. In: 2012 IEEE Conference on Computer Vision and Pattern Recognition, pp. 733–740. IEEE (2012)
16. Singh, S.K., Dhawale, C.A., Misra, S.: Survey of object detection methods in camouflaged image. IERI Procedia **4**, 351–357 (2013)
17. Sun, Y., Chen, G., Zhou, T., Zhang, Y., Liu, N.: Context-aware cross-level fusion network for camouflaged object detection. arXiv preprint arXiv:2105.12555 (2021)
18. Sun, Y., Wang, S., Chen, C., Xiang, T.Z.: Boundary-guided camouflaged object detection. arXiv preprint arXiv:2207.00794 (2022)
19. Wei, J., Wang, S., Huang, Q.: F^3Net: fusion, feedback and focus for salient object detection. In: Proceedings of the AAAI Conference on Artificial Intelligence, vol. 34, pp. 12321–12328 (2020)
20. Xie, E., Wang, W., Wang, W., Ding, M., Shen, C., Luo, P.: Segmenting transparent objects in the wild. In: Vedaldi, A., Bischof, H., Brox, T., Frahm, J.-M. (eds.) ECCV 2020. LNCS, vol. 12358, pp. 696–711. Springer, Cham (2020). https://doi.org/10.1007/978-3-030-58601-0_41
21. Zhai, Q., Li, X., Yang, F., Chen, C., Cheng, H., Fan, D.P.: Mutual graph learning for camouflaged object detection. In: Proceedings of the IEEE/CVF Conference on Computer Vision and Pattern Recognition, pp. 12997–13007 (2021)
22. Zhang, J., et al.: UC-Net: uncertainty inspired RGB-D saliency detection via conditional variational autoencoders. In: Proceedings of the IEEE/CVF Conference on Computer Vision and Pattern Recognition, pp. 8582–8591 (2020)
23. Zhang, M., Xu, S., Piao, Y., Shi, D., Lin, S., Lu, H.: PreyNet: preying on camouflaged objects. In: Proceedings of the 30th ACM International Conference on Multimedia, pp. 5323–5332 (2022)
24. Zhang, P., Wang, D., Lu, H., Wang, H., Ruan, X.: Amulet: aggregating multi-level convolutional features for salient object detection. In: Proceedings of the IEEE International Conference on Computer Vision, pp. 202–211 (2017)
25. Zhao, J.X., Liu, J.J., Fan, D.P., Cao, Y., Yang, J., Cheng, M.M.: EGNet: edge guidance network for salient object detection. In: Proceedings of the IEEE/CVF International Conference on Computer Vision, pp. 8779–8788 (2019)
26. Zhao, T., Wu, X.: Pyramid feature attention network for saliency detection. In: Proceedings of the IEEE/CVF Conference on Computer Vision and Pattern Recognition, pp. 3085–3094 (2019)
27. Zheng, Y., Zhang, X., Wang, F., Cao, T., Sun, M., Wang, X.: Detection of people with camouflage pattern via dense deconvolution network. IEEE Signal Process. Lett. **26**(1), 29–33 (2018)
28. Zhou, T., Zhou, Y., Gong, C., Yang, J., Zhang, Y.: Feature aggregation and propagation network for camouflaged object detection. IEEE Trans. Image Process. **31**, 7036–7047 (2022)
29. Zhu, H., et al.: I can find you! Boundary-guided separated attention network for camouflaged object detection. In: Proceedings of the AAAI Conference on Artificial Intelligence, vol. 36, pp. 3608–3616 (2022)

Saliency Driven Monocular Depth Estimation Based on Multi-scale Graph Convolutional Network

Dunquan Wu[1] and Chenglizhao Chen[1,2,3]($\boxtimes$)

[1] College of Computer Science and Technology, Qingdao University, Qingdao, China
cclz123@163.com
[2] Shandong Provincial Key Laboratory for Distributed Computer Software Novel Technology, Jinan, China
[3] College of Computer Science and Technology, China University of Petroleum (East China), Qingdao, China

Abstract. Monocular depth estimation is a fundamental and crucial task in computer vision that enables scene understanding from a single image. This paper proposes a novel approach for saliency-driven monocular depth estimation based on a multi-scale Graph Convolutional Network (GCN). Our method utilizes saliency information to guide the depth estimation process and employs a multi-scale GCN to capture local and global contextual cues. The proposed framework constructs a graph structure using RGB images to represent the relationships between image regions. We designed a multi-scale feature fusion module called DS Fusion, by applying GCN at multiple scales, our method effectively integrates depth features and saliency features to predict accurate depth maps. Extensive experiments conducted on KITTI and NYU datasets demonstrate the superior performance of our approach compared to state-of-the-art techniques. Additionally, we perform indepth analysis of the network architecture and discuss the impact of saliency cues on depth estimation accuracy. Our proposed method showcases the potential of combining saliency information and GCN in monocular depth estimation, contributing to the progress of scene understanding and depth perception from a single image.

Keywords: Monocular Depth Estimation · Multi-scale Graph Convolutional Network · Saliency Detection

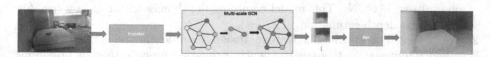

Fig. 1. The multi-scale graph convolution takes image as input and outputs depth maps and saliency maps, and then fed into a fusion network to generate the final results.

© The Author(s), under exclusive license to Springer Nature Singapore Pte Ltd. 2024
Q. Liu et al. (Eds.): PRCV 2023, LNCS 14433, pp. 445–456, 2024.
https://doi.org/10.1007/978-981-99-8546-3_36

1   Introduction

Estimating depth from a single image is a fundamental problem in the field of computer vision, applicable to areas such as autonomous navigation, augmented reality, and saliency detection [2,26]. Therefore, recovering depth information from a single image plays a crucial role in scene understanding and reconstruction. Traditional methods typically rely on handcrafted features or stereo matching techniques, which have limitations in capturing fine details and handling occlusions.

Currently, most depth estimation algorithms utilize Convolutional Neural Networks (CNN) to regress depth information [1,7,9,11,15]. However, they often overlook the characteristics of depth information, namely, each pixel value in the depth map is not only related to its neighboring pixel values but also correlated with other pixels that have the same depth value. On the other hand, Graph Convolutional Network (GCN) [12] can effectively extend the deep neural network models from the Euclidean domain implemented by CNNs to non-Euclidean domains. We also observe that within a depth map, the same object typically exhibits smooth depth values. With the rapid advancements in saliency detection, we see the promising potential of utilizing saliency information to guide depth refinement [23].

To address these issues, in this paper, we propose a novel approach for saliency-driven monocular depth estimation based on a multi-scale GCN. Our method aims to leverage the salient regions in the image to achieve accurate depth estimation. By incorporating saliency information, we can guide the depth estimation process and enhance the network's ability to capture important features. Specifically, we first extract features from RGB images using a VGG-16 backbone [20] and construct a graph structure based on the feature representation. Each node in the graph corresponds to a feature point in the input image, and the edges represent the connectivity between nodes. We apply a multi-scale GCN on the constructed graph to learn feature representations of the nodes through convolutional operations. The GCN can utilize the connectivity between nodes to acquire richer contextual information and learn more accurate depth features for the depth estimation task. By utilizing the node features learned through the multi-scale GCN, we predict the depth map and saliency map. Through the DS Fusion module, as is shown in Fig. 1, our method effectively integrates depth features and saliency features to generate accurate depth maps.

Our main contributions are as follows:

1. We propose a novel saliency-driven monocular depth estimation model based on multi-scale GCNs. This model combines depth cues with saliency cues for monocular depth estimation.
2. We designed a multi-scale feature fusion module that effectively utilizes saliency information to guide the depth estimation process, thereby providing more accurate depth estimation results.
3. Even low-quality saliency labels can drive the generation of high-quality depth maps.

4. Extensive experiments conducted on the outdoor KITTI dataset and indoor NYU dataset demonstrate that our model outperforms existing methods.

2   Related Work

2.1   Monocular Depth Estimation

Monocular depth estimation is an important problem in the field of computer vision, and its significance in downstream tasks has attracted increasing research attention. Traditional depth estimation methods mainly rely on geometric or texture features for depth inference, such as optical flow or dense reconstruction [5]. However, these methods are often limited by factors such as occlusions, texture consistency, and lighting variations, making it challenging to obtain accurate depth information. In recent years, with the emergence of deep learning, methods based on CNNs have made significant progress. These methods leverage end-to-end learning to extract depth features from large-scale datasets [1,7,9,11,15], leading to more accurate monocular depth estimation. And multi-scale networks also exhibit good advantages in depth estimation [27].

Recent studies have shown the potential of GCNs in monocular depth estimation. Graph convolution allows for effective capture of spatial relationships between pixels through convolutional operations on graph structures. This capability enables GCNs to utilize contextual information for depth estimation, thereby improving the quality of the depth map [18].

2.2   Saliency Detection

Saliency detection is an important task in computer vision, aiming to identify the most attention-grabbing regions in an image by analyzing color, texture, contrast, and spatial information, among others. In recent years, deep learning methods have made significant advancements in saliency detection. These methods leverage CNNs to learn saliency features from large-scale annotated datasets, enabling accurate detection of salient regions [25]. In our approach, saliency detection information is incorporated into monocular depth estimation to enhance the accuracy and robustness of depth estimation.

2.3   Graph Convolutional Networks

Graph Convolutional Networks (GCNs) are neural network models that effectively handle graph-structured data. GCNs perform convolution operations on graphs, allowing them to leverage the connectivity between nodes to acquire contextual information. In image processing tasks, GCNs have been successfully applied in areas such as image segmentation, object detection, and image generation [16]. In monocular depth estimation, GCNs are introduced to learn spatial relationships between pixels, thereby enhancing the accuracy of depth estimation.

3   The Proposed Method

Monocular depth estimation is an ill-posed problem because many depth cues are lost when 3D objects are projected onto a 2D plane. Inspired by the work of Johnson [8], we assume that we can obtain positional information of the observed objects in the scene. Using this clue, we can infer the distance to the objects, with the main challenge being how to handle non-Euclidean positional relationships. Therefore, we first introduce a depth topological graph as a depth clue and then utilize GCN to process the feature information from the CNN. The depth value is estimated by calculating the Adjacency matrix and multi-scale eigenvector in the depth map. The details of our model will be described in the following sections:

3.1   Overall Architecture

As shown in Fig. 2, the network consists of three main modules: the feature extraction module, the multi-scale graph convolution module, and the depth-saliency fusion module. The encoder network is mainly composed of commonly used network architectures in image recognition, such as VGG-16 [20] and ResNet-50 [13], to extract features from the RGB images. The extracted features are then fed into the multi-scale GCN, which effectively aggregates multi-scale information and filters redundant relationships. In addition, the depth-saliency fusion module combines the depth information with the saliency information to obtain the final depth map. We will discuss these modules in the following sections.

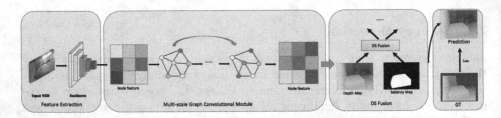

Fig. 2. Network structure of the proposed framework. After feature extraction by the encoder, the features are used to construct a graph structure, which is then fed into the multi-scale graph convolutional network. The network aggregates multi-scale information and filters out redundant relationships. Finally, the saliency information from multiple scales is combined with the depth information, resulting in the final depth.

3.2   Feature Extraction

We utilize the VGG-16 network as the backbone to extract its 2D features. For the input RGB image, we can map them into a semantically powerful 2D appearance representation $V \in R^{h \times w \times c}$. Subsequently, the multi-scale graph convolution module is introduced. To fully leverage the scene information in the images, we extract the scene features of the entire image from the image feature maps of the backbone network. The extracted features will be passed to subsequent modules for further processing.

3.3   Graph Neural Network

Saliency-driven GCN Module takes the underlying 2D appearance features V as inputs and outputs high quality depth map and saliency map.

Graph Construction. Based on the feature V obtained from the CNN encoder, we construct a graph $G = (V, E)$, where $V = v_1, ..., v_n$ represents the set of nodes and E represents the set of edges. Each node v_i corresponds to a predefined scale and edge-connected feature map and is composed of N nodes with a $(1 \times D)$ feature. Each node has its own features, and we denote the features of these nodes as an $N \times D$ matrix X, where N is the number of nodes and D is the number of input features. The node relationship can be described by an adjacency matrix A, which is a sparse $N \times N$ matrix.

Graph Convolutional Network. In general, graph convolution is defined as:

$$H^{(l+1)} = f(H^{(l)}, A) \tag{1}$$

$$f(H^{(l)}, A) = \sigma(AH^{(l)}W^{(l)}) \tag{2}$$

The input for each GCN layer consists of the adjacency matrix A and the node features H. l denotes the layer number, A is the adjacency matrix, and $f(\cdot)$ is the nonlinear function, which represents the propagation rule across layers. In Equation(2), $W^{(l)}$ represents the weight matrix of the l-th neural network layer, and $\sigma(\cdot)$ is a nonlinear activation function such as ReLU. To avoid the change in scale of feature vectors caused by the adjacency matrix, we can address this issue by enforcing self-loops in the graph, which involves adding the identity matrix to A:

$$\tilde{A} = A + I \tag{3}$$

Network Architecture. Our saliency-driven GCN module utilizes the ReLU activation function to reduce parameter dependency. To improve the quality of predicted depth maps and saliency maps, we adopt a multi-scale GCN algorithm that combines saliency information and depth information from different scales. As shown in Fig. 3, we employ a U-shaped network [6] architecture for message propagation among instance nodes. The left half of the network performs graph convolution and down-sample operations, while the right half performs up-sample and graph convolution operations. By executing graph convolution-down sample-upsample-graph convolution operations at multiple scales, we preserve high-level information propagated from coarse feature maps as well as fine local information provided by low-level feature maps, while maintaining the relationships among nodes in the depth map. The formula for our graph convolution operation is as follows:

$$\hat{X} = \sigma(AX^{(l)}W^{(l)}) \tag{4}$$

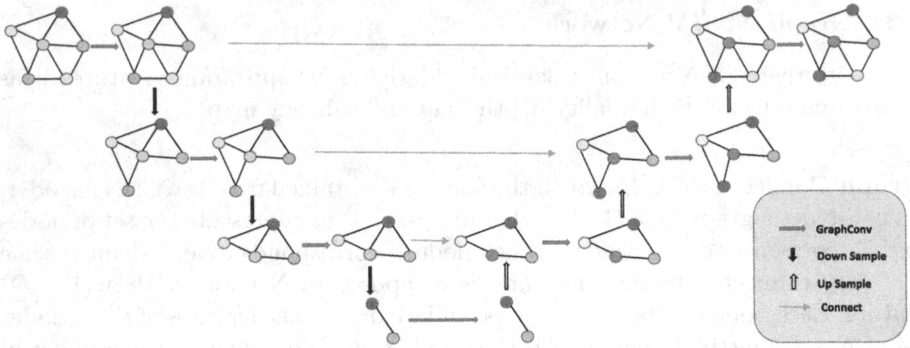

Fig. 3. Illustration of the Multi-scale Graph Convolutional Network.

where $\sigma(\cdot)$ represents the activation function, $W^{(l)}$is the learnable weight matrix of the lth neural network layer, $X^{(l)}$ denotes the input of the lth layer, and $\hat{X}$ represents the output of the graph convolution.

In the decoder network, we add depth map reconstruction layer and saliency map reconstruction layer, and introduce loss functions to measure the errors.

$$depth = sigmoid(GCN_1(\hat{X})) \tag{5}$$

$$saliency = sigmoid(GCN_2(\hat{X})) \tag{6}$$

3.4   Multi-feature Fusion Module

The "saliency-driven fusion" refers to the approach of incorporating saliency information into the depth estimation network to improve its representation of details. To effectively fuse saliency and depth information, it is crucial to leverage multi-task information to enhance depth estimation. As shown in Fig. 4, our multi-feature fusion module adopts a "fuse before expand" strategy to gradually integrate depth information and saliency information from the multi-scale graph convolutional network outputs. This module aims to capture local and global contexts for more accurate depth estimation.

At each scale, the depth map and saliency map are inputted into the fusion module. A fusion mechanism is employed to aggregate these features and create a fused representation at that scale, refining and enhancing the information. The output of each fusion layer is combined with the feature representation from the next higher scale. This process is repeated at each scale, gradually integrating information from different detail levels. Ultimately, the model predicts depth maps and saliency maps based on the fused features. Through this approach, our model effectively fuses depth and saliency information at multiple scales, achieving comprehensive contextual awareness and improving the accuracy of depth estimation.

$$F_ids = Conv(F_id \oplus F_is) \tag{7}$$

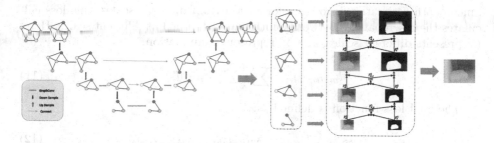

Fig. 4. Depth-Saliency Fusion Module. By integrating depth features and saliency features from multi-scales, the network becomes more attentive to the depth in salient regions, leading to the generation of higher-quality depth maps.

The fused representation at each scale, denoted as F_ids, is obtained by convolving the element-wise addition of the depth information (F_id) and the saliency information (F_is), where i $\in$ 1, 2, 3, 4. $Conv(\cdot)$ represents a convolutional layer, and $\oplus$ denotes element-wise addition. Then, we combine F_ids with depth features and saliency features using different convolutional layers, enabling interaction and fusion between depth and saliency information. By emphasizing salient areas through saliency cues, the depth estimation network focuses on salient regions and enhances the features for depth estimation. The enhanced features are then fed into the next decoder for fusion at the next scale.

3.5 Loss Function

To ensure accurate predictions of real depth values, we use the Scale-Invariant Log-arithmic Loss (SILog) [7] to supervise the training process.

$$\triangle d_i = log\hat{d}_i - logd_i^* \tag{8}$$

$$l_{depth} = \alpha\sqrt{\frac{1}{k}\sum_i \triangle d_i^2 - \frac{\lambda}{k^2}(\sum_i \triangle d_i)^2} \tag{9}$$

where d_i^* represents the ground truth depth value and $\hat{d}_i$ represents the predicted depth at pixel i. For K pixels with valid depth values in an image, the Scale-Invariant Loss is calculated as (9).

l_{weight} represents the weighted loss, where each pixel is assigned a different weight based on its saliency or uncertainty. This allows us to focus more on pixels that contribute more to depth estimation during training. The weighted loss can be calculated using the saliency map S as follows:

$$l_{weight} = S * \frac{1}{k}\sum_1^k(\hat{d}_i - d_i^*)^2 \tag{10}$$

$l_{multiscale}$ represents the multi-scale consistency loss, which aims to promote consistency by measuring the differences between depth maps at different scales.

Since depth estimation involves information from multiple scales, this loss calculates the discrepancies between depth maps obtained at different scales. Here, i represents different scales, and $\sum$ represents summation.

$$l_{multiscale} = \sum \frac{1}{k} \Sigma_1^k (\hat{d}_i - d_i^*)^2 \qquad (11)$$

The final loss function is defined as:

$$Loss = \lambda_1 l_{depth} + \lambda_2 l_{weight} + \lambda_3 l_{multiscle} \qquad (12)$$

where λ_1, λ_2, and λ_3 are weighting coefficients for each loss term.

4   Experiments

4.1   Datasets

We used the commonly used outdoor dataset KITTI [10] and indoor dataset NYU-v2 [19] as our training and testing datasets. Since the KITTI and NYU-v2 datasets do not have ground truth of saliency labels, we employed the Saliency Detection Networks to generate saliency maps as pseudo ground truth.As shown in Fig. 5.

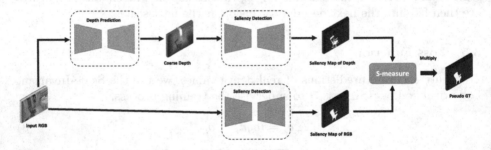

Fig. 5. Generate saliency pseudo ground truth.

4.2   Evaluations

Evaluation on KITTI Dataset: For outdoor scenes, we employed the training-testing split method proposed by Eigen et al. [7], as shown in Table 1. Quantitative comparison results with [3,4,7,9,11,15,18,21,22] demonstrate that our method outperforms previous methods, showing significant improvements in all metrics. Qualitative comparison results, as shown in Fig. 6, illustrate that our method exhibits superior visualization performance compared to other methods. The contours are clearer, and there is higher distinction between connected objects.

Evaluation on NYU-v2 Dataset: For indoor scenes, we utilized the NYU-v2 dataset and evaluated our experiments using the official split. The quantitative

Table 1. Quantitative results on KITTI.

Methods	Lower Better↓				Higher Better↑		
	Abs_Rel	RMSE	Sq_Rel	RMSE_Log	δ1	δ2	δ3
Eigen et al. [7]	0.190	7.156	1.515	0.270	0.692	0.899	0.967
DORN [9]	0.072	2.727	0.307	0.120	0.932	0.984	0.994
Monodepth2 [11]	0.109	4.960	0.879	0.209	0.864	0.948	0.975
BTS [15]	0.059	2.756	0.245	0.096	0.956	0.993	0.998
Adabins [3]	0.058	2.360	0.190	0.088	0.964	0.995	**0.999**
GCNDepth [18]	0.104	4.494	0.720	0.181	0.888	0.965	0.984
Cha et al. [4]	0.096	4.334	0.700	0.179	0.894	0.964	0.983
Tang et al. [21]	0.103	4.918	0.873	0.203	0.875	0.949	0.975
Wang et al. [22]	0.109	4.656	0.790	0.185	0.882	0.962	0.983
Ours	**0.052**	**2.077**	**0.151**	**0.078**	**0.976**	**0.997**	**0.999**

Table 2. Quantitative results on NYU-v2.

Methods	Lower Better↓					Higher Better↑		
	Abs_Rel	RMSE	Sq_Rel	RMSE_Log	log10	δ1	δ2	δ3
Eigen et al. [7]	0.215	0.907	0.212	0.285	-	0.611	0.887	0.971
DORN [9]	0.115	0.509	-	-	0.051	0.828	0.965	0.992
Monodepth2 [11]	0.156	0.561	0.123	0.211	-	0.732	0.895	0.940
BTS [15]	0.110	0.392	0.066	0.142	0.047	0.885	0.978	0.994
Adabins [3]	0.103	0.364	-	-	0.044	0.903	0.984	0.997
Jun et al. [14]	0.100	0.362	-	-	0.043	0.907	0.986	0.997
Ling et al. [17]	0.104	0.356	-	-	0.044	0.903	0.986	0.997
Hi-Net [24]	0.130	0.522	-	-	0.056	0.828	0.959	0.989
Cha et al. [4]	0.156	0.561	0.123	0.211	-	0.784	0.939	0.979
Ours	**0.093**	**0.327**	**0.044**	**0.118**	**0.040**	**0.924**	**0.990**	**0.998**

comparison results with [3,4,7,9,11,14,15,17,24] in Table 2 demonstrate that our method significantly improves the performance of all metrics. The "Abs Rel" error is reduced to below 0.1, and the data indicates the superiority of our model in enhancing depth estimation accuracy. Qualitative results in Fig. 7 show that our method can better estimate depth, particularly in challenging scenarios with cluttered environments, poor lighting conditions, and reflective surfaces.

4.3 Abalation Study

To better assess the effectiveness of each module in our approach, we conducted ablation studies to evaluate each component and present the results in Table 3.

Baseline: To validate the effectiveness of our proposed multi-scale graph convolutional network and saliency-driven approach, we designed a baseline model. In comparison to our model, the baseline model employed a CNN encoder-decoder structure and removed the saliency-driven branch and fusion module. We then sequentially added the multi-scale graph convolutional module, the uni-directional fusion saliency-driven branch, and the bi-directional fusion module, observing significant improvements in performance.

Multi-scale graph convolutional network: Building upon the baseline model, we replaced the decoder with a multi-scale graph convolutional network. This network can handle non-Euclidean domain data and capture local and global contextual cues more effectively. As evident from the data in the table, this network brings notable improvements to the depth estimation model.

Fig. 6. Qualitative results on the KITTI dataset.

Table 3. Ablation study on the Eigen split of KITTI dataset. "Ms" refers to multi-scale GCN, "Sal" refers to saliency-driven branch, and "Fu" refers to Fusion module.

Settings	Abs_Rel↓	RMSE↓	Sq_Rel↓	RMSE_Log↓	δ1↑	δ2↑	δ3↑
Baseline	0.180	5.231	1.500	0.180	0.699	0.899	0.968
+Ms	0.105	4.058	0.550	0.185	0.895	0.942	0.986
+Ms+Sal	0.068	2.964	0.352	0.108	0.951	0.965	0.992
+Ms+Sal+Fu	0.052	2.077	0.151	0.078	0.976	0.997	0.999

Saliency-driven branch: Within the same object, depth values may exhibit smooth variations. By incorporating the saliency-driven branch and adopting unidirectional fusion, the depth estimation is enhanced, resulting in depth maps with more prominent contours and clearer representations of objects.

Fusion module: Replacing the unidirectional fusion of saliency-driven depth information module with bidirectional fusion of depth and saliency information allows the network to comprehensively integrate both depth and saliency information. The improvement in predicted results is relatively limited compared

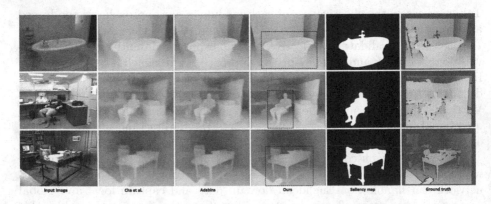

Fig. 7. Qualitative results on the NYUv2 dataset.

to the previous step, but it enhances saliency detection and provides valuable guidance for further experiments' improvements.

5 Conclusions

We propose a saliency-driven module based on a multi-scale GCN to address the problem of monocular depth estimation. To effectively utilize features at different scales and improve the accuracy of depth estimation, we employ a multi-scale GCN for multi-scale feature fusion and information propagation. Additionally, we introduce saliency information as an extra cue. By integrating saliency detection with depth estimation, our model can better understand important regions in the image and provide more accurate guidance for depth estimation. However, our research also has certain limitations. For example, the saliency detection effect of some scenes is poor, and different adjacent objects are detected as the same salient object, which will have a certain impact on the final effect. Future research directions could explore more complex network architectures and richer information fusion strategies to further enhance the performance of monocular depth estimation.

Acknowledgments. This work was supported by the National Natural Science Foundation of China (No. 62172246) and the Youth Innovation and Technology Support Plan of Colleges and Universities in Shandong Province (2021KJ062).

References

1. Agarwal, A., Arora, C.: Depthformer: multiscale vision transformer for monocular depth estimation with global local information fusion. In: 2022 IEEE International Conference on Image Processing (ICIP), pp. 3873–3877. IEEE (2022)
2. Badue, C., et al.: Self-driving cars: a survey. Expert Syst. Appl. **165**, 113816 (2021)
3. Bhat, S.F., Alhashim, I., Wonka, P.: Adabins: depth estimation using adaptive bins. In: Proceedings of the IEEE/CVF Conference on Computer Vision and Pattern Recognition, pp. 4009–4018 (2021)
4. Cha, G., Jang, H.D., Wee, D.: Self-supervised monocular depth estimation with isometric-self-sample-based learning. IEEE Rob. Autom. Lett. **8**, 2173–2180 (2022)
5. Chen, J., Yang, X., Jia, Q., Liao, C.: Denao: monocular depth estimation network with auxiliary optical flow. IEEE Trans. Pattern Anal. Mach. Intell. **43**(8), 2598–2610 (2020)
6. Chen, M., Lyu, X., Guo, Y., Liu, J., Gao, L., Song, J.: Multi-scale graph attention network for scene graph generation. In: 2022 IEEE International Conference on Multimedia and Expo (ICME), pp. 1–6. IEEE (2022)
7. Eigen, D., Puhrsch, C., Fergus, R.: Depth Map Prediction from a Single Image Using a Multi-Scale Deep Network. MIT Press, Cambridge (2014)
8. Flynn, J., Neulander, I., Philbin, J., Snavely, N.: Deep stereo: learning to predict new views from the world's imagery. In: 2016 IEEE Conference on Computer Vision and Pattern Recognition (CVPR) (2016)
9. Fu, H., Gong, M., Wang, C., Batmanghelich, K., Tao, D.: Deep ordinal regression network for monocular depth estimation. In: Proceedings of the IEEE Conference on Computer Vision and Pattern Recognition, pp. 2002–2011 (2018)

10. Geiger, A., Lenz, P., Urtasun, R.: Are we ready for autonomous driving? the kitti vision benchmark suite. In: 2012 IEEE Conference on Computer Vision and Pattern Recognition, pp. 3354–3361 (2012). https://doi.org/10.1109/CVPR.2012.6248074

11. Godard, C., Mac Aodha, O., Firman, M., Brostow, G.J.: Digging into self-supervised monocular depth estimation. In: Proceedings of the IEEE/CVF International Conference on Computer Vision, pp. 3828–3838 (2019)

12. Goyal, P., Ferrara, E.: Graph embedding techniques, applications, and performance: a survey. Knowl.-Based Syst. **151**, 78–94 (2018)

13. He, K., Zhang, X., Ren, S., Sun, J.: Deep residual learning for image recognition. IEEE (2016)

14. Jun, J., Lee, J.H., Lee, C., Kim, C.S.: Depth map decomposition for monocular depth estimation. In: Computer Vision-ECCV 2022: 17th European Conference, Tel Aviv, Israel, 23–27 October 2022, Proceedings, Part II, pp. 18–34. Springer, Heidelberg (2022). https://doi.org/10.1007/978-3-031-20086-1_2

15. Lee, J.H., Han, M.K., Ko, D.W., Suh, I.H.: From big to small: Multi-scale local planar guidance for monocular depth estimation (2019)

16. Li, C., Liu, F., Tian, Z., Du, S., Wu, Y.: DAGCN: dynamic and adaptive graph convolutional network for salient object detection. IEEE Trans. Neural Netw. Learn. Syst. (2022)

17. Ling, C., Zhang, X., Chen, H., Zhu, X., Yan, W.: Combing transformer and CNN for monocular depth estimation. In: 2022 China Automation Congress (CAC), pp. 3048–3052. IEEE (2022)

18. Masoumian, A., Rashwan, H.A., Abdulwahab, S., Cristiano, J., Asif, M.S., Puig, D.: Gcndepth: self-supervised monocular depth estimation based on graph convolutional network. Neurocomputing **517**, 81–92 (2023)

19. Silberman, N., Hoiem, D., Kohli, P., Fergus, R.: Indoor segmentation and support inference from RGBD images. In: Fitzgibbon, A., Lazebnik, S., Perona, P., Sato, Y., Schmid, C. (eds.) ECCV 2012. LNCS, vol. 7576, pp. 746–760. Springer, Heidelberg (2012). https://doi.org/10.1007/978-3-642-33715-4_54

20. Simonyan, K., Zisserman, A.: Very deep convolutional networks for large-scale image recognition. arXiv preprint arXiv:1409.1556 (2014)

21. Tang, S., Ye, X., Xue, F., Xu, R.: Cross-modality depth estimation via unsupervised stereo RGB-to-infrared translation. In: ICASSP 2023–2023 IEEE International Conference on Acoustics, Speech and Signal Processing (ICASSP), pp. 1–5. IEEE (2023)

22. Wang, G., Zhong, J., Zhao, S., Wu, W., Liu, Z., Wang, H.: 3D hierarchical refinement and augmentation for unsupervised learning of depth and pose from monocular video (2021)

23. Wang, Y., Ma, L.: Saliency guided depth prediction from a single image (2019)

24. Wu, G., Li, K., Wang, L., Hu, R., Guo, Y., Chen, Z.: Hi-net: boosting self-supervised indoor depth estimation via pose optimization. IEEE Rob. Autom. Lett. **8**(1), 224–231 (2022)

25. Zhang, Y., Wang, H., Yang, G., Zhang, J., Gong, C., Wang, Y.: CSNET: a convnext-based siamese network for RGB-D salient object detection. Visual Comput. 1–19 (2023)

26. Zhou, T., Fan, D.P., Cheng, M.M., Shen, J., Shao, L.: RGB-D salient object detection: a survey. Comput. Visual Media **7**, 37–69 (2021)

27. Zuo, Y., Wang, H., Fang, Y., Huang, X., Shang, X., Wu, Q.: MIG-NET: multi-scale network alternatively guided by intensity and gradient features for depth map super-resolution. IEEE Trans. Multimedia **24**, 3506–3519 (2021)

Mask-Guided Joint Single Image Specular Highlight Detection and Removal

Hao Chen[1] [iD], Li Li[1,2,3]([✉]) [iD], and Neng Yu[1] [iD]

[1] School of Computer Science and Artificial Intelligence, Wuhan Textile University,
Wuhan 430200, China
`lli@wtu.edu.cn`
[2] Engineering Research Center of Hubei Province for Clothing Information,
Wuhan 430200, China
[3] Hubei Provincial Engineering Research Center for Intelligent Textile and Fashion,
Wuhan 430200, China

Abstract. Detecting and removing specular highlights is a highly challenging task that can effectively enhance the performance of other computer vision tasks in real-world scenarios. Traditional highlight removal algorithms struggle to accurately distinguish the differences between pure white or nearly white materials and highlights, while recent deep learning-based highlight removal algorithms suffer from complex network architectures, lack of flexibility, and limited object adaptability. To address these issues, we propose an end-to-end framework for single-image highlight detection and removal that utilizes mask guidance. Our framework adopts an encoder-decoder structure, with EfficientNet serving as the backbone network for feature extraction in the encoder. The decoder gradually restores the feature map to its original size through upsampling. In the highlight detection module, the network layers are deepened, and residual modules are introduced to extract more feature information and improve detection accuracy. In the highlight removal module, we introduce the Convolutional Block Attention Module, which dynamically learns the importance of each channel and spatial position in the input feature map. This helps the model better distinguish foreground and background and improve adaptability and accuracy in complex scenes.Experimental results demonstrate that our proposed method outperforms existing methods, as evaluated on the public SHIQ dataset through comparative experiments. Our method achieves superior performance in highlight detection and removal.

Keywords: Highlight detection · Highlight removal · Attention · End to End · Encoder-decoder

1 Introduction

Specular highlights are a common physical phenomenon that typically appear as bright spots on the surface of smooth objects, altering the color of the object

Q. Liu et al. (Eds.): PRCV 2023, LNCS 14433, pp. 457–468, 2024.
https://doi.org/10.1007/978-981-99-8546-3_37

surface, disrupting its contour, and obscuring its texture. The presence of specular highlights can interfere with various computer vision and image processing tasks, such as image segmentation, edge detection, and precision measurement, thereby reducing the overall performance of visual systems. Therefore, effective detection and removal of specular highlights can greatly benefit a variety of visual tasks in real-world scenarios.

As an important research topic in the field of computer vision, many methods have been proposed in the past few decades to address the challenges of specular highlight detection and removal. These methods can be roughly divided into two categories: model-based methods and deep learning-based methods. Model-based methods mostly refer to traditional approaches such as optimization-based methods [3], clustering [15], and filtering [23]. However, these methods have limitations in that they lack semantic information of the image, making it difficult to remove highlights and pure white colors semantically from complex real-world scenes.In recent years, the majority of methods for specular highlight detection and removal have shifted towards deep learning-based approaches, which have achieved significant progress [5,18,24]. However, these methods also face some common challenges. Firstly, most of them rely on synthetic datasets, which limits their ability to remove highlights from real-world scenes. Secondly, the transition between highlight and non-highlight regions in the removal results is often not seamless enough, leading to the appearance of artifact regions and reduced realism in the highlight removal results.

Our research proposes an end-to-end framework for detecting and removing specular highlights in single images using a mask-guided approach. The framework consists of an encoder-decoder structure, with the encoder utilizing EfficientNet as the feature extraction backbone network to extract features from the input RGB image. These features are then used to generate a predicted mask through the decoder, while also being fed into a highlight detection module (HLDM) to generate another predicted mask. The input RGB image and predicted mask are then processed through a highlight removal module (HLRM) to generate a sequence of predicted RGB images with gradually reduced highlights. By sharing features between HLDM and HLRM, the framework can focus on the global and local features of highlight images, leading to more realistic highlight removal results. Additionally, the framework utilizes the CBAM attention mechanism and the lightnessAdjustModule module to better learn input information, improving the model's accuracy and adaptability for complex scenes and highlight removal tasks.

Our contributions in this network are as follows:

1. We obtain two different-scale specular highlight masks from the decoder's second and fourth layers, respectively, and then restore them to the original image size through interpolation. This can better capture the details of different scales in the image and improve the performance and effectiveness of the model;

2. We deepen the network layers in the highlight detection module and introduce ResidualBlocks to extract more feature information, thereby improving the accuracy of highlight detection;

3. We use the UNet [11] structure as a highlight removal module and introduce the CBAM attention mechanism in the UNetConvBlock base module. This mechanism captures important information in the input feature map through channel attention and spatial attention modules, and combines them to improve the model's ability to capture features at different scales and judge their importance, thereby improving the performance and effectiveness of the model.

Based on the aforementioned three points, our approach effectively helps the model learn input image information better, thereby improving the adaptability and accuracy of the highlight removal task.

2 Related Work

In this section, we will review related methods for highlight detection in Sect. 2.1 and highlight removal in Sect. 2.2.

2.1 Highlight Detection

Most early studies were based on various thresholding schemes. In most cases, only a small portion of the image contains highlight regions. Zhang et al. [25] defined the concept of highlight detection as non-negative matrix factorization (NMF). Fu et al. [4] designed a deep learning-based SHDNet for mirror highlight detection, which utilized multi-scale contextual contrast features to accurately detect mirror highlights of different scales, making it easier to detect highlights of various sizes. The following year, Fu et al. [5] designed a multi-task network, JSHDR, for joint mirror highlight detection and removal. To accurately detect and remove highlights of different sizes, they further proposed an extended Dilated Spatial Contextual Feature Aggregation (DSCFA) module to extract and aggregate multi-scale spatial contextual information to deal with different highlight regions. Wang et al. [17] proposed a new deep supervised network for highlight detection and introduced a self-attention module, PSCA, to detect highlights through contextual features. Additionally, they proposed a new highlight detection fine-tuning method that introduces highlight contours to constrain the highlight detection mask. They also designed a new HMC loss function to improve detection performance. However, these approaches have some common limitations in terms of generality, scalability, and high computational complexity, making it difficult to meet real-time requirements.

2.2 Highlight Removal

Currently, there are many methods for highlight removal, which can be broadly divided into two categories: traditional methods based on physical models and data-driven deep learning methods.

Traditional methods based on physical models are typically based on the basic principles of optical physics and geometry. By simulating the physical processes of highlight reflection formation, the corresponding reflection model or constraint conditions are established to achieve the effect of highlight removal. Yang [21] proposed a method based on the H-S color space, which separates the reflection components by adjusting the saturation of mirror pixels to the value of only diffuse pixels with the same diffuse chromaticity. Recently, Ramos et al. [10] removed highlights by matching histograms in the YCbCr color space based on a two-color reflection model. Feng et al. [2] innovatively combined the advantages of polarization properties and light field imaging, established the relationship between image grayscale and polarization angle based on the Stokes parameters of the light field, removed mirror highlights, and restored the output image using a color transfer algorithm. These methods typically require accurate optical parameters and geometric information as data support. Therefore, under complex lighting conditions, the actual effect of traditional methods based on physical models will be greatly discounted. In summary, existing traditional algorithms cannot semantically remove highlights and pure white from complex real-world scenes.

Data-driven deep learning methods have achieved great success in highlight removal. These methods use deep convolutional neural networks (CNN) to learn the mapping relationship of highlight removal from a large amount of annotated image or video data. Through end-to-end learning, these methods can automatically extract features from images and generate high-quality highlight removal results. Hou et al. [7] designed a two-stage network that includes a highlight detection network and a highlight removal network, using the output of the highlight detection network to guide the subsequent highlight removal network to remove highlights. Fu et al. [5] created a dataset containing 16k real images and proposed a multi-task network for joint highlight detection and removal, but it cannot effectively remove highlights in colored lighting. Wu et al. [19] obtained a highlight area mask through an encoder-decoder framework and used the mask to guide highlight removal. However, this method is sensitive to area size and color, and the actual effect on images with large-area highlights or highlights with text is not satisfactory.

3   Methods

In this section, we first present the overall architecture of our proposed method introduced in Sect. 3.1. Then, we describe our highlight detection module in Sect. 3.2 and highlight removal module in Sect. 3.3. Finally, the loss function will be provided in detail in Sect. 3.4.

3.1   Network Architecture

In the network structure proposed in this paper, the input is an RGB image with highlights, and the output is a mask of the highlight image and the corresponding image without highlights. The detailed network diagram is shown in Fig. 1.

Specifically, we first use a pre-trained EfficientNet [16] as the feature extraction module and input the features of the feature extraction module into the highlight detection module (HLDM) and the highlight removal module (HLRM) to obtain the final image without highlights.

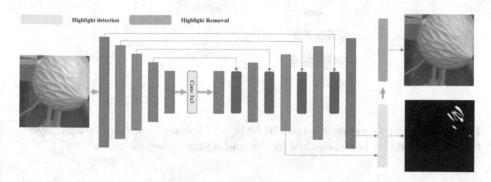

Fig. 1. Overview of our proposed network. From the figure, we can see that the input of the network is an RGB image with highlights, and the output is the corresponding RGB image without highlights and the mask image of the highlight image. The overall network architecture adopts an encoder-decoder structure. First, we use a pre-trained EfficientNet [16] as a feature extractor to obtain multi-level encoded features, which are simultaneously input into the highlight detection module (detailed structure in Sect. 3.2) and the highlight removal module (detailed structure in Sect. 3.3) to obtain the predicted RGB image without highlights and the mask image of the highlight image.

3.2 Highlight Detection Module (HLDM)

The Highlight Detection Module (HDLM) is mainly used to generate a highlight mask for high-light images. It consists of convolutional layers, batch normalization layers, residual blocks, and a sigmoid activation function. The specific structure diagram is shown in Fig. 2, and the generated mask can be used to guide the subsequent Highlight Removal Module (HLRM) for highlight removal operations.

Specifically, the first layer of the network is a 'nn.Conv2d' convolutional layer used for feature extraction. Then, the network uses two 'ResidualBlock' modules for feature extraction, each containing two convolutional layers and a residual connection. Each residual block uses batch normalization layers and LeakyReLU activation functions. Finally, the network uses a 'nn.Conv2d' convolutional layer to output the predicted highlight mask, which is then non-linearly mapped through a 'nn.Sigmoid' activation function. The combination of these layers is simple and efficient, and achieves high accuracy and robustness.

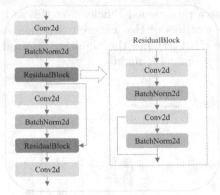

Fig. 2. Highlight Detection Module (HLDM), which consists of convolutional layers, batch normalization layers, residual blocks, and a sigmoid activation function.

3.3   Highlight Removal Module (HLRM)

The Highlight Removal Module (HLRM) model consists of a convGRU combination composed of multiple lightnessAdjustModules, which are used to process the input of the image and the mask and generate the output RGB image. The model takes an RGB image and a mask (from the HLDM module) as input and outputs a series of RGB images with highlights removed. Each convGRU module in the model has a hidden state to store the internal state of the module. In each iteration, the model takes the current predicted image and mask as input, then uses the convGRU module to process the input and generate output. The output of the model is used as input for the next iteration, gradually achieving the effect of removing highlights. The lightnessAdjustModule model includes:

1. **Unet optimization module:** is used to extract image features. The encoder consists of four UNetConvBlocks, each block containing two convolutional layers, a batch normalization layer, ReLU activation function, and a CBAM attention module. The detailed information is shown in Fig. 3(a). The CBAM attention mechanism adjusts the feature map in both channel and spatial dimensions to enhance its representational power. UNetConvBlock is repeatedly used in each resolution layer of the U-Net encoder and decoder, contributing to accurate semantic segmentation and image processing tasks. The CBAM attention mechanism dynamically learns the importance of each channel and spatial position in the input feature map, better distinguishing foreground from background. It also performs effective channel and spatial scaling of the input feature map, expanding the receptive field, which helps the model to better capture global contextual information and improve segmentation performance.
2. **ConvGRUCell** [14] is a gated recurrent neural network module that performs multiple recursive calculations on features to enhance their representation ability. It takes features as input and performs recursive calculations

at multiple time steps, with each module having a hidden state to store its internal state. In this experiment, ConvGRUCell is used to process the input of current predicted image and mask, and generate output. These outputs are used as input for the next iteration to gradually achieve the de-highlighting effect. The detailed structure of ConvGRUCell is shown in Fig. 3(b). The calculation formula of ConvGRU is shown in the Eq. (1–4):

$$R_t = \sigma(X_t * W_{xr} + H_{t-1} * W_{hr}) \tag{1}$$

$$Z_t = \sigma(X_t * W_{xz} + H_{t-1} * W_{hz}) \tag{2}$$

$$\widetilde{H}_t = tanh(X_t * W_{xh} + (R_t \odot H_{t-1}) * W_{hh}) \tag{3}$$

$$H_t = Z_t \odot H_{t-1} + (1 - Z_t) \odot \widetilde{H}_t \tag{4}$$

where σ is the activation function; $*$ denotes convolution operation; $\odot$ denotes element-wise multiplication; R_t is the reset gate; Z_t is the update gate; x_t is the input of the network layer at time t; H_{t-1} is the hidden state at time $t-1$; and $\widetilde{H}_t$ is the candidate set.

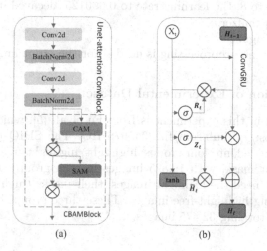

(a)   (b)

Fig. 3. (a) shows the basic structure of the Unet optimization module, and (b) shows the internal structure of the ConvGRU.

3.4 Loss Function Design

The overall loss function consists of two parts: the loss of the segmentation mask and the loss of the highlight-removed image. In this experiment, we use the same loss function for both parts, which is shown in the following Eq. (5–8):

$$SSIM_{loss} = 1 - SSIM(input, target) \tag{5}$$

$$smooth_{l1} = \begin{cases} 0.5 \cdot x^2 & , |x| < 1 \\ |x| - 0.5 & , otherwise \end{cases} \quad (6)$$

$$Loss = \lambda_1 \cdot smooth_{l1} + \lambda_2 \cdot ssim_{loss} \quad (7)$$

The overall loss function is:

$$Total_{Loss} = \beta_1 \cdot mask_{loss} + \beta_2 \cdot rgb_{loss} \quad (8)$$

Through experimental optimization and comparative analysis, it was found that the highlight removal effect is optimal when $\lambda_1 = 0.3$, $\lambda_2 = 0.7$, $\beta_1 = \beta_2 = 0.5$.

4 Experiment

4.1 Experimental Environment

The experiments in this paper were implemented in PyTorch and conducted on a single NVIDIA RTX 2080Ti-11G GPU.

Training settings:

During training, we uniformly resized the input images to 200*200 (pixels), set the batch size to 8, the learning rate to 0.000125, decayed it per epoch, and used the Adam optimizer.

Testing settings:

During testing, no preprocessing is needed for the input images.

4.2 Introduction of Experimental Dataset

The dataset used in this experiment is from Fu et al. [5], which can be downloaded from https://github.com/fu123456/SHIQ. The SHIQ dataset contains 43,300 images (divided into four groups: highlight, highlight-free, highlight intensity, and highlight mask, with 10,825 images in each group). For our research purposes, we only need the highlight images, their masks (highlight mask), and their corresponding highlight-free images. These three groups of images constitute our dataset, totaling 32,475 images.

4.3 Results

Quantitative Analysis. In this section, We used PSNR and SSIM values for quantitative analysis of the highlight removal task. In Table 1, we presented the quantitative comparison results of our method with the latest and widely recognized highlight removal methods on the SHIQ dataset. The best result is highlighted in bold in Table 1, which shows the detailed PSNR and SSIM values of different methods on the SHIQ dataset. The results indicate that our network is slightly inferior to Fu [5] in terms of PSNR, but has a significant improvement in SSIM.

Qualitative Analysis. In this section, we conducted a qualitative analysis on a subset of images from the dataset, and Fig. 4 shows the results of various specular highlight removal methods. From the results, it can be observed that traditional methods based on optimization and color analysis cannot effectively remove specular highlights or may cause color distortion. In contrast, our method can successfully remove specular reflections, preserve the details and structural information of the image, and restore the overall color balance, making the image more realistic, natural, and visually comfortable. These results demonstrate the superiority of our method in specular highlight removal tasks, and the qualitative analysis validates the effectiveness and robustness of our approach, providing a reliable foundation for subsequent experimental results.

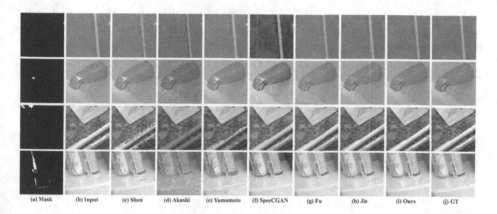

(a) Mask   (b) Input   (c) Shen   (d) Akashi   (e) Yamamoto   (f) SpecCGAN   (g) Fu   (h) Jie   (i) Ours   (j) GT

Fig. 4. Visual comparison of our method with state-of-the-art specular highlight removal methods on real-world images.

4.4 Ablation Experiments

In order to verify the effectiveness of the main components of our network, we made corresponding modifications to the network to construct baselines as follows:

- N_1:Replace the original simple two convolutional networks with a residual network (as described in Sect. 3.2 above) only in the highlight detection module (HLDM) to improve the accuracy and robustness of the model.
- N_2:Add CBAM attention only in the highlight removal module (HLRM).

The PSNR and SSIM results of our network framework and two baselines are presented in Table 2. The results indicate that our network outperforms N_1 and N_2 in terms of both SSIM and PSNR .Specifically, in the ablation experiments, we introduced a residual module only in our final HDLM module for N_1, while excluding the CBAM module in the HLRM. Through experimental comparisons, we obtained a PSNR of 32.71 and SSIM of 0.9473, which still have some distance from our final experimental metrics. This effectively demonstrates the effectiveness of introducing the CBAM attention mechanism in the HLRM. For N_2, we added CBAM attention only in our final HLRM module, while replacing the residual part in HDLM with a simple two-layer convolutional network. Through experimental comparisons, we obtained a PSNR of 32.59 and SSIM of 0.9568, which still have some distance from our final experimental metrics. This also effectively demonstrates the effectiveness of introducing the residual network in HDLM. In summary, for the ablation experiments, we employed a subtraction operation on the final network. By comparing the experimental metrics before and after removing the added module, we verified the effectiveness of each module.

Table 1. Quantitative comparison of our method with state-of-the-art and well-accepted specular highlight removal methods on the SHIQ dataset. The best result is highlighted in bold.

Methods	PSNR$_{uparrow}$	SSIM$_{uparrow}$
Tan [8]	11.04	0.40
Shen [12]	13.90	0.42
Yang [22]	14.31	0.50
Akashi [1]	14.01	0.52
Guo [6]	17.18	0.58
Yamamoto [20]	19.54	0.63
Shi [13]	18.21	0.61
Yi [24]	21.32	0.72
Fu [5]	**34.13**	0.86
Jie [9]	30.12	0.93
Our	33.67	**0.95**

Table 2. Analysis of the highlight removal performance metrics on the SHIQ dataset, with the best result highlighted in bold.

Methods	Ours	Fu [5]	N_1	N_2
PSNR↑	33.67	**34.13**	32.71	32.59
SSIM ↑	**0.9578**	0.86	0.9473	0.9568

5   Conclusion

The proposed framework for joint highlight detection and removal in single-image based on Mask guidance utilizes an encoder-decoder structure and EfficientNet as the feature extraction backbone network and attention mechanism, which can better distinguish foreground and background and improve adaptability and accuracy for complex scenes in highlight removal tasks. Although experimental results show that the method achieves significant effects in highlight removal tasks, making images more realistic and natural while retaining details and structural information, highlight removal is still a complex task, and there are still situations where highlights cannot be completely removed or cause loss of image details. Therefore, future research can explore more refined feature representations and attention mechanisms to improve the accuracy of identifying and removing highlight regions, combine multimodal information such as depth maps and normal maps to further enhance the robustness and generalization ability of the algorithm, introduce methods such as Generative Adversarial Networks (GAN) to generate more realistic highlight removal results and improve image quality, and apply the method to real-world scenarios, such as deploying the algorithm for real-time highlight removal on terminals, to further improve the effectiveness of highlight removal and expand its application areas. In summary, these research directions have the potential to further improve the effectiveness of highlight removal and apply it to a wider range of fields and scenarios.

References

1. Akashi, Y., Okatani, T.: Separation of reflection components by sparse non-negative matrix factorization. In: Cremers, D., Reid, I., Saito, H., Yang, M.-H. (eds.) ACCV 2014. LNCS, vol. 9007, pp. 611–625. Springer, Cham (2015). https://doi.org/10.1007/978-3-319-16814-2_40
2. Feng, W., Cheng, X., Sun, J., Xiong, Z., Zhai, Z.: Specular highlight removal and depth estimation based on polarization characteristics of light field. Optics Commun. **537**, 129467 (2023)
3. Feng, W., Li, X., Cheng, X., Wang, H., Xiong, Z., Zhai, Z.: Specular highlight removal of light field based on dichromatic reflection and total variation optimizations. Opt. Lasers Eng. **151**, 106939 (2022)
4. Fu, G., Zhang, Q., Lin, Q., Zhu, L., Xiao, C.: Learning to detect specular highlights from real-world images. In: Proceedings of the 28th ACM International Conference on Multimedia, pp. 1873–1881 (2020)
5. Fu, G., Zhang, Q., Zhu, L., Li, P., Xiao, C.: A multi-task network for joint specular highlight detection and removal. In: Proceedings of the IEEE/CVF Conference on Computer Vision and Pattern Recognition, pp. 7752–7761 (2021)
6. Guo, J., Zhou, Z., Wang, L.: Single image highlight removal with a sparse and low-rank reflection model. In: Proceedings of the European Conference on Computer Vision (ECCV), pp. 268–283 (2018)
7. Hou, S., Wang, C., Quan, W., Jiang, J., Yan, D.-M.: Text-aware single image specular highlight removal. In: Ma, H., Wang, L., Zhang, C., Wu, F., Tan, T., Wang, Y., Lai, J., Zhao, Y. (eds.) PRCV 2021. LNCS, vol. 13022, pp. 115–127. Springer, Cham (2021). https://doi.org/10.1007/978-3-030-88013-2_10
8. Ikeuchi, K., Miyazaki, D., Tan, R.T., Ikeuchi, K.: Separating reflection components of textured surfaces using a single image. In: Digitally Archiving Cultural Objects, pp. 353–384 (2008)
9. Jie, L., Zhang, H.: MGRLN-NET: mask-guided residual learning network for joint single-image shadow detection and removal. In: Proceedings of the Asian Conference on Computer Vision, pp. 4411–4427 (2022)

10. Ramos, V.S., Júnior, L.G.D.Q.S., Silveira, L.F.D.Q.: Single image highlight removal for real-time image processing pipelines. IEEE Access **8**, 3240–3254 (2019)
11. Ronneberger, O., Fischer, P., Brox, T.: U-net: convolutional networks for biomedical image segmentation. In: Navab, N., Hornegger, J., Wells, W.M., Frangi, A.F. (eds.) MICCAI 2015. LNCS, vol. 9351, pp. 234–241. Springer, Cham (2015). https://doi.org/10.1007/978-3-319-24574-4_28
12. Shen, H.L., Zheng, Z.H.: Real-time highlight removal using intensity ratio. Appl. Opt. **52**(19), 4483–4493 (2013)
13. Shi, J., Dong, Y., Su, H., Yu, S.X.: Learning non-lambertian object intrinsics across shapenet categories. In: Proceedings of the IEEE Conference on Computer Vision and Pattern Recognition, pp. 1685–1694 (2017)
14. Shi, X., et al.: Deep learning for precipitation nowcasting: a benchmark and a new model. Adv. Neural Inf. Process. Syst. **30** (2017)
15. Souza, A.C., Macedo, M.C., Nascimento, V.P., Oliveira, B.S.: Real-time high-quality specular highlight removal using efficient pixel clustering. In: 2018 31st SIBGRAPI Conference on Graphics, Patterns and Images (SIBGRAPI), pp. 56–63. IEEE (2018)
16. Tan, M., Le, Q.: Efficientnet: rethinking model scaling for convolutional neural networks. In: International Conference on Machine Learning, pp. 6105–6114. PMLR (2019)
17. Wang, C., Wu, Z., Guo, J., Zhang, X.: Contour-constrained specular highlight detection from real-world images. In: The 18th ACM SIGGRAPH International Conference on Virtual-Reality Continuum and its Applications in Industry, pp. 1–4 (2022)
18. Wu, S., et al.: Specular-to-diffuse translation for multi-view reconstruction. In: Proceedings of the European conference on computer vision (ECCV), pp. 183–200 (2018)
19. Wu, Z., Guo, J., Zhuang, C., Xiao, J., Yan, D.M., Zhang, X.: Joint specular highlight detection and removal in single images via unet-transformer. Comput. Visual Media **9**(1), 141–154 (2023)
20. Yamamoto, T., Nakazawa, A.: General improvement method of specular component separation using high-emphasis filter and similarity function. ITE Trans. Media Technol. Appl. **7**(2), 92–102 (2019)
21. Yang, J., Liu, L., Li, S.: Separating specular and diffuse reflection components in the HSI color space. In: Proceedings of the IEEE International Conference on Computer Vision Workshops, pp. 891–898 (2013)
22. Yang, Q., Tang, J., Ahuja, N.: Efficient and robust specular highlight removal. IEEE Trans. Pattern Anal. Mach. Intell. **37**(6), 1304–1311 (2014)
23. Yang, Q., Wang, S., Ahuja, N.: Real-time specular highlight removal using bilateral filtering. In: Daniilidis, K., Maragos, P., Paragios, N. (eds.) ECCV 2010. LNCS, vol. 6314, pp. 87–100. Springer, Heidelberg (2010). https://doi.org/10.1007/978-3-642-15561-1_7
24. Yi, R., Tan, P., Lin, S.: Leveraging multi-view image sets for unsupervised intrinsic image decomposition and highlight separation. In: Proceedings of the AAAI Conference on Artificial Intelligence, vol. 34, pp. 12685–12692 (2020)
25. Zhang, W., Zhao, X., Morvan, J.M., Chen, L.: Improving shadow suppression for illumination robust face recognition. IEEE Trans. Pattern Anal. Mach. Intell. **41**(3), 611–624 (2018)

CATrack: Convolution and Attention Feature Fusion for Visual Object Tracking

Longkun Zhang, Jiajun Wen$^{(\boxtimes)}$, Zichen Dai, Rouyi Zhou, and Zhihui Lai

College of Computer Science and Software Engineering, Shenzhen University,
Shenzhen 518060, China
enjoy_world@163.com

Abstract. In visual object tracking, information embedding and feature fusion between the target template and the search have been hot research spots in the past decades. Linear convolution is a common way to perform correlation operations. The convolution operation is good at processing local information, while ignoring global information. By contrast, the attention mechanism has the advantages of innate global information modeling. To model the local information of the target template and the global information of the search area, we propose a convolution and attention feature fusion module (CAM). Thus, the efficient information embedding and feature fusion can be achieved in parallel. Moreover, a bi-directional information flow bridge is constructed to realize information embedding and feature fusion between the target template and the search area. Specifically, it includes a convolution-to-attention bridge module (CABM) and an attention-to-convolutional bridge module(ACBM). Finally, we present a novel tracker based on convolution and attention (CATrack), which combines the advantages of convolution and attention operators, and has enhanced ability for accurate target positioning. Comprehensive experiments have been conducted on four tracking benchmarks: LaSOT, TrackingNet, GOT-10k and UAV123. Experiments show that the performance of our CATrack is more competitive than the state-of-the-art trackers.

Keywords: Visual object tracking · Attention learning · Feature fusion

1 Introduction

Visual object tracking is a fundamental and challenging task in the field of computer vision. Even though great progress have been made by the recent studies, the difficulties such as occlusions, out-of-view, deformation, background cluttering and other variations, still exist.

In recent years, the mainstream tracking framework is based on Siamese network [1,13,18,19,27,32]. These trackers use a backbone network of shared network parameters to extract the features of the target template and search area. Then, the features of the target template and the search area are fused to achieve sufficient information embedding, which provides a reliable response

Q. Liu et al. (Eds.): PRCV 2023, LNCS 14433, pp. 469–480, 2024.
https://doi.org/10.1007/978-981-99-8546-3_38

map for the subsequent tracking head network to decode the target information. The state-of-the-art Siamese trackers [1,18,19,32] use cross-correlation convolution network layers on the deep feature maps to achieve feature fusion and information embedding. By using the target template as the convolution kernel to linearly slide on the feature maps of the search area, the local features can be extracted. However, global information can hardly be detected and extracted. Recently, the attention mechanism in vision Transformer [9] has shown great advantage in processing global information and achieves remarking performance over CNNs.

In this paper, we propose a convolution operation and attention feature fusion module (CAM) to realize feature extraction and feature fusion of the two branches of the target template and the search area. Meanwhile, the collaboration of local information processing and global information modeling can be realized. In order to achieve sufficient information embedding, a bi-directional bridge of information flows is established between the convolution operation of the target template and the attention operation of the search area. The size of the target template is smaller than that of the search area. In this paper, we use convolution operation to model the target template features, and use attention mechanism to process global information and model the search area. By introducing the convolution operation and attention combined module, we build a novel tracker, named Convolution and Attention Feature Fusion for Visual Object Tracking (CATrack). The main contributions of our work are as follows:

- We design a convolution and attention feature fusion module (CAM), which realizes the coordination of local information processing and global information modeling, and establish a bi-directional bridge between the convolution operation of the target template and the attention operation of the search area to achieve full feature fusion.
- We propose a novel tracker CATrack, which is simple but efficient. Combining with the advantages of convolution operator and attention mechanism, CATrack is more adaptable and robust to complex scenes compared to previous trackers.
- The comprehensive experiments on several public datasets show that our tracker exhibits more competitive performance than the state-of-the-art trackers.

2   Related Work

Siamese Tracking. In recent years, Siamese network based trackers [1,13, 14,18,19,27,32,36] have gained great attention due to their excellent performance. These trackers are mainly composed of three parts: a backbone network to extract the features of the target template and search area, a similarity matching module for information embedding and feature fusion, and a tracking head to decode the fused features. These trackers formulate visual tracking as a cross-correlated problem. To generate a similarity map from the Siamese branch, they train two neural network branches that contain the object template and the search area. To further optimize the tracking model of Siamese network, some researchers [18,35] mainly focus on the feature extraction of the backbone network and the design of better tracking heads to obtain more accurate bounding box regression. However, less attention has been paid to the feature fusion between the target template and the search area.

Transformers and Attention. Recently, Transformers [26] have been proposed for the task of machine text translation. The attention mechanism shows strong global information processing ability. Later, DETR [3] and ViT [9] were proposed to open the Pandolai box of the attention mechanism in computer vision field. Due to the good ability of attention mechanism to model long-order dependencies, ViT and its variants have demonstrated impressive and superior performance on multiple vision tasks. In order to control the computational complexity and achieve multi-scale network architectures, Swin Transformer [21] has been proposed, which proposes window-based attention operations to achieve state-of-the-art in multiple vision tasks.

Convolution and Attention Combination. Several recent works [8,12] demonstrate the superiority of combining convolutions and attention. Attention mechanism offers advantages for long-term sequence dependency modeling and global information fusion extraction. To realize the full feature fusion for visual object tracking, we combine the advantages of convolution operation for modeling local information and attention operation for modeling long-order dependencies and global information. The attention mechanism is used to make up for the weakness of the long-order modeling of the convolution operation, while retaining the strength of its local information fusion, so as to achieve a more efficient and robust target tracker.

3   Method

3.1   Convolution and Attention Feature Fusion

The convolution operation is a linear operation, which has a good advantage of local information modeling. The attention mechanism has the ability to capture the global information so as to extract the global features. After the attention was proposed in the transformer framework, it has been widely used in feature extraction due to its advantages as mentioned above [26]. Combining these operations in parallel enables a full use of their advantages to realize local and global information embedding and feature modeling.

In this paper, the convolution operation is used for the feature extraction of the target template, which can model the local information of the feature map of the target template. The attention operation model the search area and capture global information for feature extraction in a large-scale search area.

Specifically, as shown in Fig. 1, pointwise (PW) convolution mainly realizes the alignment operation of the feature map and depthwise (DW) convolution realizes the feature extraction of the target template. Attention operation are

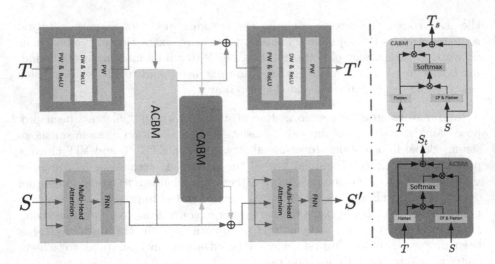

Fig. 1. Convolution and attention feature fusion module (CAM). The convolution part consists of pointwise (PW) and depthwise (DW) convolutions to model the local information of the target template. The attention part consists of multi-head attention and FNN to model the global information of the search area. The bi-directional information flow bridge includes convolution-to-attention bridge module (CABM) and attention-to-convolution bridge module (ACBM). The proposed two modules realize information embedding and feature fusion between the target template and the search area.

used to establish global dependencies on the large-scale search area and global features are extracted to locate the target on the search area. The feature extraction of the target template and the search area are performed in parallel under the convolution operation and the attention operation, respectively. This feature fusion scheme achieves the optimal modeling of the local information of the target template and the efficient modeling of the global information of the search area, which provides the semantic feature expression for the target positioning.

3.2   Feature Fusion Bi-directional Bridge

After sufficient feature extraction of the target template and the search area, it is also necessary to perform sufficient feature fusion. We build a bi-directional bridges between the target template and the search area to perform feature extraction and feature fusion.

To transfer the embedded information from the target template to the search area, we build a convolution to attention bridge module(CABM). First, the query vectors are obtained after the feature vectors of the search area are convolutional linear projected. For the feature vector of the target template, we directly use it without space mapping. The feature vectors of the target template are adjusted to form the key vector and value vector. We use multi-head attention, thus the CABM can be summarized as

$$T_s = Concat(H_1^s, ..., H_n^s)W_s^O \tag{1}$$

$$H_i^s = softmax\left(\frac{Flatten(CP(S))Flatten(T)}{\sqrt{C_\varphi}}\right)Flatten(T) \qquad (2)$$

where **Concat**, **Faltten** and **CP** represent the concatenation, flattening and convolution projection, respectively. The search area is more important for feature extraction than the target template in visual object tracking, so we do a linear mapping of the convolution projection. So the discriminant ability of the network can be enhanced by bridging the feature gap between the target template and the search area. We build attention to convolution bridge modules (ACBM) to perform information embedding based on the search area and the target template. The feature maps extracted by convolution operation are used as the query vectors without performing space mapping. For the feature vectors of the search area, we use convolution projection to obtain two new feature space vectors as the key vectors and the value vectors. The feature maps of the target template will converge more expressive features after the fusion of the feature maps of the search area. Thus, we can formulate the ACBM as

$$S_t = Concat(H_1^t, ..., H_n^t)W_t^O \qquad (3)$$

$$H_i^t = softmax\left(\frac{Flatten(T)Flatten(CP(S))}{\sqrt{C_\varphi}}\right)Flatten(CP(S)) \qquad (4)$$

By building bi-directional bridges of two information flows, we realize the information interaction and the deep feature fusion between the target template and the search area while the convolution and attention operation jointly extract the target template and search area features. We achieve the synchronization of feature extraction and feature fusion. Moreover, the collaborative interaction is realized between convolution operation for local feature extraction and attention operation for global information embedding.

3.3   Overall Tracking Architecture

We build our object tracker (CATrack) shown in Fig. 2. The whole object tracker adopts the Siamese network architecture.

Backbone Network for Feature Extraction. Feature extraction of our CATrack based on siamese network contains two branches which share the weight parameters of the convolutional neural network. We adopt ResNet-50 [15] as the backbone network to extract the features of the object template and the search area. We only adopt the first four stages of the ResNet-50 as our backbone network. In the fourth stage, the down-sampling step of the convolution operation is modified from 2 to 1 to increase the spatial size of the feature maps. This modification increases the resolution of the output features to improve the spatial expressiveness for object localization. The entire backbone network takes image pairs as input: one image as the target template $T_0 \in \mathbb{R}^{3 \times H_{t0} \times W_{t0}}$ and other image as the search area $S_0 \in \mathbb{R}^{3 \times H_{s0} \times W_{s0}}$.

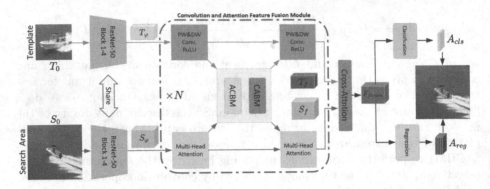

Fig. 2. The architecture of our proposed tracker CATrack. ResNet-50 is used as the backbone network to extract the features of the object template and the search area. The convolution and attention combination module performs information embedding and feature fusion, which includes a bi-directional information flow bridge between the target template and the search area. The prediction head consists of a classification branch and a regression branch.

Feature Fusion Modules. After extracting the features of the target template and the search area through the backbone network ResNet-50, our CATrack need to realize the information interaction between the target and the search area. The input of the whole feature fusion module consists of two parts generated by the backbone feature extraction network: one is the feature maps $T_\varphi \in \mathbb{R}^{C_\varphi \times H_t \varphi \times W_t \varphi}$ and the other is the feature maps $S_\varphi \in \mathbb{R}^{C_\varphi \times H_s \varphi \times W_s \varphi}$. While the convolution operation and the attention operation cooperate in parallel to model local information and global information, we use CABM and ACBM for information embedding to achieve full feature fusion. The entire module works with convolution operation, attention operation and the bi-directional bridge, iterating N times to generate features $T_f \in \mathbb{R}^{C_{tf} \times H_{tf} \times W_{tf}}$ and $S_f \in \mathbb{R}^{C_{sf} \times H_{sf} \times W_{sf}}$. After experiments, the effect is best when N is 3. Too many layers may have parameter redundancy, resulting in poor learning effect, and too few layers cannot carry out sufficient feature extraction and fusion. Then a cross-attention operation is performed on T_f and S_f to obtain a fused feature $F_{fusion} \in \mathbb{R}^{C_f \times H_f \times W_f}$, which provides a strong feature expression ability for target positioning.

Prediction Head. The prediction head contains two branches: one for classification and the other for regression. For both branches, we stack three MLP layers. Except for the last layer after each MLP layer, the activation function ReLu is immediately followed. The classification result $A_{cls} \in \mathbb{R}^{d_{cls} \times H_f \times W_\upsilon}$ of foreground and background will be generated after the classification branch head. The regression branch head generates a normalized coordinate vector $A_{reg} \in \mathbb{R}^{d_{reg} \times H_f \times W_\upsilon}$ whose size corresponds to that of F_{fusion}. The prediction head of the tracker directly predicts the normalized coordinates of the target in the search area without adopting the prior knowledge of anchor points or anchor boxes, which makes the whole tracker more streamlined.

4   Experiments

4.1   Implementation Details

In our experiments, we used the modified ResNet-50 [15] as backbone feature extraction network, where ResNet-50 is pre-trained on the ImageNet [25] dataset. The training datasets on which we trained our tracker includ LaSOT [10], GOT-10k [16], TrackingNet [23], and COCO [20] datasets. The backbone's learning rate was set to 1e-5, and other parameters' learning rates were set to 1e-4. The total sample pairs of each epoch are 1000 and the total epoch is 1000. The weight decay was set to 1e-4 and the learning rate decreased by factor 10 after 500 epochs. The loss function consists of a classification branch and a regression branch. Refer to [5] for more details. Finally, the proposed tracker can speed up to 39 FPS on a single GeForce RTX 3090 GPU.

4.2   Comparison with the State-of-the-Art Methods

To verify the performance of CATrack we conducted extensive experiments on five benchmarks: LaSOT [10], TrackingNet [23], GOT-10k [16], UAV123 [22].

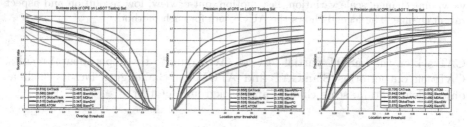

Fig. 3. Success, precise and normalized precise plots comparisons on LaSOT dataset. The proposed tracker was compared with nine other existing state-of-the-art trackers on 280 videos.

LaSOT. LaSOT is a large-scale dataset with high-quality annotations, with a total of 1400 long videos. We evaluated our approach on the test split consisting of 280 challenging videos. CATrack is compared with the advanced trackers including SiamFC [1], SiamDW [35], MDNet [24], SiamMask [30], SiamRPN++ [18], ATOM [7], GlobalTrack [17], DaSiamRPN [37] and DiMP [2]. As shown in Fig. 3, CATrack achieves the best performance in terms of Success, Precision and Normalized Precision.

In Fig. 4, we visualize the success polts of different trackers under 14 challenge factors on LaSOT. These factors evaluate the performance of the tracking algorithm in different ways. The experimental results show that our CATrack performs well on these challenging factors and demonstrate its superiority. As shown in Fig. 5, we visualize the tracking results of six different trackers, including SiamFC [1], ATOM [7], SiamGAT [13], Ocean [36], SiamRPN++ [19] and

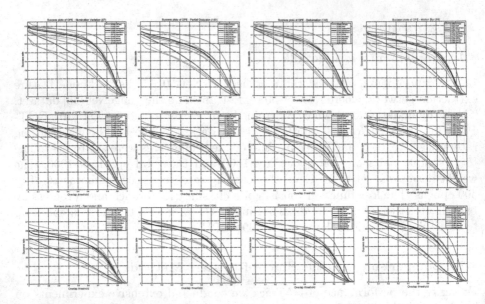

Fig. 4. The success plots on LaSOT [10] dataset for 14 challenging factors: illumination variation, partial occlusion, deformation, motion blur, rotation, background clutter, viewpoint change, scale variation, fast motion, out-of-view, low resolution and aspect ration change. The legend shows the AUC score for each method.

CATrack on the LaSOT partial test set. According to the experimental results, CATrack outperforms SiamGAT and ATOM on challenging factors such as object deformation, scale change, and background clutter.

Table 1. Comparison with the state-of-the-art methods on TrackingNet test set.

	SiamFC [1]	SiamRPN++ [18]	SiamFC++ [32]	SiamIRCA [4]	AutoMatch [34]	MCSiam [33]	DualMN [28]	TrDiMP [29]	TREG [6]	CATrack
Succ. (%)	57.1	73.3	75.4	75.7	76.0	76.5	77.8	78.4	78.5	**79.8**
Prec. (%)	53.3	69.4	70.5	71.7	72.6	73.1	72.8	73.1	75.0	**77.4**
Norm. Prec. (%)	66.3	80.0	80.0	81.4	–	82.0	83.2	83.3	83.8	**84.7**

TrackingNet. TrackingNet is another large-scale dataset, and its test set consists of 511 video sequences. The test set is not publicly available, so the test results need to be submitted to the official server to evaluate the performance metrics. We compared our CATrack with state-of-the-art trackers, including SiamFC [1], SiamRPN++ [18], SiamIRCA [4], SiamFC ++ [32], MCSiam [33], AutoMatch [34], DualMN [28], TrDiMP [29], and TREG [6]. As shown in Table 1, Our CATrack achieves the best performance with a Success score 79.8%, a precision score 77.4% and a normalization precision score 84.7%. These experimental results indicate that our CATrack is very robust.

GOT-10k. The training set of GOT-10k contains 10k video sequences and the test set contains 180 sequences. The test results need to be submitted to the

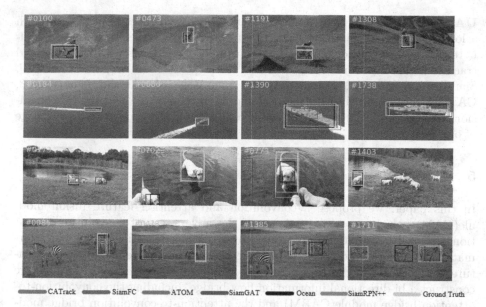

Fig. 5. Comparison with other four state-of-the-art trackers such as SiamFC, ATOM, SiamGAT, Ocean, SiamRPN++ in four different challenging sequence in LaSOT dataset. The video sequences are from top to down are bicycle-18, boat-17, dog-19 and zebra-10, respectively.

Table 2. Comparison with the state-of-the-art methods on GOT-10k dataset.

	SiamFC [1]	SiamRPN++ [18]	SiamFC++ [32]	SiamAGN [31]	SiamIRCA [4]	SiamGAT [13]	SiamR-CNN [27]	AutoMatch [34]	**CATrack**
AO(%)	34.8	51.7	59.5	60.5	62.3	62.7	64.9	65.2	**66.7**
$SR_{0.5}$(%)	35.3	61.6	69.5	70.2	73.6	74.3	72.8	76.6	**77.7**
$SR_{0.75}$(%)	9.8	32.5	47.7	49.2	–	48.8	59.7	54.3	**59.8**

official evaluation server for evaluation. It requires the trackers to be trained only on the GOT-10k training split. We follow this protocol. We compared our CATrack with the advanced trackers including SiamFC [1], SiamRPN++ [18], SiamFC++ [32], SiamIRCA [4], SiamGAT [13], SiamAGN [31], SiamR-CNN [27] and AutoMatch [34]. As shown in Table 2, our CATrack achieves the best performance with an AO(average overlap) score 66.7%, a $SR_{0.5}$ (success rates) score 77.7% and a $SR_{0.75}$ score 59.8%.

Table 3. Comparison with the state-of-the-art methods on UAV123 dataset.

	SiamFC [1]	SiamRPN++ [18]	MCSiam [33]	SiamGAT [13]	SiamR-CNN [27]	STMTracker [11]	SiamAGN [31]	**CATrack**
AUC(%)	48.5	61.0	62.5	64.6	64.9	64.7	65.3	**66.6**
P(%)	69.3	80.3	79.1	83.4	–	–	82.2	**83.4**

UAV123. The UAV123 contains 123 aerial video sequences, captured through a low-altitude UAV platform. Success and precision metrics were used to evaluate the performance of the tracker. We compared CATrack with the advanced trackers including SiamFC [1], SiamRPN++ [18], MCSiam [33], SiamGAT [13], SiamR-CNN [27], STMTracker [11] and SiamAGN [31]. As shown in Table 3, CATrack performs well, with the best success rate of 66.6% and the best precision rate of 83.4%. These experimental results show the competitive advantage of our CATrack.

5   Conclusion

In this paper, we propose a convolution and attention feature fusion module(CAM), which can make full use of the advantages of convolution operation for local information processing and attention operation for global information processing. Moreover, to achieve the information embedding and feature fusion between the target template and the search area in parallel, we construct a bi-directional information flow bridge, including the convolution-to-attention bridge module(CABM) and the attention-to-convolution bridge module(ACBM). Finally, we propose a novel and efficient tracker CATrack with accurate target positioning ability. The comprehensive experiments on four challenging tracking benchmarks validate the superiority of the proposed method in target positioning, tracking robustness and efficiency compared with the state-of-the-art methods under complex environments.

Acknowledgement. This work was supported by the Natural Science Foundation of China under Grant 61976145, and in part by the Guangdong Basic and Applied Basic Research Foundation under Grant 2021A1515011318, and 2314050002242, and the Shenzhen Municipal Science and Technology Innovation Council under Grant JCYJ20190808113411274 and JCYJ20220818095803007.

References

1. Bertinetto, L., Valmadre, J., Henriques, J.F., Vedaldi, A., Torr, P.H.S.: Fully-convolutional siamese networks for object tracking. In: Hua, G., Jégou, H. (eds.) ECCV 2016. LNCS, vol. 9914, pp. 850–865. Springer, Cham (2016). https://doi.org/10.1007/978-3-319-48881-3_56
2. Bhat, G., Danelljan, M., Gool, L.V., Timofte, R.: Learning discriminative model prediction for tracking. In: Proceedings of the IEEE/CVF International Conference on Computer Vision, pp. 6182–6191 (2019)
3. Carion, N., Massa, F., Synnaeve, G., Usunier, N., Kirillov, A., Zagoruyko, S.: End-to-End object detection with transformers. In: Vedaldi, A., Bischof, H., Brox, T., Frahm, J.-M. (eds.) ECCV 2020. LNCS, vol. 12346, pp. 213–229. Springer, Cham (2020). https://doi.org/10.1007/978-3-030-58452-8_13
4. Chan, S., Tao, J., Zhou, X., Bai, C., Zhang, X.: Siamese implicit region proposal network with compound attention for visual tracking. IEEE Trans. Image Process. **31**, 1882–1894 (2022)

5. Chen, X., Yan, B., Zhu, J., Wang, D., Yang, X., Lu, H.: Transformer tracking. In: Proceedings of the IEEE/CVF Conference on Computer Vision and Pattern Recognition, pp. 8126–8135 (2021)

6. Cui, Y., Jiang, C., Wang, L., Wu, G.: Target transformed regression for accurate tracking. arXiv preprint arXiv:2104.00403 (2021)

7. Danelljan, M., Bhat, G., Khan, F.S., Felsberg, M.: Atom: accurate tracking by overlap maximization. In: Proceedings of the IEEE/CVF Conference on Computer Vision and Pattern Recognition, pp. 4660–4669 (2019)

8. d'Ascoli, S., Touvron, H., Leavitt, M.L., Morcos, A.S., Biroli, G., Sagun, L.: Convit: improving vision transformers with soft convolutional inductive biases. In: International Conference on Machine Learning, pp. 2286–2296. PMLR (2021)

9. Dosovitskiy, A., et al.: An image is worth 16×16 words: transformers for image recognition at scale. arXiv preprint arXiv:2010.11929 (2020)

10. Fan, H., et al.: Lasot: a high-quality benchmark for large-scale single object tracking. In: Proceedings of the IEEE/CVF Conference on Computer Vision and Pattern Recognition, pp. 5374–5383 (2019)

11. Fu, Z., Liu, Q., Fu, Z., Wang, Y.: Stmtrack: template-free visual tracking with space-time memory networks. In: Proceedings of the IEEE/CVF Conference on Computer Vision and Pattern Recognition, pp. 13774–13783 (2021)

12. Graham, B., et al.: Levit: a vision transformer in convnet's clothing for faster inference. In: Proceedings of the IEEE/CVF International Conference on Computer Vision, pp. 12259–12269 (2021)

13. Guo, D., Shao, Y., Cui, Y., Wang, Z., Zhang, L., Shen, C.: Graph attention tracking. In: Proceedings of the IEEE/CVF Conference on Computer Vision and Pattern Recognition, pp. 9543–9552 (2021)

14. Guo, Q., Feng, W., Zhou, C., Huang, R., Wan, L., Wang, S.: Learning dynamic siamese network for visual object tracking. In: Proceedings of the IEEE International Conference on Computer Vision, pp. 1763–1771 (2017)

15. He, K., Zhang, X., Ren, S., Sun, J.: Deep residual learning for image recognition. In: Proceedings of the IEEE Conference on Computer Vision and Pattern Recognition, pp. 770–778 (2016)

16. Huang, L., Zhao, X., Huang, K.: Got-10k: a large high-diversity benchmark for generic object tracking in the wild. IEEE Trans. Pattern Anal. Mach. Intell. 43(5), 1562–1577 (2019)

17. Huang, L., Zhao, X., Huang, K.: Globaltrack: a simple and strong baseline for long-term tracking. In: Proceedings of the AAAI Conference on Artificial Intelligence, vol. 34, pp. 11037–11044 (2020)

18. Li, B., Wu, W., Wang, Q., Zhang, F., Xing, J., Yan, J.: Siamrpn++: evolution of siamese visual tracking with very deep networks. In: Proceedings of the IEEE/CVF Conference on Computer Vision and Pattern Recognition, pp. 4282–4291 (2019)

19. Li, B., Yan, J., Wu, W., Zhu, Z., Hu, X.: High performance visual tracking with siamese region proposal network. In: Proceedings of the IEEE Conference on Computer Vision and Pattern Recognition, pp. 8971–8980 (2018)

20. Lin, T.-Y., et al.: Microsoft COCO: common objects in context. In: Fleet, D., Pajdla, T., Schiele, B., Tuytelaars, T. (eds.) ECCV 2014. LNCS, vol. 8693, pp. 740–755. Springer, Cham (2014). https://doi.org/10.1007/978-3-319-10602-1_48

21. Liu, Z., et al.: Swin transformer: Hierarchical vision transformer using shifted windows. In: Proceedings of the IEEE/CVF International Conference on Computer Vision, pp. 10012–10022 (2021)

22. Mueller, M., Smith, N., Ghanem, B.: A benchmark and simulator for UAV tracking. In: Leibe, B., Matas, J., Sebe, N., Welling, M. (eds.) ECCV 2016. LNCS, vol. 9905, pp. 445–461. Springer, Cham (2016). https://doi.org/10.1007/978-3-319-46448-0_27
23. Muller, M., Bibi, A., Giancola, S., Alsubaihi, S., Ghanem, B.: Trackingnet: a large-scale dataset and benchmark for object tracking in the wild. In: Proceedings of the European conference on computer vision (ECCV), pp. 300–317 (2018)
24. Nam, H., Han, B.: Learning multi-domain convolutional neural networks for visual tracking. In: Proceedings of the IEEE Conference on Computer Vision and Pattern Recognition, pp. 4293–4302 (2016)
25. Russakovsky, O., et al.: Imagenet large scale visual recognition challenge. Int. J. Comput. Vision **115**, 211–252 (2015)
26. Vaswani, A., et al.: Attention is all you need. Adv. Neural Inf. Process. Syst. **30** (2017)
27. Voigtlaender, P., Luiten, J., Torr, P.H., Leibe, B.: SIAM R-CNN: visual tracking by re-detection. In: Proceedings of the IEEE/CVF Conference on Computer Vision and Pattern Recognition, pp. 6578–6588 (2020)
28. Wang, J., Zhang, H., Zhang, J., Miao, M., Zhang, J.: Dual-branch memory network for visual object tracking. In: Pattern Recognition and Computer Vision: 5th Chinese Conference, PRCV 2022, Shenzhen, China, 4–7 November 2022, 2022, Proceedings, Part IV, pp. 646–658. Springer, Heidelberg (2022). https://doi.org/10.1007/978-3-031-18916-6_51
29. Wang, N., Zhou, W., Wang, J., Li, H.: Transformer meets tracker: exploiting temporal context for robust visual tracking. In: Proceedings of the IEEE/CVF Conference on Computer Vision and Pattern Recognition, pp. 1571–1580 (2021)
30. Wang, Q., Zhang, L., Bertinetto, L., Hu, W., Torr, P.H.: Fast online object tracking and segmentation: a unifying approach. In: Proceedings of the IEEE/CVF Conference on Computer Vision and Pattern Recognition, pp. 1328–1338 (2019)
31. Wei, B., Chen, H., Ding, Q., Luo, H.: Siamgn: siamese attention-guided network for visual tracking. Neurocomputing **512**, 69–82 (2022)
32. Xu, Y., Wang, Z., Li, Z., Yuan, Y., Yu, G.: Siamfc++: towards robust and accurate visual tracking with target estimation guidelines. In: Proceedings of the AAAI Conference on Artificial Intelligence, vol. 34, pp. 12549–12556 (2020)
33. Zhang, J., Wang, J., Zhang, H., Miao, M., Cai, Z., Chen, F.: Multi-level cross-attention siamese network for visual object tracking. KSII Trans. Internet Inf. Syst. **16**(12), 3976–3989 (2022)
34. Zhang, Z., Liu, Y., Wang, X., Li, B., Hu, W.: Learn to match: automatic matching network design for visual tracking. In: Proceedings of the IEEE/CVF International Conference on Computer Vision, pp. 13339–13348 (2021)
35. Zhang, Z., Peng, H.: Deeper and wider siamese networks for real-time visual tracking. In: Proceedings of the IEEE/CVF Conference on Computer Vision and Pattern Recognition, pp. 4591–4600 (2019)
36. Zhang, Z., Peng, H., Fu, J., Li, B., Hu, W.: Ocean: object-aware anchor-free tracking. In: Vedaldi, A., Bischof, H., Brox, T., Frahm, J.-M. (eds.) ECCV 2020. LNCS, vol. 12366, pp. 771–787. Springer, Cham (2020). https://doi.org/10.1007/978-3-030-58589-1_46
37. Zhu, Z., Wang, Q., Li, B., Wu, W., Yan, J., Hu, W.: Distractor-aware siamese networks for visual object tracking. In: Proceedings of the European Conference on Computer Vision (ECCV), pp. 101–117 (2018)

SText-DETR: End-to-End Arbitrary-Shaped Text Detection with Scalable Query in Transformer

Pujin Liao[1,3] and Zengfu Wang[1,2,3(✉)]

[1] University of Science and Technology of China, Hefei, China
liaopj23@mail.ustc.edu.cn, zfwang@ustc.edu.cn
[2] Institute of Intelligent Machines, Chinese Academy of Sciences, Beijing, China
[3] National Engineering Research Center of Speech and Language Information Processing, Hefei, China

Abstract. Recently, Detection Transformer has become a trendy paradigm in object detection by virtual of eliminating complicated post-processing procedures. Some previous literatures have already explored DETR in scene text detection. However, arbitrary-shaped texts in the wild vary greatly in scale, predicting control points of text instances directly might achieve sub-optimal training efficiency and performance. To solve this problem, this paper proposes Scalable Text Detection Transformer (SText-DETR), a concise DETR framework using scalable query and content prior to improve detection performance and boost training process. The whole pipeline is built upon the two-stage variant of Deformable-DETR. In particular, we present a Scalable Query Module in the decoder stage to modulate position query with text's width and height, making each text instance more sensitive to its scale. Moreover, Content Prior is presented as auxiliary information to offer better prior and speed up the training process. We conduct extensive experiments on three curved text benchmarks Total-Text, CTW1500, and ICDAR19 ArT, respectively. Results show that our proposed SText-DETR surpasses most existing methods and achieves comparable performance to the state-of-art method.

Keywords: Detection transformer · Text detection · Scalable query

1 Introduction

Reading text in the wild is an active area in the computer vision community, due to its wide range of practical applications, such as image understanding, intelligent navigation, document analysis, and autonomous driving. As a prerequisite of scene text reading, text detection which aims at locating the region of text instances has been studied extensively. However, texts in the wild have large aspect ratios, various scales and curved shapes, making arbitrary-shaped text detection remain challenging.

Q. Liu et al. (Eds.): PRCV 2023, LNCS 14433, pp. 481–492, 2024.
https://doi.org/10.1007/978-981-99-8546-3_39

Previous arbitrary-shaped text detection methods can be mainly divided into two categories: segmentation-based and contour-based. Segmentation-based methods, taking advantage of results at a pixel level, can easily describe texts in various scales and curved shapes. However, most segmentation methods require heuristic and complex post-processing for assembling pixels into text instances, resulting in undesirable time cost in the inference stage. A point in case, PSENet [26] utilizes a progressive scale expansion algorithm to improve detection performance, while the inference time is much longer than other methods. On the contrary, contour-based methods directly predict control points of each text instance, clearing out complex post-processing procedures and contributing a more concise framework. ABCNet [14] creatively models curved text by Bezier Curve, and Bezier Align is presented to transform arbitrary-shaped texts into regular ones. Although contour-based methods are more straightforward, they rely on unique contour representations (such as the bezier curve) for texts, and specially designed operations (*e.g.* Bezier Align) are also closely coupled with these representations.

Transformer [24] has achieved great success in natural language processing. The pioneering work, DETR [2] seminally introduces Transformer into generic object detection. It discards several hand-designed processes, forming a concise end-to-end framework, which also attracts some researchers to investigate its potential in arbitrary-shaped text detection. [21] is the first one to attempt DETR in scene text detection, while it is dedicated to oriented texts rather than curved texts. TESTR [30] enables Deformable-DETR [32] to predict bounding box proposals, then generates polygon control points or bezier curve control points by proposed box-to-polygon process. Taking a further step on TESTR, DPText-DETR [28] improves the performance and training efficiency by EPQM and EFSA. However, these methods rarely take text scale information into account. They assume texts in the wild are isotropic, which is inconsistent with reality. Arbitrary-shaped texts vary a lot in scales and aspect ratios, texts with very large scale and with very small scale may exist in one image, it is difficult for the model to learn a unified representation at the same time. Based on these observations, this paper proposes Scalable Text Detection Transformer to solve this problem. In particular, we present Scalable Query Module (SQM) to deal with scale information and Content Prior as auxiliary information to improve training efficiency. Our main contribution can be summarized as follows:

- We propose SText-DETR, an end-to-end DETR framework for arbitrary-shaped text detection, in which Scalable Query Module is presented to use scale information to improve the detection performance.
- We investigate the impact of Content Prior, experimental results show that content prior is of great importance in speeding up the training process, which is seriously ignored in previous research.
- On three representative arbitrary-shaped text detection benchmarks Total-Text, CTW1500 and ICDAR19 ArT, SText-DETR surpasses most existing methods and achieves comparable performance to the state-of-art method.

2 Related Works

2.1 Text Detection

Scene text detection can be roughly classified into two categories: segmentation-based and contour-based. Segmentation-based methods usually learn semantic segmentation at a pixel level, compared to contour-based methods, segmentation-based methods are easier to describe text in arbitrary shapes while highly relying on complicated post-processing steps to form text instances. EAST [31] learns a shrink text region and directly regresses multi-oriented quadrilateral boxes from text pixels. PixelLink [6] predicts connections between adjacent pixels and localizes text region by separating the links belonging to different text instances, PSENet [26] gradually expands predefined kernels to obtain the final detection. On the other hand, contour-based methods directly regress control points of text contour, removing complicated post-processing procedures. For instance, ABCNet [14] predicts control points for a bezier curve to adaptively fit arbitrary-shaped texts. FCEnet [33] and Text-ray [25] model text contour in Fourier domain and Polar system respectively, PCR [5] evolves initial text proposal to arbitrarily shaped contours in a progressive way with convolution network.

2.2 Detection Transformer

DETR [2] introduces a transformer into object detection, it models object detection as a set prediction problem supervised by Hungarian loss [10]. In this way, DETR discards hand-craft anchor generation and some complex post-processing procedures (*e.g.* Non-Maximum Suppression), contributing to a very concise framework. However, computation complexity in vanilla DETR is quadratic to the feature size, resulting in slow training convergence and inefficient usage of high resolution, many DTER followers devote efforts to solving these problems. For instance, Deformable DETR [32] attends sparse positions in multi-scale feature layers with a specially designed multi-scale deformable attention module. Spatial conditional query is presented in Conditional-DETR [19] to restrain the bounding box coordinates and boost the training process. DAB-DETR [13] explicitly decouples query into the content query and positional query, and validates that positional query is the key to speed up training convergence.

2.3 DETR-Based Text Detection

DETR provides a simple and effective solution for generic object detection, which attracts researchers to explore its potential in arbitrary text detection. [21] firstly applies DETR into scene text detection. With its proposed rotated-box-giou loss, [21] achieves good performance on oriented texts but leaves unsatisfied results on curved texts. To leverage multi-scale features, TESTR [30] and DPText-DETR [28] use Deformable-DETR as their basic scheme. TESTR presents box-to-polygon operation to model arbitrary-shaped texts, proposals from the Transformer encoder are encoded as the position queries by sinusoidal function to guide the control points regression. Nevertheless, DPText-DETR emphasizes that box information shared by all control points is coarse for target points in detection. To better model the control point query, DPText-DETR presents Explicit Positional Query Modeling (EQPM) method which can encode each point query separately. Different from both two methods, FSG [23] harnesses multi-scale features by sampling a few representative feature points and uses Transformer encoder layers to aggregate them, then, control points for the bezier curve are predicted by a detection head.

3   Methodology

In this section, we firstly describe our overall pipeline, then detail the key components: Scalable Query Module and Content Prior in our method.

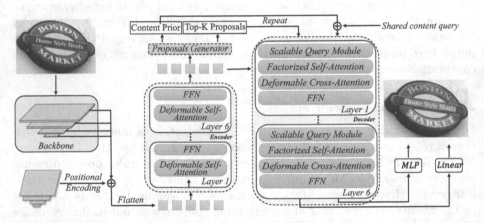

Fig. 1. The pipeline of SText-DETR. Firstly, flattened features are sent to encoder to perform muti-scale deformable attention. Top-K proposals along with content prior are generated by proposals generator, and then fed into decoder to further mine the relationship in texts. Finally, a linear layer and MLP are used to predict the class scores and control points of text.

3.1   Overview

As illustrated in Fig. 1, our model adopts a two-stage scheme of Deformable-DETR as the basic architecture. Given a text image, a CNN backbone (ResNet50 [8]) is first applied to extract multi-scale features. With 2D positional embedding, flattened features are sent to the Transformer encoder to interact with each other and mine their relationship. After the last encoder layer, we generate top-K proposals by a Linear layer and MLP module. At the same time, we select top-K features with respect to the top-K proposals, termed as content prior. Content prior then is added on the top of shared content queries to form content queries. In the decoder stage, proposals from the encoder are firstly encoded by a sinusoidal function as positional embedding. Content queries modulate positional embedding in Scalable Query Module (SQM). The output of SQM is the sum of modulated positional queries and content queries, which are called composite queries in our paper. Composite queries are then sent to Factorized Self-Attention Module proposed in TESTR [30] to capture the relationship between different text instances and between different sub-queries within a single text instance in a structural way. After that, composite queries are fed into the deformable cross-attention module and FFN sequentially. Similar to [13], we update control point coordinates layer by layer, refining them in a progressive way. Finally, a linear layer and an MLP module are used to generate a class score and control points coordinates. And we follow TESTR [30] to calculate the loss for classification and control point coordinates.

3.2   Content Prior

DAB-DETR [13] offers an insight that query can be decoupled into content part and positional part, the content query is responsible for the object class, and the positional query is responsible for object position. In the original two-stage scheme of Deformable-DETR, positional queries can be generated by top-k proposals, while content queries are randomly initialized as a learnable variable. Now that we can use the top-k proposals to generate the initial positional queries, features with respect to these proposals are supposed to contain key information of the text. We gather these features and term them content prior, this process is shown in Fig. 2. Then, content prior and shared content queries are summed to form content queries. Intuitively, content queries from the second decoder layer to the last decoder layer are the weighted sum of encoder features, where the weights are produced by multi-scale deformable attention operation. However, original content queries for the first decoder layer are just randomly initialized learnable variables, which causes a misalignment to some extent. If these learnable queries are initialized improperly during the training stage, the model might take a lot of time to converge. Fortunately, our content prior is able to alleviate this issue. In fact, content prior provides auxiliary information. With the help of this information, the model still shows fast convergence efficiency even if the randomly initialized learnable query is not good enough.

3.3   Scalable Query Module

Scale information is of great importance to text detection. TESTR [30] uses scale information by directly encoding the top-K proposals coordinates into the 256-dimension vector, in which the point embedding and scale embedding are separated from each other. And all control points in a text instance share the same position prior, We conjecture that is coarse for control points regression.

DAB-DETR [13] proposes a way to modulate attention with scale information, which inspires us to apply it in text detection. However, DAB-DETR is designed for generic object detection, which only needs to predict bounding boxes (x, y, ω, h) for objects. While text detection needs to predict a series of control points, that means there is no explicit width and height of text instances. So can we predict control points as well as the text's weight and height at the same time? The answer is yes. We notice that TESTR directly uses the bounding boxes information to predict control points, insight behind that is all control points belonging to one text can share the same position prior. Based on that, we regard a proposal as a control point, and then replicate it for N times to generate the initial control points for a text instance, N is the predefined number of text control points. In this way, each point in each text has explicit scale information. Now, we can elaborate on our Scalable Query Module.

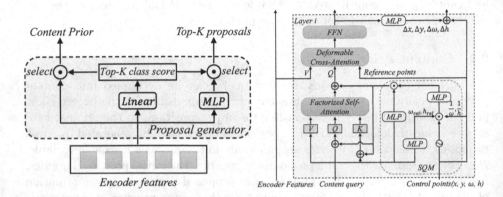

Fig. 2. Proposal Generator   **Fig. 3.** Scalable Query Module

Figure 3 shows Scalable Query Module in a decoder layer. First of all, the input of SQM includes two parts: content queries and control points. As discussed before, content queries are the sum of content prior and shared content queries, control points (also called reference points in the deformable attention module) are obtained by replicating proposals N times. We denote P_i^j as i-th control point of j-th text instance, so the initial control points can be described as follows:

$$P_i^j = (x^j, y^j, \omega^j, h^j) j = 1, 2, ..., K, i = 1, 2, ..., N \tag{1}$$

where K denotes the number of proposals, N denotes the number of control points. When given a control point $P = (x, y, \omega, h)$ the positional embedding P is generated by:

$$Embed_{pos} = PE(P(x, y)) \tag{2}$$

where PE means positional encoding which generates sinusoidal embedding from float numbers. $P(x, y)$ means we only encode the point coordinates. At the same time, content queries are sent a MLP followed by a sigmoid function to generate reference width and height for texts.

$$\omega_{ref}, h_{ref} = \sigma(MLP(Q_{content})) \tag{3}$$

Now, we modulate the positional queries by dividing relative width and height from their x part and y part separately.

$$Q_{pos}^{mod} = MLP((\boldsymbol{PE}(x)\frac{\omega_{\text{ref}}}{\omega}, \boldsymbol{PE}(y)\frac{h_{\text{ref}}}{h})) \qquad (4)$$

In this way, the scale information is explicitly injected on the top of positional queries. In fact, as initial positional embedding is generated by a sinusoidal function, we scale this embedding with the text's width and height, not only introducing the scale information, but also limited the searching space of the control points, which can help the model converge faster. After that, a MLP module is applied to transform the modulated positional embedding into position queries. Furthermore, we generate spatial conditional query described in Conditional-DETR.

$$Q_{pos}^{cond} = MLP(Q_{content}) \odot Q_{pos}^{mod} \qquad (5)$$

The spacial conditional queries and content queries are then summed to form the composite queries, which are sent to Factorized Self-Attention Module [30] to mine their relationship, finally fed into deformable cross-attention module and FFN.

As we model control points as a quaternion, we can update control points layer by layer with predicting relative position offsets $(\Delta x, \Delta y, \Delta \omega, \Delta h)$ by a prediction head. By this method, we can refine the reference points in a progressive way.

4   Experiments

4.1   Datasets

SynthText 150K [14] is a synthesized dataset for arbitrary-shaped scene text, which contains 94,723 images with multi-oriented text and 54327 images with curved text. **Total-Text** [4] is a popular curved text benchmark that consists of 1255 training images and 300 test images. **CTW1500** [15] is another important curved text benchmark, containing 1000 training images and 500 testing images. Different from Total-Text, both English and Chinese texts exist in this dataset. **ICDAR19 ArT** [3] is a large-scale multi-lingual arbitrary-shaped text detection benchmark. It consists of 5603 training images and 4563 test images.

We adopt standard evaluation metric as the ICDAR challenges for all experiments, including precision, recall, and F-score.

4.2   Implementation Details

We use ResNet-50 [8] as the feature extraction backbone for all experiments. Multi-scale feature maps are directly extracted from the last three stages of ResNet without FPN [12]. As for the deformable attention module, we follow the original settings in Deformable-DETR [32], using 8 heads and 4 sampling points. The number of both encoder layers and decoder layers is set to 6. And the number of top-K proposals (composite queries) K is 100, control points of each text instance are 16. Following TESTR [30], in encoder loss, weighting factors are $\lambda_{cls} = 2.0$, $\lambda_{coord} = 5.0$, $\lambda_{gIoU} = 2.0$, in decoder loss, weighting factors are $\lambda_{cls} = 2.0$, $\lambda_{coord} = 5.0$. And we set $\alpha = 0.25$, $\gamma = 2$ for the focal loss. The final loss is a sum of encoder loss and decoder loss. Furthermore, for a fair comparison to DPText-DETR, we use its positional label form [28] in our experiments. The model is trained on 4 RTX3090 GPUS (24 GB) with a batch size of 8.

Table 1. Quantitative results on benchmarks. "P", "R" and "F" denote precision, recall and F-score, respectively. "†" means results on ICDAR19 ArT are collected from the official website [3], others are collected from their original papers. Best results are shown in bold, and the underlined values are the second high results.

model	backbone	Total-Text			CTW1500			ICDAR19 ArT		
		P	R	F	P	R	F	P	R	F
TextSnake [17]	VGG16	82.7	74.5	78.4	67.9	85.3	75.6	–	–	–
PAN [26]	Res18	89.3	81.0	85.0	86.4	81.2	83.7	–	–	–
CRAFT [1]†	VGG16	87.6	79.9	83.6	86.0	81.1	83.5	77.2	68.9	72.9
TextFuseNet [27]†	Res50	87.5	83.2	85.3	85.8	85.0	85.4	82.6	69.4	75.4
DBNet [11]	Res50-DCN	87.1	82.5	84.7	86.9	80.2	83.4	–	–	–
PCR [5]	DLA34	88.5	82.0	85.2	87.2	82.3	84.7	84.0	66.1	74.0
ABCNet-V2 [16]	Res50	90.2	84.1	87.0	85.6	83.8	84.7	–	–	–
I3CL [7]	Res50	89.2	83.7	86.3	87.4	84.5	85.9	82.7	71.3	76.6
TextBPN++ [29]	Res50	91.8	85.3	88.5	87.3	83.8	85.5	81.1	71.1	75.8
FSG [23]	Res50	90.7	85.7	88.1	88.1	82.4	85.2	–	–	–
SwinTextSpotter [9]	Swin	–	–	88.0	–	–	88.0	–	–	–
TESTR-polygon [30]	Res50	93.4	81.4	86.9	92.0	82.6	87.1	–	–	–
DPText-DETR [28]	Res50	91.8	86.4	89.0	91.7	86.2	**88.8**	83.0	73.7	**78.1**
SText-DETR (ours)	Res50	91.4	86.9	**89.1**	92.2	85.6	**88.8**	74.0	82.4	78.0

Data Augmentations. We follow previous works to conduct data augmentations during training, including instance-aware random crop, random blur, brightness adjusting, and color change. Also, a multi-scale training strategy is applied with the shortest edge ranging from 480 to 832, and the longest edge is kept within 1600.

Training Process. The whole training process consists of two stages: pre-training stage and finetuning stage. To keep the same with TESTR [30] and DPText-DETR [28], we pre-train the model on a mixture of SynthText 150K, MLT [20] and Total-Text for 360k iterations. The basic learning rate (lr) is set to 1×10^{-4} and is decayed at the 280k-th iteration by a factor of 0.1. Then, we both adopt 20k iterations to finetune the model on Total-Text and CTW1500 respectively, during finetuning stage, lr for Total-Text is set 5×10^{-5} and is divided by 10 at 16k, lr for CTW1500 is set for 2×10^{-5} and is divided by 10

Table 2. Ablation results on Total-Text and CTW1500. "SQM" and "CP" denote Scalable Query Module and Content Prior respectively.

(a) ablation results on Total Text

method	SQM	CP	P	R	F
baseline			87.95	82.07	84.91
	✓		89.21	81.80	85.34
		✓	88.44	83.24	85.76
	✓	✓	89.84	82.25	85.88

(b) ablation results on CTW1500

method	SQM	CP	P	R	F
baseline			82.19	84.45	83.30
	✓		86.00	84.52	85.25
		✓	84.62	84.52	84.57
	✓	✓	86.98	84.75	85.85

at 15k. As for ICDAR19 ArT, SynthText 150K, MLT, ArT and LSVT [22] are used as a mixture to pre-train the model for 400k iterations. lr is set to 1×10^{-4} and is decayed to 1×10^{-5} at 320k. Then, we finetune it on ArT for another 50k iterations, with 5×10^{-5} lr which is divided by 10 at 40k. In ablation experiments, we do not pre-train the model as the pre-training process is time-consuming, all ablation experiments are only trained for 40k iterations, lr is set to 1×10^{-4} and is decayed to 1×10^{-5} at 28k. AdamW [18] is used as the optimizer, with $\beta_1 = 0.9$, $\beta_2 = 0.999$ and weight decay is 10^{-4}.

4.3 Experimental Results

Quantitative results on Total-Text, CTW1500, and ICADR19 ArT are shown in Table 1. First of all, we can obviously see that SText-DETR outperforms TESTR [30] by 2.2% on Totaltext and 1.7% on CTW1500 respectively in terms of the F-score metric. Compared with other existing methods, SText-DETR still achieves considerable improvements. For example, SText-DETR outperforms FSG [23] by 1%, 3.6% in terms of F-score on Total-Text and CTW1500. It also surpasses I3CL [7] by 2.8%, 2.9%, 1.4% on all benchmarks. Moreover, compared with Swin-TextSpotter [9], which improves performance by the synergy between text detection and text recognition, SText-DETR still leads 1.1%, 0.8% higher F-score on Total-Text and CTW1500. Furthermore, SText-DETR achieves almost the same performance in contrast with the latest state-of-art method: DPText-DETR [28], which firmly verifies its effectiveness on arbitrary-shaped text detection. Overall, SText-DETR outperforms most existing text detectors by a large margin and is comparable to the state-of-art method.

4.4 Ablation Studies

In this section, extensive ablation experiments are conducted on Total-Text and CTW1500 to demonstrate the effectiveness of two key components in our model: Scalable Query Module and Content Prior. Ablation results on Total-Text and CTW1500 are mainly presented in Table 2.

Table 2(a) shows the ablation results on Total-Text. Compared with the baseline model, F-score is improved by 0.43% when SQM is used alone and it further increases by 0.54% when content prior is added, resulting in 0.97% F-score gain

Table 3. Ablation results on decoder layers.

(a) Baseline model

method	layers	P	R	F
baseline	6	87.95	82.07	84.91
	4	83.65	83.20	83.42
	2	76.32	76.29	76.30
	1	65.90	6 6.94	66.41

(b) baseline model with content prior

method	layers	P	R	F
baseline+CP	6	88.44	83.24	85.76
	4	87.79	83.11	85.38
	2	86.81	82.07	84.37
	1	87.03	81.21	84.02

in total. The same situation can also be observed in CTW1500. Comparison between line 1 and line 2 in Table 2(b) demonstrates SQM can bring 1.95% improvement on F-score, results in line 1 and line 3 indicate content prior can improve the F-score by 1.27%, a total improvement of F-score in CTW1500 is up to 2.55%. On the whole, results on two benchmarks consistently confirm the effectiveness of the Scalable Query Module and Content Prior, moreover, the two components synergize with each other to further improve detection performance. Further more, We plotted the F-score (metric) curves during the training stage in Fig. 4, it can be observed that model with CP shows faster convergence than the baseline model at the early training stage. We attribute these to two aspects: (1) SQM modulates position embedding with the scale information, explicitly incorporating the width and height of text into the control points, this scale information limits the search space of control points, contributing to faster convergence. And control points update layer by layer in the decoder, refining the results in a progressive way, which further improves the detection performance. (2) Content Prior contains a lot of information about the text, these auxiliary information offers good guidance at the early training stage, which is beneficial for the model to stabilize training and boost convergence.

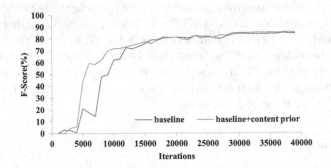

Fig. 4. F-score Curves during training stage.

More Discussion on Content Prior. To verify the impact of auxiliary information in content prior, we train our model with less decoder layers. Results are shown in Table 3. With the number of decoder layers decreasing, F-score on baseline model drops heavily, while performance nearly keeps unchanged when content prior is used. When we only use 1 decoder layer, it still achieves 84.02% in terms of F-score, which is almost same as baseline model with 6 layers. These experimental results demonstrate that content prior is enough for model to learn the relationship in texts. With the help of content prior, we can reduce decoder layers with slightly drops on performance, achieving the balance between model complexity and performance, and we leave this for the future work.

5 Conclusion

In this paper, we propose SText-DETR, a simple and effective DETR framework for arbitrary-shaped text detection. Scalable Query Module is presented to explicitly modulate position embedding, so that control points can be more sensitive to its scale information. Besides, we investigate the impact of content prior in training process, which is seriously neglected by previous works, we hope content prior can be exploited more in the future work. Experimental results on challenging arbitrary-shaped text benchmarks, Total-Text, CTW1500, ICDAR19 ArT, demonstrate the state-of-art performance of SText-DETR.

References

1. Baek, Y., Lee, B., Han, D., Yun, S., Lee, H.: Character region awareness for text detection. In: CVPR, pp. 9365–9374 (2019)
2. Carion, N., Massa, F., Synnaeve, G., Usunier, N., Kirillov, A., Zagoruyko, S.: End-to-End object detection with transformers. In: Vedaldi, A., Bischof, H., Brox, T., Frahm, J.-M. (eds.) ECCV 2020. LNCS, vol. 12346, pp. 213–229. Springer, Cham (2020). https://doi.org/10.1007/978-3-030-58452-8_13
3. Chng, C.K., et al.: ICDAR 2019 robust reading challenge on arbitrary-shaped text-RRC-ART. In: ICDAR, pp. 1571–1576. IEEE (2019)
4. Ch'ng, C.K., Chan, C.S., Liu, C.L.: Total-text: toward orientation robustness in scene text detection. IJDAR **23**(1), 31–52 (2020)
5. Dai, P., Zhang, S., Zhang, H., Cao, X.: Progressive contour regression for arbitrary-shape scene text detection. In: CVPR, pp. 7393–7402 (2021)
6. Deng, D., Liu, H., Li, X., Cai, D.: Pixellink: detecting scene text via instance segmentation. In: Proceedings of the AAAI Conference on Artificial Intelligence, vol. 32 (2018)
7. Du, B., Ye, J., Zhang, J., Liu, J., Tao, D.: I3CL: intra-and inter-instance collaborative learning for arbitrary-shaped scene text detection. Int. J. Comput. Vision **130**(8), 1961–1977 (2022)
8. He, K., Zhang, X., Ren, S., Sun, J.: Deep residual learning for image recognition. In: CVPR, pp. 770–778 (2016)
9. Huang, M., et al.: Swintextspotter: scene text spotting via better synergy between text detection and text recognition. In: CVPR, pp. 4593–4603 (2022)
10. Kuhn, H.W.: The Hungarian method for the assignment problem. Naval Res. Logist. Q. **2**(1–2), 83–97 (1955)
11. Liao, M., Wan, Z., Yao, C., Chen, K., Bai, X.: Real-time scene text detection with differentiable binarization. In: AAAI, vol. 34, pp. 11474–11481 (2020)

12. Lin, T.Y., Dollár, P., Girshick, R., He, K., Hariharan, B., Belongie, S.: Feature pyramid networks for object detection. In: CVPR, pp. 2117–2125 (2017)
13. Liu, S., et al.: DAB-DETR: dynamic anchor boxes are better queries for detr. In: ICLR (2022)
14. Liu, Y., Chen, H., Shen, C., He, T., Jin, L., Wang, L.: ABCNET: real-time scene text spotting with adaptive bezier-curve network. In: CVPR, pp. 9809–9818 (2020)
15. Liu, Y., Jin, L., Zhang, S., Luo, C., Zhang, S.: Curved scene text detection via transverse and longitudinal sequence connection. PR **90**, 337–345 (2019)
16. Liu, Y., et al.: Abcnet v2: adaptive bezier-curve network for real-time end-to-end text spotting. IEEE Trans. Pattern Anal. Mach. Intell. **44**(11), 8048–8064 (2021)
17. Long, S., Ruan, J., Zhang, W., He, X., Wu, W., Yao, C.: Textsnake: a flexible representation for detecting text of arbitrary shapes. In: ECCV, pp. 20–36 (2018)
18. Loshchilov, I., Hutter, F.: Decoupled weight decay regularization. In: ICLR (2017)
19. Meng, D., et al.: Conditional detr for fast training convergence. In: CVPR, pp. 3651–3660 (2021)
20. Nayef, N., et al.: ICDAR 2019 robust reading challenge on multi-lingual scene text detection and recognition-RRC-MLT-2019. In: ICDAR, pp. 1582–1587. IEEE (2019)
21. Raisi, Z., Naiel, M.A., Younes, G., Wardell, S., Zelek, J.S.: Transformer-based text detection in the wild. In: CVPR, pp. 3162–3171 (2021)
22. Sun, Y., et al.: ICDAR 2019 competition on large-scale street view text with partial labeling-RRC-LSVT. In: ICDAR, pp. 1557–1562. IEEE (2019)
23. Tang, J., et al.: Few could be better than all: feature sampling and grouping for scene text detection. In: CVPR, pp. 4563–4572 (2022)
24. Vaswani, A., et al.: Attention is all you need. Adv. Neural Inf. Process. Syst. **30** (2017)
25. Wang, F., Chen, Y., Wu, F., Li, X.: Textray: contour-based geometric modeling for arbitrary-shaped scene text detection. In: ACM MM, pp. 111–119 (2020)
26. Wang, W., et al.: Efficient and accurate arbitrary-shaped text detection with pixel aggregation network. In: ICCV, pp. 8440–8449 (2019)
27. Ye, J., Chen, Z., Liu, J., Du, B.: Textfusenet: scene text detection with richer fused features. In: IJCAI, vol. 20, pp. 516–522 (2020)
28. Ye, M., Zhang, J., Zhao, S., Liu, J., Du, B., Tao, D.: DPTEXT-DETR: towards better scene text detection with dynamic points in transformer. arXiv preprint arXiv:2207.04491 (2022)
29. Zhang, S.X., Zhu, X., Yang, C., Yin, X.C.: Arbitrary shape text detection via boundary transformer. arXiv preprint arXiv:2205.05320 (2022)
30. Zhang, X., Su, Y., Tripathi, S., Tu, Z.: Text spotting transformers. In: CVPR, pp. 9519–9528 (2022)
31. Zhou, X., et al.: East: an efficient and accurate scene text detector. In: CVPR, pp. 5551–5560 (2017)
32. Zhu, X., Su, W., Lu, L., Li, B., Wang, X., Dai, J.: Deformable detr: deformable transformers for end-to-end object detection. In: ICLR (2021)
33. Zhu, Y., Chen, J., Liang, L., Kuang, Z., Jin, L., Zhang, W.: Fourier contour embedding for arbitrary-shaped text detection. In: CVPR, pp. 3123–3131 (2021)

SSHRF-GAN: Spatial-Spectral Joint High Receptive Field GAN for Old Photo Restoration

Duren Wen, Xueming Li$^{(\boxtimes)}$, and Yue Zhang

Beijing University of Posts and Telecommunications, Beijing, China
{dorren,lixm,zhangyuereal}@bupt.edu.cn

Abstract. Old photo restoration is a challenging problem since it suffers from mixed degradation like noise, blurriness, scratch, dots, or color fading during its preserving or digitalizing process. The common practice of old photo restoration is to apply single degradation restoring methods sequentially, but it cannot restore the old images at one time. The existing mixed degradation restoring methods are designed in a complex architecture or training paradigm. To tackle this problem in a simple way, we propose SSHRF-GAN: Spatial-Spectral joint High Receptive Field GAN. It contains three independent modules for the mixed degradation that appeared in old photos: (1) High Receptive Field Fast Convolution block for noise and blurriness restoration; (2) Partial convolution block for scratch, dots, and cracks inpainting and (3) Image-wise histogram attention for color tone mapping. We evaluate our methods on synthesized datasets and gather real old data. The result shows that our simple network has a comparative performance with those sequential applied single degradation restoration models and previously proposed complex models.

Keywords: Image restoration · Combined degradation · Spatial-spectral joint convolutional network

1 Introduction

During the years the camera was invented, lots of precious photos (or videos) were shot. However, due to the storage medium and digitalization process, the old photos suffered from many kinds of degradation during their preservation. Thanks to mobile cameras and scanners, many old photos and movies have been turned into digital forms making it possible to be repaired by modern techniques.

In the early years, old photos were restored in manufacturing ways, the one who specialized in photo editing may take hours to restore a piece of an old photo. After that many digital image processing methods [2,5] were used to

---

This work was supported by National Key R&D Program "Cultural Technology and Modern Service Industry" Key Special Project(#2021YFF0901700).

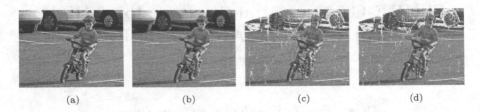

(a)   (b)   (c)   (d)

Fig. 1. The example of (b) unstructured degradation only; (c) structured degradation only and (d) mixed degradation. (a) stands for the image before degraded

help restore degraded images in those photography toolkits, but unfortunately, the exact degradation function for those real old photos is unknown, so these methods cannot effectively handle all kinds of degradation.

In recent years, since the boosting development of machine learning in computer vision, there are many deep learning-based methods for denoising [6,10,13,27], deblurring [7,9,14] and inpainting [11,19,24]. However, these methods are mainly designed for single degradation, which means one model was only designed or trained for single degradation. However, old photos often suffer from multiple defects and for different images, the degradation types and levels are all different. Thus, by using those single degradation restoration models, the old photo restoration is a complex multi-stage task. Although there is some awesome research on old photo restoration these years [21,23], they all designed a complex structure and difficult training strategy.

According to the image degradation theory and above mentioned methods, we sum up the key points for old photo restoration as follows:

First, the mixed degradation of the old images could be roughly divided into two classes: unstructured degradation(like noise, blurriness, and color fading), and structured degradation (like scratch, hole, and spots, there is an example in Fig. 1). Since these two kind of degradation would appear in one image at the same time, the model should have the ability to handle both of them. Besides, we found that color fading has an image-wise characteristic, so it should be considered independently.

Second, it is a good idea to combine image restoration with image inpainting together to better handle the mixed degradation. When it comes to deep image restoration and inpainting, a high receptive field is important for both of them [9,13,19,26]. Since the deep image restoration and inpainting models are mostly based on generative models like Variational Auto Encoder (VAE) or Generative Adversarial Nets(GAN), it is important to model long-range pixel dependencies to generate reasonable results since

In this paper, we propose a simple network trained in the GAN paradigm for old photo restoration that handles the structured and unstructured degradation at the same time. First, we use Fast Fourier Convolution(FFC) to achieve the Spatial-Spectral joint High Receptive Field(SSHRF) generator which effectively restores those unstructured degradation; Second, we add a partial convolution block to inpaint the structured degradation. Third, we propose a reference image-based color shifting module, image-wise histogram attention, to control the restored images' color tone since the color fades in different ways for different images. At last, to stabilize the training process both in the spatial and spectral domain, we introduce a Spatial-Spectral joint Feature Matching Loss (SSFMLoss).

2 Related Work

2.1 Image Restoration

Images can be made up of both useful information and degradation, the latter one reduces the clarity of the image. Image restoration aims to restore useful information from those degraded images. The degradation of the images could be divided into two classes, 1) unstructured degradation like noise, blurriness, and color fading; 2) structured degradation like holes, scratches, and spots. Since the different mathematical modeling, the approach of restoration has differences.

Unstructured Degradation Restoration. The unstructured degradation is modeled as additive or multiplicative noise for the high-quality images. In the early times, restoration methods worked on the prior of degradation, like the median filter for salt-and-pepper noise. Since the convolutional neural network(CNN) has impressing performance on vision, it has become the main trend of image restoration, like [10,13,27] for denoising, [7,14] for debluring. Recently, many transformer-based methods like [9,26] have surpassed the CNN-based method and applied for different degradation restoration by training on different datasets.

Structured Degradation Restoration. As structured degradation is often caused by physical reasons, mostly the image content destroyed irreversibly, it modeled an inpainting problem. Image inpainting aims to repair the inner content of a given mask [4]. Traditional methods inpaint the masked area according to the pixels outside the hole or search for the most similar patch. As generative models like GAN and VAE have great success in image generation, there emerged many generative network-based inpainting models like [8,11,17,19,24].

Mixed Degradation Restoration. Images often suffer from more than one kind of degradation during their capturing and preserving process. [25] proposed an agent and toolbox framework, the agent calculates the most possible crack at time step t and the toolbox takes different actions to restore the input image till the stopping action. [18] achieved the action selection through an attention mechanism to restore unknown combined distortions. The above-mentioned method concentrates on unstructured mixed degradation.

2.2   Old Photo Restoration

Old photo restoration is a standard mixed degradation restoration task, [21] first proposed a combined degradation restoration model based on latent space translation. The model consists of two VAEs and a mapping model. The mapping model includes a non-local partial block responsible for structured defects. [23] focused on end-to-end old photo reconstruction and colorization, they used three sub-nets to handle image restoration, color transition, and colorization. As for the reconstruction net, they used a multi-scale dense net encoder to encode the photos to achieve high receptive field, and there is no extra strategy to handle structured defects. [12] proposed a two-stage GAN for old photo restoration, global restoration, and local restoration. They divide the degradation into four classes and first restore the degradation in latent space using latent class-attribute information, and then restore the local detail by a dynamic condition-guided restoration module.

3   Method

Old photo restoration is a mixed degradation restoration problem, we formulate structured degradation and unstructured degradation in different ways. The structured degradation is formulated as $x_d = x \odot m$, where m is a binary mask and $m = 0$ stands for damaged areas like scratches or dots in old images. As for the unstructured degradation, although the degradation function is often unknown, we can simply assume it follows $x_d = x + \epsilon$, where ϵ follows a particular distribution.

To restore the old photos that suffered from both structured and unstructured degradation by a single model, we propose Spatial-Spectral joint High Receptive Field GAN (SSHRF-GAN). The SSHRF-GAN is trained on the synthesized old images and fine-tuned on real old images to achieve better performance.

Given the degraded image x_d and binary mask m, SSHRF-GAN recovers the clean image $x' = G_{SSHRF}(x_d, m)$, more details in Sect. 3.1. The image-wise histogram attention maps the restored image' color tone to the reference one by fusing the reference image's spatial histogram feature map, see in Sect. 3.2. Then the reconstruction loss calculated on x' and x, and the discriminator discriminates whether the input is clean or degraded to calculate adversarial loss(see more about loss function in Sect. 3.3).

3.1   Spatial-Spectral Joint High Receptive Field Generator

Our spatial-spectral joint high receptive field generator is a fully convolutional network consisting of downsampling layers, PConv-FFC residual blocks, and upsampling layers. The downsampling layers simply apply vanilla convolution on input images and decrease the size of feature maps slightly. The PConv-FFC residual block separately operates global features on the spectral domain and local features on the spatial domain and fuses them together at last. The upsampling layers do the opposite operation with the downsampling layer.

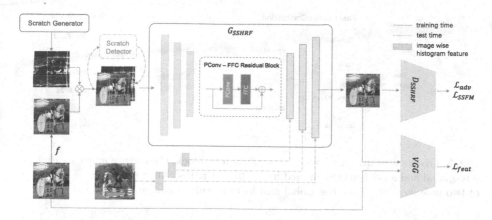

Fig. 2. SSHRF-GAN consists of a spatial-spectral joint high receptive field generator (G_{SSHRF}) and corresponding discriminator (D_{SSHRF}). During the training time, the degraded images are synthesized by a scratch generator and unstructured degradation function f, and then fed into G_{SSHRF} and computed the combined loss between the clean images and restored images. The training loss consists of the weighted sum of the adversarial loss, SSFM loss, and perceptual loss. At test time, the old image was first fed into a scratch detector(pretrained segmentation model in UNet structure) to get a binary mask that indicates the structured degradation area, similar to that in [21]. Also, the image-wise histogram attention is applied to control the color tone of the restored image during the test time.

PConv-FFC Residual block consists of PConv block and FFC block, where PConv is responsible for the structured degradation mask updating and FFC mainly for unstructured degradation restoration.

Due to the combination of FFC, our PConv-FFC residual block takes 75% of the features as global features and the rest as local features. Global features are operated on the spectral domain and local features on the spatial domain (see in Fig. 2). The PConv block first applies partial convolution on the local features and updates the binary mask. Then FFC block separately operates the feature maps on spatial and spectral domains.

The PConv block ignores the degraded pixels during convolution and updates the binary mask at every step. Denote the input feature map at ith residual block as F_i, and the mask as M_i, the partial convolution operation could be formally expressed as

$$F_i' = \begin{cases} W^T(F_i \odot M_i)\frac{sum(1)}{sum(M_i)} + b, & if\ sum(M_i) > 0 \\ 0, & other \end{cases} \qquad (1)$$

where $\odot$ denotes Hadamard product and **1** denotes the valid values in binary mask. When F_i' is known, the mask updating function is

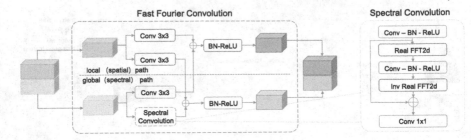

Fig. 3. Overview of the Fast Fourier Convolution [1] block, the feature maps are divided into global and local features and fed into different paths. After the separate operation of two independent paths, the global and local features fused and combined together.

$$M_i' = \begin{cases} 1, & if\ sum(F_i') > 0 \\ 0, & other \end{cases} \tag{2}$$

The output of the PConv block is then fed into the FFC block. FFC [1] is a convolution block designed for the convolutional networks to have the global receptive field in a shallow layer. The global receptive field of FFC comes from Fourier Theory: a point-wise update in the spectral domain globally affects all input features involved in the Fourier transform.

FFC consists of two inter-connected paths: a local path that conducts spatial convolution on a part of input features and a global path that operates on the other parts in the spectral domain(see Fig. 3). Each path captures information from different receptive fields and at last fused together. Formally, the whole procedure of the FFC is as follow:

$$\begin{cases} Y^l & = Y^{l \to l} + Y^{g \to l} = f_l(X^l) + f_{g \to l}(X^g), \\ Y^g & = Y^{g \to g} + Y^{l \to g} = f_g(X^g) + f_{l \to g}(X^l) \end{cases} \tag{3}$$

In general, the PConv-FFC residual block plays an important role in the whole framework. The PConv block is mainly responsible for mask updating which is important for structured degradation restoration and FFC extracts deep features to restore the input images due to its global receptive field.

Compared to the custom ways for enlarging the receptive field for CNN, like making networks deeper, using larger convolution kernels, or operating on multi-scale images, FFC has fewer memory footprint and calculation wastage, most importantly it provides global context information in shallow layers. Also, spectral filters like low bandpass filter was a popular DIP method applied to denoising, thus we assume that by using fast Fourier convolution, we got a series of learnable filters both on the spatial and spectral domain.

3.2 Image Wise Spatial Histogram Attention

Since there is no unified color fading mode for degraded images, we introduce image-wise spatial histogram attention similar to [23] to restrict the color of the restored image with a reference image. The reference image could be a target color distribution or the input image itself. The spatial histogram describes the probability of a pixel falling in each bin by Gaussian Expansion. The probability of a pixel at position (i, j) falls into the kth bin is expressed as follow:

$$h(i, j, k) = \frac{exp(-(D_{ij} - u_k)^2/2\sigma^2)}{\sum_{k=1}^{K} exp(-(D_{ij} - u_k)^2/2\sigma^2)} \tag{4}$$

where D_{ij} is the value of kth channel of the reference picture at spatial coordinate (i, j), $sigma$ is fixed as 0.1, and u_k is the mean value of the kth channel.

Note that the structured degradation area would be set to 0 to ensure it doesn't affect the output. After that, the spatial histogram resized to (H, W, C) and fused to the upsampling features similar to UNet skip connection.

3.3 Loss

Spatial-Spectral Joint Feature Matching Loss. Feature matching loss, or known as discriminator-based perceptual loss, is known to stabilize training [22]. It is a perceptual loss on the features of the discriminator network, here we introduce Spatial-Spectral joint Feature Matching Loss based on our FFC-based discriminator. Denote the ith layer feature extractor of the discriminator D as $D^{(i)}$, denote the local (spatial) path features as $D^{(i,l)}$, global (spectral) path features as $D^{(i,g)}$, the spatial-spectral joint feature matching loss function is as follow:

$$\mathcal{L}_{SSFM} = E_{x_d, x} \sum_{i=1}^{T} \frac{1}{N_i} \left[\left\| D_k^{(i,l)}(x_d, x) - D_k^{(i,l)}(x_d, G(x_d, m)) \right\|_1 \right.$$
$$\left. + \left\| D_k^{(i,g)}(x_d, x) - D_k^{(i,g)}(x_d, G(x_d, m)) \right\|_1 \right] \tag{5}$$

Total Loss. Except for the above-mentioned SSFM loss, we also use the adversarial loss to ensure the model restores both structured and unstructured degradation. The discriminator works on the local patch to discriminate whether the input is "real" or "fake", "real" as clean, and "fake" as structured degraded or unstructured degraded. We use Least Squares GAN Loss [15] to minimize the gap between the distribution of degraded images and clean images. Besides we use perceptual loss [3] to require the generator to reconstruct the ground truth precisely. The final loss function is in the form of

$$\mathcal{L} = \alpha \mathcal{L}_{Adv} + \beta \mathcal{L}_{percaptral} + \gamma \mathcal{L}_{SSFM} \tag{6}$$

4 Experiment

4.1 Dataset and Implementation Detail

Dataset. Since it is difficult to gather repaired real old data to supervise the training, we use PASCAL VOC dataset and follow [21] to synthesize a set of paired data for supervised training. As for the structured degradation, we randomly choose the following two approaches: (1) random walk crack generation; (2) texture-based crack generation, the texture images are mainly from [23].

Implementation Detail. We use Adam optimizer and set exponential decay rates as 0.9 and 0.999. The initial learning rate is set to 5e–4 and then reduce to 1e–5 at epoch 100. We use PyTorch framework with 1 24G NVIDIA 4090 GPU. To avoid the mode crash, we accomplish the upsampling approach by 1×1 convolution and upsampling layer rather than deconvolution.

4.2 Comparison

We compare our model with published old photo restoration methods [21,23] potential generation network [20,22], and a sequential applied single image restoration methods. To fairly compare the methods, we train above-mentioned methods on synthesized PASCAL VOC old images and evaluate on DIV2K synthesized old images. Test time scratch mask was detected by a pretrained UNet mentioned in [21]. Also, we compared the impact on real old images collected from the internet and old photo datasets from [21,23].

Quantitative Comparison. We test different models on synthesized images from DIV2K dataset and calculate the peak signal-to-noise ratio (PSNR), structural similarity index (SSIM), and learned perceptual image patch similarity (LPIPS) metrics between restored output and corresponding ground truth. The results are reported in Table 1.

The table is organized by whether the model is trained or inferred in a complex way. The training and inferring paradigm of three complex models are as follows:

Table 1. Quantitative results on the DIV2K dataset. The upward arrow(↑) indicates that a higher score denotes a good image restoration. We highlight the top 2 scores for each measure.

	Method	PSNR↑	SSIM↑	LPIPS↓
Complex Model	DIP [20]	22.59	0.57	0.54
	Sequential [2,16]	22.71	0.60	0.49
	Wan [21]	**23.33**	**0.69**	0.25
End-to-end Model	pix2pix [22]	22.18	0.62	**0.23**
	pikfix [23]	22.22	0.60	0.42
	Ours	**23.15**	**0.78**	**0.24**

Input   GT/manupulated   Pix2Pix[22]   Latent space   Ours
                                            translation[21]

Fig. 4. Qualitative comparison against state-of-the-art methods. It shows that our method has comparative impact with [21]

- **DIP** [20] first pre-trained on a large scale of clean images, then optimized thousands of steps to generate clean images from single degradation. It has been proven powerful in denoising, super-resolution, and blind inpainting.
- **Sequential** sequentially applies denoising method in [2] and inpainting method in [16]. One can apply more restoration methods to get a higher score, but all models need pretrain.
- **Wan** [21] first train the VAE A for old image reconstruction and VAE B for clean image reconstruction respectively, then train a mapping network from VAE A to VAE B. After that there are some refinement processes.

In contrast, [22] and [23] just take the degraded photos as input and produce clean images, which is the same as ours in the training and inferring process. A scratch detection process is needed by all methods to indicate the structural defect area.

| Input | Reference Image | Vanilla Conv | FFC | FFC w PConv | FFC w PConv&HA |

Fig. 5. Ablation study on FFC block, PConv block, and image-wise histogram attention. It shows that PConv is important for structured degradation restoration and image-wise histogram attention (denoted as HA) maps the color tone of the restored image to reference image

Qualitative Comparison. Figure 4 provides a qualitative comparison of DIV2K synthesized images (the upper two rows) and real old images. SSHRF-GAN robustly produced vivid and realistic images both on synthesized datasets and real old images. [22] can restore unstructured degradation to some extent, but have no idea for structured degradation since there is no mask indicating degradation area. [21] provides clean and sharp results for real old images, but the color tone seems not very realistic and there exists a global shift on that.

4.3 Ablation Study

We conducted experiments on the synthesized DIV2K dataset to evaluate the effectiveness of FFC block, PConv block, and image-wise spatial histogram attention. The visual results are in Fig. 5 and the quantitative results in Table 2. It shows that the FFC block generates a clearer image than the Vanilla Conv block and PConv mainly contributes to structured degradation inpainting. The image-wise spatial histogram influences the color tone of the output image slightly according to the selected image. All the above-mentioned blocks together help SSHRF-GAN perform impressively.

Table 2. The evaluation result of ablation study for FFC block, PConv block, and histogram attention (denote as HA).

Setting	PSNR↑	SSIM↑	LPIPS↓
FFC, w/o PConv, w/o HA	15.86	0.51	0.49
FFC, w PConv, w/o HA	21.28	0.74	0.26
FFC, w PConv, w HA	**23.15**	**0.78**	0.24
Vanilla Conv, w PConv, wHA	22.18	0.62	**0.23**

5   Conclusion

We proposed a simple network for old photo restoration, a complex mixed degradation restoration problem. We simply apply different modules for different degradation to achieve this target and have a comparative impact with the previous proposed complex models. Although there is a gap between the model restored and manipulated images (see Fig. 4 GT/manipulated), our model could restore the images from those low-level degradations simply and effectively. We believe it could provide much convenience for further manufacturing. Due to the complex and various degradation functions of the old images, it is hard to synthesize all categories of degradation. Thus our model also has bad performance on those rarely appeared degradation which do not fulfill our synthesis function. To further improve the performance on those odd degradations, one can increase the diversity of the synthesized dataset by learning those odd degradations with another neural network.

References

1. Chi, L., Jiang, B., Mu, Y.: Fast fourier convolution. Adv. Neural. Inf. Process. Syst. **33**, 4479–4488 (2020)
2. Dabov, K., Foi, A., Katkovnik, V., Egiazarian, K.: Image denoising by sparse 3-d transform-domain collaborative filtering. IEEE Trans. Image Process. **16**(8), 2080–2095 (2007). https://doi.org/10.1109/TIP.2007.901238
3. Dosovitskiy, A., Brox, T.: Generating images with perceptual similarity metrics based on deep networks. Adv. Neural Inf. Process. Syst. **29**, 1–9 (2016)
4. Elharrouss, O., Almaadeed, N., Al-Maadeed, S., Akbari, Y.: Image inpainting: a review. Neural Process. Lett. **51**, 2007–2028 (2020)
5. Giakoumis, I., Nikolaidis, N., Pitas, I.: Digital image processing techniques for the detection and removal of cracks in digitized paintings. IEEE Trans. Image Process. **15**(1), 178–188 (2005)
6. Ho, J., Jain, A., Abbeel, P.: Denoising diffusion probabilistic models. Adv. Neural. Inf. Process. Syst. **33**, 6840–6851 (2020)
7. Kupyn, O., Martyniuk, T., Wu, J., Wang, Z.: Deblurgan-v2: deblurring (orders-of-magnitude) faster and better. In: Proceedings of the IEEE/CVF International Conference on Computer Vision, pp. 8878–8887 (2019)
8. Li, W., Lin, Z., Zhou, K., Qi, L., Wang, Y., Jia, J.: Mat: mask-aware transformer for large hole image inpainting. In: Proceedings of the IEEE/CVF Conference on Computer Vision and Pattern Recognition, pp. 10758–10768 (2022)
9. Liang, J., Cao, J., Sun, G., Zhang, K., Van Gool, L., Timofte, R.: SwinIR: image restoration using swin transformer. In: Proceedings of the IEEE/CVF International Conference on Computer Vision, pp. 1833–1844 (2021)
10. Liu, D., Wen, B., Fan, Y., Loy, C.C., Huang, T.S.: Non-local recurrent network for image restoration. Adv. Neural Inf. Process. Syst. **31** (2018)
11. Liu, G., Reda, F.A., Shih, K.J., Wang, T.C., Tao, A., Catanzaro, B.: Image inpainting for irregular holes using partial convolutions. In: Proceedings of the European Conference on Computer Vision (ECCV), pp. 85–100 (2018)
12. Liu, J., Chen, R., An, S., Zhang, H.: CG-GAN: class-attribute guided generative adversarial network for old photo restoration. In: Proceedings of the 29th ACM International Conference on Multimedia, pp. 5391–5399 (2021)

13. Liu, P., Zhang, H., Zhang, K., Lin, L., Zuo, W.: Multi-level wavelet-CNN for image restoration. In: Proceedings of the IEEE Conference on Computer Vision and Pattern Recognition Workshops, pp. 773–782 (2018)
14. Mao, X., Liu, Y., Shen, W., Li, Q., Wang, Y.: Deep residual fourier transformation for single image deblurring. arXiv preprint arXiv:2111.11745 (2021)
15. Mao, X., Li, Q., Xie, H., Lau, R.Y., Wang, Z., Paul Smolley, S.: Least squares generative adversarial networks. In: Proceedings of the IEEE International Conference on Computer Vision (ICCV) (2017)
16. Nazeri, K., Ng, E., Joseph, T., Qureshi, F.Z., Ebrahimi, M.: Edgeconnect: generative image inpainting with adversarial edge learning. arXiv preprint arXiv:1901.00212 (2019)
17. Quan, W., Zhang, R., Zhang, Y., Li, Z., Wang, J., Yan, D.M.: Image inpainting with local and global refinement. IEEE Trans. Image Process. **31**, 2405–2420 (2022)
18. Suganuma, M., Liu, X., Okatani, T.: Attention-based adaptive selection of operations for image restoration in the presence of unknown combined distortions. In: Proceedings of the IEEE/CVF Conference on Computer Vision and Pattern Recognition (CVPR) (2019)
19. Suvorov, R., et al.: Resolution-robust large mask inpainting with fourier convolutions. In: Proceedings of the IEEE/CVF Winter Conference on Applications of Computer Vision, pp. 2149–2159 (2022)
20. Ulyanov, D., Vedaldi, A., Lempitsky, V.: Deep image prior. In: Proceedings of the IEEE Conference on Computer Vision and Pattern Recognition, pp. 9446–9454 (2018)
21. Wan, Z., Zhang, B., Chen, D., Zhang, P., Wen, F., Liao, J.: Old photo restoration via deep latent space translation. IEEE Trans. Pattern Anal. Mach. Intell. **45**(2), 2071–2087 (2022)
22. Wang, T.C., Liu, M.Y., Zhu, J.Y., Tao, A., Kautz, J., Catanzaro, B.: High-resolution image synthesis and semantic manipulation with conditional gans. In: Proceedings of the IEEE Conference on Computer Vision and Pattern Recognition (CVPR) (2018)
23. Xu, R., et al.: PIK-FIX: restoring and colorizing old photos. In: Proceedings of the IEEE/CVF Winter Conference on Applications of Computer Vision, pp. 1724–1734 (2023)
24. Yu, J., Lin, Z., Yang, J., Shen, X., Lu, X., Huang, T.S.: Free-form image inpainting with gated convolution. In: Proceedings of the IEEE International Conference on Computer Vision, pp. 4471–4480 (2019)
25. Yu, K., Dong, C., Lin, L., Loy, C.C.: Crafting a toolchain for image restoration by deep reinforcement learning. In: Proceedings of the IEEE Conference on Computer Vision and Pattern Recognition, pp. 2443–2452 (2018)
26. Zamir, S.W., Arora, A., Khan, S., Hayat, M., Khan, F.S., Yang, M.H.: Restormer: efficient transformer for high-resolution image restoration. In: Proceedings of the IEEE/CVF Conference on Computer Vision and Pattern Recognition, pp. 5728–5739 (2022)
27. Zhang, K., Zuo, W., Chen, Y., Meng, D., Zhang, L.: Beyond a gaussian denoiser: residual learning of deep CNN for image denoising. IEEE Trans. Image Process. **26**(7), 3142–3155 (2017)

Author Index

Q. Liu et al. (Eds.): PRCV 2023, LNCS 14433, pp. 505–507, 2024.
https://doi.org/10.1007/978-981-99-8546-3

Printed in the United States
by Baker & Taylor Publisher Services